Biology Now

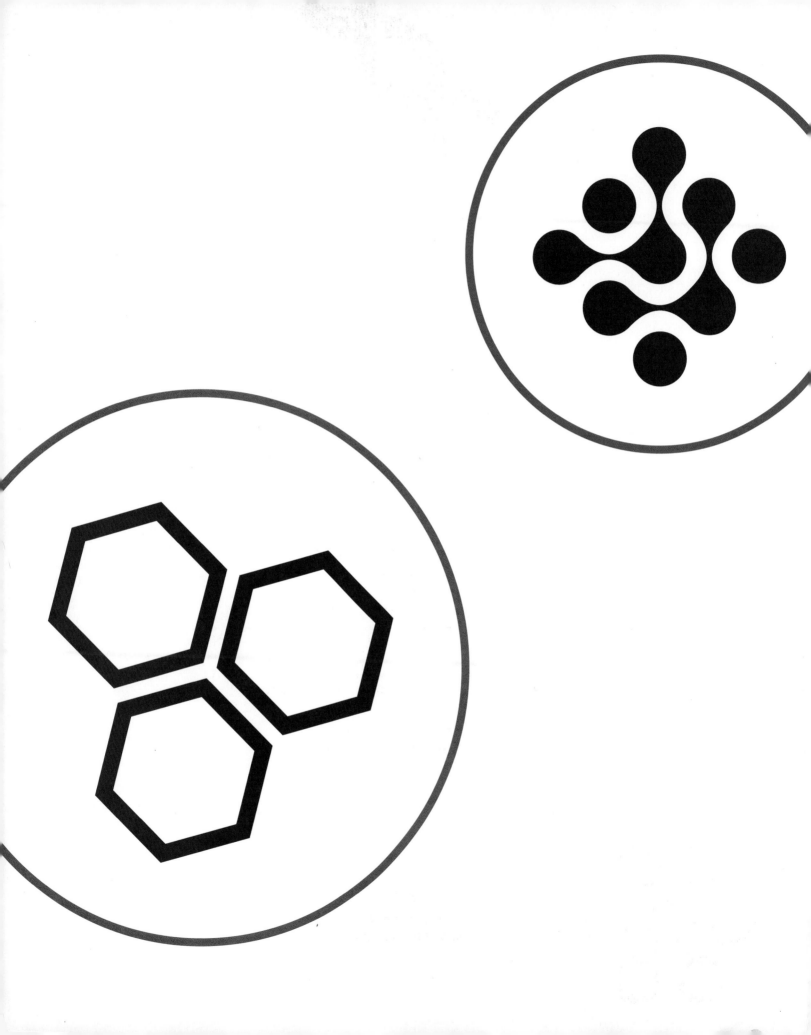

Biology Now

ANNE HOUTMAN
CALIFORNIA STATE UNIVERSITY, BAKERSFIELD

MEGAN SCUDELLARI
SCIENCE JOURNALIST

CINDY MALONE
CALIFORNIA STATE UNIVERSITY, NORTHRIDGE

ANU SINGH-CUNDY
WESTERN WASHINGTON UNIVERSITY

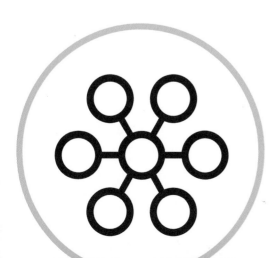

W. W. NORTON
NEW YORK · LONDON

W. W. Norton & Company has been independent since its founding in 1923, when William Warder Norton and Mary D. Herter Norton first published lectures delivered at the People's Institute, the adult education division of New York City's Cooper Union. The firm soon expanded its program beyond the Institute, publishing books by celebrated academics from America and abroad. By midcentury, the two major pillars of Norton's publishing program—trade books and college texts— were firmly established. In the 1950s, the Norton family transferred control of the company to its employees, and today— with a staff of four hundred and a comparable number of trade, college, and professional titles published each year—W. W. Norton & Company stands as the largest and oldest publishing house owned wholly by its employees.

Editor: Betsy Twitchell
Developmental Editor: Andrew Sobel
Senior Project Editor: Thomas Foley
Editorial Assistants: Katie Callahan and Courtney Shaw
Copy Editor: Stephanie Hiebert
Managing Editor, College: Marian Johnson
Managing Editor, College Digital Media: Kim Yi
Associate Director of Production, College: Benjamin Reynolds
Media Editor: Karl Bakeman
Associate Media Editors: Cailin Barrett Bressack and Callinda Taylor
Media Project Editor: Danielle Belfiore
Media Editorial Assistant: Victoria Reuter
Marketing Manager, Biology: Meredith Leo
Design Director: Hope Miller Goodell
Photo Editor: Evan Luberger
Photo Researcher: Fay Torresyap
Permissions Manager: Megan Jackson
Permissions Clearer: Elizabeth Trammell
Composition and Illustrations by Precision Graphics
Project manager at Precision Graphics: Mimi Polk
Manufacturing: Courier—Kendallville, IN

Permission to use copyrighted material is included in the backmatter of this book.

ISBN 978-0-393-91892-2

ISBN **978-0-393-91894-6** (enhanced ebook)
ISBN **978-0-393-91895-3** (PDF ebook)

W. W. Norton & Company, Inc., 500 Fifth Avenue, New York, NY 10110-0017
wwnorton.com

W. W. Norton & Company Ltd., Castle House, 75/76 Wells Street, London W1T 3QT

1 2 3 4 5 6 7 8 9 0

Brief Contents

Contents

INTRODUCTION

UNIT 1: CELLS

UNIT 2: GENETICS

CHAPTER 6: Patterns of Inheritance   90

Dog Days of Science

CHAPTER 7: Chromosomes and Human Genetics   108

A Deadly Inheritance

 UNIT 3: EVOLUTION

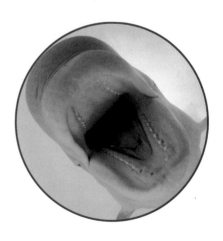

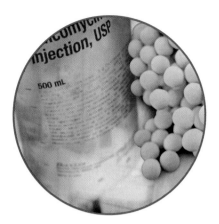

UNIT 4: ECOLOGY

 # UNIT 5: PHYSIOLOGY

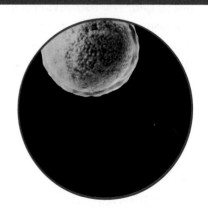

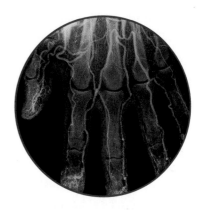

CONCLUSION

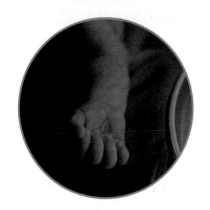

About the Authors

ANNE HOUTMAN is Dean of the School of Natural Sciences, Mathematics, and Engineering at California State University, Bakersfield. Anne has over 20 years of experience teaching non-majors biology at a variety of private and public institutions, which gives her a broad perspective and a wide view of the education landscape. She is strongly committed to evidence-based, experiential education and has been an active participant in the national dialogue on STEM (science, technology, engineering, and math) education for almost 20 years. Anne's research interests are in the ecology and evolution of hummingbirds. She grew up in Hawaii, received her doctorate in zoology from the University of Oxford, and conducted postdoctoral research at the University of Toronto.

MEGAN SCUDELLARI is an award-winning freelance science writer and journalist based in Boston, Massachusetts, specializing in the life and environmental sciences. She has contributed to *Newsweek*, *Scientific American*, *Discover*, *Nature*, and *Technology Review*, among others. For five years she worked as a correspondent and later as a contributing editor for *The Scientist* magazine. In 2013, she was awarded the prestigious Evert Clark/Seth Payne Award in recognition of outstanding reporting and writing in science. She has also received accolades for investigative reporting on traumatic brain injury and a feature story on prosthetics bestowed with a sense of touch. Megan received an MS from the Graduate Program in Science Writing at the Massachusetts Institute of Technology, a BA from Boston College, and worked as an educator at the Museum of Science, Boston.

CINDY MALONE began her scientific career wearing hip-waders in a swamp behind her home in Illinois. She earned her BS in Biology at Illinois State University and her PhD in Microbiology and Immunology at UCLA. She continued her postdoctoral work at UCLA in Molecular Genetics. Dr. Malone is currently a distinguished educator and an Associate Professor at California State University, Northridge, where she is the Director of the CSUN-UCLA Bridges to Stem Cell Research Program funded by the California Institute for Regenerative Medicine. Dr. Malone's research is aimed at training undergraduates and masters degree candidates to understand how genes are regulated through genetic and epigenetic mechanisms that alter gene expression. She has been teaching non-majors biology for over 15 years and has won curriculum enhancement and teaching awards at CSUN.

ANU SINGH-CUNDY received her PhD from Cornell University and did postdoctoral research in cell and molecular biology at Penn State. She is an associate professor at Western Washington University, where she teaches a variety of undergraduate and graduate courses, including organismal biology, cell biology, plant developmental biology, and plant biochemistry. She has taught introductory biology to non-majors for over 15 years and is recognized for pedagogical innovations that communicate biological principles in a manner that engages the non-science student and emphasizes the relevance of biology in everyday life. Her research focuses on cell-cell communication in plants, especially self-incompatibility and other pollen-pistil interactions. She has published over a dozen research articles and has received several awards and grants, including a grant from the National Science Foundation.

Preface

A good biology class has the potential to substantially improve the quality of students' lives. Biology will be a part of so many decisions that students will make as individuals and as members of society. It helps parents to see the value of vaccinating a child because they will understand what viruses are and how the immune system works. It can assist homeowners on New Jersey's beaches as they decide how to respond to the ongoing cleanup from 2012's Hurricane Sandy because they understand how an ecosystem functions. The examples are endless. Making informed decisions on these real-world issues requires students to be comfortable with scientific concepts and the process of scientific discovery.

So how do we achieve that? The last decade has seen an explosion of research on how students learn best. In a nutshell, they learn best when they see the relevance of a subject to their lives, when they are actively engaged in their learning, and when they are given opportunities to practice critical thinking.

In addition, most faculty who teach non-majors biology would agree that our goal is to introduce students to both the key concepts of biology (e.g., cells, DNA, evolution) and the tools to think critically about biological issues. Many would add that they want their students to leave the class with an appreciation for the value of science to society, and with an ability to distinguish between science and the non-science or pseudoscience that they are bombarded with on a daily basis.

How can a textbook help combine the ways students learn best with the goals of a non-majors biology class? At the most basic level, if students don't read the textbook, they can't learn from it. When they do read, traditional textbooks are adept at teaching key concepts, and have recently begun to emphasize the relevance of biology to students' lives. But students may be intimidated by the length of chapters and the amount of difficult text, and they often cannot see the connections between the story and the science. More importantly, textbooks have not been successful at helping students become active learners and critical thinkers, and none emphasize the process of science. It was our goal to make *Biology Now* relevant and interactive, and to be sure that it emphasized the process of science in short chapters students *want* to read, while still covering the essential content found in other non-majors biology textbooks.

Each chapter in our book covers a current news story about people *doing* science, reported first-hand by Megan, an experienced journalist who specializes in reporting scientific findings in a compelling and accurate way. The process of writing each chapter begins when Anne and Cindy use their decades of teaching experience to select the essential biological concepts that students need to learn. Megan then weaves these science concepts into the news story, giving the biology a context that students will engage with and remember. Anne and Cindy also support the text with pedagogical features developed in their classrooms, such as the three thought questions that accompany almost every figure. Colorful, dynamic art and engaging, data-driven infographics round out our book, which we sincerely hope you—and your students—enjoy.

Anne Houtman
Megan Scudellari
Cindy Malone
Anu Singh-Cundy

The perfect balance of science and story

Every chapter is structured around a story about people doing science that motivates students to read and stimulates their curiosity about biological concepts.

Dynamic chapter-opening spreads inspired by each chapter's story draw students in to the material.

"After reading this chapter you should be able to" learning outcomes provide a preview of the concepts presented in each chapter.

Cast of Character Bios

highlight the scientists, researchers, and professors at the center of each story.

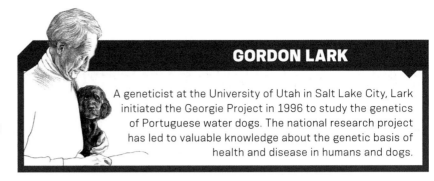

GORDON LARK

A geneticist at the University of Utah in Salt Lake City, Lark initiated the Georgie Project in 1996 to study the genetics of Portuguese water dogs. The national research project has led to valuable knowledge about the genetic basis of health and disease in humans and dogs.

J. G. M 'HANS' THEWISSEN

Paleontologist and embryologist J.G.M. 'Hans' Thewissen is a professor and whale expert at Northeast Ohio Medical University in the Department of Anatomy and Neurobiology. He and his lab study ancestral whale fossils and modern whale species.

LISA COOPER

Lisa Cooper is an assistant professor at Northeast Ohio Medical University in the Department of Anatomy and Neurobiology. She earned her PhD in Thewissen's lab.

MICHAEL HELLBERG AND CARLOS PRADA

Michael Hellberg (right) is an evolutionary biologist at Louisiana State University who studies how species evolve in marine environments. Carlos Prada (left) is a graduate student in Hellberg's lab at LSU. Prada initiated and conducted a study about how corals form new species in the ocean.

XU XING

Xu Xing is a paleontologist at the Chinese Academy of Sciences in Beijing. He has discovered more than 60 species of dinosaurs and specializes in feathered dinosaurs and the origins of flight.

An inquiry-based approach that builds science skills—asking questions, thinking visually, and interpreting data.

Most **figures** in the book are accompanied by three questions that promote understanding and encourage engagement with the visual content. Answers are provided at the back of the book, making the questions a useful self-study tool.

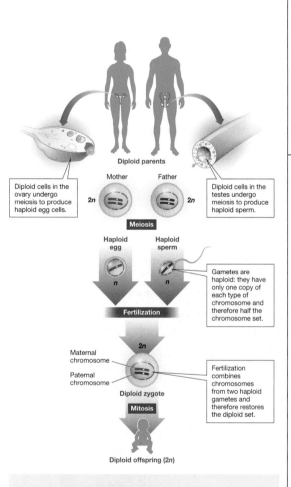

Q1: Is a zygote haploid or diploid?

Q2: What cellular process creates a baby from a zygote?

Q3: If a mother or father is exposed to BPA prior to conceiving a child, how does that explain potential birth defects in the fetus?

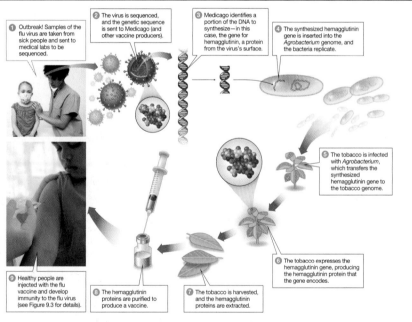

Q 1: In which of the steps illustrated here does DNA replication occur? In which steps does gene expression occur?

Q2: Why do vaccine producers not simply replicate the entire viral genome? Why do they instead isolate the gene for one protein and replicate only that gene?

Q3: What role do the bacteria play in this process? Why are they needed?

Engaging data-driven **infographics** appear in every chapter. Topics range from global renewable energy consumption (Chapter 4) to the size and geographic distribution of food-poisoning outbreaks (Chapter 8) and many more. The infographics expose students to scientific data in an engaging way.

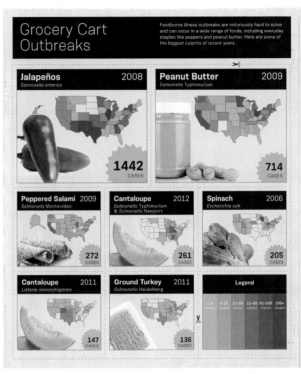

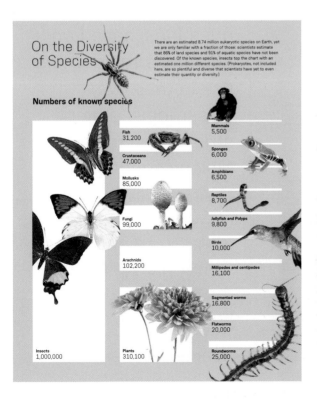

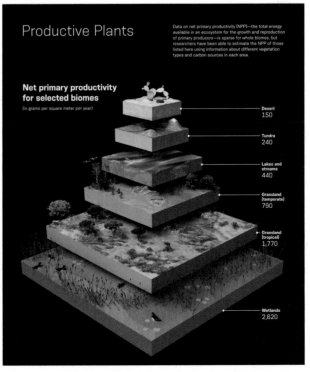

Extensive end-of-chapter review ensures that students see the forest for the trees.

Reviewing the Science identifies each chapter's key science concepts, providing students with a guide for studying.

End-of-chapter questions follow Bloom's Taxonomy, moving from review—The Basics, to synthesis—Try Something New, to application—Leveling Up.

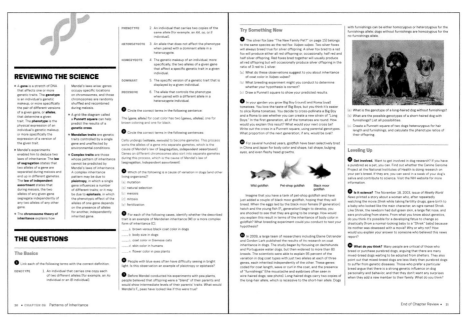

Leveling Up questions, based on questions the authors use in their own classrooms, prompt students to relate biology concepts to their own lives. The questions focus on one of the following themes: "What do *you* think?," "Life Choices," "Looking at data," "Is it Science?," "Doing science," and "*Write now* biology."

Leveling Up Rubrics in the coursepack coincide with the questions at the end of each chapter in the book. Extensive grading rubrics for instructors make these critical-thinking, application-based questions easier to assign and grade.

Chapter 6, *Leveling Up* 13 TOTAL SCORE_____

Is it science? *The November 18, 2003, issue of Weekly World News printed a story about a woman who, after repeatedly watching the movie Shrek while taking fertility drugs, gave birth to a baby who looked like the main character, an ogre named Shrek. Like Shrek, the newborn had dull green skin, a large flat nose, and ears protruding from stems. From what you know about genetics, do you think it's possible for a developing fetus to change so drastically (from a normal-looking baby to a "Shrek" baby) because its mother was obsessed with a movie? Why or why not? How would you explain your answer to someone who believed this news report?*

Level	Opinion on Possibility of "Shrek Baby" and Explanation	Quality of Writing
Excellent	Opinion demonstrates thorough understanding of the genetic basis of inheritance. Explanation is clear, complete and accurate, and would be convincing to a person who believes the news report.	Writing style is clear and concise and flows well. Close-to-perfect spelling, grammar, and language usage.
Adequate	Opinion demonstrates understanding of the genetic basis of inheritance. Explanation is mostly clear, complete, and accurate. May contain minor errors or some lack of clarity.	Clearly written, with only a few errors in spelling, grammar, and language usage.
Flawed	Opinion demonstrates a flawed understanding of the genetic basis of inheritance. Explanation contains major errors **or** is very unclear. Explanation would **not** be convincing to a person who believes the news report.	Occasional lack of clarity in writing. Several errors in spelling, grammar, and language usage.
Deficient	Opinion is not explained **or** is completely unintelligible.	Writing is difficult to follow. Serious difficulties with misspelling, errors in grammar, misuse of language.

Chapter 6, *Leveling Up* 14 TOTAL SCORE_____

What do you think? *Many people are critical of those who breed or purchase purebred dogs, arguing that there are many mixed-breed dogs waiting to be adopted from shelters. They also point out that mixed-breed dogs are less likely than purebred dogs to suffer from genetic diseases. Those who prefer a particular breed argue that there is a strong genetic influence on dog personality and behavior, and that they don't want any surprises when they add a new member to their family. What do you think?*

Level	Opinion on Purebred versus Mixed-Breed Dogs	Quality of Writing
Excellent	Opinion is thoughtful, clear, and well-supported by data and/or logic. Acknowledges and argues against the opposing viewpoint.	Writing style is clear and concise and flows well. Close-to-perfect spelling, grammar, and language usage.
Adequate	Opinion is clear, and supported by data and/or logic. Doesn't necessarily address the opposing viewpoint **or** argument may contain some weaknesses.	Clearly written, with only a few errors in spelling, grammar, and language usage.
Flawed	Opinion is not well supported by logic **or** argument is unclear.	Occasional lack of clarity in writing. Several errors in spelling, grammar, and language usage.
Deficient	Opinion is not supported at all **or** is completely unintelligible.	Writing is difficult to follow. Serious difficulties with misspelling, errors in grammar, misuse of language.

Powerful resources for teaching and assessment

The Ultimate Guide helps instructors bring *Biology Now's* inquiry-based approach into the classroom through a wealth of resources, including activities useful in a variety of classroom sizes and setups, suggested online videos with discussion questions, clicker questions, sample syllabi, and suggested lecture outlines. Created by Julie Harless (Lone Star College) with the help of dozens of contributing instructors from around the country, *The Ultimate Guide* provides instructors with an array of resources for enriching class time.

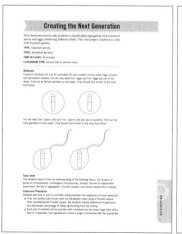

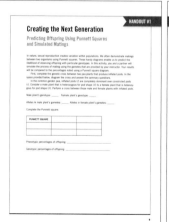

Other presentation tools for instructors:

InQuizitive—A new, formative, adaptive-learning tool that personalizes quiz questions for each student and preserves valuable lecture and lab time by building knowledge outside of class (see image below). Engaging, game like elements motivate students as they learn. Students wager points on each question based on their confidence level, win additional points for hot streaks and bonus questions, and can improve their grade by continuing to work on the assignment.

A variety of question types prepare students for lecture, quizzes, and exams, while performance-specific feedback helps students understand their mistakes. Animations, videos, and other resources built into InQuizitive allow students to review core concepts as they answer questions. Instructors can assign InQuizitive out of the box or use intuitive tools to customize material for the learning objectives they want students to work on. Students can access InQuizitive on computers, tablets, and smartphones, making it easy to study on the go.

SmartWork—*Biology Now's* emphasis on inquiry is reinforced through personalized, flexible assignments, high quality questions, and answer-specific feedback that help students apply, analyze, and evaluate key concepts. SmartWork's instructor tools provide teachers with the actionable student performance data they need so they can do what they do best: teach.

Coursepack—Norton's free coursepacks offer a variety of concept-based opportunities for assessment and review. Coursepacks include Leveling Up Rubrics that coincide with the questions in each end-of-chapter review section. These critical-thinking, application-based, writing activities are accompanied by grading rubrics, making them easier to assign. Also included are an optional ebook, chapter summaries and quizzes, flashcards, animations, and infographic quizzes that help students build skills in the interpretation of charts and graphs, as well as testing them on the information in the graphic itself.

Test Bank—The test bank is based on an evidence-centered design that was collaboratively developed by some of the brightest minds in educational testing. Each chapter is structured around the learning objectives from the textbook and conforms to Bloom's taxonomy. Questions are further classified by section and difficulty, and they are provided in multiple-choice, fill-in-the-blank, and short answer form.

Ebook—Norton ebooks allow students to access the entire book and much more. Students can search, highlight, and take notes with ease, as well as collaborate and share their notes with instructors and classmates. The *Biology Now* ebook can be viewed on any device—laptop, tablet, phone, even a public computer—and will stay synced between devices.

Animations—Key concepts and processes are clearly explained through high-quality, ADA-compliant animations developed from the meticulously designed art in the book. These animations are available for lecture presentation in PowerPoint and the coursepack as well as within our ebook, InQuizitive, and SmartWork.

Art Files—All art and photos from the book are available, in presentation-ready resolution, as both JPEGs and PowerPoints for instructor use.

Lecture Slides—Complete lecture PowerPoints thoroughly cover chapter concepts and include images and clicker questions to encourage student engagement.

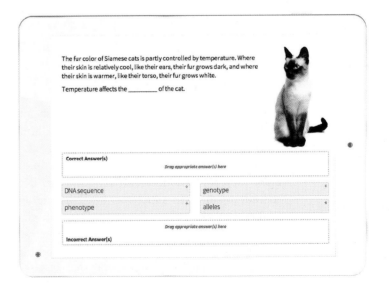

The fur color of Siamese cats is partly controlled by temperature. Where their skin is relatively cool, like their ears, their fur grows dark, and where their skin is warmer, like their torso, their fur grows white.

Temperature affects the _____ of the cat.

Correct Answer(s)

Drag appropriate answer(s) here

DNA sequence | genotype
phenotype | alleles

Drag appropriate answer(s) here

Incorrect Answer(s)

Acknowledgments

Creating this textbook would not have been possible without the enthusiasm and hard work of many people. First and foremost, we'd like to thank our indefatigable editor, Betsy Twitchell, for her keen eye to the market, terrific visual sense, and endless author wrangling skills. Andrew Sobel has done far more than a developmental editor ought to in order to ensure that our book is both accurate and readable (not to mention his tireless work on the eye-catching infographics you'll see in these pages), and for that he has our eternal gratitude.

Thank you to our supremely focused and talented project editor, Thom Foley, for creating such a superior layout and for keeping our chapters moving even when the schedule got tight. Thank you to our talented copy editor, Stephanie Hiebert, for being so meticulous with our manuscript, and so pleasant to work with.

We are grateful to photo researcher Fay Torresyap for her reliable and creative work and to Evan Luberger for managing the photo process. Production manager Ben Reynolds skillfully oversaw the process of translating our raw material into the beautiful book you hold in your hands; he too has our thanks. Special thanks to the cover designer, Carl De Torres for creating such a unique and truly gorgeous cover image.

Media editor Robin Kimball and associate editors Cailin Barrett-Bressack and Callinda Taylor worked tirelessly to create the instructor and student resources accompanying our book. Their determination, creativity, and positive attitude resulted in supplements of the highest quality that will truly make an impact on student learning. Danielle Belfiore and Kristin Sheerin's commitment to quality as media project editors ensured that every element of the resource package meets Norton's high standards. Likewise, editorial assistant Katie Callahan and media assistant Victoria Reuter contributed in a myriad of ways, large and small, and for that they have our thanks.

We appreciate the tireless enthusiasm of marketing manager Meredith Leo and her colleagues, director of marketing Steve Dunn and associate director of marketing Nicole Netherton. We thank director of sales Michael Wright and every single one of Norton's extraordinary sales people for spreading the word about our book. Finally, we thank Marian Johnson, Julia Reidhead, Roby Harrington, Drake McFeely, and everyone at Norton for believing in our book.

Thank you to our accuracy reviewers Michelle Cawthorn and Michael Zierler. We would be remiss to not also thank all of our colleagues in the field who gave their time and expertise in reviewing, class testing, and contributing to *Biology Now* and all of its supplements and resources. Thank you all.

Reviewers

Joseph Ahlander, Northeastern State University

Stephen F. Baron, Bridgewater College

David Bass, University of Central Oklahoma

Erin Baumgartner, Western Oregon University

Cindy Bida, Hentry Ford Community College

Charlotte Borgeson, University of Nevada Reno

Bruno Borsari, Winona State University

Ben Brammell, Eastern Kentucky University

Christopher Butler, University of Central Oklahoma

Stella Capoccia, Montana Tech

Kelly Cartwright, College of Lake County

Emma Castro, Victor Valley College

Michelle Cawthorn, Georgia Southern University

Jeannie Chari, College of the Canyons

Jianguo Chen, Claflin University

Beth Collins, Iowa Central Community College

Angela Costanzo, Hawai'i Pacific University

James B. Courtright, Marquette University

Danielle DuCharme, Waubonsee Community College

Julie Ehresmann, Iowa Central Community College

Laurie L. Foster, Grand Rapids Community College

Teresa Golden, Southeastern Oklahoma State University

Sue Habeck, Tacoma Community College

Janet Harouse, New York University

Olivia Harriott, Fairfield University

Tonia Hermon, Norfolk State University

Glenda Hill, El Paso Community College

Vicki J. Huffman, Potomac State College of West Virginia University

Carl Johansson, Fresno City College

Victoria Johnson, San Jose State University

Anthony Jones, Tallahassee Community College

Hinrich Kaiser, Victor Valley College

Vedham Karpakakunjaram, Montgomery College

Dauna Koval, Bellevue College

Maria Kretzmann, Glendale Community College

MaryLynne LaMantia, Golden West College

Brenda Leady, The University of Toledo

Lisa Maranto, Prince George's Community College

Roy B. Mason, Mt. San Jacinto College

Gabrielle L. McLemore, Morgan State University

Paige Mettler-Cherry, Lindenwood University

Rachel Mintell, Manchester Community College

Kiran Misra, Edinboro University of Pennsylvania

Lori Nicholas, New York University

Louise Mary Nolan, Middlesex Community College

Fran Norflus, Clayton State University

Brian Paulson, California University of Pennsylvania

Carolina Perez-Heydrich, Meredith College

Ashley Ramer, The University of Akron

Nick Reeves, Mt. San Jacinto College

Tim Revell, Mt. San Antonio College

Eric Ribbens, Western Illinois University

Kathreen Ruckstuhl, University of Calgary

Michael L. Rutledge, Middle Tennessee State University

Brian Sato, University of California, Irvine

Malcolm D. Schug, The University of North Carolina at Greensboro

Craig M. Scott, Clarion University of Pennsylvania

J. Michael Sellers, The University of Southern Mississippi

Marieken Shaner, The University of New Mexico

David Sheldon, St. Clair County Community College

Jack Shurley, Idaho State University

Daniel Sigmon, Alamance Community College

Molly E. Smith, South Georgia State College-Waycross

Lisa Spring, Central Piedmont Community College

Steven R. Strain, Slippery Rock University of Pennsylvania

Jeffrey L. Travis, University at Albany, SUNY

Suzanne Wakim, Butte College

Mark E. Walvoord, University of Oklahoma

Sherman Ward, Virginia State University

Lisa Weasel, Portland State University

Jennifer Wiatrowski, Pasco-Hernando State College

Rachel Wiechman, West Liberty University

Bethany Williams, California State University, Fullerton

Satya M. Witt, The University of New Mexico

Donald A. Yee, The University of Southern Mississippi

Focus Group Participants

Michelle Cawthorn, Georgia Southern University
Marc Dal Ponte, Lake Land College
Kathy Gallucci, Elon University
Tamar Goulet, The University of Mississippi
Sharon Gusky, Northwestern Connecticut Community College
Krista Henderson, California State University, Fullerton
Tara Jo Holmberg, Northwestern Connecticut Community College
Brenda Hunzinger, Lake Land College
Jennifer Katcher, Pina Comminity College
Cynthia Kay-Nishiyama, California State University, Northridge
Kathleen Kresge, Northampton Community College
Sharon Lee-Bond, Northampton Community College
Suzanne Long, Monroe Community College
Boriana Marintcheva, Bridgewater State University
Roy B. Mason, Mt. San Jacinto College
Gwen Miller, Collin College
Kimo Morris, Santa Ana College
Fran Norflus, Clayton State University
Tiffany Randall, John Tyler Community College
Gail Rowe, La Roche College
J. Michael Sellers, The University of Southern Mississippi
Uma Singh, Valencia College
Patti Smith, Valencia College
Bethany Stone, University of Missouri
Willetta Toole-Simms, Azusa Pacific University
Bethany Williams, California State University, Fullerton

Class Test Participants

Bruno Borsari, Winona State University
Jessica Brzyski, Georgia Southern University
Beth Collins, Iowa Central Community College
Christopher Collumb, College of Southern Nevada
Jennifer Cooper, The University of Akron
Julie Ehresmann, Iowa Central Community College
Michael Fleming, California State University, Stanislaus
Susan Holecheck, Arizona State University
Dauna Koval, Bellevue College
Kiran Misra, Edinboro University of Pennsylvania
Marcelo Pires, Saddleback College
Michael L. Rutledge, Middle Tennessee State University
Jack Shurley, Idaho State University
Uma Singh, Valencia College
Paul Verrell, Washington State University
Daniel Wetzel, Georgia Southern University
Rachel Wiechman, West Liberty University

Instructor and Student Resource Contributors

Holly Ahern, SUNY Adirondack
Steven Christenson, Brigham Young University-Idaho
Beth Collins, Iowa Central Community College
Julie Ehresmann, Iowa Central Community College
Jenny Gernhart, Iowa Central Community College
Julie Harless, Lone Star College
Janet Harouse, New York University
Vedham Karpakakunjaram, Montgomery College
Dauna Koval, Bellevue College
Brenda Leady, The University of Toledo
Boriana Marintcheva, Bridgewater State University
Paige Mettler-Cherry, Lindenwood University
Lori Nicholas, New York University
Christopher Osovitz, University of South Florida

Tiffany Randall, John Tyler Community
    College
Lori Rose, Sam Houston State University
Suzanne Wakim, Butte College
Bethany Williams, California State
    University, Fullerton

Anne would like thank her husband, Will, and children Abi and Ben, who remained supportive and loving throughout the long process of writing this book. Megan's husband, Ryan, bore, with great patience, many dinner conversations about bats, algae, wolves, and more, and for that he has her thanks. To Megan's children, May and Parker; may you read this book someday and share your mother's joy about all things biology. Cindy thanks her husband, Mike, children Ben and Lily, and their numerous pets for the chaotic lifestyle that inspired her to step up her game. Also, Cindy thanks her friends and students who laugh at her jokes and keep her grounded in reality.

Perhaps most of all, we are indebted to the many scientists and individuals who shared their time and stories for these chapters. To the men and women we interviewed for this book, we cannot thank you enough. Your stories will inspire the next generation of biologists.

Caves of Death

Scientists scramble to identify a mysterious scourge decimating bat populations.

After reading this chapter you should be able to:

- Determine whether something is living or nonliving based on the characteristics of living things.
- Caption a diagram of the scientific method, identifying each step in the process.
- Given an observation, develop a hypothesis and suggest one or more predictions based on that hypothesis.
- Design an experiment using appropriate variables, treatments, and controls.
- Distinguish between correlation and causation, and give a clear example of each.
- Give specific examples of a scientific fact and a scientific theory.
- Create a graphic showing the levels of biological organization.

CHAPTER
01

THE NATURE
OF SCIENCE

Every spring for 30 years, Alan Hicks laced up his hiking boots, packed his camera, and set out to count bats in caves in upstate New York. A biologist with the New York State Department of Environmental Conservation, Hicks leads one of the few efforts in the country to collect annual data on bat populations. Since 1980, he had never missed the annual cave trip— until March 17, 2007.

"That day, of all days in my entire career, I stayed at my desk," recalls Hicks, who had remained behind to write a report for his supervisor. A couple of hours after his crew left to inspect some local caves, 15 miles from the Albany office, Hicks's cell phone rang.

"Hey, Al. Something weird is going on here," said a nervous voice. "We've got dead bats. Everywhere."

The line went quiet. "What are we talking here?" asked Hicks. "Hundreds of dead bats?"

"No," said the voice. "Thousands."

At first, Hicks conjectured that the bats had died in a flood, which had happened in that particular cave before. But the next day, a young volunteer who had been out with the team told Hicks to check his e-mail. The volunteer had sent him a picture taken the day before of eight little brown bats (*Myotis lucifugus*) hanging from a cave outcropping. Each one had a fuzzy white nose. This was a surprise because little brown bats do not have white noses.

Hicks e-mailed the picture to every bat researcher he knew. The fuzzy white material looked like a fungus, but there was no previous record of a fungus killing bats. As scientist after scientist looked at the picture, they all replied the same way: "What is that?" Whatever the white material was, no scientist had ever seen it before.

Hicks resolved to find out what was killing the bats and how the white fuzz was involved. As a biologist, Hicks took a scientific view of the world—logical, striving for objectivity, and valuing evidence over other ways of discovering the truth. **Science** is a body of knowledge about the natural world, but it is much more than just a mountain of data. Science is an evidence-based process for acquiring that knowledge.

- Science deals with the natural world, which can be detected, observed, and measured.

- Science is based on evidence that can be demonstrated through observations and/or experiments.

- Science is subject to independent validation and peer review.

- Science is open to challenge by anyone at any time on the basis of evidence.

- Science is a self-correcting enterprise.

To gather knowledge, Hicks would apply the **scientific method** (**Figure 1.1**). The scientific method is not a set recipe that scientists follow in a rigid manner. Instead, the term is meant to capture the core logic of how science works. Some people prefer to speak of the **process of science** rather than the scientific method. Whatever we call it, the practices that produce scientific knowledge can be applied across a broad range of disciplines—including bat biology.

Keep in mind that, as powerful as the scientific method is, it is restricted to seeking natural causes to explain the workings of our world. There are other areas of inquiry that science cannot address. The scientific method cannot tell us what is morally right or wrong. For example, science can inform us about how men and women differ physically, but it cannot identify the morally correct way to act on that information. Science also cannot speak to the existence of God or any other supernatural being. Nor can it tell us what is beautiful or ugly, which poems are most lyrical, or which paintings are most inspiring. So, although science exists comfortably alongside different belief systems—religious, political, and personal—it cannot answer all questions.

But science is the best way to answer questions about the natural world. The first two steps of the scientific method are to *gather observations* and *form a hypothesis*. Hicks didn't waste a moment of time before applying the scientific method to the question of the white fuzz. Bats were dying. "Bats are part of the planet and vital members of the ecosystem," says Hicks. "They play an important role in the environment in which we live."

ALAN HICKS

Alan Hicks is a retired bat specialist who began the investigation of a mysterious bat illness while working for the New York Department of Environmental Conservation.

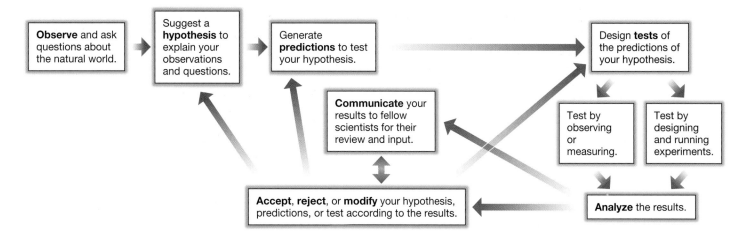

Figure 1.1

The scientific method
The scientific method is a logical process that helps us learn more about the natural world.

Bat Crazy

On March 18, the day after the first dead bats were discovered, Hicks entered the cave to make observations, a key part of the scientific process. An **observation** is a description, measurement, or record of any object or phenomenon. Hicks's team observed that the sick bats had not only white noses but also depleted fat reserves, meaning that the bats did not have enough stored energy to get through the winter. The bats also had white fuzz on their wings, scarred and dying wing tissue, and were behaving abnormally, waking up early from hibernation and leaving the cave when it was still too cold outside to hunt. Hicks's team also observed that the illness cut across species—many different types of bats were getting sick—and the bats exhibited a high rate of death: in some cases, up to 97 percent of infected bats died. Hicks and others began to call the illness white-nose syndrome (WNS). But they still didn't know what caused the syndrome, although from its characteristics, they assumed that the cause was a living organism (See "The Characteristics of Living Organisms" on page 6).

"For the first few years, we were just sleuthing," says Paul Cryan, a research biologist with the U.S. Geological Survey (USGS), and one of the scientists who received the original e-mailed picture from Hicks. From that first picture, Cryan was involved in trying to pinpoint the cause. "We were trying to understand something that had never happened before in a group of animals that was poorly understood."

In the caves, Hicks began collecting dead bats and sending them to laboratories around the nation. In those labs, technicians scraped samples from the bats' noses and wings, rubbed the samples into petri dishes—shallow glass or plastic plates containing a nutrient solution used to grow microorganisms—and watched to see if the white fuzz would grow. Time after time, many different types of bacteria and fungi grew on the dishes, speckling them with dots of different-colored colonies, but none of the samples were unusual. Nothing special or dangerous appeared to be present on the bats.

One researcher, a young microbiologist named David Blehert, decided to try something different. Blehert worked at the USGS National

Wildlife Health Center in Madison, Wisconsin. In December 2007, Hicks called Blehert. Blehert listened carefully as Hicks described how WNS was spreading. "He said, 'We have a major problem on our hands,'" recalls Blehert. "It turns out he was 100 percent right."

Hicks described to Blehert the conditions under which the bats lived during hibernation—caves in upstate New York, where the temperature was often between 30°F and 50°F. Blehert realized that most of the laboratories, including his, were trying to grow the samples from the bats at room temperature, a method conducive to the growth of many fungi. But in the caves, any living thing would have to grow at cold temperatures, so Blehert and his technicians took samples from dead bats, put them on petri dishes, and placed the dishes in the fridge.

At the same time, Melissa Behr, an animal disease specialist at the New York State Health Department, accompanied Hicks on a trip to a local cave (**Figure 1.2**). Behr swabbed a sample of the white fuzz directly from a bat in the cave, immediately spread it onto a glass slide and looked at it under a microscope. A unique fungus was on the plate. The fungus was visible in little white fuzzy patches of cells, and up close, the individual spores of the fungus appeared

The Characteristics of Living Organisms

All living things share certain features that characterize life.

1. *They are composed of one or more cells.* The **cell** is the smallest and most basic unit of life; all organisms are made of one or more cells. Larger organisms are made up of many different kinds of specialized cells and are known as *multicellular organisms*.

2. *They reproduce using DNA.* All living organisms are able to **reproduce**, to make new individuals like themselves. **DNA** is the genetic material that transfers information from parents to offspring. A segment of DNA that codes for a distinct genetic characteristic is called a *gene*. Life, no matter how simple or how complex, uses this inherited genetic code to direct the structure, function, and behavior of every cell.

3. *They obtain energy from the environment to support metabolism.* All organisms need **energy** to survive. Organisms use a wide variety of methods to capture this energy from their environment. The capture, storage, and use of energy by living organisms is known as *metabolism*.

4. *They sense the environment and respond to it.* Living organisms **sense** many aspects of their external environment, from the direction of sunlight to the presence of food and mates. All organisms gather information about the environment by sensing it, and then respond appropriately.

5. *They maintain a constant internal environment.* Living organisms sense and respond to not only the external environment, but also their internal conditions. All organisms maintain constant internal conditions—a process known as **homeostasis**.

6. *They can evolve as groups.* **Evolution** is a change in the genetic characteristics of a group of organisms over generations. When a characteristic becomes more or less common across generations, evolution has occurred within the group.

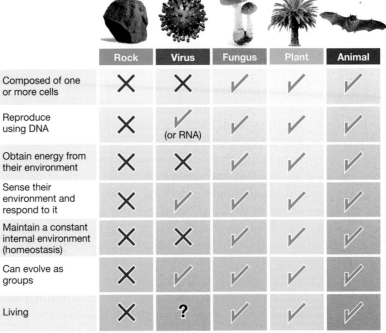

	Rock	Virus	Fungus	Plant	Animal
Composed of one or more cells	X	X	✓	✓	✓
Reproduce using DNA	X	✓ (or RNA)	✓	✓	✓
Obtain energy from their environment	X	X	✓	✓	✓
Sense their environment and respond to it	X	✓	✓	✓	✓
Maintain a constant internal environment (homeostasis)	X	X	✓	✓	✓
Can evolve as groups	X	✓	✓	✓	✓
Living	X	?	✓	✓	✓

crescent-shaped—different from all the other "normal" microbes growing on the bats' skin, and different from any fungus known to the researchers.

But Behr's single observation wasn't enough evidence to convince anyone that the strange-looking fungus was the cause of WNS. To be of any use in science, an observation must be repeatable, preferably by multiple techniques. Independent observers should be able to see or detect the same object or phenomenon, at least some of the time.

Luckily, Blehert was able to reproduce Behr's results by an independent technique. After letting his plates sit in the fridge for a few weeks, Blehert removed them and observed white patches of the same strange, crescent-shaped fungal spore. "Okay, we now have in laboratory culture what Melissa captured when she collected white material in the caves," thought Blehert. "We've got it."

Prove Me Wrong

In science, just as in everyday life, observations lead to questions, and questions lead to potential explanations. For example, if you flip on a light switch but the light does not turn on, you wonder why, and then look for an explanation: Is the lamp plugged in? Has the lightbulb burned out? You then identify one of these explanations as the most likely hypothesis for why the light did not turn on. A scientific **hypothesis** (plural "hypotheses") is an informed, logical, and plausible explanation for observations of the natural world. From the start, Hicks hypothesized that a new, cold-loving fungus was the primary cause of death in the bats. After observing the unique crescent-shaped fungal spores, Behr and Blehert agreed with his hypothesis. "It was the simplest solution," says Blehert. "We had bats with a white fungus that nobody had ever seen before growing on them, so that was the most likely thing that was doing it."

But other scientists disagreed. A fungus itself is rarely deadly to a mammal; more often a fungus causes an annoying, but not lethal, skin infection or is a secondary response after an animal gets sick from a viral or bacterial infection. So scientists proposed other hypotheses for the cause of WNS. Some suggested the fungus was a

Figure 1.2

Researchers prepare to enter the bat cave
Scientists suit up to collect more observations on the infected bats and the environmental conditions in the bats' roosting cave.

Q1: Which step(s) in the scientific method does this photograph illustrate?

Q2: What types of environmental data might the researchers have collected?

Q3: Why do you think the researchers are wearing protective gear?

secondary effect of an underlying condition, such as a viral infection. Others hypothesized that an environmental contaminant, such as a pesticide, was the cause of death. "There were so many different hypotheses," says Cryan. "But that's what is beautiful about the scientific process. You

DAVID BLEHERT

A microbiologist at the National Wildlife Health Center, David Blehert studies a variety of fungal and bacterial pathogens that are harmful to bats, humans, and other species.

observe as much as you can, and from those observations you can form multiple hypotheses. Science doesn't proceed by just landing on the right hypothesis the first time."

One of the joys, and challenges, of the scientific method is that after scientists suggest competing hypotheses, they then test their own hypotheses against those of others. A scientific hypothesis must be constructed in such a way that it is potentially **falsifiable**, or refutable. In other words, it must make predictions that can be clearly determined to be true or false, right or wrong (**Figure 1.3**). A well-constructed hypothesis is precise enough to make predictions that can be expressed as "if … then" statements.

For example, *if* WNS is caused by a transmissible fungus, *then* healthy bats that hibernate in contact with affected bats should develop the condition. *If* the fungus is secondary to an underlying condition, *then* the infection will only occur in bats after the primary underlying condition is present. *If* an environmental contaminant is the cause, *then* bats with WNS symptoms will have elevated levels of that contaminant in their blood or on their skin.

In each "if … then" case, it is possible to design tests able to demonstrate that a prediction is right or wrong. Although predictions can be shown to be true or false, the same is not true of hypotheses. Hypotheses can be *supported*, but no amount of testing can *prove* a hypothesis is correct with complete certainty (**Figure 1.4**). This is because there may be another, unmeasured or unobserved, reason that the prediction is true. For example, consider the first prediction stated in the previous paragraph—that healthy bats hibernating in contact with affected bats will develop the condition. If this is true, the reason might be that the healthy bats were infected by a fungus from their neighbor, supporting the hypothesis that the disease is caused by a transmissible fungus. Alternatively, related bats may tend to hibernate together in the same cave, and the disease, or at least vulnerability to the disease, might be genetically based. The hypothesis that the disease is fungal is *supported* but not *proved* by the correctness of this prediction.

Blehert set out to test the hypothesis that he, Behr, and Hicks had put forward—that a unique, cold-loving fungus was the primary cause of death in the bats. One can test a hypothesis through observational studies or experimental studies. Blehert's first study was observational. Observational studies can be purely **descriptive**—reporting information (**data**) about what is found in nature. Observational studies can also be **analytical**—looking for (analyzing) patterns in the data and addressing how or why those patterns came to exist. The tools of **statistics**—a branch of mathematics that can quantify the reliability of data—help scientists determine how well those patterns support a hypothesis. Observational studies usually rely on both descriptive and analytical methods to test predictions made by a hypothesis.

In 2009, Blehert, Behr, and Hicks published a scientific paper in which they described the results from inspecting 117 dead bats. They identified microscopic damage caused by a specific kind of fungus in 105 of the bats, and isolated and identified the fungus from a subset of 10 of them. It was a type of cold-loving fungus belonging to a group of fungi called *Geomyces*. They named this new species *Geomyces destructans*.

Their results revealed a correlation. **Correlation** means that two or more aspects of the natural world behave in an interrelated manner: if one shows a particular value, we can predict a value for the other aspect. From the physical

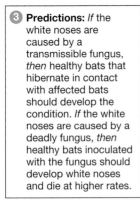
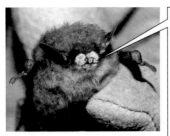

❶ **Observations and questions:** Bats are observed with white noses. What is causing the white fuzz? These bats are dying at higher rates than bats without white noses. Why?

❷ **Hypothesis:** Bats with white noses are infected with a fungus, and this fungus is causing death.

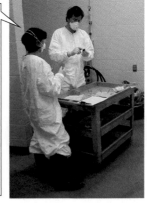

❸ **Predictions:** *If* the white noses are caused by a transmissible fungus, *then* healthy bats that hibernate in contact with affected bats should develop the condition. *If* the white noses are caused by a deadly fungus, *then* healthy bats inoculated with the fungus should develop white noses and die at higher rates.

Figure 1.3

From observation to hypothesis to testable prediction

Figure 1.4

Hypotheses are supported or not supported, but never proved
Although this vintage advertisement for cigarettes is amusing, "science" is still used to sell products today. Most Americans see thousands of advertisements every day, and many of these make "scientific" claims that are exaggerated or inaccurate.

Q1: State the hypothesis that this advertisement is claiming was scientifically tested.

Q2: State a prediction that comes from this hypothesis. Is it testable? Why or why not?

Q3: Explain in your own words why the hypothesis cannot be "proved."

characteristics of the dead bats, Blehert predicted that the white fungus was correlated with the illness. But correlation does not establish *causation*. Observational studies suggest possible causes for a phenomenon, but they do not establish a cause-effect relationship. To demonstrate that the fungus was actually causing the illness—and not just correlated with it—Blehert designed

and conducted an experiment. Testing scientific hypotheses often involves both observational and experimental approaches (**Figure 1.5**).

Catching the Culprit

An **experiment** is a repeatable manipulation of one or more aspects of the natural world. Blehert's experiment was to take healthy bats into his laboratory and expose them to the fungus. Like analytical observational studies, experimental studies use statistics to determine whether the experimental results support or refute the hypothesis being tested.

In studying nature, whether through observations, experiments, or both, scientists focus on **variables**—characteristics of any object or individual organism that can change. In a scientific experiment, a researcher typically manipulates a single variable, known as the **independent variable**. In this case, Blehert's independent variable was fungal exposure. Some bats were exposed; others were not. A **dependent variable** is any variable that responds, or could potentially respond, to the changes in the independent variable. Blehert's dependent variable was any sign of WNS on the healthy bats. If we think of the independent variable as the cause, then the dependent variable is the effect. In the most basic experimental design, a researcher manipulates a single independent variable and tracks how that manipulation changes the value of a dependent variable. Blehert manipulated his independent variable—exposing some bats to the fungus but not others—and then tracked his dependent variable, whether the bats showed symptoms of WNS.

MELISSA BEHR

Melissa Behr, formerly with the New York Department of Health, is now a veterinary pathologist at the University of Wisconsin (UW), Madison. She conducts research on the pathology and biology of bats and teaches at the UW veterinary school.

Descriptive: Looking for physical evidence of WNS.

Analytical: Measuring the weight of each bat.

Experimental: Injecting a fungicide to determine whether it will protect a bat from WNS.

One research team; three approaches.

Figure 1.5

Testing hypotheses using multiple approaches
Scientists set up an underground laboratory in Tennessee's New Mammoth Cave to test hypotheses about white-nose syndrome (WNS) using descriptive, analytical, and experimental approaches.

Q1: Give a possible hypothesis that could be tested by weighing the bats.

Q2: State the hypothesis being tested in the photo on the bottom right.

Q3: Explain in your own words why an experimental study is the only way to show a cause-effect relationship.

Blehert made sure his experiment was a controlled experiment. A **controlled experiment** measures the value of the dependent variable for two groups of subjects that are comparable in all respects, except that one group is exposed to a change in the independent variable and the other group is not. In this case, healthy bats were either exposed to the fungus or not exposed. Typically, a researcher obtains a sufficiently large sample of study subjects and assigns them randomly to two groups. Randomization helps ensure that the two groups are comparable to start with.

One group, the **control group**, is maintained under a standard set of conditions with no change in the independent variable. Blehert had 34 healthy bats in his control group; he kept these bats in the laboratory, under the same conditions as the other bats, but he did not expose them to *G. destructans*.

The other group, known as the experimental or **treatment group**, is maintained under the same standard set of conditions as the control group, but the independent variable is manipulated. Blehert exposed 83 healthy bats to the fungus. Of these, 36 were exposed to the fungus

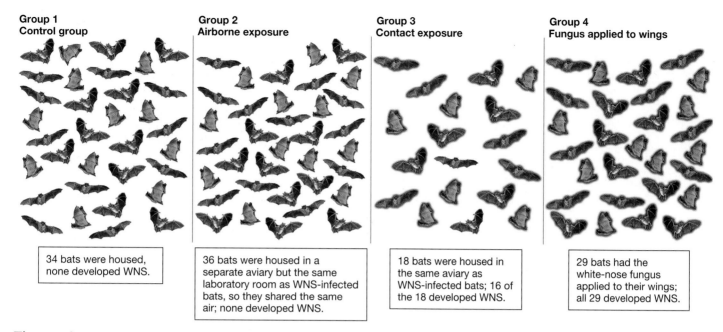

Group 1
Control group

34 bats were housed, none developed WNS.

Group 2
Airborne exposure

36 bats were housed in a separate aviary but the same laboratory room as WNS-infected bats, so they shared the same air; none developed WNS.

Group 3
Contact exposure

18 bats were housed in the same aviary as WNS-infected bats; 16 of the 18 developed WNS.

Group 4
Fungus applied to wings

29 bats had the white-nose fungus applied to their wings; all 29 developed WNS.

Figure 1.6

Blehert's experimental design
Blehert and his colleagues captured 117 healthy bats and brought them into the laboratory. They divided the bats into control and treatment groups and observed them for 102 days.

Q1: Which is the control group in this experiment, and what are the three treatment groups?

Q2: What are the hypotheses being tested in this experiment?

Q3: In one or two sentences, state the conclusions you can draw from the experiment. Were the hypotheses supported? Why or why not?

through the air, 18 were put into close contact with naturally infected bats, and 29 had the fungus applied directly to their wings (**Figure 1.6**).

After watching both the control group and the treatment group, Blehert saw what he had hypothesized: physical exposure to the fungus caused white-nose syndrome, but exposure through the air did not. Healthy bats that had fungus applied directly to their wings or were caged with naturally infected bats had high rates of WNS by the experiment's end. It was the first direct evidence that the fungus was the primary cause of white-nose syndrome.

As noted earlier, when scientists test a prediction of a hypothesis and find it upheld, the hypothesis is said to be supported. Scientists can be relatively confident in a supported hypothesis, but they cannot say that the hypothesis has been proved true. Even well-established scientific ideas

can be overturned if new evidence against the prevailing view comes to light. Albert Einstein is famously reported to have said, "No amount of experimentation can ever prove me right; a single experiment can prove me wrong."

When a prediction is not upheld, the hypothesis is reexamined and changed, or it is discarded. Over the years, other hypotheses about the cause of white-nose syndrome have not been upheld. For example, scientists were not able to identify a single environmental contaminant at elevated levels in infected bats. A follow-up study by Blehert and others showed that the fungus not only leads to symptoms of WNS in bats, but is sufficient to cause death (**Figure 1.7**).

The story of the bats with white noses is just a single example of the scientific process. One of the greatest strengths of science is that scientific knowledge is tentative and therefore open

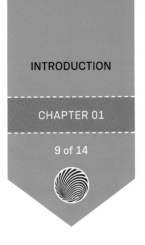

to challenge at any time by anyone. An absolute requirement of the scientific method is that evidence be based on observations or experiments, or both. Furthermore, the observations and experiments that furnish the evidence must be subject to testing by others: Independent researchers should be able to make the same observations, or obtain the same results, if they use the same conditions. In addition, the evidence must be collected in as objective a fashion as possible—that is, as free of personal or group bias as possible. Blehert's experiment fit all these conditions.

The main mechanism for policing personal or group bias and even outright fraud in science is **peer-reviewed publication**. Peer-reviewed publications are found in scientific journals that publish original research after it has passed the scrutiny of experts who have no direct involvement in the research under review. Before

Blehert's research was published, it was reviewed by numerous scientists who did not participate in the experiment. If reviewers have concerns during the peer-review process, such as whether the evidence is strong enough to support a hypothesis, they can ask the paper's authors to address those concerns (for example, by gathering additional evidence) and to resubmit the paper. Blehert's paper passed the peer-review process and was published in the scientific journal *Nature* in 2011. At that point, says Blehert, the evidence that *G. destructans* causes WNS was sufficient enough that "I think we'd convinced most people." (In 2013, scientists renamed *G. destructans* to *Pseudogymnoascus destructans* based on a reclassification of the fungus's genus.)

But identifying the cause of WNS did not stop the disease from spreading. By March 2008, just a year after Hicks's team found the thousands of dead bats near Albany, more bats were found dead and dying in caves across Vermont, Massachusetts, and Connecticut. By 2009, the disease had spread as far as Tennessee and Missouri.

The spread of WNS is a fact: Bats around the United States are dying. In casual conversation, we typically use the term "fact" to mean a thing that is known to be true. A scientific **fact** is a direct and repeatable observation of any aspect of the natural world.

Scientific fact should not be confused with scientific **theory**. Outside of science, people often use the word "theory" to mean an unproven explanation. In science, a theory is a hypothesis, or a group of related hypotheses, that has received substantial confirmation through diverse lines of investigation by independent researchers. Scientific theories have such a high level of certainty that we base our everyday actions on them. For example, the *germ theory of disease*, formally verified by Robert Koch in 1890, is the basis for treating infections and maintaining hygiene in the modern world (**Figure 1.8**).

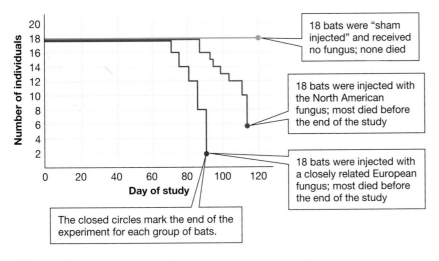

Figure 1.7

Experiments support the hypothesis that the fungus causes WNS in bats

The experiment that resulted in this graph was conducted by some of the same researchers interviewed for this story.

Q1: What is the control group in this experiment, and what are the two treatment groups?

Q2: At day 40, approximately how many individuals were alive in each treatment group? At day 80? At day 100?

Q3: In one or two sentences, state the conclusions you can draw from the experiment. Was the hypothesis supported? Why or why not?

No End in Sight

According to the U.S. Fish and Wildlife Service, white-nose syndrome has infected bats in at least 25 U.S. states and five Canadian provinces since 2007. Almost all species of bats that hibernate in these regions have been affected, including little brown bats and endangered Indiana bats, both of which have been particularly hard hit.

The fungus appears to be related to a type of fungus common in caves in Europe, so a human traveler from Europe most likely carried it across the Atlantic and into the Albany cave, where it infected its first bat in the United States. Researchers continue to explore exactly how the fungus kills the bats. It appears to wake them from hibernation too many times during the winter, so the bats use up their fat reserves too soon and do not survive the months of cold weather. The fungus also damages bats' delicate wings, which are important not only for flight but also for maintaining healthy levels of water, oxygen, and carbon dioxide in the bats' bodies. It is a powerful example of how a microorganism can affect many levels of life, from individual tissues and organs up to whole populations, communities, and even the ecosystem itself (**Figure 1.9**).

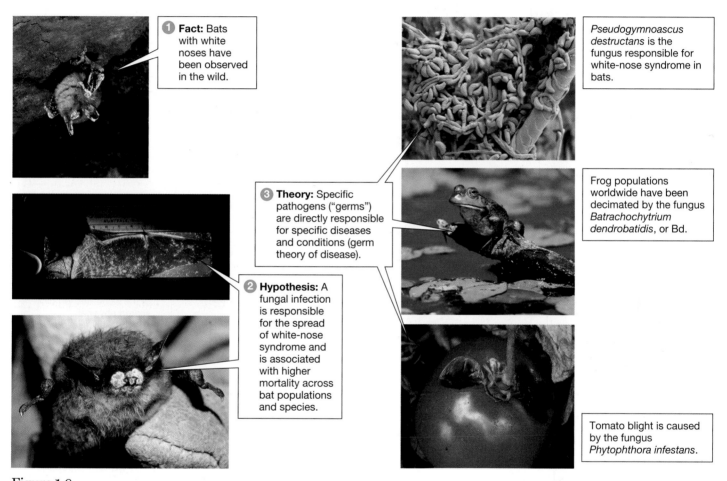

1 **Fact:** Bats with white noses have been observed in the wild.

2 **Hypothesis:** A fungal infection is responsible for the spread of white-nose syndrome and is associated with higher mortality across bat populations and species.

3 **Theory:** Specific pathogens ("germs") are directly responsible for specific diseases and conditions (germ theory of disease).

Pseudogymnoascus destructans is the fungus responsible for white-nose syndrome in bats.

Frog populations worldwide have been decimated by the fungus *Batrachochytrium dendrobatidis*, or Bd.

Tomato blight is caused by the fungus *Phytophthora infestans*.

Figure 1.8

Facts, hypotheses, and theories

It is important to distinguish between facts, hypotheses, and theories when thinking and talking about science.

Q1: Give one *fact* about bats that you learned from this chapter.

Q2: What is another example of evidence for the *germ theory of disease*? (*Hint:* Think about human diseases.)

Q3: Explain in your own words the difference between a fact and a hypothesis, and between a hypothesis and a theory.

Today, WNS continues to spread, and scientists have not found a way to stop it, says Hicks. Every winter, bats continue to die across America. Now, Hicks worries that students, hikers, and tourists will visit caves in the United States, not see any bats, and think that's normal. "In 2006, in one big cave in the Adirondacks, we counted over 185,000 bats. Anywhere you shined a light, there was a bat," says Hicks. "You go in now, and there's not a bat in sight."

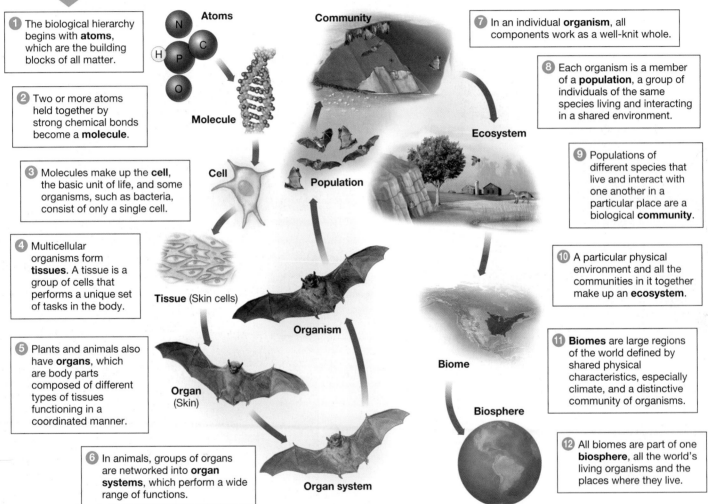

① The biological hierarchy begins with **atoms**, which are the building blocks of all matter.

② Two or more atoms held together by strong chemical bonds become a **molecule**.

③ Molecules make up the **cell**, the basic unit of life, and some organisms, such as bacteria, consist of only a single cell.

④ Multicellular organisms form **tissues**. A tissue is a group of cells that performs a unique set of tasks in the body.

⑤ Plants and animals also have **organs**, which are body parts composed of different types of tissues functioning in a coordinated manner.

⑥ In animals, groups of organs are networked into **organ systems**, which perform a wide range of functions.

⑦ In an individual **organism**, all components work as a well-knit whole.

⑧ Each organism is a member of a **population**, a group of individuals of the same species living and interacting in a shared environment.

⑨ Populations of different species that live and interact with one another in a particular place are a biological **community**.

⑩ A particular physical environment and all the communities in it together make up an **ecosystem**.

⑪ **Biomes** are large regions of the world defined by shared physical characteristics, especially climate, and a distinctive community of organisms.

⑫ All biomes are part of one **biosphere**, all the world's living organisms and the places where they live.

Atoms · **Molecule** · **Cell** · **Tissue** (Skin cells) · **Organ** (Skin) · **Organ system** · **Organism** · **Population** · **Community** · **Ecosystem** · **Biome** · **Biosphere**

Figure 1.9

The biological hierarchy

The **biological hierarchy** is a way to visualize the breadth and scope of life, from the smallest structures to the broadest interactions between living and nonliving systems.

Q1: Give examples of other kinds of organs that mammals such as bats have. (*Hint:* Think of the organs in your own body.)

Q2: Are bats in California part of the community of bats in upstate New York, if they are of the same species? Why or why not?

Q3: Is the soil within a cave in which bats live a part of the bats' population, community, or ecosystem? Explain your reasoning.

Bats in the Barn

Bats are skilled natural exterminators, consuming billions upon billions of insects each year, including crop pests. Because of this, bats play a critical role in agriculture. The loss of bat populations to white-nose syndrome would have a significant impact on farms and forests around the country.

1.3
MILLION

A single colony of **150 big brown bats** (*Eptesicus fuscus*) in Indiana is estimated to eat nearly **1.3 million pest insects each year.**

4–8
GRAMS

A single little brown bat (*Myotis lucifugus*) can consume **4 to 8 grams of insects each night** during the active season.

660–1320
METRIC TONS

Extrapolating the diet of a single bat to one million bats estimated to have died from WNS, between **660 and 1320 metric tons of insects** are no longer being consumed each year in WNS-affected areas.

$22.9
BILLION

If bats disappeared entirely from the United States, it would cost the agricultural industry roughly **$22,900,000,000 per year** to save crops by dealing with the insects no longer being eaten by bats.

REVIEWING THE SCIENCE

- **Science** is both a body of knowledge about the natural world and an evidence-based process for generating that knowledge.

- The **scientific method** represents the core logic of the process by which scientific knowledge is generated. The scientific method requires that we (1) make **observations**, (2) devise a **hypothesis** to explain those observations, (3) generate predictions from that hypothesis, (4) test those predictions, and (5) share the results of the tests, for **peer review** by fellow scientists.

- A hypothesis cannot be proved true; it can only be supported or not supported. If the predictions of a hypothesis are not supported, the hypothesis is rejected or modified. If the predictions are upheld, the hypothesis is supported.

- We can test hypotheses by making further observations or by performing **experiments** (controlled, repeated manipulations of nature) that will either uphold the predictions or show them to be incorrect.

- In a scientific experiment, the **independent variable** is manipulated by the investigator. Any variable that can potentially respond to the changes in the independent variable is called the **dependent variable**.

- **Correlation** means that two or more aspects of the natural world behave in an interrelated manner: if one shows a particular value, we can predict a particular value for the other aspect. However, correlation between two variables does not necessarily mean that one is the cause of the other.

- A scientific **fact** is a direct and repeatable observation of any aspect of the natural world. A scientific **theory** is a major idea that has been supported by many observations and experiments.

- The **biological hierarchy** refers to the many levels at which life can be studied: atom, molecule, cell, tissue, organ, organ system, organism, population, community, ecosystem, biome, biosphere.

- All living organisms have the following characteristics in common: (1) They are built of **cells**; some are single-celled and some are multicellular. (2) They **reproduce**, using **DNA** to pass genetic information from parent to offspring. (3) They take in **energy** from their environment. (4) They **sense** and respond to their environment. (5) They exhibit **homeostasis**, maintaining constant internal conditions. (6) They can **evolve** as groups.

THE QUESTIONS

The Basics

1 Which of the following is a characteristic of life? Circle all that apply.

(a) metabolism

(b) movement

(c) evolve as a group

(d) cellular

(e) hair or fur

2 Which of the following is a living organism? Circle all that apply. If it is not living, which criterion, or criteria, does it not meet?

(a) an oak tree

(b) an influenza virus

(c) the fungus that causes white-nose syndrome in bats

(d) a diamond

(e) your teacher

3 Which of the following statements is a scientific hypothesis (that is, it makes testable predictions)? Choose only one.

(a) Even though no one else can see him, the ghost of my dog lives in my back yard.

(b) The Atkins diet helps people lose more weight and keep it off than Weight Watchers does.

(c) People born under the sun sign Aquarius are kinder and cuter than those born under Scorpio.

(d) It is unethical to text while driving.

(e) none of the above

4 Consider an experiment in which subjects are given a pill to test its effectiveness in reducing the duration of a cold. Which of the following is the best way to treat the control group?

(a) Give the control group two pills instead of one.

(b) Do nothing with the control group.

(c) Give the control group a pill that looks like the test pill but does nothing.

(d) Let the control group choose whether or not to take any pills.

(e) Expose the control group to the cold virus.

5 When scientists use the word "theory," they mean

(a) an educated guess.

(b) an overarching explanation of an interrelated set of observations.

(c) wild speculation.

(d) an experimental prediction.

(e) a fact proved by many experiments.

6 Circle the correct terms in the following sentence:
The process of science begins with a(n) (**prediction, observation**) about the natural world. A scientist then proposes a (**hypothesis, prediction**), which is the basis of one or more testable (**observations, predictions**).

7 Place the following steps of the scientific method in the correct order by numbering them from 1 to 7.

_____ a. Make observations about the natural world.

_____ b. Test the prediction(s) by designing an experiment or collecting observational data.

_____ c. Run the experiment and analyze the results.

_____ d. Generate predictions to test the hypothesis.

_____ e. Share the results with fellow scientists so that they can review and evaluate them.

_____ f. Develop a hypothesis to explain the observation.

_____ g. Accept, reject, or modify your hypothesis depending on the results.

8 Identify the level of biological organization for each of the following.

_____ a. the kidney of a bat

_____ b. an oak tree outside a cave in upstate New York

_____ c. bats in a cave in upstate New York

_____ d. the physical and biological components of a cave in upstate New York

_____ e. the respiratory system of a bat

_____ f. all the species living and interacting within a cave in upstate New York

Try Something New

9 Using the scientific method as outlined in this chapter, describe the steps you would take if you made the observation, "My computer is not working." What are the experimental variables, treatment group, and control group?

10 Mad cow disease, or bovine spongiform encephelopathy, appears to be caused by a novel infectious agent: a protein that replicates by causing related proteins to modify their structure from a harmless shape to a dangerous one. These prions (short for proteinaceous infectious particles) also appear to be the cause of several other diseases, such as scrapie in sheep, and kuru and Creutzfeldt-Jakob disease in humans. Are prions alive? What information would you need to make this determination?

11 Describe one observation, one hypothesis, and one experiment from the white-nose syndrome research discussed in this chapter.

Leveling Up

12 **Looking at data** Review the map shown here (an up-to-the-minute map can be found at http://whitenosesyndrome. org/resources/map). What do you hypothesize about where WNS is

spreading? From your hypothesis, where do you predict that WNS will next be observed?

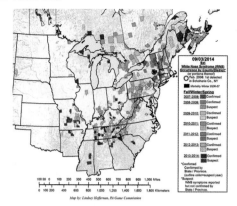

Map by: Lindsey Heffernan, PA Game Commission

13 **What do *you* think?** Insectivorous bats save the agriculture industry billions of dollars each year because they eat insects that damage crops. Other species of bats pollinate some crops. What would you predict about crop yields in areas that are experiencing heavy bat mortality from WNS? Does this affect your level of concern about WNS? Why or why not?

14 **Doing science** Although you are an expert on white-nose syndrome in bats, you have been asked to contribute your scientific expertise to understanding the fungal infection that is decimating frog populations worldwide: *Batrachochytrium dendrobatidis*, a.k.a. Bd.

(a) Read this May 2013 *National Geographic* article on how Bd has been spread around the world: http://news.nationalgeographic. com/news/2013/13/130515-chytrid-fungus-origin-african-clawed-frog-science.

(b) The graph provided here, based on data from a scientific paper published in 2013, shows that frog species with higher average body temperatures tend to have lower Bd infection rates. It also shows that females within a species tend to have higher average body temperatures than males, and also to have lower Bd infection rates than males. Is this observational study descriptive or analytical? Explain your answer.

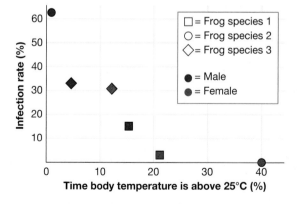

(c) Propose a hypothesis to explain the results depicted in the figure, and then identify a testable prediction from that hypothesis. Design an experiment to test the prediction, identifying your independent and dependent variables, the control group, and the treatment conditions. Create a graph to show the results you expect to find (1) if your hypothesis is supported and (2) if your hypothesis is not supported.

Ingredients for Life

A dusty box, rediscovered after 50 years, sparks a scientific treasure hunt for the origins of life on Earth.

After reading this chapter you should be able to:

- Label an illustration of an atom.
- Compare and contrast the types of bonds used to form molecules.
- Diagram the hydrogen bonds that form between water molecules and explain how those bonds produce the unique properties of water.
- Explain the difference between hydrophilic and hydrophobic molecules, and between acidic and basic molecules.
- Describe the chemical qualities of carbon that make it the basis of life on Earth.
- Create a graphic depicting how each of the four main classes of macromolecules and their subunits are related.

CHAPTER
02

CHEMISTRY
OF LIFE

Jim Cleaves hauled yet another box to the dumpster. The young, dark-haired researcher at UC San Diego had the unenviable task of cleaning out the laboratory of his PhD adviser, the recently retired chemist Stanley Miller. It was 2003. Cleaves tossed box after box—20 years' worth of chemical samples from experiments done by Miller and his former students. As the lab emptied, a small cardboard box high on a shelf caught Cleaves's eye. It was labeled "Electric Discharge Samples."

"That box looked like something I'd really regret throwing out," says Cleaves. So instead of a trip to the trash, Cleaves gave the box to another former student and friend of Miller's, the marine chemist Jeffrey Bada at the Scripps Institution of Oceanography. The box sat on Bada's shelf, unopened, for years. Neither he nor Cleaves had any idea there was a scientific treasure trove inside, preserved for them by Miller.

In 1952, Miller was a graduate student at the University of Chicago looking for a thesis idea. He approached Harold Urey, a Nobel Prize–winning chemist who, a year and a half earlier, had proposed a radical idea to the scientific community, a hypothesis about Earth's early atmosphere and the origins of life. Scientists knew that several key types of **matter** (anything that has mass and occupies a volume of space) existed on the early Earth, but they debated which types were necessary for the emergence of life between three and four billion years ago. One type of matter is an **element**, a pure substance that has distinct physical and chemical properties, and that cannot be broken down into other substances by ordinary chemical methods. There are 98 natural elements known to us, and another 20 have been created in laboratories.

An **atom** is the smallest unit of an element that retains the element's distinctive properties. Atoms make up all common materials, including this book, the air, and you. Every atom has a dense core called a **nucleus** (plural "nuclei")

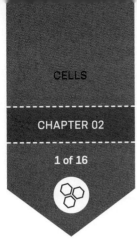

STANLEY MILLER

Stanley Miller was an American chemist who designed the first experiment to mimic Earth's early atmosphere and pioneered the study of the origins of life. He died in 2007 at the age of 77.

made up of positively charged **protons** and electrically neutral **neutrons**. A cloud of negatively charged **electrons** surrounds the nucleus (**Figure 2.1**). Electrons have significantly less mass than protons and neutrons: if an electron weighed as much as a bowling ball, a proton or neutron would be as heavy as a car.

The number of protons in an atom's nucleus is called its **atomic number** and is unique to that element. **Isotopes** of an element have the same number of protons but different numbers of neutrons. The sum of the number of protons and the number of neutrons is the **atomic mass number** of an isotope. The atomic mass number is how much mass is in an element, or in other words, how much it weighs. For example, the most common isotope of carbon has 6 protons and 6 neutrons, giving it an atomic mass number of 12, and we call it carbon-12 (^{12}C). The isotope of carbon with 6 protons and 8 neutrons is carbon-14 (^{14}C).

Atoms interact with other atoms via electrons; they can donate electrons, accept electrons, and even share electrons. When two atoms share electrons, they form a **covalent bond**. Atoms linked by covalent bonds form **molecules**. Molecules that include at least one carbon atom are referred to as **organic molecules**. Urey and others suspected that gas molecules in early Earth's atmosphere combined to form the earliest organic molecules, which later assembled into the first living organism.

Urey had proposed that Earth's early atmosphere resembled that of other planets in our solar system, including Jupiter, Saturn, and Uranus. He suggested that, like the atmospheres of those planets, our atmosphere was once rich with important **chemical compounds**, molecules that contain atoms from two or more different elements. He proposed that Earth's early atmosphere was made up of mainly four compounds: the gases methane (CH_4), ammonia (NH_3), hydrogen (H_2), and water vapor (H_2O).

The young Miller was intrigued by Urey's hypothesis, and he asked Urey if he could perform an experiment to test it, to see if such a combination of gases could be used to create other simple compounds. Urey dismissed the idea, saying that it was too difficult for a student's thesis and that, if it didn't work, Miller would have nothing to show in order to graduate. But Miller pestered him until at last Urey agreed to let Miller work on

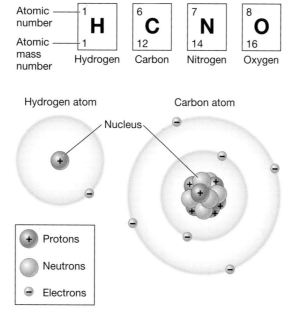

Atomic number

Atomic mass number

1	6	7	8
H	**C**	**N**	**O**
1	12	14	16
Hydrogen	Carbon	Nitrogen	Oxygen

Hydrogen atom   Carbon atom

Nucleus

+ Protons

Neutrons

– Electrons

Figure 2.1

Atomic structure

The electrons, protons, and neutrons of these hydrogen and carbon atoms are shown greatly enlarged in relation to the size of the whole atom.

Q1: How many protons, neutrons, and electrons does the hydrogen atom shown here have?

Q2: What are the atomic number and the atomic mass number of the carbon isotope shown?

Q3: Nitrogen-11 is an isotope of nitrogen that has 7 protons and 4 neutrons. What are the atomic number and atomic mass number of nitrogen-11?

the project for six months, a year at the most. As it turned out, all Miller needed was a few weeks.

Miller mixed the four gases together in a large glass apparatus. He then zapped the swirling cloud of compounds with a continuous electrical spark to mimic lightning strikes on early Earth, which he surmised would break apart the compounds (**Figure 2.2**). The process of breaking existing chemical bonds and creating new ones is known as a **chemical reaction**. The **reactants** (the gases, in this case) undergo a chemical change and form new molecules, called **products**. Some chemical reactions, like the one Miller devised, require energy in order to proceed. Others release energy.

After the first day of the experiment, a pink liquid pooled in the bottom of the apparatus. By the end of a week, it had turned "deep red and turbid," Miller reported. The color changes, he later discovered, reflected the formation of new products through chemical reactions.

To Miller and Urey's pleased surprise, the resulting red broth contained several biologically significant compounds, including five **amino acids**, small molecules important to life. There are hundreds of known amino acids, but only 20 are the building blocks for proteins, a major class

HAROLD UREY

Harold Urey was an American chemist who won the Nobel Prize in Chemistry in 1934 for the discovery of an isotope of hydrogen. He was the first to speculate that Earth's atmosphere was composed of ammonia, methane, hydrogen, and water vapor.

of molecules found in the cells of living things. Of the five amino acids that Miller isolated, three were part of that group of 20. Thus, Miller had indeed created some, but not all, of the amino acids used in life on Earth from a simple mixture of gases.

At a crowded seminar, Miller presented his results while Urey sat in the front row. Colleagues at the meeting recalled that after the presentation, the famed physicist Enrico Fermi turned to Urey and said, "I understand that you and Miller have demonstrated that this is one path by which

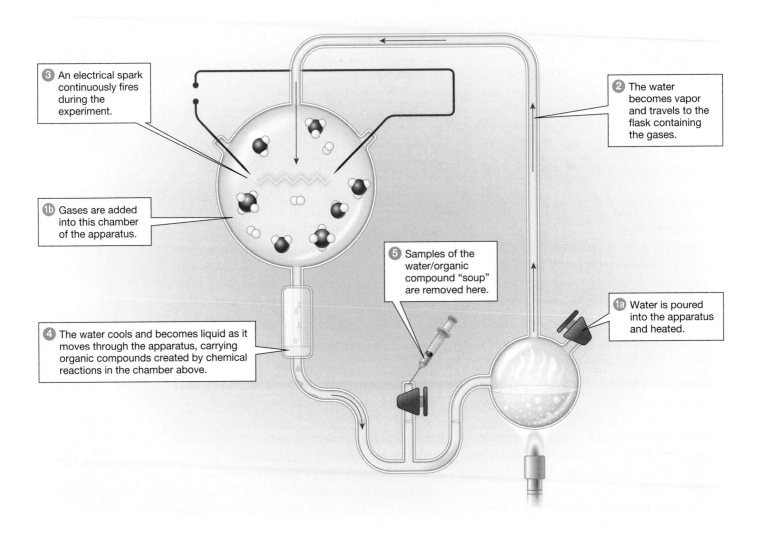

Figure 2.2

Miller's original spark discharge experiment

When Stanley Miller zapped a mixture of methane (CH_4), ammonia (NH_3), hydrogen (H_2), and water vapor (H_2O) with electrical sparks, he produced compounds called amino acids.

Q1: Before the experiment was run, the apparatus was sterilized and then carefully sealed. Why was this an important thing to do?

Q2: Why is inclusion of methane in the gas flasks an essential part of the hypothesis that complex organic molecules were formed in the early atmosphere of Earth? (*Hint:* What makes a molecule organic?)

Q3: Answer this question after reading about Miller's "steam injection" experiments: Where was the steam injected in the experimental apparatus?

life might have originated. Harold, do you think it was the way?" Urey looked at Fermi and replied, "Let me put it this way, Enrico. If God didn't do it this way, he overlooked a good bet!"

Miller and Urey's experiment, which became an instant classic in the scientific community, demonstrated that the basic chemicals for life could arise under natural conditions. Miller continued to perform spark discharge experiments, tweaking the experimental conditions and types of gases in the hope of producing more amino acids. But for an unknown reason, perhaps lack of time, Miller never published or followed up on many of his results. That is, until the mysterious box sitting on Jeffrey Bada's shelf was finally opened.

One Picture, a Thousand Experiments

In 2007, Bada was visiting the University of Texas to give a lecture. His talk was scheduled immediately after a talk by another close friend of Miller's, Antonio Lazcano, a biologist at the School of Sciences at the National Autonomous University of Mexico in Mexico City. The two men agreed to review each other's lecture slides to make sure their talks didn't overlap.

One of Lazcano's slides caught Bada's eye. It was a picture of a small glass vial, labeled as containing a residue from Miller's early experiments. Bada asked Lazcano about it. Lazcano explained that during a visit to his friend, Miller had pulled a cardboard box off a shelf, lifted out the vial, and let Lazcano take a picture. Miller told Lazcano that it was one of the leftover samples of his spark discharge experiments. Miller had saved them all.

"I was flabbergasted," says Bada. "I'd known Stanley since 1965, and he never once mentioned it." It dawned on Bada that the box might still exist. He called his lab and asked if anyone had seen the box that Cleaves gave him four years earlier. As soon as Bada returned to the lab, he found and opened the box. It was a scientist's Christmas; the box was full of carefully labeled plastic boxes containing thin glass vials, many with films of dried brown, tarlike gunk in the bottom.

HENDERSON (JIM) CLEAVES

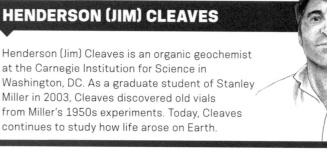

Henderson (Jim) Cleaves is an organic geochemist at the Carnegie Institution for Science in Washington, DC. As a graduate student of Stanley Miller in 2003, Cleaves discovered old vials from Miller's 1950s experiments. Today, Cleaves continues to study how life arose on Earth.

"It was on the order of 200–300 vials," says Bada (**Figure 2.3**). "It was extracts from experiments throughout the course of his life."

Luckily, Miller kept notebooks detailing the specific contents of each vial. He had performed two other experiments shortly after his original spark discharge work, using variations on the original apparatus. In one, a different method generated the spark. In another, which caught Bada's attention, hot steam was injected directly into the spark chamber.

Bada decided to analyze the contents of the vials. He sent the samples to Jason Dworkin, also one of Miller's former students, at NASA's Goddard Space Flight Center in Maryland. The Goddard Space Flight Center is home to an advanced mass spectrometer, an instrument that measures the weights of tiny amounts of matter as a way to identify them.

The results were stunning. In 1953, Miller had identified just five amino acids, three of which

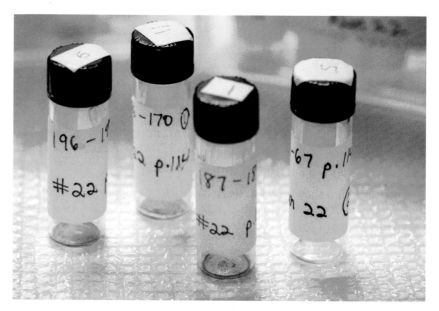

Figure 2.3

Electric discharge samples from Miller's experiments

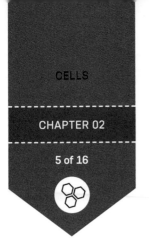

were common constituents of proteins. In 2008, analyzing the same samples with more sophisticated techniques, Bada's team identified 14 amino acids in the original experiment and a whopping 22 in the unpublished steam experiment. Bada was intrigued. Why did the second experimental apparatus result in more kinds of amino acids?

The World of Water

The major difference between the two experiments was steam—hot water vapor shooting through the gas chamber. Water is essential for life because of its unique chemical properties, which allow molecules to dissolve and interact in special ways through chemical reactions.

A water molecule is made up of two hydrogen atoms and one oxygen atom held together by shared electrons—that is, by covalent bonds. However, these electrons are not shared equally; they spend more time near the oxygen atom than near the hydrogen atoms. Because electrons are negatively charged particles, the oxygen end of a water molecule takes on a slightly negative charge, and the hydrogen ends have slightly positive charges. This lopsided electron sharing means that water is a **polar molecule** (**Figure 2.4**).

In addition to covalent bonds, two other types of **chemical bonds** attach atoms to one another: ionic bonds and hydrogen bonds. Atoms that have lost or gained electrons are called **ions**. Electrons are negatively charged, so an atom that has gained an electron is a negative ion and an atom that has lost an electron is a positive ion. When a negatively charged ion and a positively charged ion are in the same vicinity, they will chemically attract each other and form an **ionic bond**. Common table salt, sodium chloride (NaCl), is composed of sodium and chlorine held together by ionic bonds.

The third means of attaching atoms to one another is called a **hydrogen bond**. Hydrogen bonds are weak electrical attractions between a hydrogen atom with a partial positive charge and a neighboring atom with a partial negative charge. Molecules of water bind to each other through hydrogen bonds because the negatively charged oxygen end of one water molecule weakly attracts one of the positively charged hydrogen ends of another water molecule. A single hydrogen bond is about 20 times weaker than a covalent bond, but water makes up for that lack of strength with sheer quantity. The collective crosslinking of many, many water molecules through hydrogen bonds amounts to a potent force.

The polarity of water molecules and hydrogen bonding explain nearly all of the special properties of water, which were critical in Miller's experiments. The foremost of these was that water was able to break apart the compounds in the flask. As you may have noticed the last time you soaked a dirty dish, water has an incredible ability to dissolve other materials. This is because water molecules form hydrogen bonds with other polar

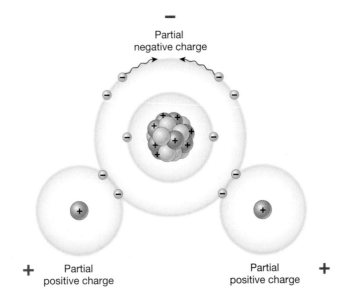

Partial negative charge

Partial positive charge

Partial positive charge

Figure 2.4

Water molecules are polar
The polarity of water molecules is the cause of water's unique properties.

Q1: Where are the covalent bonds in this figure?

Q2: This figure shows a water molecule (H_2O). A hydrogen molecule (H_2) consists of two hydrogen nuclei that share two electrons. Draw a simple diagram of a hydrogen molecule indicating the positions of the two electrons.

Q3: When table salt (sodium chloride, NaCl) dissolves in water, it separates into a sodium ion (Na^+) and a chloride ion (Cl^-). Which portion of a water molecule would attract the sodium ion, and which portion would attract the chloride ion?

molecules, like sugars or, in Miller's experiment, ammonia. The formation of hydrogen bonds with polar molecules causes those compounds to dissolve in water. Such compounds are said to be **soluble**; that is, they mix completely with the water.

A **solution** is any combination of a **solute** (a dissolved substance, such as sugar) and a **solvent** (the fluid, such as water, into which the solute has dissolved). Because of their polar nature, water molecules will not interact with uncharged or nonpolar substances, such as fat or oil. Molecules that are soluble in water (such as salt) are called **hydrophilic** ("water-loving"); molecules that don't dissolve well in water (such as oil) are called **hydrophobic** ("water-fearing"). **Figure 2.5** shows these processes in action.

When Bada and his colleagues published their results from reanalyzing Miller's vials, some scientists proposed that Miller's second experiment, in which he shot a jet of steam into the spark, resulted in more amino acids because the hot water enabled a wider variety of chemical reactions. Whether or not that was the case, water was central to Miller's success.

Water can exist in all three states of matter: liquid, solid, and gas. Hydrogen bonds explain the physical properties of water in these three phases. Though water is composed of two elements that are gases at room temperature (hydrogen and oxygen), it forms a liquid at room temperature because hydrogen bonds stick water molecules together, keeping them close. Those bonds are constantly forming and breaking in water, creating a nonstop jostling that gives water its liquid form (**Figure 2.6**, left).

When water chills, its molecules cannot move about as vigorously (**Figure 2.6**, right). As water turns into ice at 0°C, a stable network of hydrogen bonds emerges. The molecules become spaced farther apart, locked into an orderly pattern known as a crystal lattice. That spacing is why ice occupies more space than liquid water. Normal ice is nine percent less dense than liquid water (density is mass divided by volume), which explains why ice floats on water. This property of water is quite unusual; most substances are *more* dense in the solid state than in the liquid state.

Boiling water, as Miller did to inject steam into his experiments, requires the addition of a

Oil molecules are hydrophobic. They are excluded from water and tend to clump together.

Olive oil

Vinegar molecules are hydrophilic. They are held in solution by water molecules.

Vinegar

Figure 2.5

Hydrophilic substances dissolve in water, but hydrophobic substances do not

Q1: Describe what will happen to the molecules of olive oil if you shake the bottle and then leave it alone for an hour. What about the molecules of vinegar?

Q2: What would happen if you added another fat to the bottle, such as bacon grease, and shook it?

Q3: Given how sugar behaves when it is mixed into coffee or tea, would you predict that it is hydrophobic or hydrophilic?

significant amount of energy to break hydrogen bonds. When energy in the form of heat is supplied to a liquid, some molecules become energetic enough to make the transition from the

JEFFREY BADA

Jeffrey Bada is a chemist at the Scripps Institution of Oceanography at UC San Diego. With Jim Cleaves, he closely analyzed and then duplicated the Miller-Urey experiments. He is also a leading scientist studying organic compounds outside of Earth, including in meteorites and on Mars.

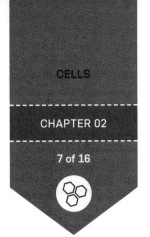

liquid to the gas state—a phenomenon known as **evaporation**.

In the case of water, a substantial amount of energy must be invested in snapping the network of hydrogen bonds before water molecules in the liquid state can move fast enough to escape as water vapor (steam) (**Figure 2.6**, bottom).

The Smell of Success

Water was key to the success of Miller's spark discharge experiments, but another compound would turn out to be almost as important. In 2011, three years after reanalyzing the first of

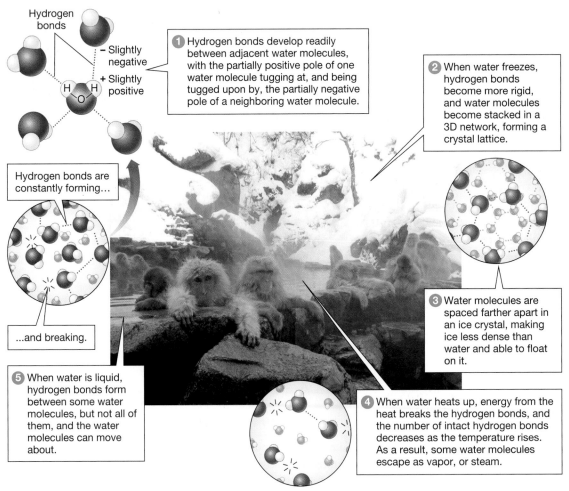

Figure 2.6

Water molecules change state as hydrogen bonds increase or decrease

Japanese snow macaques escape the cold with a daily dip in natural hot springs. Water can be seen here in its liquid, solid, and gas states.

Q1: Identify where in the picture water can be seen in its liquid, solid, and gas states.

Q2: In the gas state, water molecules move too rapidly and are too far apart to form hydrogen bonds. Compare the volumes occupied by an equal number of water molecules in the liquid, solid, and gas states.

Q3: Explain in your own words how ice floats on water.

Miller's samples, Bada, Cleaves, and a group of scientists teamed up again to analyze another set of vials tucked away in the cardboard box.

In 1958, Miller had performed additional spark discharge experiments that included a new gas: hydrogen sulfide (H_2S). You may know hydrogen sulfide gas as the cause of the awful smell released by rotten eggs, but it is also released by volcanoes and should have been present in Earth's early atmosphere (**Figure 2.7**). Miller's lab notes show that he performed the experiment and isolated the amino acids, but then never analyzed them. "Why he never analyzed it, I don't know," says Bada. In 2011, Bada's team finished the job. They identified 23 amino acids in Miller's test tube, including six that contained sulfur. One of those sulfur-containing amino acids, methionine, initiates the construction of all proteins in cells.

Miller's experiments support the hypothesis that volcanoes—a major source of hydrogen sulfide today—coinciding with lightning, which is often focused around volcanoes, may have played a role in making large and varied quantities of biologically crucial molecules, setting the stage for the evolution of life on Earth. But this is not the only hypothesis about the origins of life. There are other hypotheses, including one that has gained traction in recent years, that the building blocks for life arrived from space.

Although proteins use only 20 amino acids, scientists have identified 90 amino acids in meteorites, suggesting that the first organic materials could have come to Earth from outer space. At NASA's Ames Research Center in California, the astrophysicist Scott Sandford and his colleagues have been able to make amino acids from gases using conditions like those that exist in interstellar space.

Sandford's team has frozen different combinations of gases to simulate ices found on comets, and in interstellar clouds (where new stars and planets form) and then bombarded those ices with radiation. And what does their icy-hot recipe create? "We make amino acids all the time," says Sandford. In many cases, he and his team identified amino acids similar to those that Miller's experiment had created. "The universe seems to be an organic chemist," says Sandford. "It is hardwired to turn simple molecules into more complicated ones. The consequence of that is that

Figure 2.7

Miller's experiments were like volcanic eruptions

An erupting volcano releases all of the gases included in Miller's mixtures, and the volcanic ash contains iron and other metals. Steam is produced when the magma (superheated rock) comes in contact with groundwater, and lightning bolts lance through the cloud of gas and ash.

Q1: Suppose you were going to repeat Miller's experiments. How would you decide how much of each gas to include in the chamber?

Q2: Why did the addition of steam to the gases in Miller's second set of experiments increase the yield of amino acids?

Q3: Miller used electrical energy in his experiment. What other forms of energy were present in the early atmosphere of Earth that could have led to the formation of complex molecules?

it makes a whole host of products, some of which are biologically interesting."

On October 7, 2008, an asteroid entered Earth's atmosphere over Africa, and its impact was photographed by a NASA satellite (**Figure 2.8**). Initially 6–15 feet in diameter, the asteroid exploded over eastern Africa, and its fragments landed in Sudan. Scientists from NASA were able

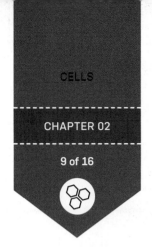

to find some of the pieces of the asteroid on the ground because the dark rocks stood out against the light-colored desert sand. These fragments are called the "Almahata Sitta" or "Station Six" meteorites, named after a train station near the location where pieces were recovered.

When the fragments were analyzed, 19 different amino acids were found. The extraterrestrial origin of these amino acids is confirmed by a particular characteristic of amino acid molecules: they have two forms, left-handed and right-handed, that are mirror images of each other. All of the amino acids produced by living organisms on Earth are of the left-handed form; right-handed amino acids are produced only in laboratories. The amino acids in the Almahata Sitta meteorites are a mixture of left-handed and right-handed forms. The presence of right-handed amino acids shows that the amino acids in the fragments really came from space, not from terrestial organic contamination after the fragments landed.

Getting the Right Mix

Miller never stopped trying to make complex molecules from simple ones. Though his 1953 experiment was originally met with fanfare, scientists later began to dispute the usefulness of his experiments. Methane and ammonia, they argued, didn't exist in large amounts on the early Earth. Instead, new evidence suggested that the atmosphere contained nitrogen gas (N_2) and carbon dioxide (CO_2). "Most people agree now [that Miller and Urey] didn't have the right composition," says Sandford.

So, in 1983, 30 years after his original experiment, Miller repeated it using nitrogen and carbon dioxide. But instead of a deep red/brown broth, the experiment produced a clear, seemingly barren liquid. The experiment seemed to have failed.

In 2007, Bada and Cleaves decided to revisit that experiment to see what had gone wrong. Instead of just reanalyzing samples, they redid the experiment and discovered that the chemical reactions between the new gases were producing chemicals called nitrites, which destroy amino acids. The solution also became acidic because of the presence of nitrous acid (HNO_2). An **acid** is a hydrophilic compound that dissolves in water and loses one or more hydrogen ions (H^+). Acids donate H^+ ions to water and therefore increase the concentration of free H^+ ions in an aqueous solution. H^+ ions are extremely reactive and can disrupt or alter other chemical reactions. The acidic solution, Bada and Cleaves realized, was preventing amino acids from forming.

To counteract that acidity, Bada needed to add a **base**. Acids and bases are chemical opposites.

A NASA satellite photographed the impact of the asteroid on October 7, 2008. The yellow arrow traces the path the asteroid followed, and the reddish-orange blob shows the point at which it exploded after entering Earth's atmosphere.

The asteroid broke into fist-sized pieces that fell into the Nubian Desert in Sudan. The dark color of the fragments made them conspicuous on the desert sand, and many of the fragments were recovered by NASA scientists in February 2009.

Figure 2.8

Asteroid 2008 TC3 contains amino acids that formed in outer space

Q1: How did the NASA scientists protect the fragments of the meteorite from contamination by Earth's amino acids when they collected and transported them to the laboratory for analysis?

Q2: What two pieces of evidence suggest that amino acids found in meteorites originated in outer space, rather than being contaminated once they came to Earth?

Q3: What is the significance of finding extraterrestrial amino acids?

What's It All Made Of?

Everything in the universe is composed of matter—from ordinary matter, made of atoms, to dark matter, which may consist of unknown types of particles. Here, we stick with what we know and describe the common elements that compose the world around us.

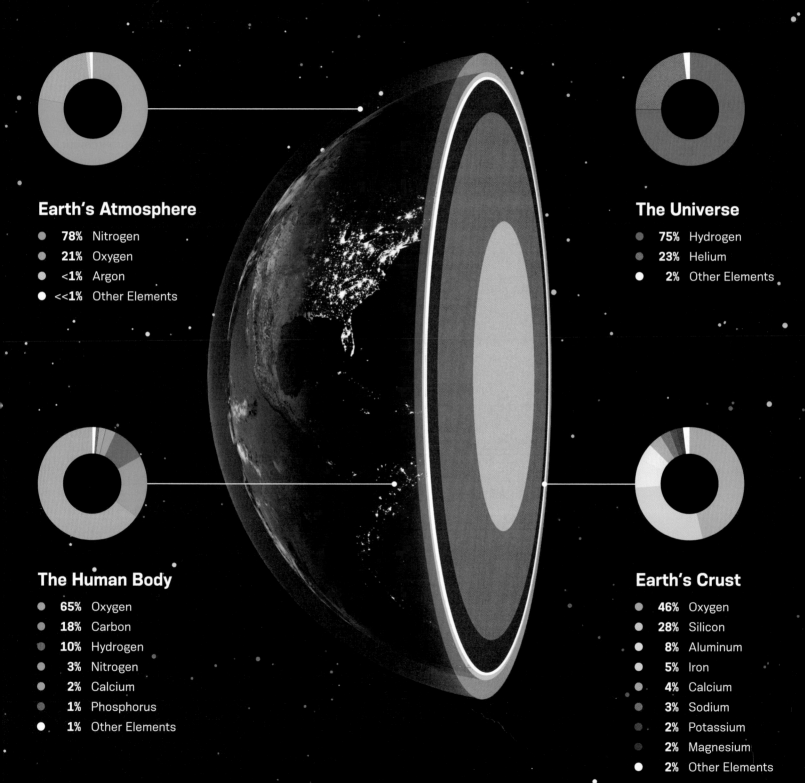

Earth's Atmosphere

- **78%** Nitrogen
- **21%** Oxygen
- **<1%** Argon
- **<<1%** Other Elements

The Universe

- **75%** Hydrogen
- **23%** Helium
- **2%** Other Elements

The Human Body

- **65%** Oxygen
- **18%** Carbon
- **10%** Hydrogen
- **3%** Nitrogen
- **2%** Calcium
- **1%** Phosphorus
- **1%** Other Elements

Earth's Crust

- **46%** Oxygen
- **28%** Silicon
- **8%** Aluminum
- **5%** Iron
- **4%** Calcium
- **3%** Sodium
- **2%** Potassium
- **2%** Magnesium
- **2%** Other Elements

All proportions are by mass except Earth's atmosphere, which is by volume

Unlike acids, bases *accept* hydrogen ions from aqueous surroundings. Because a base removes H⁺ ions from solution, it has the overall effect of *reducing* the concentration of free H⁺ ions in an aqueous solution (strong bases, like strong acids, can be dangerous because they disrupt chemical reactions important to life). Acids react with bases to have an overall neutralizing effect, reducing the concentration of reactive H⁺ ions.

Hydrogen ion concentration is commonly expressed on a scale from 0 to 14, where 0 represents an extremely high concentration of free H⁺ ions and 14 represents the lowest concentration. This scale, called the **pH scale**, is logarithmic: each pH unit represents a 10-fold increase or decrease in the concentration of hydrogen ions (**Figure 2.9**). Pure water is said to be neutral at pH 7, in the middle of the pH scale. The addition of acids to pure water raises the concentration of free hydrogen ions, making the solution more acidic and pushing the pH below the neutral value of 7. Adding a base lowers the concentration of free hydrogen ions in the solution, making the resulting solution more basic and raising the pH above 7.

Bada added a simple base, calcium carbonate, to the experiment in order to raise the pH of his solution. This time, the resulting brew was bursting with amino acids, suggesting that many of the building blocks of life could indeed have originated on Earth and not been solely delivered to Earth by meteorites or comets. Today, many scientists agree that it is likely that both processes—amino acids arriving from space and originating on Earth—contributed to life as we know it. But no matter where amino acids came from, the logical next question is one that continues to stump scientists to this day: What happened next?

Figure 2.9

The pH scale indicates hydrogen ion concentration

Basic

14

Oven cleaner (13.5)

13

12

Household ammonia (11.7)

11

Antacids (10.5)

10

Borax (9.5)

9

Baking soda (8.3)

8 — Seawater (7.5–8.3)

Human blood (7.4)

7 — Neutral – Pure water

Milk (6.5)

6

Natural rainwater (5.6)

5

Tomatoes (4.5)

4

Oranges (3.5)

3

Lemons (2.3)

2

Stomach acid (1.5–2.0)

1

0

Acidic

Values above 7 indicate basic solutions; the higher the value, the more basic the solution.

A pH of 7 means that the solution is neutral.

Values below 7 indicate acidic solutions; the lower the value, the more acidic the solution.

A solution with a pH of 10 is 100 times more basic than a solution with a pH of 8.

A solution with a pH of 3 is 10 times more acidic than a solution with a pH of 4.

Q1: Which has a higher concentration of free hydrogen ions: vinegar, pH 2.8; or milk, pH 6.5?

Q2: What happens to the concentration of free hydrogen ions in your stomach when you drink a glass of milk?

Q3: Black coffee has a pH of 5. If you add water with a pH of 7, do you increase or decrease the concentration of free hydrogen ions in the liquid?

Life's First Steps

"No one really knows how life got started," says Sandford. But researchers can agree on what life requires. If all the water in any living organism were removed, four major classes of large organic molecules, or **macromolecules**, would remain, all of them critical for living cells: **proteins, carbohydrates**, **lipids**, and **nucleic acids**.

Each of these biologically important molecules is built on a framework of covalently bonded carbon atoms. Carbon is the predominant element in living systems, partly because it can form large molecules that contain thousands of atoms. A single carbon atom can form strong covalent bonds with up to four other atoms. Carbon atoms can also bond to other carbon atoms, forming long chains, branched molecules, and even rings. No other element is as versatile as carbon in the sheer diversity of complex molecules that can be assembled from it. In fact, while there are around 4,500 known naturally occurring inorganic molecules on Earth, the number of known organic molecules is in the range of millions, and the number of those as yet uncharacterized is likely several orders of magnitude larger.

Proteins, carbohydrates, and nucleic acids are **polymers**, long strands of repeating units of small molecules called **monomers** (**Figure 2.10**). Amino acids are the monomers making up proteins, simple sugars are the monomers in carbohydrates, and nucleotides are the basis of nucleic acid polymers. Lipids are macromolecules but are not polymers, because their structure, while repeating, is not composed of a chain of monomers.

Though amino acids abound in the experiments of Miller, Bada, and other scientists, these investigators have not yet been able to identify any macromolecules (**Figure 2.11**) in their prebiotic soups. "It's still very much of an unknown how we go from simple to complex molecules," says Bada. Lipids and carbohydrate-like compounds, however, have been found in meteorites, and Sandford's team at NASA is now attempting to create carbohydrates using its space ice simulations. Other scientists are trying to re-create lipids and nucleic acids in the lab (see Chapter 3).

Fifty More Years

Sadly, Miller died in 2007, never having seen the results of Bada's reanalysis. Today, researchers continue to perform spark discharge experiments with equipment that is essentially the same as in Miller's original design. "We're not done with it," says Jim Cleaves. "There are just so many variations to look at—different combinations of gases, changing the pH, adding metals, et cetera."

"This experiment lives on," agrees Bada. "Not many people can say that an experiment done over 50 years ago is still being investigated today, and finding a wealth of new information." And with better and better instruments to detect the results, who knows what scientists might discover over the next 50 years?

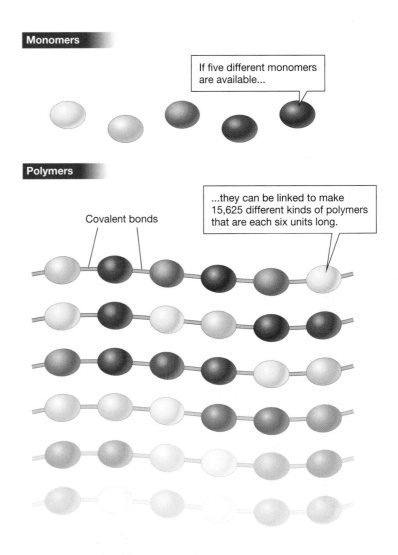

Figure 2.10

Polymers are long chains made from repeating units known as monomers

A handful of monomers (mono = one) can be assembled into a great variety of polymers (poly = many).

Q1: How many different four-unit polymers can be formed by three different monomers?

Q2: Why are monomers so important to living organisms?

Q3: Name three common kinds of organic molecules that are polymers.

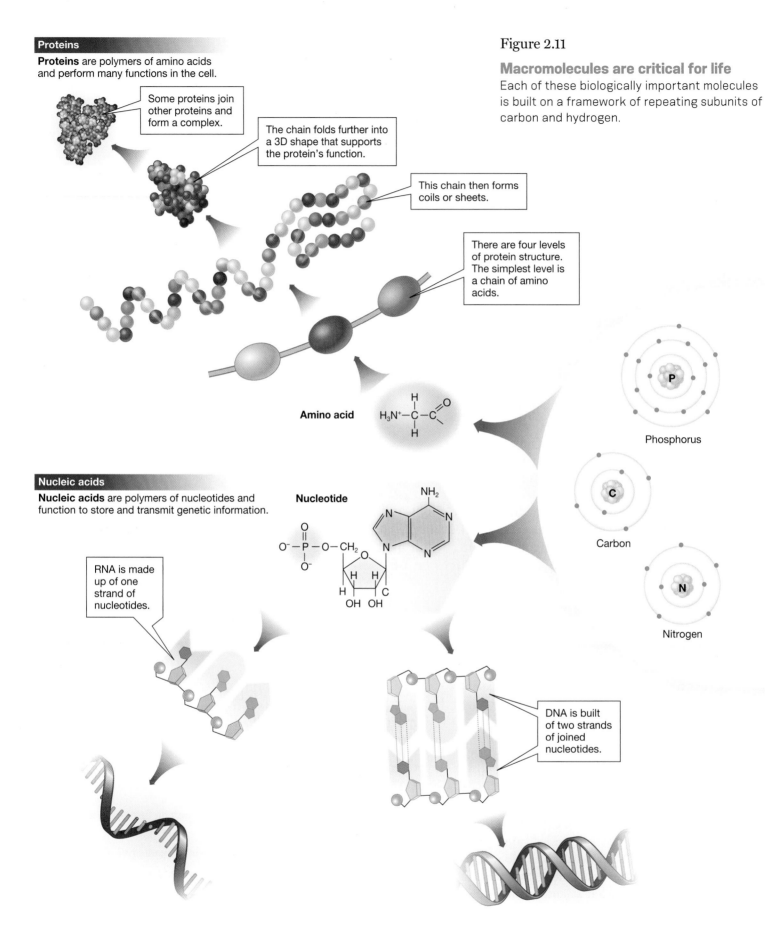

Proteins

Proteins are polymers of amino acids and perform many functions in the cell.

Some proteins join other proteins and form a complex.

The chain folds further into a 3D shape that supports the protein's function.

This chain then forms coils or sheets.

There are four levels of protein structure. The simplest level is a chain of amino acids.

Amino acid

Figure 2.11

Macromolecules are critical for life
Each of these biologically important molecules is built on a framework of repeating subunits of carbon and hydrogen.

Phosphorus

Carbon

Nitrogen

Nucleic acids

Nucleic acids are polymers of nucleotides and function to store and transmit genetic information.

Nucleotide

RNA is made up of one strand of nucleotides.

DNA is built of two strands of joined nucleotides.

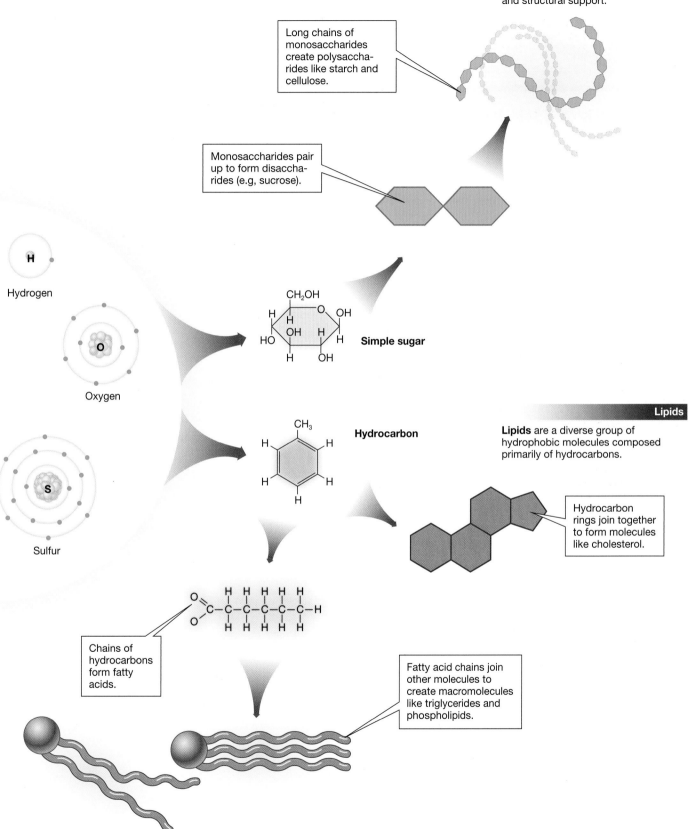

Carbohydrates are polymers of simple sugars, and are used for energy storage and structural support.

Long chains of monosaccharides create polysaccharides like starch and cellulose.

Monosaccharides pair up to form disaccharides (e.g, sucrose).

H

Hydrogen

O

Oxygen

CH_2OH

H

H

O

OH

HO

OH

H

H

H

OH

Simple sugar

S

Sulfur

CH_3

Hydrocarbon

H

H

H

H

H

H

Lipids are a diverse group of hydrophobic molecules composed primarily of hydrocarbons.

Hydrocarbon rings join together to form molecules like cholesterol.

O

C — C — C — C — C — C — H

O

H H H H H

H H H H H

Chains of hydrocarbons form fatty acids.

Fatty acid chains join other molecules to create macromolecules like triglycerides and phospholipids.

Ingredients for Life ■ 33

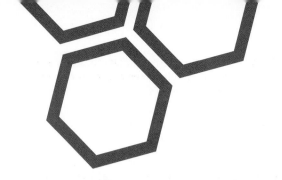

REVIEWING THE SCIENCE

- The physical world is composed of **matter**, which is anything that has mass and occupies space. There are 98 distinct **elements** found naturally on Earth. An **atom** is the smallest unit of an element that maintains its unique properties; it contains positively charged **protons**, uncharged **neutrons**, and negatively charged **electrons**.

- The chemical interactions that cause atoms to associate with each other are known as **chemical bonds**. When an atom loses or gains electrons, it becomes, respectively, a positively or negatively charged **ion**. Ions of opposite charge are held together by **ionic bonds**.

- **Covalent bonds** are formed by the sharing of electrons between atoms. A **molecule** contains at least two atoms that are held together by covalent bonds.

- **Hydrogen bonds** are weak associations between two molecules such that a partially positive hydrogen atom within one molecule is attracted to a partially negative region of the other molecule.

- Partial electrical charges result from the unequal sharing of electrons between atoms, giving rise to **polar molecules**.

- The polarity of individual water molecules and hydrogen bonding across water molecules explain nearly all of the special properties of water.

- Ions and polar molecules are **hydrophilic**; they readily dissolve in water. Nonpolar molecules cannot associate with water and are therefore **hydrophobic**.

- In **chemical reactions**, bonds between atoms are formed or broken. The participants in a chemical reaction (**reactants**) are modified to give rise to new ions or molecules (**products**).

- The concentration of free hydrogen ions in water is expressed by the **pH scale**.

- **Chemical compounds** are molecules that contain atoms from at least two different elements. Carbon atoms can link with each other and with other atoms to generate a great diversity of chemical compounds called **organic molecules**. The four main types of large organic molecules, or **macromolecules**, are **proteins**, **carbohydrates**, **lipids**, and **nucleic acids**.

THE QUESTIONS

The Basics

1 Matter is anything that

(a) is visible.

(b) can be measured.

(c) has mass and occupies a volume of space.

(d) none of the above

2 An element is matter

(a) that cannot be broken down to a simpler substance.

(b) whose composition is known.

(c) that has a combination of two or more different atoms.

(d) that contains molecules.

3 The atomic number of an atom is determined by the number of _____ in the atom.

(a) protons

(b) neutrons

(c) electrons

(d) electrons plus neutrons plus protons

4 The atomic mass number of an atom is determined by the sum of the number of _____ in the atom.

(a) protons plus electrons plus neutrons

(b) protons plus electrons

(c) neutrons plus electrons

(d) neutrons plus protons

5 A chemical compound is composed of

(a) atoms of two or more elements.

(b) atoms that are joined by chemical bonds.

(c) atoms that are present in specific ratios.

(d) all of the above

6 Link each of the following terms with the correct definition.

ION	1. The smallest subunit of an element.
MOLECULE	2. A molecule made up of repeating subunits.
ATOM	3. An atom that has gained or lost an electron.
POLYMER	4. An association of atoms bonded to each other through shared electrons.

7 Which of the following is the correct definition of an organic compound?

(a) a molecule that is found in a living organism

(b) a macromolecule

(c) a molecule that includes at least one carbon atom

(d) a molecule that contains carbon, hydrogen, and oxygen

8 Circle the correct terms in the following sentences: Proteins are (**polymers, monomers**) of amino acids. A common carbohydrate is (**sugar, fat**). Nucleic acids are composed of (**nucleotides, DNA**). Lipids (**are, are not**) polymers. All of these organic molecules contain (**carbon, nitrogen**).

Try Something New

9 Explain why life on Earth is carbon-based rather than, for example, hydrogen- or oxygen-based.

10 Laundry detergent molecules have a short polar end and a long nonpolar end, enabling them to bind to water (on the polar end) and oils (on the nonpolar end). Suppose you spill some salad dressing on your shirt. Explain how washing your shirt with detergent will help remove the dressing.

11 Can NASA scientists determine with absolute certainty that there was no contamination of asteroid 2008 TC3 by amino acids from Earth? If contamination had occurred, would that weaken the conclusion that the asteroid contains amino acids that were formed in space? Explain.

12 Why did Miller's addition of hydrogen sulfide (H_2S) to the mixture of molecules in the spark discharge experiments increase the number of amino acids he found?

Leveling Up

13 **What do *you* think?** Read the articles "Two Miles Underground, Strange Bacteria Are Found Thriving" in *News at Princeton* (http://www.princeton.edu/main/news/archive/S16/13/72E53/index.xml?section=newsreleases) and "Martian Underground Could Hold Clues to Life's Origins" (http://www.nhm.ac.uk/about-us/news/2013/january/martian-underground-could-hold-clues-to-lifes-origins118124.html).

(a) How are bacteria able to stay alive below the Earth's surface or (potentially) below the surface of Mars?

(b) How does the discovery of bacteria living 2 miles beneath Earth's surface relate to the importance of the McLaughlin Crater on Mars?

(c) What research do you think should next occur to answer the questions brought up by these studies?

(d) From this, what do you think is the probability of life on other planets? What would you expect that life to look like?

Engineering Life

In 2003, scientists began trying to build an artificial cell from scratch. Today, they're closer than you might think.

By the end of this chapter you should be able to:

- Explain cell theory and why it is central to the study of life.
- Describe the differences in the structures of viruses, prokaryotes, and eukaryotes.
- Diagram a plasma membrane, showing how the structure allows some substances in and keeps others out.
- Compare and contrast passive and active transport of materials into and out of cells.
- Describe the role of any given organelle in a eukaryotic cell.
- Identify the main differences between a plant cell and an animal cell.

CHAPTER
03

LIFE IS
CELLULAR

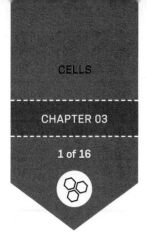
The sky was still dark when Daniel Gibson hurried into the J. Craig Venter Institute (JCVI) in La Jolla, California, at 5:00 a.m. on Monday, March 29, 2010. His footsteps echoed through the empty halls of the building. He reached a laboratory door and quickly went inside. There, Gibson peered into a warm incubator, his eyes scanning rows of palm-sized petri dishes. His stomach was in knots. For three months the experiment had failed. Would this day be any different?

Gibson is part of a team at the JCVI with a single, audacious goal: to create life. For more than a decade, this team of scientists and engineers has attempted to build a synthetic, or human-made, **cell** (**Figure 3.1**). Cells are the smallest and most basic unit of life—microscopic, self-contained units enclosed by a water-repelling membrane (**Figure 3.2**). The human body is composed of approximately 100 trillion (10^{14}) cells. That fateful day in 2010, however, the JCVI was trying to synthesize just one cell—a single-celled bacterium.

Gibson's team had sequenced a bacterium's complete genetic information, its **genome**; then built a synthetic version of that genome using basic laboratory chemicals; and finally, replaced the natural DNA of another species of bacterium with the synthetic DNA. **DNA** (deoxyribonucleic acid) is a large and complex molecule that acts as a set of instructions for building an organism, like a blueprint. Almost every cell of every living organism contains DNA. DNA transfers information from parents to offspring, which is why it is essential for reproduction. Life, no matter how simple or how complex, uses this inherited genetic code to direct the structure, function, and behavior of every cell. DNA is made up of many nucleotides held together in a structure called the double helix, a ladderlike assembly twisted into a spiral along its length (see Figure 2.11).

Gibson's boss, the famous geneticist J. Craig Venter, worked for more than 15 years and spent millions of dollars to have his team construct a synthetic DNA helix from chemicals in the laboratory, but Gibson and the team had been unable to get that synthetic DNA to work inside a cell. Every Friday for three months, they transplanted the synthetic DNA into a bacterial cell whose own DNA had been removed. The synthetic DNA included a gene to make the cells bright blue, so every Monday, Gibson hurried to the incubator and checked the petri dishes for a colony of blue cells. But Monday after Monday, the dishes were barren. "We did the genome transplantation again and again," he recalls, "but nothing was working."

Then, in March of 2010, Gibson identified an error in a single gene in the synthetic DNA. A **gene** is a segment of DNA that codes for a distinct genetic characteristic, such as having O-type blood or a dimpled chin. In Gibson's bacterium, the gene with the error was responsible for DNA replication. When it wasn't working, the bacterium couldn't replicate its DNA, and it died. So in late March, Gibson fixed the DNA error, transplanted the genome yet again, and waited.

Life, Rewritten

The JCVI is just one institution among a large group of universities and companies pursuing synthetic biology, a field that aims to design and construct new biological entities with novel and useful functions. Current research projects include algae that digest trash and produce energy, microbes that use light and water to create hydrogen gas, and bacteria that produce

Figure 3.1

Daniel Gibson and the JCVI team

Daniel Gibson (back row, second from right) is an associate professor specializing in synthetic biology at the J. Craig Venter Institute in San Diego, California. As a postdoctoral researcher in 2004, Gibson led the work of a large team of researchers to synthesize two complete bacterial genomes, culminating in the creation of the first synthetic cell.

new kinds of antibiotics to treat infections. With synthetic biology "we can harness what nature has made, but repurpose it," says James Collins, a synthetic biologist at Boston University. "We can reprogram organisms and endow them with novel functions." In addition to useful tools, the pursuit of artificial life allows scientists to explore the very origins of life. "It's going to be a big challenge to create a totally synthetic cell," says Collins, "but it's fundamentally intriguing to explore how life may have arisen on the planet."

To make their own cell, scientists are pushing the boundaries of **cell theory**, one of the unifying principles of biology. Cell theory has two main parts: Every living organism is composed of one or more cells, and all cells living today came from a preexisting cell. By trying to engineer a cell in the laboratory, Venter, Gibson, and others are challenging the second part of the definition.

Starting Small

The JCVI's first step toward a synthetic cell was a small one. In 2003, Venter's team flexed its scientific muscles by synthesizing the 11-gene, 5,386-base-pair genome of phiX174, a virus that infects bacteria. A **virus** is a small, infectious agent that can replicate only inside a living cell. Most viruses are little more than stripped-down genetic material wrapped in proteins, yet these pathogens attack and devastate organisms in every kingdom of life, from bacteria to plants and animals. Though the JCVI team successfully created a virus with a synthetic genome, it was not considered the first synthetic life, because scientists debate whether viruses are alive. (For more on this debate, see "Viruses—Living or Not?" on page 40.)

Next, Venter and his colleagues moved up to a bacterium, a living organism that consists of a single cell. In 2010, they sequenced and built the genome of a bacterium called *Mycoplasma mycoides* (*M. mycoides*), an organism that can cause the mammary glands of goats to swell. They constructed the genome—a 1.1-million-base-pair DNA sequence—using four necessary ingredients: the nucleotides **adenine** (**A**), **thymine** (**T**), **guanine** (**G**), and **cytosine** (**C**), the building blocks for DNA. A, T, G, and C, organized in different combinations, carry all the instructions for everything a cell does.

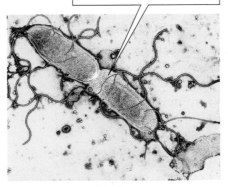

Salmonella is a single-celled bacteria that is a common cause of food poisoning.

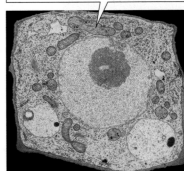

This is one cell of the multicellular plant *Arabidopsis*, which is used extensively in genetic studies.

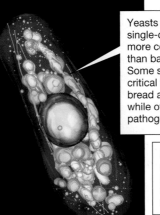

Yeasts are single-celled, but more complex than bacteria. Some species are critical for making bread and beer, while others are pathogens.

Humans are multicellular animals with many specialized cells, such as these neurons within the central nervous system.

Figure 3.2

An individual organism may consist of a single cell or many cells

All of these photos of cells were taken using electron microscope technology. Color has been added to the images to differentiate structures within the cells.

The team used a machine to read the nucleotide sequence of the *M. mycoides* genome and then "print out" little bits of that code, creating strands of DNA about 50–80 bases long. They then strung these pieces together using living cells as factories, inserting the short segments into yeasts and

J. CRAIG VENTER

J. Craig Venter is an American biologist and founder and CEO of the J. Craig Venter Institute. He led a team to fully sequence and publish the human genome in 2001, and initiated the effort to create the first cell constructed with synthetic DNA.

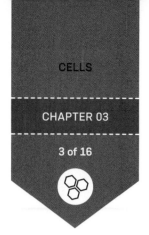
Escherichia coli—small, single-celled organisms. These organisms interpreted the strands as broken pieces of DNA and stitched them together, creating longer and longer sequences. To the researchers, it was like building the Eiffel Tower from a massive box of Legos, constructing a single support beam at a time. The effort—with many mistakes along the way—took years. "It was very complex," said Venter. "It was a long, involved process."

Congratulations, It's a ... Cell

Once the DNA sequence of *M. mycoides* was complete and intact, it was up to the JCVI team to transfer it into another species and make it work. This was the experiment that almost drove Gibson crazy. The researchers removed the DNA from a cell of a closely related bacterium, *Mycoplasma capricolum*, and replaced it with the *M. mycoides* synthetic DNA.

After months of trying, on that Monday morning at 5:00 a.m., Gibson cautiously scanned the petri dishes. There, on a single dish, was a group of bright blue cells—proof that the *M. capricolum* cell had "booted up" the *M. mycoides* DNA and transformed itself into an *M. mycoides* cell (**Figure 3.3**). Gibson was ecstatic. Moments later, he sent a text message to Venter, waking him up. Within the hour, Venter was in the lab with a camera, taking pictures of the tiny blue dollop in the dish. "How does it feel to create life?" Venter asked Gibson. They opened champagne and toasted their success.

Over the following weeks, Gibson repeated the experiment hundreds of times to make sure the blue cells were not an accident or a fluke, confirming that they only contained *M. mycoides* DNA. Every time, the cells with a synthetic genome survived. The team had done it—created the first synthetic cell. "They are living cells," Venter told *The Scientist* magazine when the research was published two months later. "The only difference is that they have no natural history. Their parents were the computer."

Viruses—Living or Not?

A virus is a microscopic, noncellular infectious particle. Like living organisms, viruses reproduce and evolve. Yet they lack some of the key characteristics of life, which is why most scientists today regard viruses as nonliving. For one thing, viruses are not made up of cells. A virus is much simpler than a cell, usually consisting of a small piece of genetic material (for example, DNA) that is wrapped in a protein coat. Some viruses also have an envelope, an oily layer rather like the plasma membrane, enclosing the central core of genetic material and protein.

Another difference, compared to living organisms, is that viruses lack the many structures within cells that are necessary for critical cellular functions such as homeostasis, reproduction, and metabolism. To gain these functions, viruses use their genetic material to make the cells of the organisms they infect do the work for them. They accomplish this feat by invading cells, releasing their genetic material into the cell interior, and "hijacking" the host cell's machinery.

But What Does It Mean?

Once the research was published, the reaction from the academic community, captured in *Nature* magazine, was swift and divided. Some called it a significant advance: "We now have an unprecedented opportunity to learn about life," said Mark Bedau, a professor of philosophy at Reed College in Oregon. Arthur Caplan, a bioethicist at the University of Pennsylvania, said, "Venter's achievement would seem to extinguish the argument that life requires a special force or power to exist. In my view, this makes it one of the most important scientific achievements in the history of mankind."

Others were more hesitant. "Has [Venter] created 'new life'?" asked George Church, a prominent geneticist at Harvard Medical School. "Not really ... Printing out a copy of an ancient text isn't the same as understanding the language."

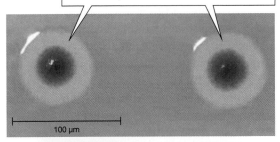

Two colonies of *M. mycoides*, transformed from *M. capricolum*.

100 μm

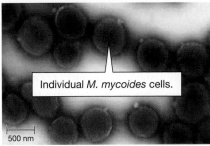

Individual *M. mycoides* cells.

500 nm

Figure 3.3

The first synthetic organism

A colony of bacteria was transformed from *M. capricolum* by the insertion of synthetic DNA of a closely related species of bacterium, *M. mycoides*. The investigators inserted a gene into the synthetic DNA that codes for blue pigment.

Q1: What was the purpose of inserting the gene that codes for blue pigment into the synthetic DNA?

Q2: What part of the transformed bacterium is synthetic?

Q3: Did this experiment create life?

Gibson and Venter agree that they did not create life from scratch but argue that they did create new life from existing life.

A Different Approach

While Venter and Gibson were making headlines, other scientists were quietly pursuing a different approach to building a cell, working from the bottom up rather than from the top down. One young scientist in California decided to start by building one of the simplest components of a cell, but also one of the most vital: the layer of molecules that surrounds a cell.

In the chemistry department at UC San Diego, assistant professor Neal Devaraj was fascinated by the idea of building life, but he approached synthetic biology from the perspective of chemistry. "Most molecular biologists study what exists," says Devaraj, "but when you're a chemist and constantly make new compounds, you want to engineer something from scratch." So instead of taking a cell apart and determining how it works, Devaraj decided to try to build a cell artificially, using materials not typically found in nature. "If you want to really understand the principles by which life operates and evolves, the best way to do so is to build a cell from the ground up," he says.

Scientists suspect that one of the first events at the beginning of life on Earth was the formation of a **plasma membrane**, a barrier separating a cell from its external environment. A plasma membrane is made of two layers of **phospholipids**, organic molecules with a water-loving, or hydrophilic, head, and a water-fearing, or hydrophobic, tail. In water, these molecules form a double layer with heads out and tails in, a barrier that separates the contents of the cell from what lies outside the cell. Thus, the membrane is a **phospholipid bilayer**, a mostly impermeable barrier. When a phospholipid bilayer forms a sphere, or **liposome**, the fluid inside the liposome can have a different composition from the fluid outside (**Figure 3.4**). The ability to maintain an internal environment separate from the external environment is one of the most critical functions of the plasma membrane of a cell.

Given a container of phospholipids, anyone can make a membrane, says Devaraj. "Making membranes is almost a trivial thing," he says. "You take natural or synthetic phospholipids, add water, and they form membranes." Yet researchers had been unable to form a membrane from scratch—without using preexisting

NEAL DEVARAJ

Neal Devaraj is a biochemist at UC San Diego who is working to make an artificial cell from the bottom up, starting with the membrane and then building other organelles.

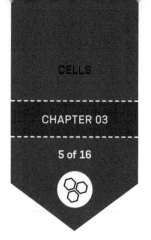
phospholipids. In nature, new phospholipids are created by enzymes embedded in the cell.

Instead of trying to engineer new phospholipids, Devaraj wanted to start with something simpler. He worked with graduate student Itay Budin, then at Harvard University, and now a postdoctoral researcher at UC Berkeley, to create a self-assembling membrane. They first mixed together oil and a detergent. Then they added copper, a metal ion, as a catalyst to spark a chemical reaction. With the addition of copper, sturdy membranes begin to bud off the oil; these were self-assembling structures. "There's no equivalent whatsoever in nature," says Devaraj. "Our goal was simply to mimic biology."

Through the Barrier

Devaraj admits that his artificial membrane is far simpler than a real cell's plasma membrane, which is dotted with numerous proteins, including transport proteins. **Transport proteins** are gates, channels, and pumps that allow molecules to move into and out of the cell, making it selectively permeable (**Figure 3.5**). **Selective permeability** means that some substances can cross the membrane, others are excluded, and still others can pass through the membrane when they are aided by transport proteins.

All movement of substances through the plasma membrane occurs by either active or passive transport. Some transport proteins facilitate **active transport**, the movement of a substance that requires an input of energy (**Figure 3.5**, left). Molecules move across the plasma membrane by active transport when they need to move from a region of *lower* concentration to a region of *higher* concentration. In contrast, **passive transport** is the movement of a substance without the addition of energy (**Figure 3.5**, middle and right). Movement via passive transport is spontaneous.

A primary type of passive transport is **diffusion**, the movement of a substance from a region of *higher* concentration to a region of *lower* concentration (**Figure 3.6**). Water, oxygen, and carbon dioxide usually enter and leave cells by **simple diffusion**: these small, uncharged molecules slip

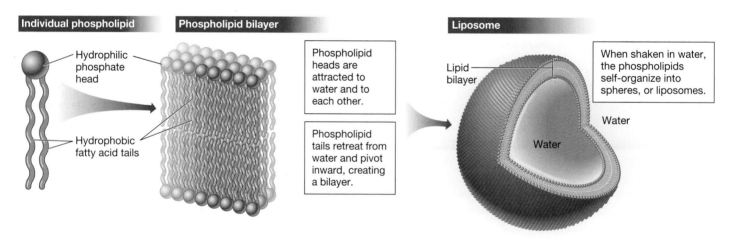

Individual phospholipid

Hydrophilic phosphate head

Hydrophobic fatty acid tails

Phospholipid bilayer

Phospholipid heads are attracted to water and to each other.

Phospholipid tails retreat from water and pivot inward, creating a bilayer.

Liposome

When shaken in water, the phospholipids self-organize into spheres, or liposomes.

Lipid bilayer

Water

Water

Figure 3.4

Liposomes form when phospholipids and water are shaken together

When you shake a mixture of phospholipids and water, the phospholipid bilayers bend and link together to form spheres called liposomes. This simple structure is remarkably similar to the basic structure of a cell.

Q1: Why is it important that the phosphate head of a phospholipid is hydrophilic?

Q2: What essential component of a cell do liposomes lack, and why is that omission important?

Q3: Could the tendency of phospholipid bilayers to spontaneously form spheres have played a role in the origin of life? (*Hint*: Refer to "The Characteristics of Living Organisms" on page 6 of Chapter 1.)

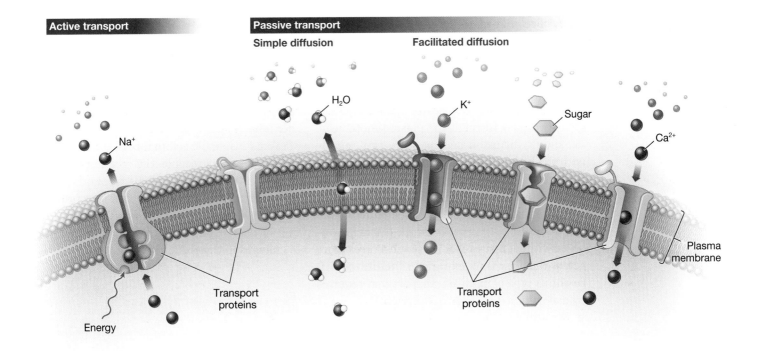

Active transport

Passive transport

Simple diffusion

Facilitated diffusion

H_2O

K^+

Sugar

Na^+

Ca^{2+}

Plasma membrane

Transport proteins

Transport proteins

Energy

Figure 3.5

The plasma membrane is a barrier and a gatekeeper
The plasma membrane moves substances in a highly selective fashion, determined in large part by the types of membrane proteins embedded in the phospholipid bilayer.

Q1: In what ways is the plasma membrane a barrier, and in what ways is it a gatekeeper?

Q2: Why can't ions (such as Na^+) cross the plasma membrane without the help of a tranport protein?

Q3: If no energy were available to the cell, what forms of transport would not be able to occur? What forms of transport could occur? (*Hint:* Look ahead at Figures 3.6 and 3.7.)

between the large molecules in the phospholipid bilayer without much hindrance.

The movement of water in and out of cells (and compartments inside cells) occurs by **osmosis**. Osmosis is a form of simple diffusion because the water molecules are moving from areas of higher concentration to areas of lower concentration (**Figure 3.7**). Osmosis is critical for cellular processes because most cells are at least 70 percent water, and nearly all cellular processes take place in a water-rich environment.

Cells must maintain a stable internal water concentration to function properly, but the concentration of water in most cells changes from moment to moment. For example, when salt molecules move into a cell, the concentration of water in the cell decreases because additional molecules have been added. In response, water molecules immediately move by osmosis—that is, they diffuse—across the plasma membrane into the cell until the concentration of water inside the cell is the same as the concentration on the outside. On the other hand, when salt molecules move out of a cell, the water molecules become more concentrated, and osmotic movement of water out of the cell then restores the concentration of water in the cell to its correct level.

You can imagine then, that the concentration of solutes within a cell in relation to the concentration *outside* the cell is of critical importance. When cells are surrounded by fluids with the same solute concentration as the cell interior, the extracellular and intracellular environments are said to be **isotonic** (iso = equal) to each other. If the extracellular environment has a higher solute concentration, it is **hypertonic** (hyper = more) to the cell interior. If it has a lower solute concentration than the cell's interior, it is said to be **hypotonic** (hypo = less) to it.

Most hydrophobic molecules, even fairly large ones, can pass through the plasma membrane via simple diffusion because they mix readily with the hydrophobic tails that form the core of the phospholipid bilayer. But hydrophilic substances such as sodium ions (Na^+), hydrogen ions (H^+), and larger molecules, including sugars and amino acids, cannot cross the plasma membrane without assistance. These substances move across the plasma membrane by **facilitated diffusion**, a type of passive transport that requires transport proteins (**Figure 3.5**, right).

Devaraj's artificial membrane does not currently contain any transport proteins, he says, so it is impermeable to large molecules. But small molecules such as water can pass through his membrane via simple diffusion.

The plasma membrane also contains **receptor proteins**, which are sites where molecules released by other cells can bind. When a molecule binds to a receptor protein, it starts a chain of events inside the cell that causes the cell to do something. For example, the receptors in nerve cells receive molecular signals from other nerve cells that cause the nerves to fire. Receptor proteins are key components of a cell's communication system, enabling it to respond appropriately to changes in the organism.

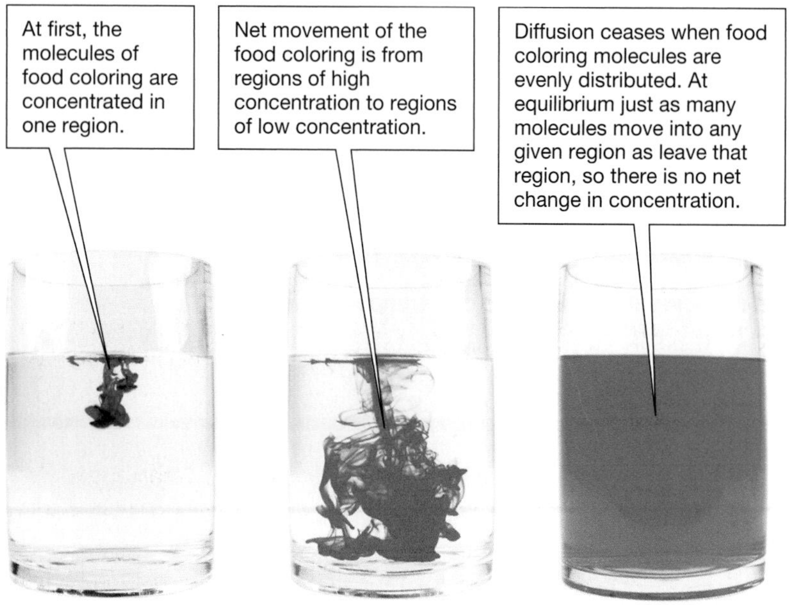

Another Way Through

In addition to transport proteins, there is another way that molecules move into and out of a cell. Sections of the plasma membrane can bulge inward or outward to form packages called **vesicles**. Vesicles move molecules from place to place inside a cell but also transport substances into and out of the cell (**Figure 3.8**).

Cells expel materials in vesicles via **exocytosis**. The substance to be exported from the cell is packaged into a vesicle, and as the vesicle approaches the plasma membrane, a portion of the vesicle's membrane fuses with the plasma membrane. The inside of the vesicle then opens to the exterior of the cell, discharging its contents.

Endocytosis is the opposite of exocytosis. In this process, a section of plasma membrane bulges inward to form a pocket around extracellular fluid, molecules, or particles. The pocket deepens until the opening in the membrane pinches off and the membrane breaks free as a closed vesicle, now wholly contained within the cell. Endocytosis can be nonspecific or specific. In nonspecific endocytosis, all of the material in the immediate area is surrounded and enclosed; in specific endocytosis, one particular type of molecule is enveloped and imported.

There are three types of endocytosis. **Receptor-mediated endocytosis** is a form of endocytosis in which receptor proteins embedded in the membrane recognize specific surface characteristics of substances to be incorporated into the cell. For example, our cells use

At first, the molecules of food coloring are concentrated in one region.

Net movement of the food coloring is from regions of high concentration to regions of low concentration.

Diffusion ceases when food coloring molecules are evenly distributed. At equilibrium just as many molecules move into any given region as leave that region, so there is no net change in concentration.

Figure 3.6

Food coloring in water illustrates diffusion

Q1: Is the dye at equilibrium in any of these glasses? Describe how the first glass will look when the dye is at equilibrium with the water.

Q2: Will diffusion mix the molecules of dye evenly through the water, or is it necessary to shake the container to get a uniform mixture?

Q3: Will diffusion mix the dye faster in hot water than in cold water? Why or why not? (*Hint:* Review the discussion of the behavior of water molecules at different temperatures in Chapter 2.)

receptor-mediated endocytosis to take up cholesterol-containing packages called low-density lipoprotein (LDL) particles. **Phagocytosis**, or "cellular eating," is a large-scale version of endocytosis in which particles considerably larger than macromolecules are ingested. Specific cells in the immune system use phagocytosis to ingest an entire bacterium or virus. **Pinocytosis** is a form of endocytosis that is often described as "cellular drinking" because it involves the capture of fluids. However, the cell does not attempt to collect particular solutions. Pinocytosis is nonspecific: the vesicle budding into the cell contains whatever happened to be dissolved in the fluid when the cell "drank."

There is a long way to go before an artificial plasma membrane will perform processes like endocytosis and exocytosis, says Neal Devaraj. "But just because a research problem is difficult, doesn't mean we shouldn't tackle it," he adds.

Prokaryotes Versus Eukaryotes

Membranes are important not only because they form the structure of a cell, but because they enclose organelles within eukaryotic cells. Depending on the fundamental structure of their cells, all living organisms can be sorted into one of two groups: **prokaryotes** or **eukaryotes** (**Figure 3.9**). *M. mycoides* and all other bacteria are prokaryotes, but virtually all the organisms you see every day, including all plants and animals, are eukaryotes.

Eukaryotic cells are larger and more complex than prokaryotic cells: they are roughly 10 times wider, with a cell volume about a thousand times greater. Unlike prokaryotic cells, eukaryotic cells have a membrane-enclosed **nucleus** (plural "nuclei") that contains the organism's DNA, and they have a variety of membrane-enclosed subcellular compartments called **organelles**. Through specialization and division of labor, these organelles act like cubicles in a large office, allowing the cell to localize different processes in different places. In contrast, prokaryotic cells are like an open floor plan: they lack a cell nucleus or any membrane-encased organelles.

Though the first fully artificial cell will most likely be a simple prokaryotic cell, synthetic

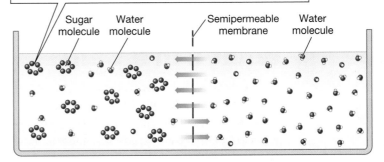

Sugar has just been added to a beaker of water. This beaker is divided by a semipermeable membrane—that is, a membrane with pores large enough to allow water molecules to pass through, but too small for sugar molecules to pass.

Sugar molecule | Water molecule | Semipermeable membrane | Water molecule

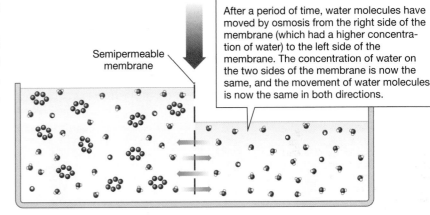

Semipermeable membrane

After a period of time, water molecules have moved by osmosis from the right side of the membrane (which had a higher concentration of water) to the left side of the membrane. The concentration of water on the two sides of the membrane is now the same, and the movement of water molecules is now the same in both directions.

Figure 3.7

In osmosis, water diffuses across a semipermeable membrane
Osmotic movement of water between a cell and the external environment is critical for maintaining a constant water concentration in the cell, which it needs to function properly.

Q1: What would the second diagram look like if the pores in the semipermeable membrane were too small to allow water molecules to pass through?

Q2: What would the second diagram look like if the pores were large enough to let both water molecules and sugar molecules through?

Q3: The osmotic concentration of the fluid in an IV bag is the same as the osmotic concentration of blood. What change would you see in the red blood cells of a patient if a bag of distilled water was used in error?

biologists aspire to build a eukaryotic cell. "We're not quite there yet, but it's interesting to think about making complex structures like organelles," says Devaraj. Having achieved a self-assembling artificial membrane, his team is working to create membranes inside preexisting vesicles, mimicking organelles like mitochondria that are made up of membranes within membranes. It is the first step toward creating the variety of organelles inside eukaryotic cells.

What's in a Cell?

The **nucleus** is the control center of the cell. It contains most of the cell's DNA and may occupy up to 10 percent of the space inside the cell (**Figure 3.10**, center top). Inside the nucleus, long strands of DNA are packaged with proteins into a remarkably small space. The boundary of the nucleus, called the **nuclear envelope**, is made up of two concentric phospholipid bilayers. The nuclear envelope is speckled with thousands of small openings called **nuclear pores**. These pores allow chemical messages to enter and exit the nucleus.

The **endoplasmic reticulum** (**ER**) is an extensive and interconnected network of sacs made of a single membrane that is continuous with the outer membrane of the nuclear envelope (**Figure 3.10**, left middle). The membranes of the ER are classified into two types based on their appearance: smooth and rough. Enzymes associated with the surface of the **smooth ER** manufacture lipids and hormones. In some cell types, smooth-ER membranes also break down toxic compounds. **Ribosomes** embedded in the **rough ER** give it the knobby appearance from

Exocytosis

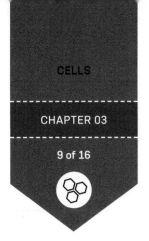

Exocytosis is used to eject substances from the cell. Here a cell ejects waste material into the outside environment.

Endocytosis

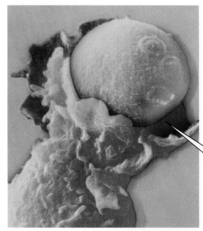

Endocytosis is the reverse of exocytosis and brings material from the outside of the cell to the inside, enclosed in vesicles. Endocytosis may be nonselective, as shown here, drawing in any and all molecules near the opening of the vesicle.

Here a white blood cell engulfs a yeast cell through phagocytosis, a form of nonspecific endocytosis.

Receptor-mediated endocytosis

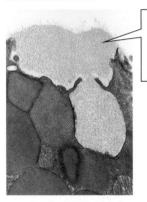

Receptor-mediated endocytosis is a selective process in which only certain molecules bind to receptor proteins. Here, low-density lipoprotein (LDL) particles bind to LDL receptors and are transported to the cell interior.

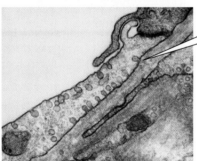

Cells lining blood vessels have created fluid-filled vesicles through pinocytosis, non-specific endocytosis of fluid.

Figure 3.8

Substances move into cells by endocytosis and out of cells by exocytosis

Q1: If endocytosis itself is nonspecific, how does receptor-mediated endocytosis bring only certain molecules into a cell?

Q2: What sorts of molecules could be moved by endocytosis or exocytosis, but not by diffusion?

Q3: How does the fluid that enters a cell via pinocytosis differ from the fluid that enters by osmosis?

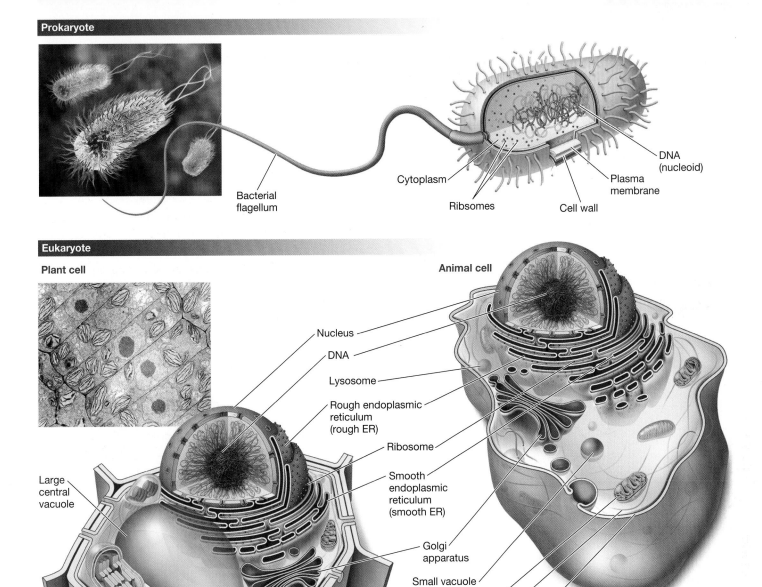

Prokaryote

Bacterial flagellum

Cytoplasm

Ribsomes

Cell wall

DNA (nucleoid)

Plasma membrane

Eukaryote

Plant cell

Animal cell

Nucleus

DNA

Lysosome

Rough endoplasmic reticulum (rough ER)

Ribosome

Smooth endoplasmic reticulum (smooth ER)

Golgi apparatus

Small vacuole

Mitochondrion

Cytoskeleton

Plasma membrane

Large central vacuole

Chloroplast

Cell wall

Figure 3.9

Prokaryotic and eukaryotic cells

A prokaryote cell, like all cells, contains DNA, cytoplasm, ribosomes, and a plasma membrane. Many prokaryotes also have a cell wall that serves as a kind of exoskeleton. To enable movement, some bacteria possess a flagellum, or several flagella. The components of eukaryotic cells are described in the text and in Figure 3.10.

Q1: What structures do prokaryotic and eukaryotic cells have in common?

Q2: What cellular processes occur in both prokaryotic and eukaryotic cells?

Q3: Both plants and animals are eukaryotes, but there are differences in their cellular structure. What are those differences?

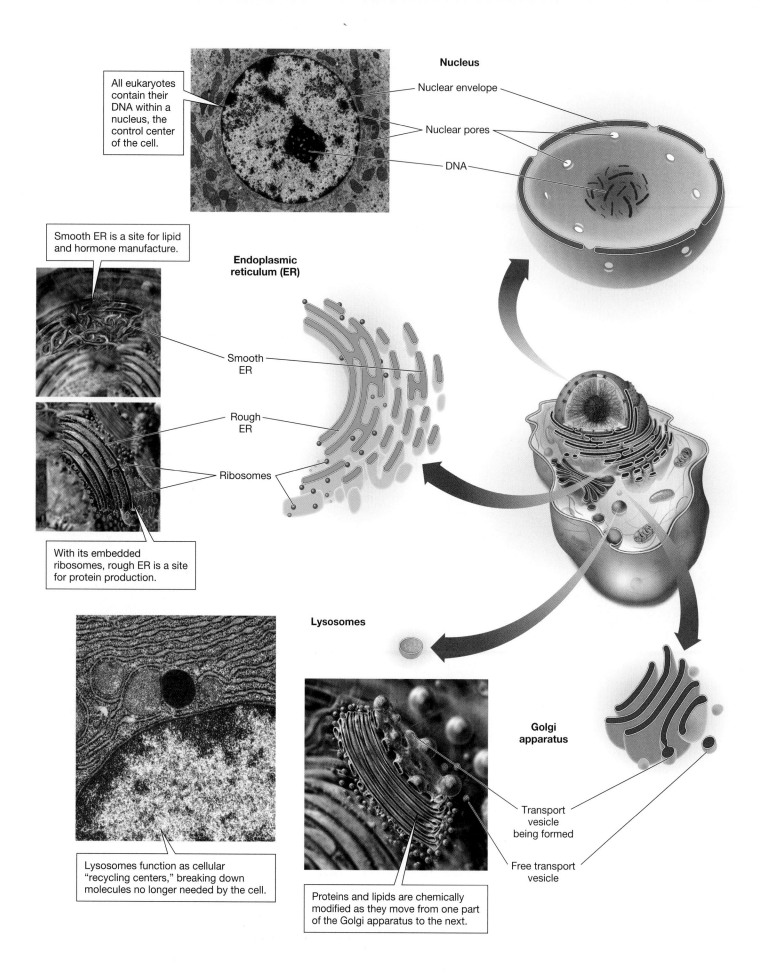

Nucleus

All eukaryotes contain their DNA within a nucleus, the control center of the cell.

Nuclear envelope

Nuclear pores

DNA

Smooth ER is a site for lipid and hormone manufacture.

Endoplasmic reticulum (ER)

Smooth ER

Rough ER

Ribosomes

With its embedded ribosomes, rough ER is a site for protein production.

Lysosomes

Golgi apparatus

Lysosomes function as cellular "recycling centers," breaking down molecules no longer needed by the cell.

Proteins and lipids are chemically modified as they move from one part of the Golgi apparatus to the next.

Transport vesicle being formed

Free transport vesicle

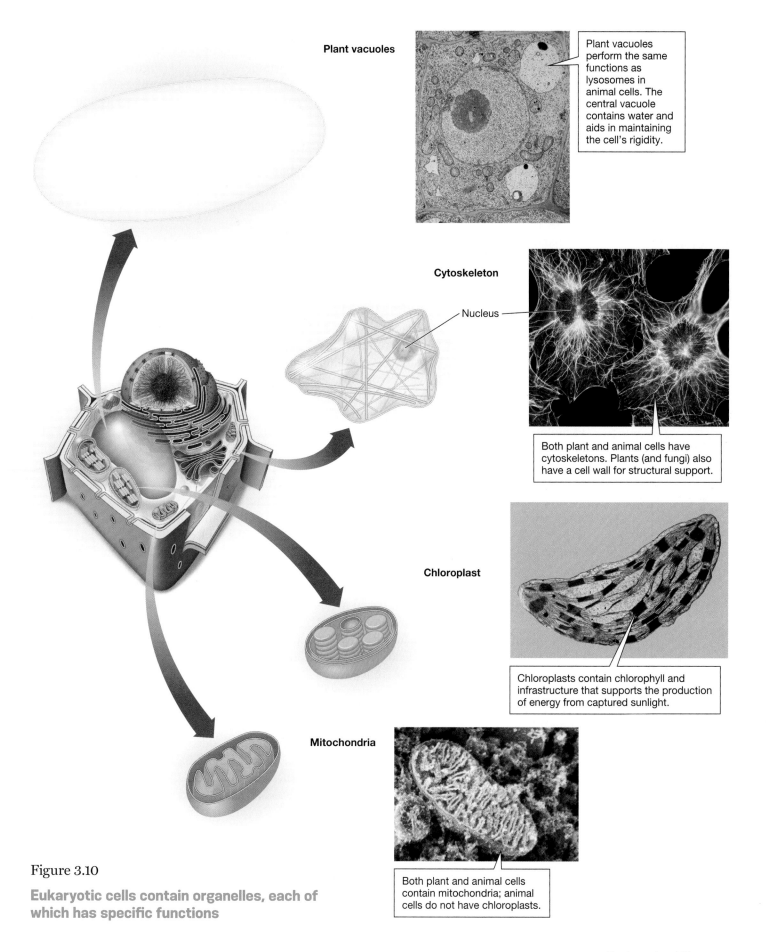

Plant vacuoles

Plant vacuoles perform the same functions as lysosomes in animal cells. The central vacuole contains water and aids in maintaining the cell's rigidity.

Cytoskeleton

Nucleus

Both plant and animal cells have cytoskeletons. Plants (and fungi) also have a cell wall for structural support.

Chloroplast

Chloroplasts contain chlorophyll and infrastructure that supports the production of energy from captured sunlight.

Mitochondria

Both plant and animal cells contain mitochondria; animal cells do not have chloroplasts.

Figure 3.10

Eukaryotic cells contain organelles, each of which has specific functions which it gets its name. Ribosomes on the rough ER assemble proteins that will be inserted into the cell's plasma membrane or its organelles.

Resembling a pile of flattened balloons, the **Golgi apparatus** is like a post office, packaging and directing proteins and lipids produced by the ER to their final destinations either inside or outside the cell (**Figure 3.10**, left bottom). Each molecule is first packaged into a transport vesicle. The transport vesicle buds off from the ER membrane and delivers the cargo to its destination by fusing with the membrane of the target compartment. The Golgi apparatus targets these destinations by adding a specific chemical tag to each molecule it receives, like attaching a shipping label to a package.

In animal cells, transport vesicles bring large molecules to be discarded to **lysosomes**, organelles that act as garbage and recycling centers (**Figure 3.10**, left middle). Lysosomes contain a variety of enzymes that degrade macromolecules and release the breakdown products into the cell interior to be discarded or reused. In plant cells, **vacuoles** perform similar functions as lysosomes, plus a few additional functions, such as water storage (**Figure 3.10**, right top). Some plant vacuoles stockpile noxious compounds that can deter herbivores from eating plants.

In most eukaryotic cells, the main source of energy is the **mitochondrion** (plural "mitochondria"), a tiny power plant that fuels cellular activities (**Figure 3.10**, right bottom). Mitochondria are made up of double membranes—a smooth external membrane and a folded internal membrane forming a mazelike interior. Mitochondria use chemical reactions to transform the energy of food molecules into ATP (adenosine triphosphate), the universal cellular fuel, in a process called *cellular respiration*. Mitochondria provide ATP to all eukaryotic cells (both plant and animal), but the cells of plants and some protists have additional organelles called **chloroplasts** that capture energy from sunlight and use it to manufacture food molecules via *photosynthesis* (**Figure 3.10**, right middle). We will explore both cellular respiration and photosynthesis in Chapter 4.

A network of protein cylinders and filaments collectively known as the **cytoskeleton** forms the framework of a cell (**Figure 3.10**, right middle). The cytoskeleton organizes the interior of a eukaryotic cell, supports the intracellular movement of organelles such as transport vesicles, and enables whole-cell movement in some cell types. It also gives shape to wall-less cells. Fungi and plants separately evolved cell walls to maintain cell structure, but animals rely only on the cell's cytoskeleton.

Life Goes On

From Venter's complex synthetic genome to Devaraj's simple self-assembling membrane, scientists debate whether the top-down or bottom-up approach will be more successful in the effort to build an artificial cell. But the two can be complementary, says synthetic biologist James Collins. "I'm not sure that one will win out. I think they bring different things to the table," he notes. Gibson agrees: "I would like to one day be able to combine all of a cell's parts from nonliving components, including the genome, and incubate them, and see if we can get life out of those nonliving components," he says. "It would help us better understand how cells work."

Today, the work on both ends continues. While Devaraj makes his membranes more complex, Gibson and his team are working to simplify the *M. mycoides* genome. The JCVI team's new goal is to determine the smallest set of genes needed to maintain life. "Now that we can put all the pieces together to make a genome, we'll go back and start removing sections and see if we still get colonies—whittling away at the genome one gene at a time," says Gibson. Once his team has a core set of genes necessary for life, it will be easier to build other synthetic cells with specific functions, he adds.

"There's not a single organism in the world where we understand what every gene does," he notes. If they can strip the *M. mycoides* genome down to just its essential genes, he concludes, "this would be the first example of that."

Sizing Up Life

Cells dramatically range in size, and viruses are even smaller. The amoeba, a single-celled eukaryote that eats smaller unicellular organisms, is visible under a light microscope. But to observe a virus, one would have to use an electron microscope, which relies on a high-voltage beam of electrons to magnify an image.

|← 500 μm →|

Amoeba proteus

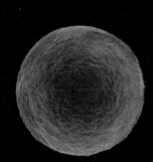

|← 130 μm →|

Human egg cell
(shown next to an Amoeba proteus)

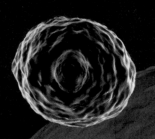

|← 30 μm →|

Skin cell
(shown next to a human egg cell)

|← 8 μm →|

Red blood cell
(shown next to a skin cell)

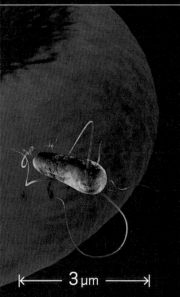

|← 3 μm →|

E. coli bacterium
(shown next to a red blood cell)

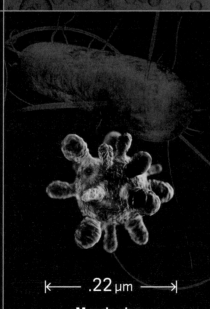

|← .22 μm →|

Measle virus
(shown next to an E. coli bacterium)

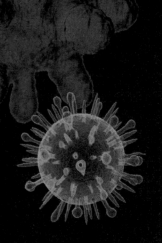

|← .13 μm →|

Influenza virus
(shown next to a measle virus)

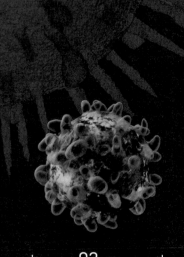

|← .03 μm →|

Rhinovirus
(shown next to an influenza virus)

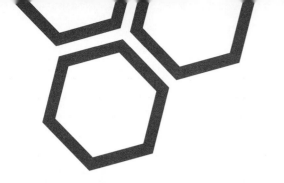

REVIEWING THE SCIENCE

- **Cells** are the basic units of all living organisms. **Cell theory** states that all living things are composed of one or more cells, and that all cells living today came from a preexisting cell.

- Every cell is surrounded by a **plasma membrane** that separates the chemical reactions inside the cell from the surrounding environment.

- The plasma membrane is a **phospholipid bilayer** that is **selectively permeable** and embedded with proteins that perform a variety of functions.

- In **passive transport**, substances move across the plasma membrane without the direct expenditure of energy. **Active transport** by cells requires energy.

- **Diffusion** is the passive transport of a substance from a region where it is at a higher concentration to a region where it is at a lower concentration.

- **Osmosis** is the diffusion of water across a selectively permeable membrane. In a **hypotonic** solution, a cell gains water. In a **hypertonic** solution, a cell loses water. In an **isotonic** solution, there is no net movement of water. Cells can actively balance their water content through osmosis.

- Cells export materials by **exocytosis** and import materials by **endocytosis**.

- In **receptor-mediated endocytosis**, **receptor proteins** in the plasma membrane recognize and bind the substance to be brought into the cell.

- **Prokaryotes** are single-celled organisms lacking a nucleus and complex internal compartments. **Eukaryotes** may be single-celled or multicellular, and their cells typically possess many membrane-enclosed compartments.

- By volume, eukaryotic cells can be a thousand times larger than prokaryotic cells. They require internal compartments, or **organelles**, that concentrate and organize cellular chemical reactions.

- The **nucleus** of a eukaryotic cell contains DNA. It is bounded by the **nuclear envelope**, which has **nuclear pores** that allow communication between the nucleus and the cell interior.

- Lipids are made in the **smooth endoplasmic reticulum**. Some proteins are manufactured in the **rough endoplasmic reticulum**.

- The **Golgi apparatus** receives proteins and lipids, sorts them, and directs them to their final destinations.

- In animals, **lysosomes** break down large organic molecules such as proteins into simpler compounds that can be used by the cell. Plant **vacuoles** are similar to lysosomes but also store ions and molecules and lend physical support to plant cells.

- **Mitochondria** produce chemical energy for eukaryotic cells in the form of ATP.

- **Chloroplasts** harness the energy of sunlight to make sugars through photosynthesis.

- Eukaryotic cells depend on the **cytoskeleton** for structural support, and for the ability to move and change shape. Plants and fungi also have a cell wall that provides structural support.

THE QUESTIONS

The Basics

1 To be able to recognize a colony of bacteria that had grown from cells of *Mycoplasma capricolum* in which the DNA had been replaced by a synthetic DNA of *Mycoplasma mycoides*, Daniel Gibson added a _____ that coded for a blue pigment.

(a) cell

(b) chromosome

(c) gene

(d) bacterium

2 A virus is _____ a cell.

(a) larger than

(b) smaller than

(c) the same size as

(d) none of the above

3 A phospholipid is a macromolecule composed of a phosphate group that is bonded to two lipid chains. Which of the following correctly describes the nature of those two components?

(a) The phosphate group is hydrophobic, and the lipid chains are hydrophilic.

(b) Both the phosphate group and the lipid chains are hydrophilic.

(c) Both the phosphate group and the lipid chains are hydrophobic.

(d) The phosphate group is hydrophilic, and the lipid chains are hydrophobic.

4 How does the phospholipid bilayer of a liposome differ from the phospholipid bilayer of the plasma membrane of a cell?

(a) The phospholipid bilayer of a liposome contains only phospholipids, without the proteins that are inserted into the plasma membrane of a cell.

(b) The phospholipid bilayer of a liposome contains two bilayers of phospholipid molecules, whereas the plasma membrane of a cell contains only one bilayer of phospholipid molecules.

(c) The phospholipid bilayer of a liposome completely envelops the liposome, whereas the plasma membrane of a cell does not completely envelop the cell.

(d) The phospholipid molecules in the phospholipid bilayer of a liposome are oriented with the lipid ends on the outside of the bilayer and the phosphate groups on the inside.

5 Which of the following statements about transport through the plasma membrane of a cell is correct?

(a) Both passive and active transport require the input of energy.

(b) Passive transport does not require the input of energy, and active transport does require energy input.

(c) Passive transport does require the input of energy, and active transport does not require energy input.

(d) Neither active nor passive transport requires the input of energy.

6 Link each of the following processes with the correct definition.

RECEPTOR-MEDIATED ENDOCYTOSIS 1. A cell ingests a large particle, such as a bacterial cell.

PHAGOCYTOSIS 2. Receptor proteins embedded in the membrane recognize specific surface characteristics of substances.

PINOCYTOSIS 3. A transport vesicle inside the cell approaches the plasma membrane of the cell, fuses with it, and releases its contents to the outside of the cell.

EXOCYTOSIS 4. A vesicle containing whatever molecules are in solution outside the cell bulges inward, pinches off, and enters the cell.

7 Link each of the following structures with the correct function.

CHLOROPLAST 1. Location of the cell's DNA.
GOLGI APPARATUS 2. Site of protein synthesis.
LYSOSOME 3. Site of lipid synthesis.
MITOCHONDRION 4. Adds chemical tags to newly synthesized proteins to direct them to their correct location.
NUCLEUS 5. Breaks down macromolecules by enzymatic action.
ROUGH ENDOPLASMIC RETICULUM 6. Site of cellular respiration.
SMOOTH ENDOPLASMIC RETICULUM 7. Site of photosynthesis.

8 Place the names of the following cellular components in the table below to indicate which structures are found in each type of cell. Items may be used more than once.

Plasma membrane   Cellulose cell wall

Nucleus   Endoplasmic reticulum

Golgi apparatus   Ribosomes

Cytoskeleton   Mitochondria

Chloroplasts

	Eukaryotes	
Prokaryotes	Animals	Plants

Try Something New

9 Picture a beaker divided by a semipermeable membrane like the one in Figure 3.7. Imagine that you put 5 grams of sugar in the left side of the beaker and 10 grams of sugar in the right side, and then add water to bring both sides to the same depth. Explain which direction water will move by osmosis and why it will move in that direction.

10 Liposomes filled with the chemical sodium cromoglycate administered in the form of a nasal spray are used for short-term relief of allergies such as hay fever. Explain how sodium cromoglycate in the liposomes could enter the cells of the nasal passages.

11 Liposomes filled with anticancer drugs are used to deliver the drugs to cancerous tissues. How could you modify a therapeutic liposome so that it could be injected into the body and would deliver its contents only to the target cells?

Leveling Up

12 **What do *you* think?** Viruses display many of the characteristics of living organisms. In particular, they reproduce, creating new virus particles. During reproduction, viruses make copies of their genetic material, and some of the copies contain mutations that are beneficial to the virus. For example, HIV (human immunodeficiency virus), the virus that causes the disease AIDS (acquired immunodeficiency syndrome) mutates so often that its surface proteins change faster than we can develop antiviral drugs. New drug-resistant strains of HIV are appearing constantly. However, viruses can reproduce only after they have entered a living cell of an organism because viruses hijack the cell's machinery and use it to produce new viruses. Where does that combination of characters place viruses on the scale of nonlife to life? Are viruses living organisms? Nonliving? If neither of those categories fits the properties of viruses, how should they be classified?

13 Did Venter create "new life" by inserting synthetic DNA copied from the genome of *Mycoplasma mycoides* into a cell of *Mycoplasma capricolum* from which the DNA had been removed? Be prepared to support your opinion in a class discussion.

14 **Doing Science** Go to the National Science Foundation website (www.nsf.gov), and read "Biologists Replicate Key Evolutionary Step in Life on Earth" (Press Release 12-009, January 16, 2012). State the hypothesis that the biologists were testing, describe their experimental design, and explain their results. Was their hypothesis supported? What experiment do you think they should try next? State a hypothesis that you would test using their multicellular yeast samples.

Pond Scum to Jet Fuel

Are algae the fuel of the future?

After reading this chapter you should be able to:

- Explain how photosynthesis and cellular respiration support life on Earth.
- Describe the importance of enzymes in metabolic pathways.
- Differentiate between anabolic and catabolic metabolic processes.
- Detail the role of ATP in a cell.
- Compare and contrast photosynthesis and cellular respiration.
- Label a diagram of the light reactions and the Calvin cycle of photosynthesis.
- Define the three stages of cellular respiration and explain the function of each.

CHAPTER
04

HOW CELLS
WORK

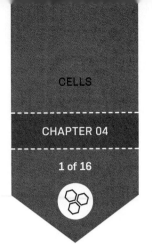

When Kim Jones landed a teaching job at Brunswick Community College on the rural coast of North Carolina, she brought a few billion friends with her—large flasks filled with millions upon millions of tiny green cells. Jones, a professional volleyball player turned scientist, was researching chemicals from algae that kill fish when she got the job at Brunswick. She needed to care for her algae every day, and luckily the college didn't mind that she brought them along.

Soon, Jones's new students discovered the cultures and offered to help maintain them by adding nutrients to the flasks, checking temperatures, and transferring the tiny organisms to new containers to grow. "They loved it, so a colleague and I applied for a grant to open an algae culturing lab at the college," says Jones, a tall, brown-haired researcher with a hint of a southern twang. She and the students spent hours and hours cultivating algae in the new lab, and soon an idea dawned on her. "At the time, renewable biofuels were really taking off," says Jones, "and I realized algae were one solution."

Humans rely heavily on fossil fuels such as oil and petroleum to power our cars, trains, planes, and more. Hundreds of millions of years ago, single-celled organisms including algae used carbon dioxide and sunlight to make cell materials, which then became buried and concentrated into oil in Earth's crust. We pump that oil from the ground today, and thus are literally burning 200-million-year-old algae in our gas tanks. Burning fossil fuels unfortunately releases that ancient carbon dioxide back into the atmosphere, and a consequence of this increase in carbon dioxide is that the surface of our planet has begun to warm. This process, commonly referred to as global warming, has significant environmental consequences that we have already begun to experience, such as the melting of the ice caps at

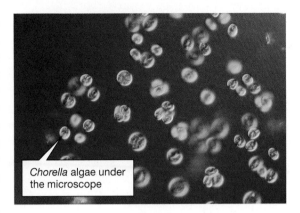

Chorella algae under the microscope

Figure 4.1

Chlorella **is a single-celled alga commonly used to produce biofuels and nutritional supplements**

the North Pole. (We will address climate change and global warming further in Chapter 15.)

To prevent further climate change and lessen our dependence on fossil fuels—many of which we pay a premium to import from other countries—scientists are exploring ways to make clean, renewable fuels that are safer for the environment and less expensive than fossil fuels. Corn and soybeans are a popular source of **biofuel**, a fuel produced by the conversion of organic matter—but some argue that using food crops for fuel is not a sustainable practice for a planet with a growing population that needs to eat.

There is at least one biofuel source that does not compete with food crops: algae (**Figure 4.1**). Microalgae, microscopic single-celled organisms (as opposed to macroalgae like seaweed), were first grown en masse on a rooftop at the Massachusetts Institute of Technology in the early 1950s. At the time, squeezing oil from algae was a futuristic, kooky idea. But, in the 1970s, oil prices spiked and alternative fuel sources became attractive. In response, the U.S. government poured $25 million into algae biofuel projects in the hope that algae could substantially contribute to the country's fuel supply. However, despite the increased research effort, algae biofuels still cost significantly more per gallon than crude oil. Research in the area dwindled.

Then, in 2008, oil prices spiked once again, and interest was renewed. "There has been a flurry of activity in the last four years," says John Hewson, a scientist studying algae biofuels at

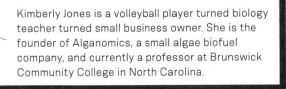

KIMBERLY JONES

Kimberly Jones is a volleyball player turned biology teacher turned small business owner. She is the founder of Alganomics, a small algae biofuel company, and currently a professor at Brunswick Community College in North Carolina.

Sandia National Laboratories in New Mexico. In January 2010, government organizations committed $78 million to commercializing algae biofuels. Private fuel companies such as Exxon, Chevron, and Shell have also poured money and time into developing an algae biofuel, but such a fuel has yet to be produced on a commercial scale and remains expensive.

Is today's push toward algae biofuels just another fuel bubble, as in the 1970s? Or could we all soon be pumping algae-derived oil into our tanks?

It Starts with Sunlight

All living cells require energy, which they must obtain from the living or nonliving components of their environment. Organisms use energy for growth, reproduction, defense, and to manufacture the many chemical compounds that make up living cells.

The sun is the ultimate source of energy for most living organisms. In the process known as **photosynthesis**, organisms capture energy from the sun and use it to create sugars from carbon dioxide and water, thereby transforming light energy into chemical energy stored in the covalent bonds of sugar molecules. This energy is then used to fuel cellular activities via **cellular respiration**, the reciprocal process to photosynthesis in which sugars are broken down into energy usable by the cell (**Figure 4.2**).

The first law of thermodynamics says that energy cannot be created or destroyed, but it can be changed from one form to another. In other words, algal cells cannot create energy from nothing, but they can take one form of energy and change it to another form. Photosynthetic algae capture energy from sunlight and convert it into chemical energy in the form of glucose, a sugar. This glucose fuels the cell's activities, and some of it is converted to fatty acids, including oil, to help build cell membranes and to store energy for future needs. The capture, storage, and use of energy by living cells is not efficient, and much of the energy in sunlight is lost during photosynthesis.

Although no organism can turn sunlight into stored energy with 100 percent efficiency, microalgae are some of the most efficient photosynthetic organisms, producing energy from sunlight at a rate several times that of corn or soybeans. They are also some of the fastest-replicating life-forms on Earth, doubling their mass many times a day. And they make oil. Researchers want to take advantage of these three qualities of microalgae to produce a cheap, renewable fuel. "We spend millions to kill algae," says Jones, "but here is an opportunity to use them in a beneficial way."

In 2008, Jones applied for and was awarded several small business grants from her local government. She used the funds to form Alganomics, a small algae biofuel company. Alganomics would focus on harvesting the oil produced by algae as a source of biodiesel. Other companies, such as Florida-based Algenol, specialize in creating ethanol from algae carbohydrates. Algae can be used to form other fuels too, including biogas and methane. At Alganomics, Jones' first task was to build a facility to grow and harvest the algae.

Algae need a lot of water to grow, so Jones identified a field behind a wastewater treatment plant where she could pump water from the plant—rich in nutrients to help the algae grow—into *bioreactors*, vessels designed to contain biological reactions (**Figure 4.3**). Her team set up two rows of long, clear plastic tubes to act as bioreactors. Each tube, about 6 feet long, was attached to the next tube by macaroni-shaped connectors, forming a ribbonlike waterway of tubing.

Once the tubes were ready and filled with water, Jones added microalgae and placed the tubes out in the field under the sun. Then she waited. The algae basked in the sunlight, capturing energy from the sun and using it to produce oil.

Energized Algae

The term **metabolism** describes all the chemical reactions that occur inside living cells, including those that store or release energy. Nearly all metabolic reactions, including photosynthesis, are facilitated by enzymes. **Enzymes** are biological catalysts—molecules that speed up chemical reactions. Without the action of enzymes—most of which are proteins—metabolism would be extremely slow, and life as we know it could not exist. Enzymes work

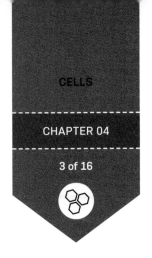

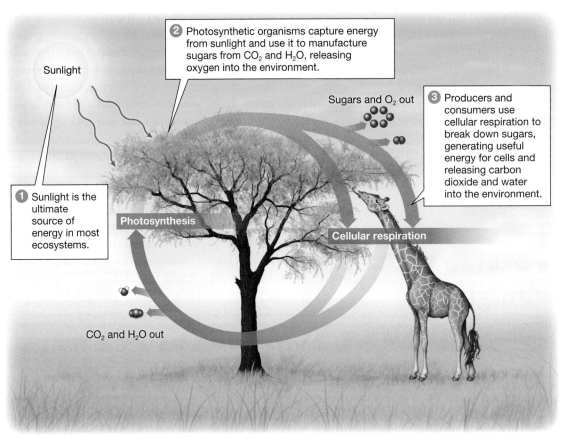

① Sunlight is the ultimate source of energy in most ecosystems.

② Photosynthetic organisms capture energy from sunlight and use it to manufacture sugars from CO_2 and H_2O, releasing oxygen into the environment.

③ Producers and consumers use cellular respiration to break down sugars, generating useful energy for cells and releasing carbon dioxide and water into the environment.

Sunlight

Sugars and O_2 out

Photosynthesis

Cellular respiration

CO_2 and H_2O out

Figure 4.2

From sunlight to usable energy

Photosynthesis transforms sunlight into sugar molecules within the cell, releasing oxygen as a by-product. Without photosynthesis, we would not have access to the sun's energy, and we wouldn't have oxygen in our atmosphere to support life. In a complementary process, cellular respiration breaks down these sugar molecules, allowing organisms to access the energy stored in them.

Q1: Why is photosynthesis called "primary production"?

Q2: How does animal life depend on photosynthesis?

Q3: Explain how photosynthesis and cellular respiration are "complementary" processes.

by positioning **substrates**—molecules that will react to form new products—in an orientation that favors the making or breaking of chemical bonds (**Figure 4.4**).

Each enzyme binds only to a specific substrate or substrates and catalyzes a specific chemical reaction. An enzyme's function is based on its chemical characteristics and the three-dimensional shape of its **active site**—the location within the enzyme where substrates bind. When molecules bind to the active site, the enzyme changes shape, a process called **induced fit**. The enzyme's shape, and therefore its activity, can be affected by temperature, pH, and salt concentration. Because the enzyme's active site is not permanently changed as reactions are catalyzed, an enzyme can be used over and over.

Most chemical reactions in a cell occur in chains of linked events known as **metabolic pathways**. Metabolic pathways produce key biological molecules in a cell, including important chemical building blocks like amino acids and nucleotides. A multistep metabolic pathway can proceed rapidly and efficiently because the required enzymes are physically close together and the products of one enzyme-catalyzed

Figure 4.3

Bioreactors similar to the ones used by Kim Jones

chemical reaction serve as the basis for the next reaction in the series.

All living cells have two main types of metabolism: anabolism and catabolism (**Figure 4.5**). **Anabolism** refers to metabolic pathways that create complex molecules from simpler compounds. Photosynthesis is an example of anabolism; as Jones's tubes sit in the noonday sun, the algae inside them perform photosynthesis, making sugars from carbon dioxide and sunlight—complex molecules made from simpler ones. **Catabolism** refers to metabolic pathways that release chemical energy in the process of breaking down complex molecules. In this case, complex molecules are broken down into simpler molecules. In the evening, when the sun has set, the algae use cellular respiration to break down the sugars made during photosynthesis.

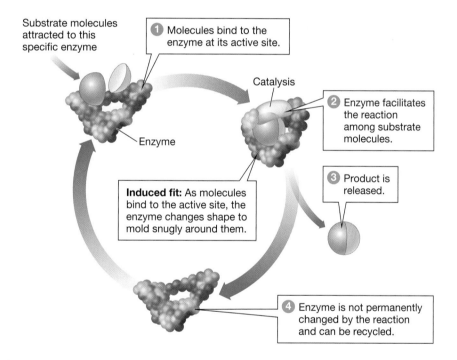

Substrate molecules attracted to this specific enzyme

1 Molecules bind to the enzyme at its active site.

Catalysis

2 Enzyme facilitates the reaction among substrate molecules.

Enzyme

3 Product is released.

Induced fit: As molecules bind to the active site, the enzyme changes shape to mold snugly around them.

4 Enzyme is not permanently changed by the reaction and can be recycled.

Figure 4.4

Enzymes are molecular matchmakers

Enzymes dramatically increase the rate of chemical reactions by positioning molecules so that they more easily form or break chemical bonds.

Q1: Why is it important that enzymes are not permanently altered when they bind with substrate molecules?

Q2: How would a higher temperature or salt concentration make it more difficult for an enzyme to function effectively?

Q3: If a cell was unable to produce a particular enzyme necessary for a metabolic pathway, describe how the absence of that enzyme would affect the cell.

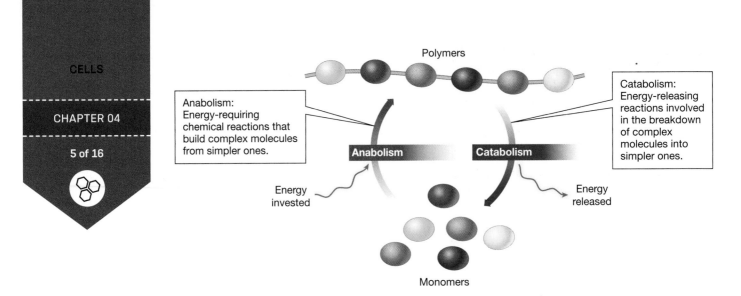

Polymers

Anabolism:
Energy-requiring
chemical reactions that
build complex molecules
from simpler ones.

Catabolism:
Energy-releasing
reactions involved
in the breakdown
of complex
molecules into
simpler ones.

Anabolism

Catabolism

Energy
invested

Energy
released

Monomers

Figure 4.5

Anabolism builds macromolecules; catabolism breaks them down
Metabolic reactions either build or break down molecules. The former (anabolism) cost energy, and the latter (catabolism) release energy.

Q1: What source of energy would algae or plants use for anabolic reactions? Would an animal use the same kind of energy?

Q2: What source of energy would algae or plants release in catabolic reactions? Would an animal release the same kind of energy?

Q3: Create a mnemonic or jingle that helps you remember the difference between anabolism and catabolism.

All of this metabolic activity requires energy, so cells need **energy carriers** to store and deliver usable energy. Every living cell uses **ATP** (adenosine triphosphate), a small, energy-rich organic molecule, to store energy and to move it from one part of a cell to another. ATP powers a variety of activities in the cell, such as moving molecules and ions in and out of the cell and moving organelles around inside the cell. In addition to powering a cell's activities, ATP fuels metabolic reactions, including photosynthesis.

Much of the usable energy in ATP is stored in its energy-rich phosphate bonds (**Figure 4.6**). Energy is released when a molecule of ATP loses its terminal phosphate group, breaking into a molecule of **ADP** (adenosine diphosphate) and a free phosphate. Converting ADP and phosphate into ATP takes metabolic energy.

ATP is not the only energy carrier that cells rely on; NADPH, NADH, and $FADH_2$ are also energy carriers. Each one is a specialist in terms of the amount of energy it carries and the types of chemical reactions to which it supplies energy and from which it receives energy. ATP, because of its phosphate bonds, carries the most energy.

Green Slime

Within days of sitting in the sun, the bioreactors at Alganomics turned bright green as they filled with algae, thanks to photosynthesis. Photosynthesis is considered by many to be the most important life process on Earth; it is the way our planet stores energy from the sun and produces oxygen for animals to breathe. In the cells of algae and all plants, photosynthesis takes place inside **chloroplasts**, organelles that look like green, oval gumballs when viewed under a light microscope (**Figure 4.7**). Chloroplasts contain an extensive network of structures called thylakoids, piled up like stacks of pancakes, that contain all the enzymes needed for photosynthesis. Also embedded in those membranes is a green pigment called **chlorophyll** that is specialized for absorbing light energy.

Photosynthesis takes place in two principal stages: the *light reactions* and the *light-independent reactions*, or the *Calvin cycle* (**Figure 4.8**).

During the **light reactions**, chlorophyll molecules absorb energy from sunlight and use that energy to split water (**Figure 4.8**, left). The splitting of water produces oxygen gas (O_2)—the oxygen that we breathe—as a by-product that is released into the atmosphere. More important for the organism, electrons and protons (H^+) from the light reactions are handed over to other molecules via the **electron transport chain**, an elaborate chain of chemical events that ultimately generates ATP and NADPH. The light reactions

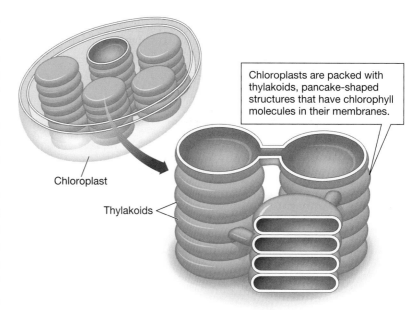

Chloroplasts are packed with thylakoids, pancake-shaped structures that have chlorophyll molecules in their membranes.

Chloroplast

Thylakoids

Figure 4.7

Chloroplasts are the site of photosynthesis in eukaryotes
Most plants and algae have multiple chloroplasts to contain their chlorophyll. Photosynthetic bacteria embed chlorophyll directly into the plasma membrane.

Q1: Is chlorophyll found only within chloroplasts?

Q2: What could be an advantage of concentrating chlorophyll molecules in the membranes of chloroplasts?

Q3: Oil from algae is now being used in beauty products and nutritional supplements. Would you use a moisturizer or eat pills made from algae? Why or why not?

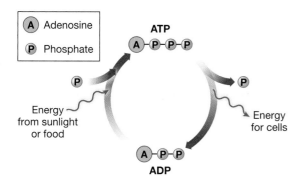

A Adenosine
P Phosphate

ATP
A-P-P-P

P

Energy from sunlight or food

Energy for cells

P

A-P-P
ADP

Figure 4.6

Cells store and deliver energy using ATP
Loading a high-energy phosphate (P) on ADP transforms the otherwise sedate molecule into the live wire that is ATP. But turning ADP and phosphate groups into ATP—the universal energy currency of cells—takes metabolic energy.

Q1: Define ATP in your own words.

Q2: How is ATP involved in anabolism and catabolism? (*Hint:* Review Figure 4.5.)

Q3: Arsenic disrupts ATP production. Why would this characteristic cause it to be a potent poison?

depend upon protein complexes embedded in the chloroplast membrane, including photosystems I and II, and ATP synthase.

In the next stage, the **light-independent reactions**, or **Calvin cycle**, a series of chemical reactions, converts carbon dioxide (CO_2) into sugar, using energy delivered by ATP and electrons and hydrogen ions donated by NADPH (**Figure 4.8**, right). These reactions are catalyzed by enzymes at each step; the enzyme needed in the first step—and the most abundant enzyme on the planet—is **rubisco**. This process is also known as **carbon fixation**. By capturing inorganic carbon atoms from CO_2 gas and converting them into glucose, the Calvin cycle reactions make carbon from the nonliving world available to the photosynthetic organisms and eventually to other living organisms, including us.

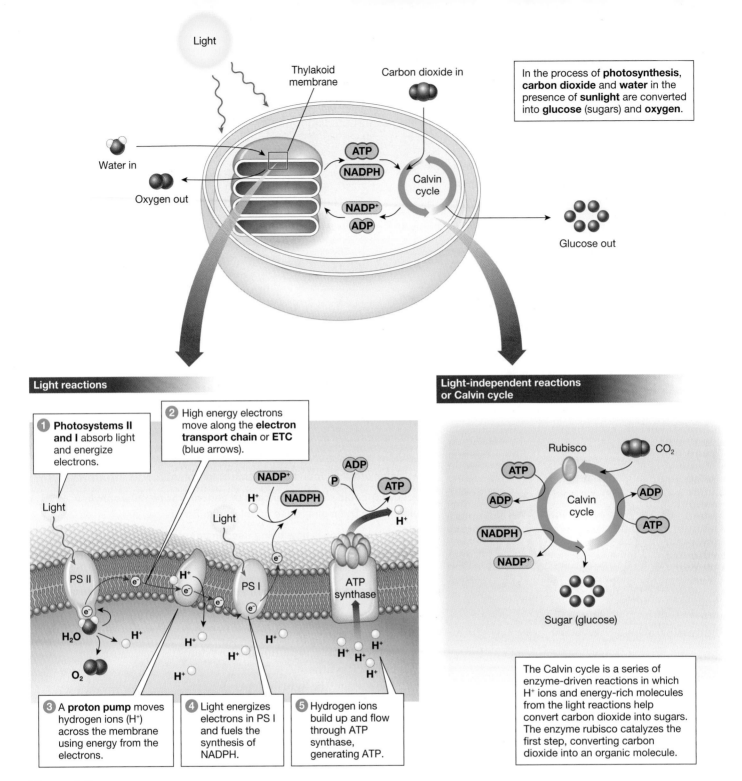

Figure 4.8

Photosynthesis occurs in two stages

The light reactions generate energy carriers; the light-independent Calvin cycle reactions create sugars. Rubisco contributes to fixing carbon during the Calvin cycle, but enzymes are critical throughout both stages of photosynthesis.

Q1: What is the source of the carbon dioxide used for photosynthesis?

Q2: What products of the light reactions of photosynthesis does the Calvin cycle use?

Q3: What are the two major products of photosynthesis?

Photosynthesis, essentially, is the production of food out of thin air. Or, in Jones's case, the production of *fuel* out of thin air. Once the algae in Jones's bioreactors had grown, her team harvested them to extract oils produced by the conversion of glucose to fatty acids in the cells. Algae could yield up to 5,000 gallons of fuel per acre per year, under the right conditions, according to some estimates. Soybeans, in comparison, which are widely used to produce fuel, yield only 50 gallons per acre per year. The production potential of algae, scientists say, is tremendous.

In addition to the oil from algae, the leftover plant matter can be dried and used as fish food for aquaculture, to create bioplastics, or for antioxidants in cosmetic products like lotions, making algae a potentially valuable crop even if they do not become a mainstream biofuel. For now, Jones is working to see whether algae biofuels can have an immediate impact on the farming community of North Carolina. After Iowa, North Carolina is the largest hog-farming state in the country. Many farmers have lagoons on their farms where waste from the hogs is poured, and those lagoons could be used to grow microalgae, says Jones. "It would create another badly needed crop for farmers here." Her team is currently experimenting with different local strains of algae to see which grow fastest and produce the most oil.

Jones's goal is simple: to demonstrate that algae biofuels can successfully be produced on a small commercial scale in a rural community. But other scientists are interested in making algae biofuels on a larger scale—a much, much larger scale.

An Ocean of Algae

In 2008, Jonathan Trent had a big idea. While working as an astrobiologist for the NASA Ames Research Center in California, Trent had spent months watching enzymes in plant systems devour plant material. He knew that algae make oil and biomass through such catabolic processes, which got him thinking about using algae to produce fuel. "I looked into the literature about using algae to make biofuel and realized immediately the problem was going to be scale. We need millions of acres of algae to meet our fuel needs, and growing that much algae on land will undoubtedly compete with agriculture, not just for huge areas of land, but also for water and fertilizer," says Trent.

Figure 4.9

NASA's OMEGA system uses large, flexible tubes as photobioreactors

Like Jones of Alganomics, Trent knew that algae can grow in wastewater and he also realized that most coastal cities dump their wastewater offshore. "It was obvious," he said. "Algae grown in wastewater. Wastewater is being pumped offshore and if we could move the algae cultivation offshore, we could solve the water, fertilizer, and land problems all at once." That kernel of an idea soon became the Offshore Membrane Enclosures for Growing Algae (OMEGA) project, a project to demonstrate the feasibility of growing algae in bioreactors in seawater, using nutrients from wastewater and a source of concentrated CO_2 to help them grow (**Figure 4.9**).

Wastewater is an important component of both Trent's and Jones's ideas because algae require nutrients to grow, especially nitrogen and phosphorus. Every amino acid in a cell contains nitrogen, and many molecules in the cell, such as plasma membrane molecules and DNA, as well as every molecule of ATP, contain phosphorus. Without these two nutrients, algal cells cannot construct these vital elements. It turns out that wastewater—liquid waste from our homes, businesses, and industries—is an excellent source of both nitrogen and phosphorus. Using wastewater has two benefits: it lowers the cost of producing the

JONATHAN TRENT

Jonathan Trent is a marine biologist and the lead scientist of NASA's Offshore Membrane Enclosures for Growing Algae (OMEGA) project. The project was established to grow oil-producing algae using municipal wastewater.

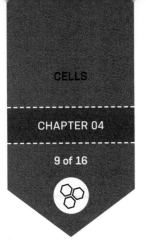
biofuel because there is no need to make or purchase fertilizer, and the algae help clean the wastewater before it is released into the environment.

With $10 million in funding from NASA, Trent and his team built a prototype system in seawater tanks at the San Francisco Southeast Wastewater Treatment Plant. Large, flexible plastic tubes filled with algae bob in a pool of seawater. Inside those tubes, microalgae perform photosynthesis, soaking up sunlight during the day and storing that energy in molecules of sugar, mainly glucose. Later, when the energy is needed, the algae perform cellular respiration to convert the energy from glucose to a usable form of cellular energy: ATP.

Making Energy the Cell Can Use

During cellular respiration, the carbon-carbon bonds in glucose molecules are broken, and each carbon atom is released into the environment in a molecule of CO_2, with water (H_2O) as a by-product. Thus, cellular respiration is the reciprocal process of photosynthesis, as illustrated earlier in Figure 4.2. In most eukaryotes, cellular respiration is a multi-step process that occurs in the cell's mitochondria (**Figure 4.10**). Cellular respiration is the main way that animal cells, which do not perform photosynthesis, obtain energy. We digest glucose from food sources, and we exhale carbon dioxide and urinate water as waste products. Algae and other photosynthetic organisms also use cellular respiration to produce ATP at night, when there is no sunlight.

The first stage of cellular respiration, glycolysis, takes place in the cytoplasm of the cell. During **glycolysis**, sugars (mainly glucose) are split to make a three-carbon compound called *pyruvate*. This process releases two molecules of ATP and two molecules of NADH for each glucose molecule that is split.

From an evolutionary standpoint, glycolysis was probably the earliest means of producing ATP from food molecules, and it is still the primary means of energy production in many prokaryotes. However, the energy yield from glycolysis is small because sugar is only partially broken down through this process. For most eukaryotes, glycolysis is just the first step in extracting

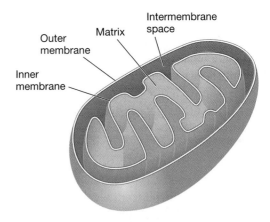

Mitochondrion

Figure 4.10

Mitochondria are the site of cellular respiration in eukaryotes

energy from sugars, and additional reactions in the mitochondria help to generate much more ATP than is possible through glycolysis alone.

Glycolysis is an **anaerobic** process, which means it does not require oxygen. Many anaerobic bacteria that live in oxygen-deficient swamps, in sewage, or in deep layers of soil, are actually poisoned by oxygen. Most anaerobic organisms extract energy from organic molecules using a *fermentation* pathway. **Fermentation** begins with glycolysis, followed by a special set of reactions whose only role is to help perpetuate glycolysis. This process enables organisms to generate ATP anaerobically—through glycolysis alone—when **aerobic** (oxygen-dependent) ATP production is constrained by low oxygen levels (**Figure 4.11**).

Some companies trying to make biofuel from algae have experimented with growing algae indoors in the presence of sugar, instead of outdoors in sunlight. This choice bypasses photosynthesis and skips straight to fermentation as a way to derive ATP. An indoor fermentation plant offers tighter controls on temperature, pressure, and other environmental conditions for the algae than those of an outdoor system like OMEGA. Still, closed indoor systems can be expensive, and relying on glycolysis alone via fermentation does not produce as much energy as the entire process of cellular respiration, of which glycolysis is only the first step.

After pyruvate is made in glycolysis, it enters the mitochondria (**Figure 4.12**), where it is broken down during the second stage of cellular

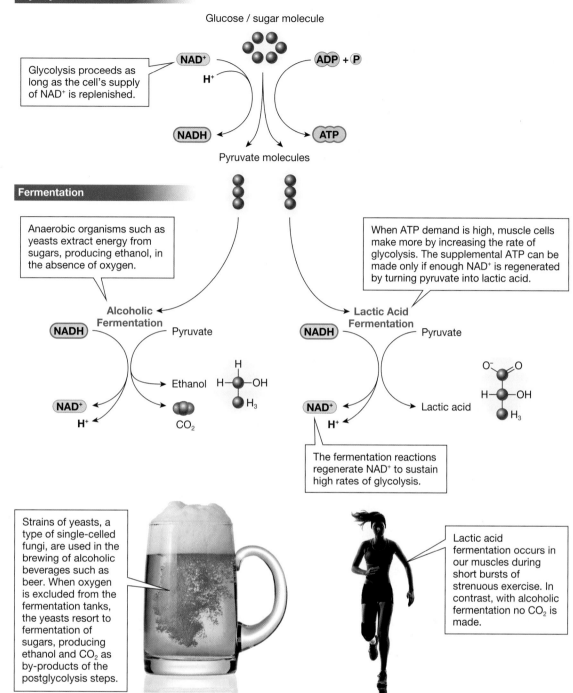

Glycolysis

Glucose / sugar molecule

NAD⁺ → NAD^+

Glycolysis proceeds as long as the cell's supply of NAD^+ is replenished.

H^+

ADP + P

$NADH$

ATP

Pyruvate molecules

Fermentation

Anaerobic organisms such as yeasts extract energy from sugars, producing ethanol, in the absence of oxygen.

When ATP demand is high, muscle cells make more by increasing the rate of glycolysis. The supplemental ATP can be made only if enough NAD^+ is regenerated by turning pyruvate into lactic acid.

Alcoholic Fermentation

$NADH$ — Pyruvate

→ Ethanol

NAD^+ ← H^+ ← CO_2

Lactic Acid Fermentation

$NADH$ — Pyruvate

→ Lactic acid

NAD^+ ← H^+

The fermentation reactions regenerate NAD^+ to sustain high rates of glycolysis.

Strains of yeasts, a type of single-celled fungi, are used in the brewing of alcoholic beverages such as beer. When oxygen is excluded from the fermentation tanks, the yeasts resort to fermentation of sugars, producing ethanol and CO_2 as by-products of the postglycolysis steps.

Lactic acid fermentation occurs in our muscles during short bursts of strenuous exercise. In contrast, with alcoholic fermentation no CO_2 is made.

Figure 4.11

Fermentation produces energy in the absence of oxygen

When the oxygen supply is inadequate to support ATP production through cellular respiration, fermentation supports ATP production through glycolysis alone.

Q1: What is produced during fermentation that accounts for the bubbles in beer?

Q2: Bakers of yeast breads also rely on fermentation, allowing bread to "rise" before baking. Describe what is occurring with the yeast as the bread rises.

Q3: Explain in your own words why lactic acid builds up in your muscles during strenuous physical activity.

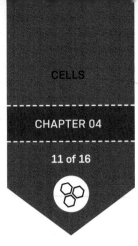

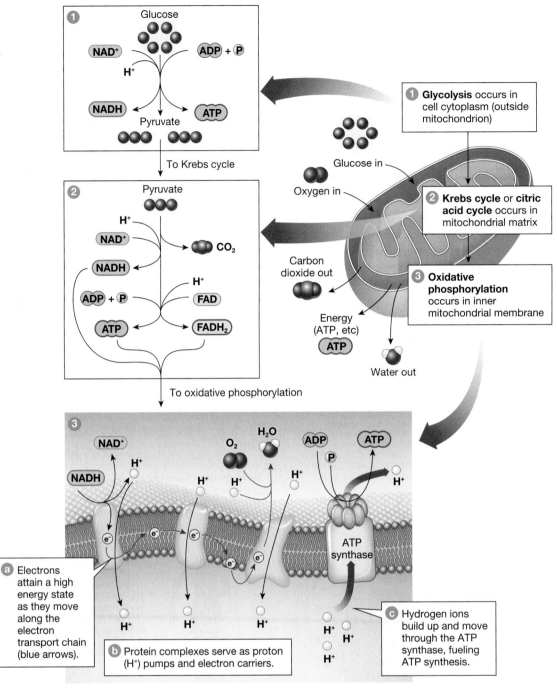

Figure 4.12

Cellular respiration is highly efficient in eukaryotes

In glycolysis, each six-carbon glucose molecule is converted into two molecules of pyruvate. The next two stages of cellular respiration require oxygen. The Krebs cycle releases carbon dioxide and generates high-energy molecules. Oxidative phosphorylation, the last stage in cellular respiration, produces more ATP than any other metabolic pathway.

Q1: What are the products of cellular respiration?

Q2: Considering the inputs and products of each process, why is cellular respiration considered the reciprocal process to photosynthesis?

Q3: Which of the three stages of cellular respiration—glycolysis, the Krebs cycle, or oxidative phosphorylation—could organisms have used four billion years ago, before photosynthesis by cyanobacteria released oxygen into the atmosphere?

Growing Demand

Renewable energy currently makes up only a small slice of the total energy resources used around the world. Here we peek into the origins of two of the renewable fuels currently in use, biodiesel and bioethanol, which together still make up less than one percent of all global energy consumed.

Total world energy consumption

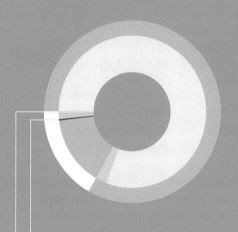

Nonrenewables

- **80.6%** Fossil Fuels
- **2.7%** Nuclear

Renewables

- **11.44%** Biomass Heating
- **3.34%** Hydropower
- **.17%** Biodiesel
- **1.25%** Other Renewables
- **.50%** Bioethanol

Biodiesel

89.2%
Vegetable oil
(including palm and soybean oils)

9%
Non-agricultural feedstock

1.8%
Jatropha

Bioethanol

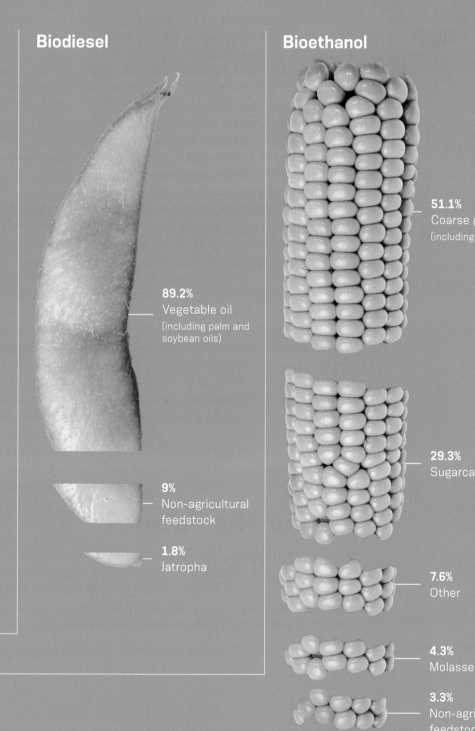

51.1%
Coarse grains
(including corn)

29.3%
Sugarcane

7.6%
Other

4.3%
Molasses

3.3%
Non-agricultural feedstock

2.2%
Wheat

2.2%
Sugar beet

respiration: a sequence of enzyme-driven reactions known as the **Krebs cycle** or **citric acid cycle**. The carbon backbone of the pyruvate molecule is taken apart, releasing carbon dioxide (**Figure 4.12**, middle). The breakdown of carbon backbones by the Krebs cycle produces a large bounty of energy carriers, including ATP, NADH, and $FADH_2$.

The largest output of ATP is generated during the third and last stage of cellular respiration: **oxidative phosphorylation**. The chemical energy of NADH and $FADH_2$ is converted into the chemical energy of ATP, while electrons and hydrogen atoms removed from NADH and $FADH_2$ are handed over to molecular O_2, creating water (H_2O). In the process, a large amount of ATP is generated (**Figure 4.12**, bottom). In fact, oxidative phosphorylation can generate 15 times as much ATP as glycolysis alone can.

ATP production in the mitochondrion is crucially dependent on oxygen; that is, the Krebs cycle and oxidative phosphorylation are strictly aerobic processes. Highly aerobic tissues, like your muscles, have high concentrations of mitochondria and a rich blood supply to deliver the large amounts of O_2 needed to support their activity.

Together, photosynthesis and cellular respiration enable cells to store and release energy from the sun. They are two sides of the same coin: The products of one reaction are the ingredients for the other. Photosynthesis, an anabolic process, requires energy (sunlight) and CO_2, and releases O_2 and glucose. Cellular respiration, a catabolic process, requires O_2 and glucose, and releases CO_2 and energy.

Fuel of the Future?

A full-scale OMEGA system, with acres and acres of algae-filled tubes, has not yet come to fruition. Despite his ambitions, getting government permission to put a large facility offshore is difficult, says Trent, as is getting public support for the project.

It is yet another hurdle in the transition from fossil fuels to sustainable alternatives. Algae biofuels have a long history, but so far no company has produced an algae biofuel on a commercial scale that is even close to being as affordable to consumers as what they currently pay for gasoline. But there has been significant progress in the last four years, says Sandia's Hewson: "There are companies now producing reasonable amounts of algae."

In September 2012, for example, Algenol, a private company based in Florida, announced that it had surpassed 8,000 gallons per acre per year in ethanol production from algae using large outdoor bioreactors. Sapphire Energy, which owns a large algae biofuel farm in New Mexico, also recently announced that the company had harvested 31 million gallons of algae to date (**Figure 4.13**). "We're proving that you can grow this like a crop," Tim Zenk, vice president of corporate affairs for Sapphire, said at a 2012 biofuels conference. "We've had a full year of continuous harvests."

"The problem now," says Hewson, "is how to make it profitable." Algae biofuels are expensive now, but when petroleum was first developed, it was very expensive too, says Patrick Hatcher, a geochemist who leads a large algae research program at Old Dominion University in Norfolk, Virginia. "It took decades to make a product that was economical," he says, and algae biofuels are currently at those early stages of development.

For algae biofuels to become a major part of the nation's fuel supply, a lot more research and investment will be needed, says Trent. But, he adds, it will be worth it. "We should be focusing on this the way we did the Manhattan Project or the Apollo program," he says. "We're burning through all this oil, leaving nothing for the next generation. We don't have to totally change our lifestyle. We have to become responsible stewards for the world."

Figure 4.13

Algaeus Prius, the first algae-run vehicle
The hybrid plug-in Prius was modified to use algae biofuel produced by Sapphire Energy.

REVIEWING THE SCIENCE

- The sun is the source of energy fueling most living organisms. Plants, algae, and some bacteria gain energy from their environment through **photosynthesis**. Most organisms use **cellular respiration** to extract usable energy from sugar molecules. In chemical terms, photosynthesis is the reverse of cellular respiration.

- All the many chemical reactions involved in the capture, storage, and use of energy by living organisms are collectively known as **metabolism**. Energy-releasing breakdown reactions like cellular respiration are **catabolism**; energy-requiring synthesis reactions like photosynthesis are **anabolism**.

- **Enzymes** are biological catalysts, usually small proteins, that speed up chemical reactions. An enzyme's function is based on its chemical characteristics and the three-dimensional shape of its **active site**.

- A **metabolic pathway** is a multistep sequence of chemical reactions, with each step catalyzed by a different enzyme.

- **Energy carriers** store energy and deliver it for cellular activities. **ATP** is found in all cells and is the most commonly used energy carrier.

- **Photosynthesis** takes place in chloroplasts and occurs in two stages. In the **light reactions**, energy is absorbed using pigment molecules that include **chlorophyll** as electrons flow along the **electron transport chain**. The light reactions create the energy carriers ATP and NADPH, splitting water molecules and releasing oxygen gas.

- The energy carriers are then used to convert carbon dioxide into sugar molecules during the light-independent **Calvin cycle** reactions. In the first of the reactions, the enzyme **rubisco** catalyzes the fixation of CO_2.

- **Cellular respiration** occurs in three stages. Small amounts of ATP and NADH are made during the first stage (**glycolysis**).

- During the next stage of cellular respiration, the **Krebs cycle**, carbon dioxide is released, and NADH, $FADH_2$, and ATP are produced.

- The final stage of cellular respiration is **oxidative phosphorylation**, during which many molecules of ATP are made in an oxygen-utilizing process.

- In the absence of oxygen, **fermentation** breaks down the products of glycolysis into alcohol and lactic acid.

THE QUESTIONS

The Basics

1 Metabolic pathways

(a) always break down large molecules into smaller units.

(b) only link smaller molecules together to create polymers.

(c) are often organized as a multistep sequence of reactions.

(d) occur only in mitochondria.

2 Enzymes

(a) catalyze reactions that would otherwise occur much more slowly.

(b) catalyze reactions that would otherwise never occur.

(c) provide energy for anabolic but not catabolic pathways.

(d) are consumed during the reactions that they speed up.

3 The chemical that is the most common energy-carrying molecule in all organisms is

(a) carbon dioxide.

(b) water.

(c) ATP.

(d) rubisco.

4 The major product of photosynthesis is

(a) lipids.

(b) sugar.

(c) amino acids.

(d) nucleotides.

5 Which of the following statements is *not* true?

(a) Glycolysis is the first stage of cellular respiration.

(b) Glycolysis can proceed under low oxygen levels with the assistance of fermentation.

(c) Glycolysis produces less ATP than the Krebs cycle or oxidative phosphorylation.

(d) Glycolysis produces most of the ATP required by aerobic organisms like us.

6 Circle the correct terms in the following sentences: Photosynthesis and cellular respiration are chemically (**identical, opposite**) processes. Cellular respiration is an example of (**anabolism, catabolism**), which (**produces, expends**) energy.

7 In the diagram of photosynthesis shown here, fill in each blank with the appropriate term: (a) Sunlight; (b) CO_2; (c) Oxygen; (d) H_2O; (e) ATP and NADPH; (f) ADP and NADP$^+$; (g) Sugar; (h) Light reactions; (i) Calvin cycle.

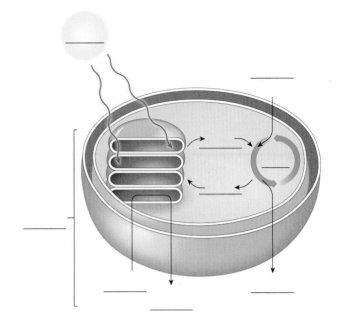

8 Place the following steps of cellular respiration in the correct order by numbering them from 1 to 4.

_____ a. The Krebs cycle produces energy carriers NADH, FADH$_2$, and ATP.

_____ b. If oxygen levels are adequate, pyruvate is transported into the mitochondrion. If oxygen levels are very low, fermentation proceeds.

_____ c. Glucose is broken down to produce ATP and NADH.

_____ d. An electron transport chain produces ATP from ADP.

Try Something New

9 The Calvin cycle reactions are sometimes called the "light-independent reactions" or "dark reactions" to contrast them with the light reactions, or light-dependent reactions. Can the Calvin cycle be sustained in algae that are kept in total darkness for several days? Why or why not?

10 In 2012, an Illinois man was killed by cyanide poisoning after he won a million dollars in the lottery. Cyanide is a lethal poison because it interferes with the electron transport chain in mitochondria. What effect would cyanide have on cellular respiration?

(a) Glycolysis, the Krebs cycle, and oxidative phosphorylation would all be inhibited.

(b) The Krebs cycle would be inhibited, but oxidative phosphorylation would not.

(c) Oxidative phosphorylation would be inhibited.

(d) Glycolysis, the Krebs cycle, and oxidative phosphorylation would all be stimulated.

11 Plants in the genus *Ephedra* have been harvested for their active substance *ephedrine* for centuries. Ephedrine is used to reduce the symptoms of bronchitis and asthma, as a stimulant and study aid, and as an appetite suppressant. It is also the main ingredient in the illegal production of methamphetamine. Since 2006, the sale of ephedrine and related substances has been limited and monitored in the United States. One effect of ingesting ephedrine is greatly increased metabolism, which has been known to kill users of ephedrine. How might an increased metabolic rate cause death?

Leveling Up

12 **What do _you_ think?** What would happen if a virus destroyed all photosynthetic organisms on Earth?

13 **_Write Now_ Biology: calorie-burning fat** Your friend has e-mailed you a link (see below) to a _New York Times_ article on "brown" fat. He has been trying to lose weight and wants to know whether you think it would be a good idea for him to spend more time in the cold, rather than continuing to exercise regularly. He is also interested that the article mentions ephedrine's ability to stimulate brown fat, and he asks if you think he should begin to take ephedrine

supplements. Compose an e-mail to your friend addressing the following points (using one to two paragraphs for each one). [_Note:_ You may need to do further reading to answer (b) and (c).]

(a) Explain in detail how brown fat burns calories when someone is chilled.

(b) Explain how ephedrine affects metabolism and what its possible side effects are.

(c) Contrast (a) and (b) with the effect of exercise on metabolism, both in the short term and in the longer term by increasing muscle mass.

(d) In your final paragraph, advise your friend as to whether he should begin spending time in the cold to increase weight loss or take ephedrine supplements, and whether he should continue to exercise regularly. Justify your opinion with data and logic.

To research your e-mail, begin with the _New York Times_ article published on April 8, 2009: "Calorie-Burning Fat? Studies Say You Have It" (http://www.nytimes.com/2009/04/09/health/research/09fat.html). Consult and reference at least two additional sources in your e-mail.

Toxic Plastic

Two ruined experiments expose the health risks of chemicals in everyday products all around us.

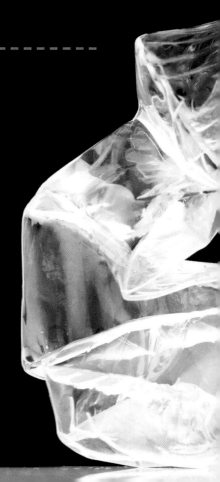

After reading this chapter you should be able to:

- Label a figure of the major stages of the cell cycle and explain the processes that occur during each of these stages.

- Compare and contrast cell division by binary fission, mitosis, and meiosis.

- Distinguish between sister chromatids and homologous chromosomes.

- Diagram, using the appropriate terms, the steps in mitosis and in meiosis.

- Explain the importance of the checkpoints in the cell cycle and the consequences of bypassing those checkpoints.

- Identify the ways in which meiosis and fertilization together produce genetically diverse offspring.

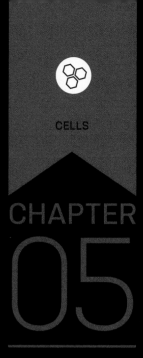
CHAPTER

05

CELL DIVISION

It began as a run-of-the-mill experiment. In 1989, biologists Ana Soto and Carlos Sonnenschein at Tufts University in Massachusetts were studying how estrogen, a hormone, regulates the growth of cells in the female reproductive system. For their research, the duo developed an experimental setup consisting of human breast tumor cells growing in plastic bottles called cell culture flasks (**Figure 5.1**, top). The flasks were filled with a liquid containing an unknown factor that prevented the cells from multiplying. But estrogen was able to cancel out that inhibition. So, when estrogen was added to the flasks, the cells grew. When estrogen was absent, they didn't.

One day, suddenly and surprisingly, cells in the flasks began growing even when estrogen hadn't been added. "What had worked for years didn't work anymore," says Sonnenschein. The two scientists immediately stopped their experiments and began searching for the cause. "It smacked of contamination," recalls Soto, as if estrogen had somehow gotten into the flasks. But after weeks of searching, Soto and Sonnenschein still couldn't identify a source of contamination. They became so paranoid that they suspected someone was entering the lab at night and secretly dripping estrogen into their flasks.

Almost 10 years later, in August 1998, geneticist Patricia Hunt at Case Western Reserve University in Ohio stared dumbfounded at another inexplicable experimental anomaly. Hunt was studying why older women are at increased risk of having children with chromosomal abnormalities, like Down syndrome, in which an individual has 47 chromosomes—the tiny, stringlike structures in cells that contain genes—instead of the usual 46 (see Chapter 7 for more on chromosomes). She hypothesized that levels of hormones have an impact on that increased risk. To test her hypothesis, she raised groups of mice with varying levels of hormones and checked their egg cells for abnormal numbers of chromosomes (**Figure 5.1**, bottom).

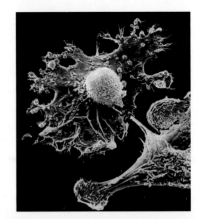

Figure 5.1

Two research models, two unusual results
Breast tumor cells (top) and mouse oocytes (bottom) grew in unexpected ways, leading scientists to study them more deeply.

The experiment was almost complete when Hunt went in to check on the control mice one last time. A control population is a necessary baseline for comparison against an experimental population; in this case, the control was a group of healthy mice whose hormone levels had not been altered. Using a light microscope, Hunt examined mouse oocytes—precursors to egg cells—at the moment just before the cells undergo a specialized type of cell division that produces the eggs. She was shocked. The cells were a mess, the chromosomes scrambled. A whopping 40 percent of the resulting eggs had chromosomal defects. "The controls were completely bonkers," says Hunt. "One week they were fine, the next week they weren't. That's when we knew something was going on."

Like Soto and Sonnenschein, Hunt scrutinized every method and

ANA SOTO AND CARLOS SONNENSCHEIN

Ana Soto is a biologist at Tufts University who studies how cell division is affected by sex steroids. Carlos Sonnenschein, also a professor at Tufts University, studies chemicals that disrupt hormone systems in mammals.

every piece of lab equipment used in the experiment, looking for the culprit. But as weeks passed, she couldn't figure out what had ruined her experiment. Soto, Sonnenschein, and Hunt didn't know it at the time, but their botched experiments would change the course of their scientific careers forever. The three would spend the next decade identifying, tracking, and investigating a potentially toxic chemical that pervades our environment.

Divide and Conquer

In Soto and Sonnenschein's experiment, the breast cells were multiplying under the wrong circumstances. In Hunt's experiment, the mouse oocytes were not producing egg cells correctly. In both cases, something was interfering with the ability of the cells to divide—disrupting the cell cycle. The **cell cycle** is a sequence of events that make up the life of a typical eukaryotic cell, from the moment of its origin to the time it divides to produce two daughter cells. The time it takes to complete a cell cycle depends on the organism, the type of cell, and the life stage of the organism. Human cells, for example, typically have a 24-hour cell cycle, while Hunt's mouse oocytes can take days to complete a cycle. Some fly embryos, on the other hand, have cell cycles that are only 8 minutes long.

There are two main stages in the cell cycle of eukaryotes: *interphase* and *cell division*—each marked by distinctive cell activities (**Figure 5.2**). **Interphase** is the longest stage of the cell cycle; most cells spend 90 percent or more of their life span in interphase. During this phase, the cell takes in nutrients, manufactures proteins and other substances, expands in size, and conducts special functions depending on the cell type. Neurons in the brain, for example, transmit electrical impulses, while beta cells in the pancreas release insulin.

Interphase can be divided into three main intervals: G_1, S, and G_2. The **G_1 phase** (for "gap 1") is the first phase in the life of a newborn cell. In cells that are destined to divide, preparations for cell division begin during the **S phase** ("S" stands for "synthesis"). A critical event during the S phase is the copying, or *replication*, of all the cell's DNA molecules, which contain the organism's genetic information. The **G_2 phase** (for "gap 2") occurs after the S phase but before the start of cell division.

Early cell biologists bestowed the term "gap" on the G_1 and G_2 phases because they believed

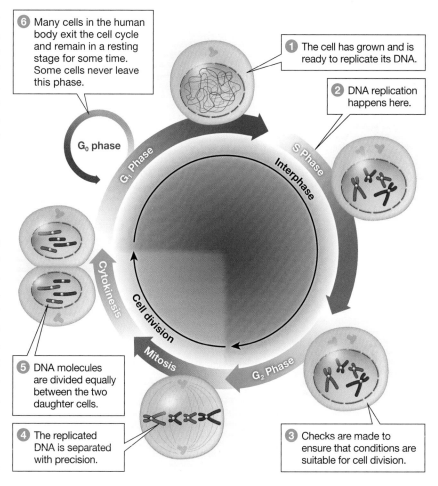

Figure 5.2

The cell cycle

The eukaryote cell cycle consists of two major stages: interphase and cell division.

6 Many cells in the human body exit the cell cycle and remain in a resting stage for some time. Some cells never leave this phase.

1 The cell has grown and is ready to replicate its DNA.

2 DNA replication happens here.

3 Checks are made to ensure that conditions are suitable for cell division.

4 The replicated DNA is separated with precision.

5 DNA molecules are divided equally between the two daughter cells.

G_0 phase

G_1 Phase

S Phase

Interphase

G_2 Phase

Mitosis

Cell division

Cytokinesis

Q1: When is DNA replicated during the cell cycle?

Q2: When in the cell cycle does DNA separate into two genetically identical daughter cells?

Q3: If a cell is not destined to separate into daughter cells, what phase does it enter? Is this part of the cell cycle?

those phases to be less significant periods in the life of a cell than are the S phase and cell division. We now know that the "gap" phases are often periods of growth during which both the size of the cell and its protein content increase. Furthermore, each "gap" phase serves as a checkpoint to prepare the cell for the phase immediately following it, ensuring that the cell cycle does not progress unless all conditions are suitable.

Cell division is the last stage in the life of an individual cell. As cell division begins, the cell

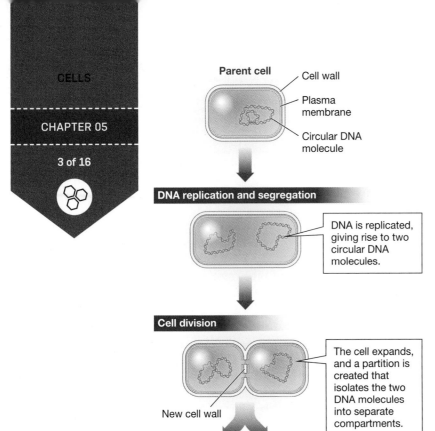

Parent cell
- Cell wall
- Plasma membrane
- Circular DNA molecule

DNA replication and segregation

DNA is replicated, giving rise to two circular DNA molecules.

Cell division

The cell expands, and a partition is created that isolates the two DNA molecules into separate compartments.

New cell wall

Cell separation

Two daughter cells

Figure 5.3

Cell division in a prokaryote
Many prokaryotes reproduce asexually in a type of cell division known as binary fission.

Q1: In what ways is cell division in prokaryotes and eukaryotes the same?

Q2: Why is binary fission referred to as "asexual reproduction"?

Q3: In what ways is cell division in prokaryotes and eukaryotes different?

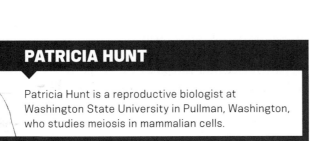

PATRICIA HUNT

Patricia Hunt is a reproductive biologist at Washington State University in Pullman, Washington, who studies meiosis in mammalian cells.

contains twice the usual amount of DNA because of DNA replication during the S phase.

Not all cells complete the cell cycle. Many types of cells—neurons and beta cells, for example—become specialized shortly after entering G_1, and they pull out of the cell cycle to enter a nondividing state called the G_0 phase. The **G_0 phase** can last for a period ranging from a few days to the lifetime of the organism.

Cells begin cell division for two basic reasons: (1) to reproduce and (2) to grow and repair a multicellular organism. Most single-celled organisms use cell division to produce offspring through **asexual reproduction**. Asexual reproduction generates *clones*, offspring that are genetically identical to the parent.

Cell division occurs in all living organisms (eukaryote and prokaryote) and involves the transfer of DNA from the parent cell to two daughter cells. Most prokaryotes carry their genetic material in just one loop of DNA and reproduce through **binary fission**, a type of cell division in which a cell simply divides into two equal halves—resulting in daughter cells that are genetically identical to each other and to the parent cell (**Figure 5.3**). Because the prokaryote chromosome is just one loop of DNA, with no surrounding nuclear membrane, cell division occurs more simply and more rapidly than in eukaryotes.

But Soto, Sonnenschein, and Hunt were studying eukaryotic cells. Cell division in eukaryotes is more complicated than binary fission because eukaryotic cells have many molecules of compacted, protected DNA that have to be replicated and equally distributed between the two daughter cells. Eukaryotic DNA lies in the nucleus, wrapped in a double layer of membranes that make up the nuclear envelope. In most eukaryotes, the nuclear envelope is disassembled in the dividing cell and then reassembled in each of the daughter cells toward the end of cell division. Eukaryotic cells undergo two types of division: asexual reproduction through *mitosis*, and sexual reproduction—the production of sperm and eggs—through *meiosis*.

Trade Secret

Back in Massachusetts, Soto and Sonnenschein spent four months trying to identify why their experiment had stopped working—how unknown estrogen was getting into their cell culture flasks and causing the cells to divide. By trial and error, they determined that a compound seemed to be shedding from the walls of the plastic tubes in which they stored the liquid that was being added to the cell culture flasks.

They called the tube manufacturer, who confirmed that an ingredient was added to make the tubes more impact resistant. But the company refused to reveal the identity of the "trade secret" ingredient. Soto and Sonnenschein spent a year purifying the secret ingredient and finally identified a compound called nonylphenol, a chemical used to make detergents and hard plastics. The reason for all their problems became clear: nonylphenol mimics the action of estrogen.

Like estrogen, nonylphenol activates **mitotic division**, a type of cell division that generates two genetically identical daughter cells from a single parent cell in eukaryotes. Mitotic division consists of two steps: *mitosis* and *cytokinesis*. The first step, **mitosis**, refers to the division of the nucleus. Mitosis is divided into four main phases: *prophase*, *metaphase*, *anaphase*, and *telophase*. Each phase is defined by easily identifiable events (**Figure 5.4**).

A parent cell sets up for an upcoming mitotic division by duplicating its DNA during the S phase of interphase, well before mitosis gets under way. Note that DNA in the nucleus is not tightly packaged during gap and synthesis phases. This is because the DNA must be accessible for replication and to conduct the business of the cell. Then, as cell division begins, each long, double-stranded DNA molecule is attached to proteins that help pack it for cell division into a more compact physical structure called a **chromosome**. This packing is necessary because every DNA molecule is enormously long, even in the simplest eukaryote cells. When a chromosome is replicated, two identical DNA molecules, called **chromatids**, are produced. These **sister chromatids** are firmly attached at a central region of the chromosome called the **centromere** and do not separate until metaphase (**Figure 5.5**).

One of the main objectives of mitosis is to separate those sister chromatids, pulling them apart at the centromere and delivering one of each to the opposite ends of the parent cell. An elaborate choreography has evolved in cells to minimize the risk of mistakes during the equal and symmetrical partitioning of the replicated genetic material. Normally, no daughter cell winds up short a chromosome, nor does it acquire

Cancer: Uncontrolled Cell Division

Cancer accounts for more than 500,000 deaths in the United States each year—one in every four deaths. Only heart disease kills more people. Over the course of a lifetime, an American male has a nearly one in two chance of being diagnosed with cancer; American women fare slightly better, with a one in three chance of developing cancer. There are more than 200 different types of cancer, but the big four—lung, prostate, breast, and colon cancers—combine to account for more than half of all cancers. More than 8 million Americans alive today have been diagnosed with cancer and are either cured or undergoing treatment. The National Cancer Institute estimates that the collective price tag for the various forms of cancer is more than $100 billion per year.

Every cancer begins with a single rogue cell that starts dividing without the checkpoints of a normal cell. This runaway cell division rapidly creates a cell mass known as a **tumor**. Tumors that remain confined to one site are **benign**. Because benign tumors can usually be surgically removed, they are generally not a threat to the patient's survival. However, an actively growing benign tumor is like a cancer-in-training. Because these tumor cells are not subject to the monitoring that occurs at checkpoints during the cell cycle of a normal cell, their descendants can become increasingly abnormal—changing shape, increasing in size, and ultimately ceasing normal cell functions. As tumor cells progress toward a cancerous state, they begin secreting substances that cause **angiogenesis**, the formation of new blood vessels. The resulting increase in blood supply to the tumor is important for delivering nutrients to it and whisking waste away from it, allowing the tumor to grow larger.

Most cells in the adult animal body are firmly anchored in one place and will stop dividing if they are detached from their surroundings—a phenomenon known as **anchorage dependence**. But some tumor cells may acquire anchorage *in*dependence, the ability to divide even when released from their attachment sites. When tumor cells gain anchorage independence and start invading other tissues, they are transformed into **cancer cells**, also known as **malignant cells**. Cancer cells may break loose from their attachment sites and enter blood or lymph vessels to emerge in distant locations throughout the body, where they form new tumors. The spread of a disease from one organ to another is known as **metastasis**. Metastasis typically occurs at later stages in cancer development. Once a cancer has metastasized to form tumors in multiple organs, it may be very difficult to fight.

Cancer cells multiply rapidly wherever they establish themselves, overrunning neighboring cells, monopolizing oxygen and nutrients, and starving normal cells in the vicinity. Without restraints on their growth and migration, cancer cells steadily destroy tissues, organs, and organ systems. The normal function of these organs is then seriously impaired, and cancer deaths are ultimately caused by the failure of vital organs.

duplicates. Unless an error occurs, each daughter cell inherits the same information that the parent cell possessed in the G_1 phase of its life.

After the DNA has been divided in two, half of it to each end of the parent cell, the cytoplasm

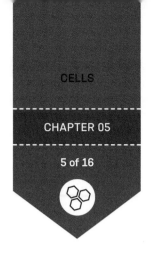

is divided by a process called **cytokinesis** ("cell movement"). Think of pulling apart a ball of Silly Putty into two halves. Cytokinesis gives rise to two self-contained daughter cells.

Mitotic division can serve both the eukaryotic organism's need to replace itself (to reproduce) and its need to add new cells to its body. Many multicellular eukaryotes use mitotic division to reproduce asexually, including most seaweed, fungi, and plants, and some animals, such as sponges and flatworms. All multicellular organisms also rely on mitotic division for the growth of tissues and organs and the body as a whole, and for repairing injured tissue and replacing worn-out cells. Mitosis is why fingernails grow and why skin closes over a cut.

Good Cells Gone Bad

Cell division is not always a good thing. Runaway cell division can turn into cancer (see "Cancer: Uncontrolled Cell Division" on page 77). In a developing organism, it can also cause an organ such as the heart or liver to form incorrectly and not function properly. It is little wonder, then, that the cell cycle is carefully controlled in healthy individuals. The decision to divide the cell is made during the G_1 phase of the cell cycle in response to internal and external signals. In humans, external signals that influence the commitment to divide include hormones and proteins called

Figure 5.4

Cell division in a eukaryote
Mitotic cell division is composed of two main stages: mitosis (with four substages) and cytokinesis.

Q1: Do all cells in an organism enter each stage of mitosis at the same time? (*Hint:* see image of onion root tip below.)

Q2: What happens between the end of interphase and early prophase that changes the appearance of the chromosomes?

Q3: Explain in your own words the role of the mitotic spindle in mitosis.

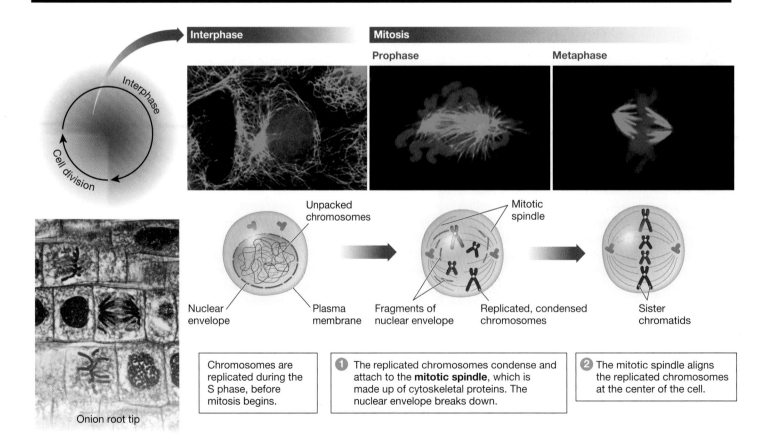

Interphase

Mitosis

Prophase

Metaphase

Interphase

Cell division

Unpacked chromosomes

Mitotic spindle

Nuclear envelope

Plasma membrane

Fragments of nuclear envelope

Replicated, condensed chromosomes

Sister chromatids

Chromosomes are replicated during the S phase, before mitosis begins.

❶ The replicated chromosomes condense and attach to the **mitotic spindle**, which is made up of cytoskeletal proteins. The nuclear envelope breaks down.

❷ The mitotic spindle aligns the replicated chromosomes at the center of the cell.

Onion root tip

growth factors. Some hormones and growth factors act like the gas pedal in a car and push a cell toward cell division; others act like a brake and hinder cell division.

Special *cell cycle regulatory proteins* are activated after a cell enters the cell cycle. These proteins "throw the switch" that enables the cell to pass through critical checkpoints and proceed from one phase of the cell cycle to the next (**Figure 5.6**). For example, upon receiving the appropriate signals, cell cycle regulatory proteins advance the cell from the G_1 phase to the S phase by triggering chromosome replication and other processes associated with it.

Cell cycle regulatory proteins also respond to negative internal or external control signals. Internal signals will pause a cell in the G_1 phase under any of the following conditions: the cell is too small, the nutrient supply is inadequate, or the cell's DNA is damaged. G_2 pauses in the same circumstances, as well as when the chromosome duplication that begins in the S phase is incomplete for any reason.

Nonylphenol interferes with G_0/G_1 checkpoints. In essence, it gives the cell a green light to enter the cell cycle at a time when the cell would not normally divide. That's why Soto and

Sonnenschein became concerned when they realized that nonylphenol enables human breast cells and rat uterine cells to divide.

In a 1991 paper detailing their discovery, Soto and Sonnenschein wrote that nonylphenol might be interfering with science experiments like theirs and that, even more important, it could be harmful to humans. "From the very beginning, we realized that this could be a health problem," says Sonnenschein.

Unequal Division

In Ohio, Patricia Hunt was having no success determining why her control mouse eggs divided abnormally, having either too many or too few chromosomes. After searching in vain for months, one day she noticed that something was wrong with the plastic mouse cages; the water bottles were leaking and the plastic cage walls were hazy.

She asked around and found out that, months earlier, a substitute janitor had used detergent with a high pH instead of the normal low pH detergent to clean the cages and bottles. With some lab work, she identified the chemical that

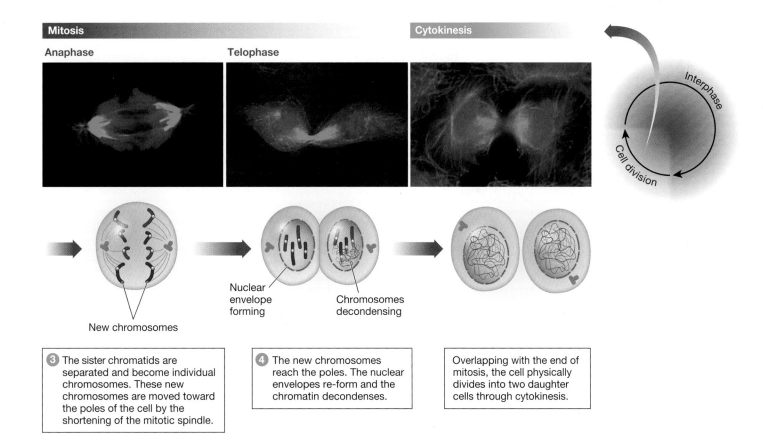

Mitosis

Anaphase

Telophase

Cytokinesis

Interphase

Cell division

New chromosomes

Nuclear envelope forming

Chromosomes decondensing

3 The sister chromatids are separated and become individual chromosomes. These new chromosomes are moved toward the poles of the cell by the shortening of the mitotic spindle.

4 The new chromosomes reach the poles. The nuclear envelopes re-form and the chromatin decondenses.

Overlapping with the end of mitosis, the cell physically divides into two daughter cells through cytokinesis.

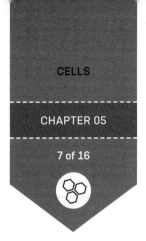

was oozing from the corroded plastic. "It took essentially one washing with the wrong detergent," says Hunt, "and once it was damaged, [the plastic] started to leach bisphenol A."

Bisphenol A, or BPA, was first synthesized in 1891. A synthetic hormone that, like nonylphenol, mimics estrogen, BPA was tested in the 1930s by clinicians seeking a hormone replacement therapy for women who needed estrogen, but it wasn't as effective as other substitutes. In the 1940s and 50s, the chemical industry found another use for BPA, as a chemical component of a clear, strong plastic. Manufacturers began to incorporate it into eyeglass lenses, water and baby bottles, the linings of food and beverage cans, and more. Unfortunately, as Hunt found out, BPA doesn't necessarily stay in those products. Not all the BPA used to make plastics gets locked into chemical bonds, and what doesn't get locked into bonds can work free—especially if the plastic is heated, such as when a baby bottle is warmed, or if the plastic is exposed to a harsh chemical, as Hunt's mouse cages were. Because of its prevalence in products and its ability to leach out of them, BPA is one of the most common chemicals we are exposed to in everyday life.

To confirm the hypothesis that BPA caused the mouse egg abnormalities, Hunt's team re-created the original event. They intentionally damaged a set of new cages and put healthy female mice into them. They also had the mice drink from damaged water bottles. Later, when they examined the eggs of these mice, they saw the same toxic effects as before: 40 percent of the eggs had abnormal chromosomes. The eggs showed errors in *meiosis*.

Meiosis is a specialized type of cell division that kicks off **sexual reproduction**, the process by which genetic information from two individuals is combined to produce offspring. Sexual reproduction has two steps: cell division through meiosis, and then fertilization. BPA was affecting the first of these steps—meiosis.

We learned earlier that mitosis produces daughter cells with the same number of chromosomes as the parent cell has. These non–sex cells are called **somatic cells**. In contrast, meiosis produces daughter cells—**gametes**—with half the chromosome count of the parent cell. In this way,

One chromosome

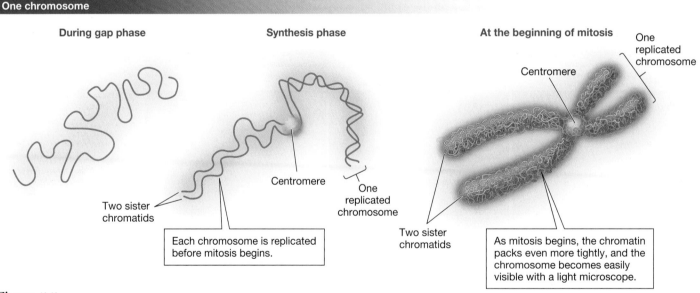

During gap phase

Synthesis phase

At the beginning of mitosis

One replicated chromosome

Centromere

Two sister chromatids

Centromere

One replicated chromosome

Each chromosome is replicated before mitosis begins.

Two sister chromatids

As mitosis begins, the chromatin packs even more tightly, and the chromosome becomes easily visible with a light microscope.

Figure 5.5

Chromosomes are copied and condensed in preparation for cell division
Chromosomes spend the majority of the cell cycle unpackaged (left). They are copied (replicated) during synthesis, then tightly packaged, or condensed, during early mitosis.

Q1: Why is it important for a chromosome to be copied before mitosis?

Q2: Are sister chromatids attached at the centromere considered to be one or two chromosomes?

Q3: Why is the chromosome's DNA tightly packed for mitosis and cytokinesis? (*Hint:* What would happen if it were unpackaged, as during interphase?)

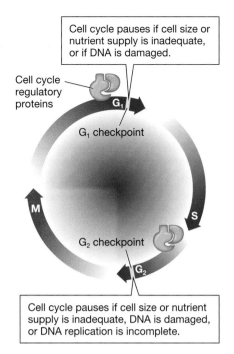

Cell cycle pauses if cell size or nutrient supply is inadequate, or if DNA is damaged.

Cell cycle regulatory proteins

G₁ checkpoint

G₂ checkpoint

Cell cycle pauses if cell size or nutrient supply is inadequate, DNA is damaged, or DNA replication is incomplete.

Figure 5.6

The cell cycle must pass checkpoints to proceed

Just two of the known cell cycle checkpoints are depicted in this diagram. Checkpoints also operate in both the S phase and the M phase.

Q1: Would the cell cycle occur more slowly or more quickly if the cell's checkpoints were disabled?

Q2: What is the advantage of stopping the cell cycle if the cell's DNA is damaged?

Q3: Which part of the cell cycle may have been influenced in Soto and Sonnenschein's breast tumor cell experiments?

the somatic cells of plants and animals have twice as much genetic information as their gametes have. The double set of genetic information possessed by somatic cells is known as the **diploid** set (represented by $2n$), and the single set possessed by gametes is called the **haploid** set (represented by n).

Fertilization, the fusion of two gametes, results in a single cell called the **zygote**. The zygote inherits a haploid set of chromosomes from each of the gametes, restoring the complete diploid set of genetic information to the offspring. Each **homologous pair** of chromosomes in the zygote consists of one chromosome received from the father and one from the mother. The zygote then divides by mitosis to create a mass of cells that will eventually develop into a mature organism (**Figure 5.7**).

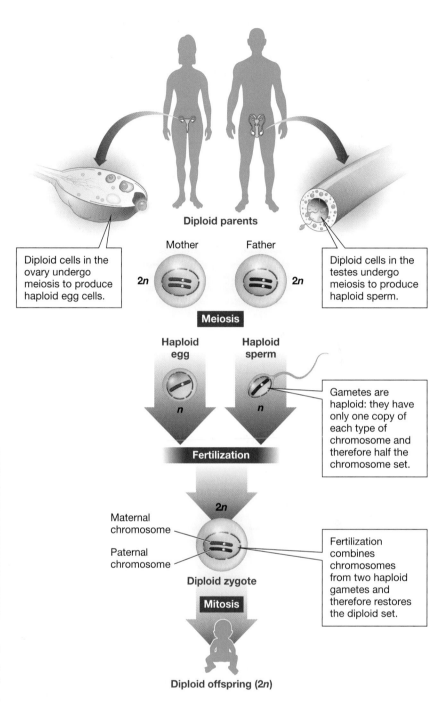

Diploid parents

Diploid cells in the ovary undergo meiosis to produce haploid egg cells.

Mother   Father

$2n$ $2n$

Diploid cells in the testes undergo meiosis to produce haploid sperm.

Meiosis

Haploid egg   Haploid sperm

n n

Gametes are haploid: they have only one copy of each type of chromosome and therefore half the chromosome set.

Fertilization

$2n$

Maternal chromosome

Paternal chromosome

Fertilization combines chromosomes from two haploid gametes and therefore restores the diploid set.

Diploid zygote

Mitosis

Diploid offspring ($2n$)

Figure 5.7

Fertilization creates a zygote from the fusion of two gametes

In species with two sexes, female gametes are *eggs* and male gametes are *sperm*. This figure shows only one of the 23 homologous pairs found in human cells.

Q1: Is a zygote haploid or diploid?

Q2: What cellular process creates a baby from a zygote?

Q3: If a mother or father is exposed to BPA prior to conceiving a child, how does that explain potential birth defects in the fetus?

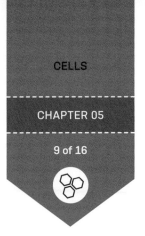
Meiosis occurs in two stages—*meiosis I* and *meiosis II*—each involving one round of nuclear division followed by cytokinesis (**Figure 5.8**). **Meiosis I** reduces the chromosome set by *separating each homologous pair* into two different daughter cells. Each homologous chromosome lines up with its partner and then separates to the two ends of the cells. **Meiosis II** *separates sister chromatids* into two new daughter cells. This time, the phases of the division cycle are almost exactly like those of mitosis: sister chromatids separate, leading to an equal segregation of chromatids into two new daughter cells. In summary, meiosis I produces two haploid cells (*n*). In meiosis II, these two haploid cells give rise to four haploid cells (*n*), each with half of the chromosome set found in the original diploid cell (2*n*) that underwent meiosis.

BPA is toxic because it disrupts the process of meiosis, hindering the ability of the chromosomes to separate into four haploid cells. Hunt realized that if BPA was disrupting meiosis in mice, it could be doing the same in humans. And if a human gamete (either the sperm or the egg) does not contain the correct number of chromosomes, fertilization typically results in a miscarriage.

Hunt was nervous about publishing the results of her experiment. "We knew we were stepping into a landmine," she says. "We knew the paper would get some press, because essentially we were publishing that this chemical—used in a wide variety of consumer products and that we are probably all exposed to—can cause an increased risk of miscarriage and babies with birth defects."

Shuffling the DNA

As a glance at a pair of parents and their biological children will tell you, the offspring resulting from sexual reproduction are similar to their parents, but—unlike the clones resulting from asexual reproduction—they are *not* identical. Because half of a sexually reproducing organism's DNA comes from a different parent, meiosis and fertilization maintain the constant chromosome number of a species while allowing for genetic diversity within the population.

Meiosis generates genetic diversity in two ways: *crossing-over* between the paternal and maternal members of each homologous pair, and *independent assortment* of the paternal and maternal chromosomes during meiosis I. **Crossing-over** is the name given to the physical exchange of chromosomal segments during meiosis I between non–sister chromatids in paired-up homologous chromosomes. These non–sister chromatids make physical contact at random sites along their length and exchange segments of DNA (**Figure 5.9**). The chromatids are said to be *recombined*, and the exchange of DNA segments is known as **genetic recombination**. Without crossing-over, every chromosome inherited by a gamete would be just the way it was in the parent cell.

The **independent assortment** of chromosomes—the random distribution of the homologous chromosomes into daughter cells during meiosis I—also contributes to the genetic variety of the gametes produced. It comes about because each homologous chromosome pair in a given meiotic cell orients itself independently when it lines up at the metaphase plate during meiosis I, leading to many possible combinations of maternal and paternal chromosomes in the daughter cells (**Figure 5.10**).

As with crossing-over, the independent assortment of chromosomes creates gametes that are likely to be different from the parent, and also from each other. Then, during fertilization, the fusion of two gametes adds a tremendous amount of genetic variation because it combines a one-in-a-million egg with a one-in-a-million sperm. These three processes together give each of us our genetic uniqueness.

Ten Years Later

Hunt and her team published their results in 2003. "We got a firestorm," she recalls with a grimace. The press covered the findings extensively, and many people and companies were upset over the allegations that BPA was toxic. Members of the plastics industry who did not agree with the paper's conclusions criticized Hunt's work. But there was more supporting research to come. Soto and Sonnenschein had also turned their attention to BPA because BPA is a far more common chemical in everyday

Meiosis I

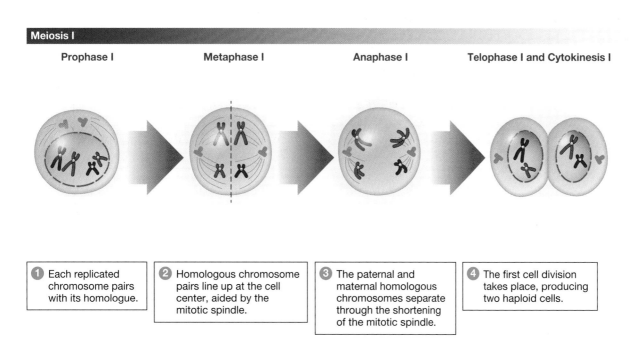

| Prophase I | Metaphase I | Anaphase I | Telophase I and Cytokinesis I |

① Each replicated chromosome pairs with its homologue.

② Homologous chromosome pairs line up at the cell center, aided by the mitotic spindle.

③ The paternal and maternal homologous chromosomes separate through the shortening of the mitotic spindle.

④ The first cell division takes place, producing two haploid cells.

Meiosis II

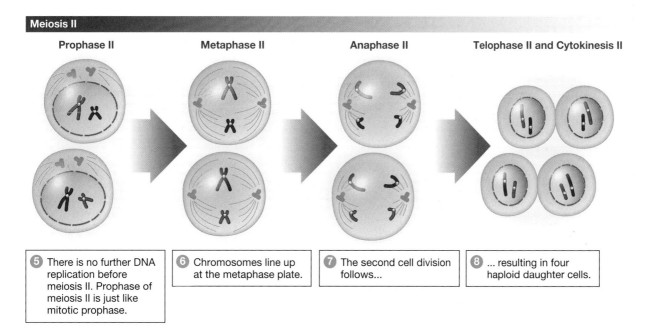

| Prophase II | Metaphase II | Anaphase II | Telophase II and Cytokinesis II |

⑤ There is no further DNA replication before meiosis II. Prophase of meiosis II is just like mitotic prophase.

⑥ Chromosomes line up at the metaphase plate.

⑦ The second cell division follows...

⑧ ... resulting in four haploid daughter cells.

Figure 5.8

Meiosis with cytokinesis creates haploid daughter cells

The homologous chromosomes inherited from each parent are paired and then separated in meiosis I. Sister chromatids are separated during meiosis II. Cytokinesis occurs at the end of both meiosis I and meiosis II.

Q1: Is a daughter cell haploid or diploid after the first meiotic division? After the second meiotic division?

Q2: What is the difference between homologous chromosomes and sister chromatids?

Q3: If the skin cells of house cats contain 38 homologous pairs of chromosomes, how many chromosomes are present in the egg cells they produce?

Prophase I

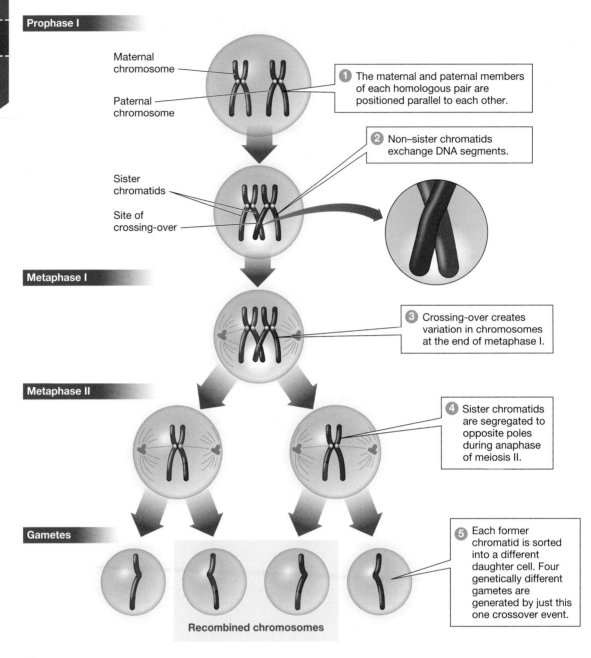

Maternal chromosome

Paternal chromosome

1 The maternal and paternal members of each homologous pair are positioned parallel to each other.

2 Non–sister chromatids exchange DNA segments.

Sister chromatids

Site of crossing-over

Metaphase I

3 Crossing-over creates variation in chromosomes at the end of metaphase I.

Metaphase II

4 Sister chromatids are segregated to opposite poles during anaphase of meiosis II.

Gametes

5 Each former chromatid is sorted into a different daughter cell. Four genetically different gametes are generated by just this one crossover event.

Recombined chromosomes

Figure 5.9

Crossing-over produces chromosomes with new combinations of DNA

Only one maternal and one paternal chromosome are depicted here, rather than the 23 pairs of homologous chromosomes found in humans.

Q1: Why is the term "crossing-over" appropriate for the exchange of DNA segments between homologous chromosomes?

Q2: At what stage of meiosis (I or II) does crossing-over occur?

Q3: What would be the effect of crossing-over between two sister chromatids?

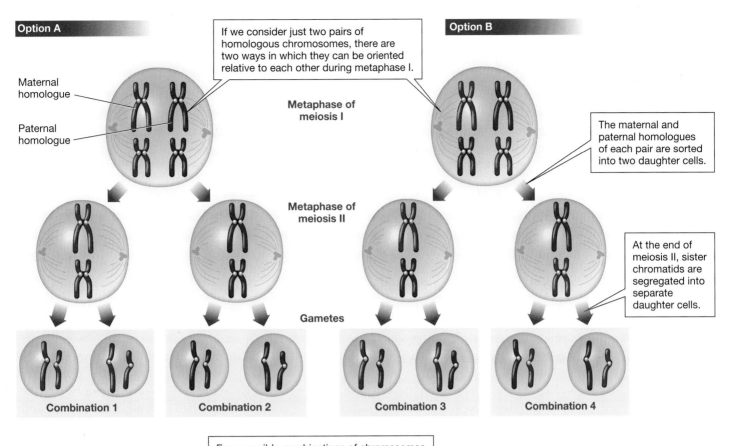

Option A

Option B

Maternal homologue

Paternal homologue

If we consider just two pairs of homologous chromosomes, there are two ways in which they can be oriented relative to each other during metaphase I.

Metaphase of meiosis I

The maternal and paternal homologues of each pair are sorted into two daughter cells.

Metaphase of meiosis II

At the end of meiosis II, sister chromatids are segregated into separate daughter cells.

Gametes

Combination 1

Combination 2

Combination 3

Combination 4

Four possible combinations of chromosomes in gametes generated by meiosis II.

Figure 5.10

The independent assortment of homologous chromosomes generates chromosomal diversity among gametes

Only two pairs of homologous chromosomes are shown here, rather than the 23 homologous pairs in human cells. Each gamete will receive either a maternal or a paternal homologue of each chromosome.

Q1: During meiosis, does random assortment occur before or after crossing-over?

Q2: What would be the effect on genetic diversity if homologous chromosomes did not randomly separate into the daughter cells during meiosis?

Q3: With two pairs of homologous chromosomes, four kinds of gametes can be produced. How many kinds of gametes can be produced with three pairs of homologous chromosomes? What does this suggest for the 23 homologous pairs of chromosomes in human cells?

products than nonylphenol and is therefore of greater concern.

At the same time that Soto/Sonnenschein and Hunt were doing their work, Frederick vom Saal at the University of Missouri found that male mice that had been exposed to BPA in utero—even at very low doses—had dramatically enlarged prostates in adulthood that were hypersensitive to hormones. This study suggests that men are also at risk of health effects from BPA.

In 2007, Hunt followed up her original work with a study that she says made the first paper look like "child's play." Her team exposed pregnant mice to BPA just as their female fetuses were producing a supply of eggs in their ovaries. When that second generation of females became

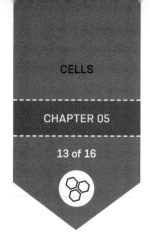

adults, their eggs were also damaged, Hunt found, demonstrating that BPA exposure affects not just adult females, but two generations of their offspring.

In the last several years, scientists all over the world have shown that BPA disrupts meiosis and mitosis and causes a plethora of health problems in mice and rats, including breast and prostate cancer, miscarriage and birth defects, diabetes and obesity, and even behavioral problems such as attention deficit hyperactivity disorder. Whether BPA is causing similar diseases in humans remains unknown, as it is difficult to study. But we do know that most people have BPA in their bodies, according to the Centers for Disease Control and Prevention. In 2008, researchers at the government health organization reported that they had found BPA in the urine of 93 percent of 2,517 people sampled. BPA has also been found in human blood, breast milk, and amniotic fluid.

Exposure aside, not everyone agrees that BPA is toxic. Numerous companies that manufacture plastics have conducted studies whose results

Figure 5.11

BPA-free bottles and cans are now widely available
If you are concerned about being exposed to BPA, check labels before you buy.

do not match Hunt and vom Saal's results. To reach a scientific consensus, on November 28, 2006, Soto, Sonnenschein, Hunt, vom Saal, and 34 other researchers from across the United States gathered at the University of North Carolina in Chapel Hill to summarize the research on BPA. The result of their two-day meeting was the "Chapel Hill Bisphenol A Consensus Statement," summarizing hundreds of studies done in vitro and in vivo over the previous 10 years. After this analysis, the group concluded firmly that BPA exposure at current levels in our environment presents a risk to human health (see "What Can You Do?"). "It was quite clear that there is a serious problem," says Soto.

Over time, many baby bottle manufacturers took BPA out of their bottles, even as government regulators were slower to respond. "As scientists, our role is to call attention to what is wrong, but it is the role of the politicians to act on it and try to straighten it out," says Sonnenschein. Then, in July 2012, the FDA banned BPA from baby bottles and children's drinking cups, though the prohibition does not apply to the use of BPA in other types of containers (**Figure 5.11**). There is still concern, however, from many scientists about the chemicals that have replaced BPA, some of which are also estrogen mimics.

Hunt, Soto, and Sonnenschein continue to explore the effects of BPA, and they recently began studying how exposure to low doses of BPA affects monkeys, a model animal that more accurately represents the human system. "We're slowly raising awareness," says Hunt, "and slowly changing things."

What Can You Do?

There are things you can do to reduce your own risk of exposure to BPA. Today, the U.S. Food and Drug Administration recommends that individuals not put hot or boiling liquid intended for consumption in plastic containers made with BPA. (Some, but not all, plastics that are marked with the recycling code 3 or 7 may be made with BPA.) The organization also recommends discarding all bottles with scratches, which may harbor bacteria and, if the plastic contains BPA, may lead to greater release of the chemical.

"Get educated about your world," says Heather Patisaul, a BPA researcher at North Carolina State University. "You can either become completely reliant on the information you get from media and government, or you can educate yourself, which is vastly more useful." If you are concerned about BPA exposure, it is possible to cut it down by making lifestyle changes like not eating canned food, not drinking bottled water, and not putting plastic in the microwave. "You can be empowered," says Patisaul. "Those types of things can effect great change."

Still, BPA and similar chemicals are ubiquitous in modern life, says Ana Soto. Ultimately, the best way to avoid them will be for government regulators to take a stand and outlaw the use of these chemicals in consumer products, she notes. She encourages individuals to contact their representatives and ask them to push legislation limiting the use of BPA in manufacturing.

Plastic by the Numbers

Plastic is one of the most ubiquitous materials in modern society, and today makes up almost 13 percent of the trash we produce (it was less than 1 percent back in 1960, according to the U.S. Environmental Protection Agency). But not all plastics are recycled equally: some, such as PETE plastics, labeled with the number 1, are the easiest to recycle. Code 7 plastics, however, are the most difficult to recycle, and most recycling centers will not accept them.

Percent of U.S. waste that is recycled

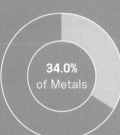

64.6% of Paper

34.0% of Metals

27.7% of Glass

8.8% of Plastics

---

 6 PS

Polystyrene
Uses include: Disposable hot cups, packing peanuts, meat trays
Can be recycled into: Plastic lumber, thermal insulation, protective packaging

.9% of waste is recycled

---

 1 PETE

Polyethylene Terephthalate
Uses include: Drink bottles (water, soft drinks, sports drinks), food jars (peanut butter, jelly)
Can be recycled into: Sleeping bag and comforter fill, carpet fibers

19.5% of waste is recycled

---

 5 PP

Polypropylene
Uses include: Yogurt containers, straws, syrup bottles, bottle caps
Can be recycled into: Plastic lumber, car battery cases

.6% of waste is recycled

---

 2 HDPE

High Density Polyethylene
Uses include: Milk jugs, butter tubs, detergent bottles, motor oil bottles
Can be recycled into: Flower pots, trash cans, traffic cones, new bottles

10.3% of waste is recycled

---

 3 V

Vinyl
Uses include: Clear food packaging, shampoo bottles
Can be recycled into: Drainage and irrigation pipes

Negligible amount of waste is recycled

---

 4 LDPE

Low Density Polyethylene
Uses include: Grocery bags, bread bags, shrink wrap, container lids
Can be recycled into: New bags

5.3% of waste is recycled

---

 7 OTHER

Other
Uses include: Ketchup bottles, 3 & 5 gallon water bottles, some juice bottles

Negligible amount of waste is recycled

daughter cells at the end of cytokinesis.

● Meiosis produces genetically diverse gametes through **crossing-over** of homologous chromosomes, leading to **genetic recombination**, and then the **independent assortment** of homologous chromosomes. Meiosis and fertilization together introduce genetic variation into populations.

REVIEWING THE SCIENCE

● The **cell cycle** is the set sequence of events over the life span of a eukaryotic cell that will divide. **Interphase** and **cell division** are the two main stages of the cell cycle. Interphase is the longest, and consists of the G_1, **S**, and G_2 phases. DNA is replicated in the S phase. Cells that will not divide exit the cell cycle and enter a G_0 **phase**.

● Cell division is necessary for growth and repair in multicellular organisms, and for **asexual** and **sexual reproduction** in all types of organisms. Many prokaryotes divide through **binary fission**, a form of asexual reproduction.

● Each **chromosome** in a cell contains a single DNA molecule compacted by packaging proteins. The **somatic cells** of eukaryotes have two of each type of chromosome, matched together in **homologous pairs**, or **homologous chromosomes**. One chromosome in each homologous pair is inherited from the mother, the other from the father. Chromosomal replication produces two identical **sister chromatids** that are held together firmly at the **centromere**.

● Eukaryotes perform cell division through **mitosis** followed by **cytokinesis**, producing daughter cells that are genetically identical to each other and to the parent cell. The four main phases of mitosis are prophase, metaphase, anaphase, and telophase. Through these phases, the chromosomes of a parent cell are condensed, positioned appropriately, and the sister chromatids are separated to opposite ends of the cell. During cytokinesis, the cytoplasm of the parent cell is physically divided to create two daughter cells.

● The cell cycle is carefully regulated. Checkpoints ensure that the cycle does not proceed if conditions are not right.

● **Meiosis** is critical for sexual reproduction. In animals, the products of meiosis are sex cells, called **gametes**, that fuse during **fertilization** to give rise to a **zygote**. Meiosis—consisting of two rounds of nuclear and cytoplasmic divisions—produces **haploid** gametes containing only one chromosome from each homologous pair. When two gametes fuse during fertilization a **diploid** zygote, with copies of chromosomes from each parent, is formed.

● During **meiosis I**, the maternal and paternal members of each homologous pair are sorted into two daughter cells. **Meiosis II** is similar to mitosis in that sister chromatids are segregated into separate

THE QUESTIONS

The Basics

1 Homologous chromosomes

(a) are the same thing as sister chromatids.

(b) are a pair of chromosomes of the same kind.

(c) are identical copies of the same chromosome.

(d) are always haploid.

2 Sister chromatids

(a) are the same thing as homologous chromosomes.

(b) are a pair of chromosomes of the same kind.

(c) are identical copies of the same chromosome, attached at the centromere.

(d) are always haploid.

3 How is the cell cycle controlled?

(a) Completion of one phase triggers the start of the next phase.

(b) A cell divides when it reaches a critical size.

(c) A cell divides when it reaches a particular age.

(d) Internal or external signals at specific checkpoints may stop the cycle or trigger the start of the next phase.

4 Which of the following is not a contributor to genetic variation?

(a) replication of sister chromatids

(b) crossing-over of homologous chromosomes

(c) random assortment of homologous chromosomes

(d) fertilization

5 Loss of cell cycle control may lead to

(a) pregnancy.

(b) cancer.

(c) fertilization.

(d) crossing-over of homologous chromosomes.

6 Link each of the following cell phases with the events that occur within it.

CYTOKINESIS   1. Each of the chromosomes in a human cell contains two sister chromatids by the end of this phase.

S PHASE   2. Most cell growth occurs during this phase.

G$_1$ PHASE   3. Cells that will never replicate leave the cell cycle and enter this phase.

G$_0$ PHASE   4. Two separate daughter cells are produced at the end of this phase.

7 Circle the correct terms in the following sentences: (**Mitosis**, **Meiosis**) produces daughter cells with half the number of chromosomes that the parent cell has. Cell division in prokaryotes is called (**mitosis, binary fission**). Meiosis I separates (**sister chromatids, homologous chromosomes**); meiosis II separates (**sister chromatids, homologous chromosomes**) into separate daughter cells.

8 Place the following events of sexual reproduction in the correct order by numbering them from 1 to 5.

_____ a. Separation of homologous chromosomes

_____ b. Separation of sister chromatids

_____ c. Mitosis within the zygote, leading to a multicellular organism

_____ d. Cytokinesis, leading to four haploid daughter cells

_____ e. Fusion of two gametes

Try Something New

9 House cat cells have a total of 38 chromosomes (versus the human 46 chromosomes). How many separate DNA molecules are present in a cat skin cell toward the end of the (a) G$_0$ phase; (b) G$_1$ phase; (c) S phase; (d) G$_2$ phase?

10 What is the difference between mitosis and cytokinesis?

11 Describe the likely consequences of bypassing the G$_1$ and G$_2$ checkpoints in the cell cycle. Why do compounds like nonylphenol lead to the multiplication of abnormal cells?

12 The consequences of mitosis are easy to see during childhood, when an individual is growing rapidly, but examples of mitosis in adult life are less conspicuous. Nonetheless, you can probably find visible evidence of mitosis in progress on your own body right now. What examples can you suggest?

Leveling Up

13 **What do *you* think?** Cancer begins with a single cell that breaks loose of normal restraints on cell division and starts dividing rapidly to establish a colony of rogue cells. As cancer cells spread through the body, they disrupt the normal functions of tissues and organs; unchecked, cancer can cause death through failure of multiple organ systems. Many cancers could be prevented by not smoking or chewing tobacco, eating less meat and processed foods, eating more fruits and vegetables, drinking alcohol only moderately if at all, exercising regularly, and maintaining a healthy weight. Only 5–10 percent of cancers are directly attributable to genetic causes.

A "sin tax" is a tax on a product or activity that has negative effects on others, as a way to offset some of those effects. Common targets of sin taxes are tobacco and alcohol because of their public health costs. Proponents of a sin tax on tobacco argue that such a tax would decrease the amount that people smoke (because of the increased cost) and could also partially fund the costs of medical care necessitated by increased rates of cancer and other diseases caused by smoking. Critics point out that sin taxes have historically triggered smuggling and black markets, and have a disproportionate effect on poor people because the wealthy can more easily afford to pay the higher prices.

What do you think? Should we institute taxes on tobacco? Would fewer people smoke, or would they smoke less, if tobacco was more expensive? What about higher health care premiums for smokers, based on their higher risk for cancer? Would increasing the premiums cause more people to stop smoking? Would you support such a policy?

14 *Write Now* **Biology: BPA effects** The studies of BPA described in this chapter use an inbred strain of mice that is known to be especially susceptible to estrogen and estrogen-like chemicals, such as BPA. The plastics industry maintains that the susceptibility of this strain of mice to estrogen renders these studies invalid as a basis for estimating the effects of BPA on humans. BPA researchers respond that the current situation, exposing millions of people to unknown levels of BPA, constitutes a massive uncontrolled experiment. They maintain that even a small risk of harm is too great to be allowed when so many people are exposed. Bills banning the use of BPA in food and beverage containers were introduced in Congress in 2009, 2011, and 2013. All failed to pass.

(a) Go to the Embryo Project Encyclopedia (http://embryo.asu.edu), an NSF-funded online repository of information about embryo research. Search for the term "BPA" to find the most up-to-date information on BPA research.

(b) Go to the website of the American Chemistry Council (http://www.americanchemistry.com), a trade organization that advocates for the chemical industry. Search for the term "BPA" to find their most up-to-date policy statement on the safety of BPA.

(c) Draft a letter urging your congressional representative to support or oppose the most recent BPA bill, explaining why you believe the weight of scientific evidence makes your position a prudent response to the situation.

Dog Days of Science

Two canine-loving researchers unravel the genetic secrets of man's best friend.

After reading this chapter you should be able to:

- Distinguish between the genotype and phenotype of a given genetic trait.
- Describe the importance of Gregor Mendel's experiments to our understanding of inheritance.
- Illustrate Mendel's laws of segregation and independent assortment.
- Create a Punnett square to predict the phenotype of offspring from parents with a known genotype—both for single genes and for two independent genes.
- Give examples of Mendelian traits and traits with complex inheritance.
- Explain how an individual's phenotype may be determined by multiple genes that interact with one another and with the environment.

CHAPTER

06

PATTERNS OF
INHERITANCE

Gordon Lark's best friend was dying. Soft and shaggy, with tousled black hair, Georgie hadn't left Lark's side in 10 years, since his daughter had first purchased her as a puppy from two kids by the side of the road. But as she aged, Georgie got sick with Addison's disease, a disorder in which her body's immune system began to attack and destroy her own tissues. Georgie passed away in 1996.

Lark, a scientist at the University of Utah in Salt Lake City, was heartbroken. To help heal the wound, he decided to adopt another dog of the same breed—a Portuguese water dog (PWD), named for its tradition of helping Portuguese fishermen with their work (**Figure 6.1**). He contacted Karen Miller, a PWD breeder on a farm in rural New York. As part of the owner screening process, Miller asked Lark about his profession. "I said I was a soybean geneticist," Lark recalls, "but all she heard was 'blah, blah, genetics, blah blah.' And she got really excited."

As a breeder, Miller was keenly interested in how dogs inherit characteristics from their parents, so Miller and Lark began having weekly phone conversations about genetics. When it came time to pick up his new puppy, Mopsa, Lark asked Miller for the bill. But Miller didn't want Lark's money. She had something else in mind: She gave him Mopsa free of charge, so that he would feel guilty and start researching dog genetics.

"That's silly," Lark told her. He wasn't a dog researcher. Lark had spent a career studying the **genetic traits** of bacteria and soybeans. A genetic trait is any inherited characteristic of an organism that can be observed or detected in some manner. Some genetic traits are **invariant**, meaning they are the same in all individuals in the species. All soybeans, for example, have pods that contain seeds. Other genetic traits are **variable**; soybean seeds occur in various sizes and colors, including black, brown, and green.

Apart from his love for canines, Lark wasn't intimately familiar with dogs' physical and biochemical traits. **Physical traits**, such as the shape of a dog's face, are easy to observe. **Biochemical traits**, on the other hand, such as a dog's susceptibility to Addison's disease, are often more difficult to observe. It is easy to collect physical and biochemical information from a field of soybeans but far more difficult to collect it from domesticated animals in homes all over the country. And to study dogs, Lark would also need data on **behavioral traits**, such as shyness and extroversion—factors he didn't have to take into account with soybeans. All of these traits—physical, biochemical, and behavioral—are influenced by genes.

Figure 6.1

The "First Dog" is a PWD
President Barack Obama plays with the family dog, Bo, a Portuguese water dog.

Getting to the Genes

A **gene** is the basic unit of information affecting a genetic trait. At the molecular level, a gene consists of a stretch of DNA on a chromosome—a thread-like molecule made of DNA and proteins, found in the nucleus of a eukaryotic cell (**Figure 6.2**). To study PWD traits, Lark would need dog DNA, which can be obtained from blood or saliva. Once he had that DNA, he could begin to search for **alleles**—different versions of a given gene—for genetic traits. Alleles of a gene arise by **mutation**, which is any change in the DNA that makes up a gene (see Chapter 8 for more on mutation). The genetic diversity in a species, whether it is soybeans or dogs or humans, comes about because the species contains many different alleles of its genes.

Dogs are the champions of genetic variation. All dogs are the same genus and species, yet a Pekingese weighs only a couple of pounds while a St. Bernard can weigh over 180 pounds. Dogs, in fact, are reported to have more variation in the size and shape of their species than any other living land mammal on Earth, with the possible exception of humans.

On a whim, Lark agreed to dabble in dog genetics, but to do so, he would need to compare the **genotypes** of individual dogs (their genetic makeup) with their **phenotypes** (the physical expression of their genetic makeup). For a particular trait, a genotype is the pair of alleles that codes for a given phenotype. To identify genes responsible for dog traits, Lark would need both types of information for each of many dogs.

Miller was already on the case. Three months after Lark received Mopsa, Miller sent him 5,000 PWD pedigrees—detailed health and breeding records for individual dogs. Lark was astonished. It was the first of many times that the enthusiasm and generosity of dog owners would contribute to his research.

"That was literally how this started," says Lark. Today their unlikely partnership has blossomed into a national research project producing valuable knowledge about the genetic basis of health and disease in both man and man's best friend. What's more, their effort has demonstrated how tiny genetic changes can create huge variation in a single species.

Pet Project

The Georgie Project, as Lark fondly named it, officially began in 1996. Lark's first task was to collect genotypes and phenotypes from PWDs. To his pleasant surprise, PWD owners were enthusiastic and began flooding him with pedigrees, blood samples, and X-rays taken by their veterinarians. In short order, Lark had DNA from more than 1,000 dogs and detailed body measurements for over 500. Then the hard work began.

Using the dogs' genotypes and phenotypes, Lark set out to pinpoint the alleles for particular traits. Some genes have alleles that are **dominant**

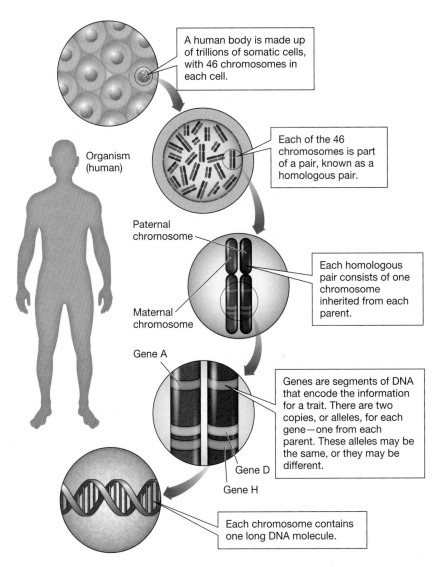

A human body is made up of trillions of somatic cells, with 46 chromosomes in each cell.

Organism (human)

Each of the 46 chromosomes is part of a pair, known as a homologous pair.

Paternal chromosome

Maternal chromosome

Each homologous pair consists of one chromosome inherited from each parent.

Gene A

Genes are segments of DNA that encode the information for a trait. There are two copies, or alleles, for each gene—one from each parent. These alleles may be the same, or they may be different.

Gene D

Gene H

Each chromosome contains one long DNA molecule.

Figure 6.2

Genes are segments of DNA that make up a genetic trait
Somatic cells (cells of the body) have two copies of most genes.

Q1: What is the physical structure of a gene?

Q2: How many copies of each gene are found in the diploid cells in a woman's body?

Q3: With 46 chromosomes in a human diploid cell, how many chromosomes are from the person's mother and how many are from her father?

when paired with another allele; that is, one allele prevents a second allele from affecting the phenotype when the two alleles are paired together. The black-fur allele (*B*), for example, is dominant in dogs. An allele that has no effect on the phenotype when paired with a dominant allele is said to be **recessive**. In dogs, the brown-fur allele (*b*) is recessive. When a gene has dominant and recessive alleles, we generally use an uppercase letter for the dominant allele and a lowercase letter for the recessive allele.

An individual who carries two copies of the same allele (such as *BB* or *bb*) is **homozygous** for that gene. An individual whose genotype consists of two different alleles for a given phenotype (*Bb*) is **heterozygous** for that gene. Having one dominant allele and one recessive allele, a heterozygous individual will show the dominant phenotype; a dog that is heterozygous for fur color (*Bb*), for example, will be black (**Figure 6.3**).

The first dog trait Lark decided to investigate was size. What makes a Great Dane large and a Chihuahua small? To find out, Lark asked for help from the "mother of all dog projects," as Lark calls her—a researcher named Elaine Ostrander, whose entry into dog research was almost as strange as Lark's.

Phenotype:

Genotype: *bb*   *BB* or *Bb*

Figure 6.3

Poodles illustrate variation in the coat color gene

These poodles, close cousins to the Portuguese water dog, may have a black coat (dominant allele *B*) or a brown coat (recessive allele *b*). Other coat colors, with different inheritance patterns, are found in poodles and other dog breeds.

> **Q1:** Which can you observe directly: the genotype or the phenotype?
>
> **Q2:** Which poodle could be heterozygous: the one with the black coat or the one with the brown coat?
>
> **Q3:** Can you identify with certainty the genotype of a black poodle? A brown poodle?

Crisscrossing Plants

In 1990, Ostrander was a young, enthusiastic researcher who had just completed her postdoctoral studies in molecular biology at Harvard University and was ready to start her own laboratory in California. But first she had to decide which organism to study. Typical choices included fruit flies, worms, or plants—organisms that are easy to grow and manipulate. Ostrander picked plants, just as Gregor Johann Mendel, an Austrian monk who later became known as

the "father of modern genetics," had done in the mid-1800s.

Mendel famously bred pea plants in a garden at his monastery. Through his work with pea plants, Mendel discovered patterns of inheritance that today form the foundation of genetics for scientists like Ostrander. "Mendel's laws," as they are now called, describe how genes are passed from parents to offspring. These laws allow us to use parental genotypes to predict offspring phenotypes.

Each time Mendel bred two pea plants together, he was performing a **genetic cross**, or just "cross" for short. A genetic cross is a controlled mating experiment performed to examine how a particular trait may be inherited. In a series of genetic crosses, the first set of parents is called the **P generation** ("P" for "parental").

For example, Mendel investigated the inheritance of flower color by crossing pea plants that had different flower colors (**Figure 6.4**). He had

GORDON LARK

A geneticist at the University of Utah in Salt Lake City, Lark initiated the Georgie Project in 1996 to study the genetics of Portuguese water dogs. The national research project has led to valuable knowledge about the genetic basis of health and disease in humans and dogs.

noticed that some plants always "bred true" for flower color; that is, the offspring always produced flowers that had the same color as the parents. He performed a genetic cross with a P generation in which one parent bred true for purple flowers (PP) and the other bred true for white flowers (pp). The first generation of offspring of a genetic cross is called the **F_1 generation** ("F" is for "filial," a word that refers to a son or daughter). When the individuals of the F_1 generation are crossed with each other, the resulting offspring are said to belong to the **F_2 generation**. Mendel allowed the F_1-generation pea plants to self-fertilize to produce the F_2 generation.

We can predict the results of an experimental cross by using a grid-like diagram called a **Punnett square** (**Figure 6.5**). A Punnett square shows all possible ways that two alleles can be brought together through fertilization. To create a Punnett square showing how a trait is inherited, list the alleles of the male genotype across the top of the grid, writing each unique allele just once. List the alleles of the female genotype along the left edge of the grid, again writing each unique allele only once. In the case of Mendel's cross of the F_1 generation, the male genotype (Pp) is mated with a female genotype (Pp).

Next, fill in each box (or "cell") in the grid by combining the male allele at the top of each column with the female allele listed at the beginning of each row. The Punnett square shows all four ways in which the two alleles in the sperm can combine with the two alleles found in the egg. The four genotypes shown within the Punnett square are all equally likely outcomes of this cross.

Using the Punnett square method, we can predict that ¼ of the F_2 generation is likely to have genotype PP, ½ to have genotype Pp, and ¼ to have genotype pp. Because the allele for purple flowers (P) is dominant, plants with PP or Pp genotypes have purple flowers, while plants with pp genotypes have white flowers. Therefore, we predict that ¾ (75 percent) of the F_2 generation will have purple flowers and ¼ (25 percent) will have white flowers—a 3:1 ratio of phenotypes. This prediction is very close to the actual results that Mendel obtained. Of a total of 929 F_2 plants that Mendel raised, 705 (76 percent) had purple flowers and 224 (24 percent) had white flowers.

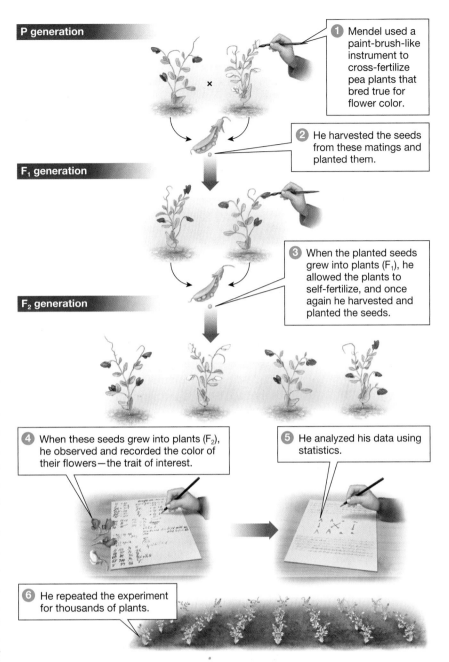

P generation

1 Mendel used a paint-brush-like instrument to cross-fertilize pea plants that bred true for flower color.

2 He harvested the seeds from these matings and planted them.

F_1 generation

3 When the planted seeds grew into plants (F_1), he allowed the plants to self-fertilize, and once again he harvested and planted the seeds.

F_2 generation

4 When these seeds grew into plants (F_2), he observed and recorded the color of their flowers—the trait of interest.

5 He analyzed his data using statistics.

6 He repeated the experiment for thousands of plants.

Figure 6.4

Mendel's careful experiments

Mendel was meticulous in conducting his research and making observations, following a very careful protocol.

Q1: What would you predict about the color of the F_1 plants' flowers?

Q2: Why was it important that Mendel begin with pea plants that he knew bred true for flower color? Why couldn't he simply cross a purple-flowered plant and a white-flowered plant?

Q3: Over the years, Mendel experimented with over 30,000 pea plants. Why did Mendel collect data on so many plants? Why didn't he study just one cross? *Hint:* Read "What Are the Odds?" on page 98 before answering.

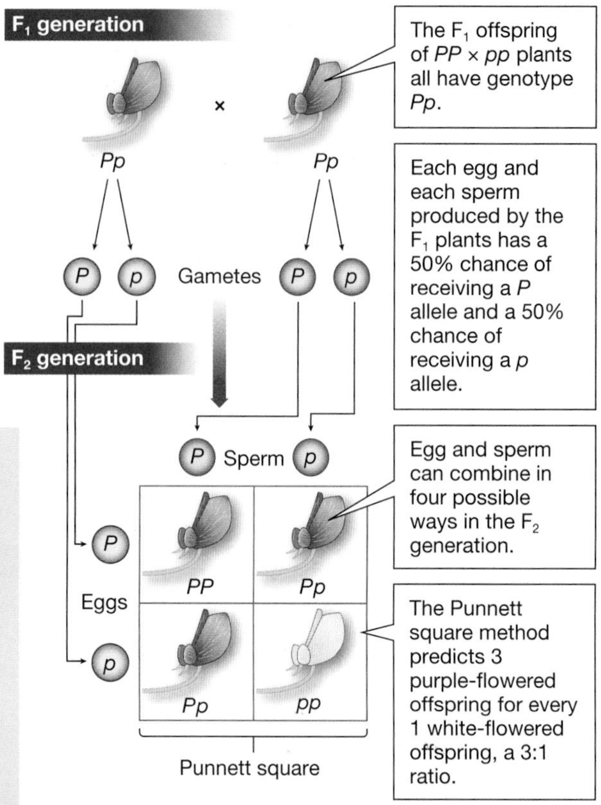

Figure 6.5

Punnett squares predict the offspring of genetic crosses

Q1: Why did Mendel's entire F$_1$ generation look the same?

Q2: The phenotype ratio in the F$_2$ generation is 3:1 purple-to-white flowers. What is the genotype ratio?

Q3: Draw a Punnett square for a genetic cross of two heterozygous, black-coated dogs. What is the phenotype ratio of their offspring? What is the offspring genotype ratio?

The F$_1$ offspring of $PP \times pp$ plants all have genotype Pp.

Each egg and each sperm produced by the F$_1$ plants has a 50% chance of receiving a P allele and a 50% chance of receiving a p allele.

Egg and sperm can combine in four possible ways in the F$_2$ generation.

The Punnett square method predicts 3 purple-flowered offspring for every 1 white-flowered offspring, a 3:1 ratio.

These results led Mendel to propose his first law, the **law of segregation**, which in modern terms (Mendel did not know about DNA) states that the two alleles of a gene are separated during meiosis, the specialized type of cell division during sexual reproduction that was discussed in Chapter 5, and end up in different gametes—egg or sperm cells. This law can be used to predict how a single trait will be inherited.

You can try this out on your own by making a Punnett square to predict the ratio of black and brown offspring that would result if two heterozygous (*Bb*) black-coated dogs were mated. It is important to understand that the predicted ratios simply give the *probability* that a particular offspring will have a certain phenotype or genotype; the actual ratio will vary (see "What Are the Odds?" on page 98).

Peas in a Pod

Mendel's research on pea seeds led to his second law, the **law of independent assortment**. This law states that when gametes form, the two copies (alleles) of any given gene segregate during meiosis independently of any two alleles of other genes. For example, pea seeds can have a round or wrinkled shape, and they can be yellow or green. Two different genes control the two different traits—the *R* gene, with alleles *R* (round) and *r* (wrinkled), controls seed shape; the *Y* gene, with alleles *Y* (yellow) and *y* (green) controls the color of the seed—but neither gene affects the inheritance of the other.

Mendel tested the idea of independent assortment in a set of experiments illustrated in **Figure 6.6**. He tracked seed shape, a trait controlled by the *R/r* alleles, and seed color, controlled by the *Y/y* alleles. The test of his hypothesis came

ELAINE OSTRANDER

The "mother of all dog projects," Elaine Ostrander is Chief of the Cancer Genetics Branch at the National Institutes of Health. She studies genes important to growth, size variation, and cancer in dogs.

when Mendel examined the phenotypes of the offspring produced by crossing the heterozygous F_1 plants (*RrYy*). As predicted by the hypothesis, two new phenotypic combinations were found among the F_2 offspring: plants with round, green seeds (*RRyy* or *Rryy*) and plants with wrinkled, yellow seeds (*rrYY* or *rrYy*). Figure 6.6 summarizes the ratios of the two parental phenotypes and the two novel, nonparental phenotypes.

Traits controlled by a single gene and unaffected by environmental conditions are called **Mendelian traits**. But when Mendel described his laws of inheritance, he had no idea what genes were made of, where they were located within a cell, or how they segregated and independently assorted. Today we know that genes are located on **chromosomes** and that these chromosomes are the basis for all inheritance. We call this the **chromosome theory of inheritance**, and it explains the mechanism underlying Mendel's laws by identifying chromosomes as the paired factors that are shuffled and recombined, and then separated randomly into sperm and egg cells during meiosis (see Figures 5.7 and 5.8 in Chapter 5). Then, during fertilization, a one-in-a-million sperm fuses with a one-in-a-million egg to create, eventually, a unique individual. That is how offspring can have genotypes and phenotypes that were not present in either parent, such as a brown puppy coming from two black dogs.

Going to the Dogs

Like Mendel, Elaine Ostrander had planned to study plants to unravel the secrets of genetics and inheritance. But when she arrived at the University of California at Berkeley to open her lab, the space was not yet available. So she wandered down the hall and into the office of Jasper Rine, a geneticist who normally studied yeast but was looking for someone to start a mammalian genome research project. Ostrander volunteered.

But which mammal to study? "I was allergic to cats, and I didn't know enough about cows or pigs or horses," she recalls, so she picked dogs. Not only was Ostrander a dog lover, but the American Kennel Club had just begun offering funding to researchers trying to identify genes associated with dog diseases.

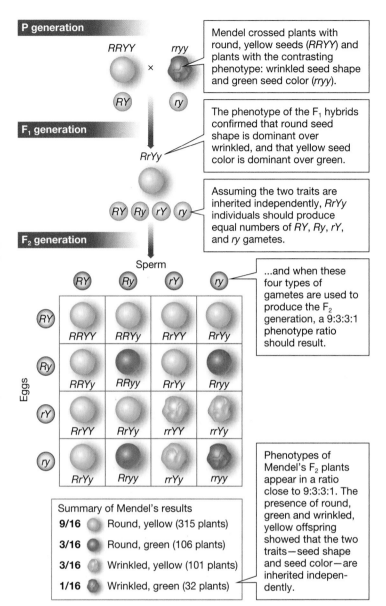

Figure 6.6

Independent assortment of pea color and shape

Mendel used two-trait breeding experiments, called **dihybrid crosses**, to test the hypothesis that the alleles of *two different genes* are inherited independently from each other.

Q1: List all the possible offspring genotypes and phenotypes.

Q2: What is the offspring phenotype ratio?

Q3: Complete a Punnett square for a genetic cross of two true-breeding Portuguese water dogs—one with a black, wavy coat (homozygous dominant, *BBWW*) and one with a brown, curly coat (homozygous recessive, *bbww*). What is the phenotype ratio of their offspring (F_1)? Now fill out another Punnett square, crossing two of the offspring. What is the phenotype ratio of the F_2 generation?

In 1993, Ostrander began identifying all the genes unique to dogs—that is, making a map of the dog **genome**. Some colleagues said she was nuts, that no one would give her money to support the research. But Ostrander was nothing if not persistent, and she knew the potential value of the research: Dogs have more than 350 inherited diseases, and up to 300 of those are similar to conditions in people, including cancer, epilepsy, heart disease, and Addison's disease, the illness that killed Lark's dog Georgie. The genetics of

bladder cancer is difficult to study in humans, for example, but the disease is quite common in Scottish terriers and could be easier to study in a dog species. By cracking the genetic code of dogs, Ostrander could possibly uncover causes and potential treatments for human diseases.

In 2005, Ostrander published the first full dog genome sequence—for a female boxer named Tasha. The achievement gained her scientific fame and raised the awareness among scientists of the importance of dog genetics to human health. "Of the more than 5,500 mammals living today, dogs are arguably the most remarkable," Ostrander's coauthor, Eric Lander, a professor of biology at the Massachusetts Institute of Technology, said when the first dog genome sequence was published. "The incredible physical and behavioral diversity of dogs—from Chihuahuas to Great Danes—is encoded in their genomes. It can uniquely help us understand embryonic development, neurobiology, human disease and the basis of evolution" (**Figure 6.7**).

But years before Ostrander completed the dog genome, she had begun a different pet project. In 2001, Ostrander received a call from a scientist in Utah who wanted to talk about dogs. It was Gordon Lark, who told her he was collecting trait information about PWDs. "The day I met Gordon was the best day of my life," says Ostrander. "I knew it was golden."

In 2002, the duo published a paper that pinpointed genes controlling dog body shape, from the tall, lanky look of a greyhound to the short, stocky frame of a pit bull. In the acknowledgments of the paper, they thanked Karen Miller and all the PWD owners who had contributed pedigree information.

In the spring of 2006, Lark and Ostrander began their second collaboration, this time to identify the genetic basis of dog size. Lark collected skeletal measurements of 92 PWDs and DNA samples from each dog. Ostrander used that genotype and phenotype information to identify a key gene for body size—*IGF1*, which controls the activity of a growth factor and is known to influence body size in mice and humans. This gene's two alleles are called *I* and *B*. Lark and Ostrander discovered that PWDs homozygous for allele *I* (*II*) were usually large dogs, and those homozygous for allele *B* (*BB*) were always small

What Are the Odds?

The *probability* of an event is the chance that the event will occur. For example, there is a probability of 0.5 that a coin will turn up "heads" when it is tossed. A probability of 0.5 is the same thing as a 50 percent chance, or ½ odds, or a ratio of one heads to one tails (1:1). If you toss the coin only a few times, the observed percentage of heads may differ greatly from 50 percent—but if you toss it many, many times, that observed percentage will be very close to 50 percent. Each toss of a coin is an independent event, in the sense that the outcome of one toss does not affect the outcome of the next toss. The probability of getting two heads in a row is a product of the separate probabilities of each individual toss: 0.5 × 0.5, which is 0.25. In our cross, the probability of getting a brown puppy is ¼, or 0.25. To go back

to a Punnett square to predict the ratio of puppies from a genetic cross of two heterozygous (*Bb*) black-coated dogs, the probability of getting a black puppy is ¾, or 0.75.

We cannot know with certainty what the actual phenotype or genotype of a particular offspring is going to be, except when true-breeding individuals are crossed. For example, two brown dogs, both of whom have a *bb* genotype, will have only *bb*-genotype, brown-phenotype offspring. Moreover, the probability that a particular offspring will display a specific phenotype is completely unaffected by how many offspring there are. The likelihood that we will see the 3 black:1 brown outcome, however, increases when we analyze a larger number of offspring, just as Mendel analyzed thousands of pea plants.

Shadow, a standard poodle, was the first dog to have its genome partially (about 80%) sequenced.

Tasha, a boxer, is the first dog to have its complete genome sequenced. Boxers are vulnerable to hip, thyroid, and heart problems. Scientists identified a gene for cardiomyopathy in boxers, a heart disorder also found in humans.

Pembroke Welsh corgis may develop a fatal neuro-degenerative condition similar to amyotrophic lateral sclerosis (ALS) in humans. The human gene mutation associated with ALS was also found in corgis with the condition.

Psychiatric disorders often have a genetic component. Doberman pinschers are susceptible to canine compulsive disorder, similar to obsessive-compulsive disorder in humans. The responsible gene in Dobermans has been linked to autism disorders in humans.

Golden retrievers are prone to cancers of the bone marrow. Ostrander's research group is analyzing the genomes of hundreds of goldens with and without cancer, hoping to identify the genes responsible.

Figure 6.7

Man's best friend

The Dog Genome Project has identified the genetic basis of several diseases and conditions in dogs, and in some cases it has been able to link the gene to a similar gene in humans.

Q1: Boxers are far more inbred than poodles. Why does that inbreeding make the former a better target for genetic studies of disease than the latter?

Q2: Explain why a geneticist interested in finding a gene linked to cancer would want to look at the DNA of senior golden retrievers with *and* without cancer?

Q3: Obsessive-compulsive disorder (OCD) in humans is characterized by obsessive thoughts and compulsive behavior, such as pacing. Canine compulsive disorder (CCD) is characterized by compulsive behavior such as "flank sucking," sometimes seen in Doberman pinschers. Would you predict that the medications given to humans with OCD would decrease compulsive behaviors in CCD dogs? Why or why not?

dogs. That single gene accounted for whether a PWD was large or small.

Interestingly, neither *IGF1* allele is dominant or recessive. Instead, heterozygous dogs, with an *IB* genotype, are medium-sized dogs. This is an example of a trait inherited by **incomplete dominance**—in which neither allele is able to exert its full effect, so a heterozygote displays an intermediate phenotype. Dogs with an *IB* genotype aren't large, and they aren't small; they are medium-sized (**Figure 6.8**).

Early in the twentieth century, geneticists identified yet another type of interaction among alleles—codominance—that Mendel had not observed among his pea plants. A pair of alleles

Genotype: *ll*
Phenotype: Large

Genotype: *BB*
Phenotype: Small

Genotype: *IB*
Phenotype: Medium

Figure 6.8

Incomplete dominance of body size alleles
Great Danes and Chihuahuas (see chapter opener) illustrate the extreme size variation found in domestic dogs. Unlike the case with Mendelian traits, dogs heterozygous for the main body size gene show an intermediate size like the Cocker Spaniel shown here.

Q1: What are the genotypes of a large and a small dog?

Q2: Is it possible to have a heterozygous large dog? Explain why or why not.

Q3: Crossing a Great Dane and a Chihuahua is likely to be unsuccessful, even though they are members of the same species (and thus have compatible sperm and egg). Why is that? What are some potential risks of such a cross?

shows **codominance** when the effect of the two alleles is equally visible in the phenotype of the heterozygote. In dogs, gum color is codominant. A dog's gums can be pink, black, or pink with black spots; in the latter case, both alleles are fully on display, and neither is diminished or diluted by the presence of the other allele (as in incomplete dominance) or suppressed by a dominant allele (as in the case of dominant and recessive alleles). In humans, blood type is a codominant trait.

It's Complicated

Many of the traits people tend to be curious about—body weight, intelligence, athleticism, and musical talent, to name a few—are yet more complicated. A **complex trait** is a genetic trait whose pattern of inheritance cannot be predicted by Mendel's laws of inheritance. Complex traits do not fit the straightforward single-gene, single-phenotype pattern discussed so far.

In some cases, a *single* gene influences a number of different traits. This is called **pleiotropy** (*pleio*, "many"; *tropy*, "change"). In PWDs, Lark found that single genes can control multiple related skeletal traits. The shape of a dog's head and the shape of its limb bones are connected by a single gene. That connection makes sense, says Lark, since a small head and long legs are advantageous for a fast dog, while a strong dog, like a pit bull, uses both its massive jaw and its short, thick legs for power.

Another good example of pleiotropy comes from a long-term breeding experiment to make Russian silver foxes tame. Researchers found that as the foxes became tamer and tamer, they also developed floppy ears instead of straight ones and had shorter legs and curlier tails than ordinary foxes have (see "The New Family Pet?" on page 102).

Patterns of inheritance can be even more complicated. Most traits are governed by the action of more than one gene. These are called **polygenic traits**. In humans, polygenic traits include eye and skin color, running speed, blood pressure, body size, and more. Of the thousands of human genetic traits, governed by an estimated 24,000 genes, fewer than 4,000 are known or suspected to be controlled by a single gene with a dominant and a recessive allele. The rest are polygenic traits.

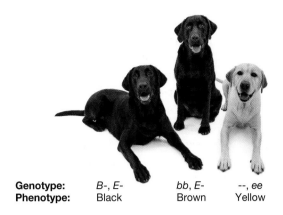

Genotype:	B-, E-	bb, E-	--, ee
Phenotype:	Black	Brown	Yellow

Figure 6.9

Epistasis in coat color

These Labrador retrievers show complex inheritance of coat color. The yellow dog carries two alleles that interfere with the deposition of melanin in hair. Both the brown and the black dogs must carry at least one allele that allows melanin deposition. A dash indicates that the allele is unknown, based on phenotype.

Q1: What are the possible genotypes (at both genes) of the black dog? The yellow dog? The brown dog?

Q2: Draw a Punnett square showing possible matings between the black dog (assuming it is heterozygous at both genes) and the yellow dog (assuming it is heterozygous at the *B* gene). List all the possible phenotypes of their offspring. (See **Figure 6.6** for an example of a Punnett square made with two traits.)

Q3: If you wanted the most variable litter possible, what colors of Labrador retrievers would you cross?

Another twist on inheritance is **epistasis**, which occurs when the phenotypic effect of the alleles of one gene depends on the presence of alleles for another, independently inherited gene. Labrador coat color, for example, is affected by epistasis (**Figure 6.9**). Dog fur, as mentioned earlier, has a dominant allele (*B*) that leads to black fur and a recessive allele (*b*) that produces brown fur. But the effects of these alleles (*B* and *b*) can be eliminated completely, depending on which allele of the expression gene (*E* or *e*) is present. Dogs with a dominant *E* allele deposit a pigment called melanin in their fur and are therefore able to express whatever fur color genotype is present. But a recessive *ee* genotype blocks the deposition of melanin in fur, so the dog is yellow, regardless of the genotype at the *B/b* gene (*BB*, *Bb*, or *bb*).

If the environment affects the phenotype, it becomes nearly impossible to predict the phenotype when given only the genotype of an individual or its parents. The effects of many genes depend on internal and external environmental conditions, such as body temperature, carbon dioxide levels in the blood, external temperature, and amount of sunlight. For example, cats have a gene that codes for an enzyme called tyrosinase, which is involved in melanin production. Siamese cats have a special C^t allele of the gene. The C^t allele codes for a tyrosinase that works well at colder temperatures ($\leq 35°C$) but does not function at warmer temperatures ($\geq 37°C$), so the production of melanin depends on the temperature of the surroundings (**Figure 6.10**). Because a cat's extremities tend to be colder than the rest of its body, melanin is produced there, and hence the paws, nose, ears, and tail of a Siamese cat tend to be dark. If a patch of light hair is shaved from the body of a Siamese cat and the skin is covered with an ice pack, the hair that grows back will be dark. Similarly, if dark hair is shaved from the tail and allowed to grow back under warm conditions, it will be light-colored.

Man's Best Friend

After describing the inheritance of size in Portuguese water dogs, Lark and Ostrander looked at the *IGF1* gene in over 350 dogs, representing 14 small breeds and 9 giant breeds. The genotype *BB* was common in small dogs and virtually nonexistent in large dogs. "All small dog breeds had them. It didn't matter when they were bred or how; they all had the exact same pattern," says Ostrander. "It was amazing," adds Lark. Breeders have, over time, been selecting for these alleles to create smaller and smaller dogs. "What mankind can do, without any genetic tools but just knowledge of heritability, is just extraordinary," he says.

After their success with dog size, Lark and Ostrander identified genes responsible for other traits—fur color, leg length, skull shape, and more. They have also identified genes related to cancer and other complex traits that might tell us something about human disease (see "Most Chronic Diseases Are Complex Traits" on page 104). In Border Collies, Ostrander's

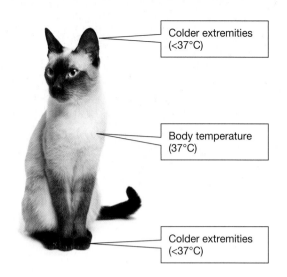

Colder extremities (<37°C)

Body temperature (37°C)

Colder extremities (<37°C)

Figure 6.10

The environment can alter the effects of genes

Coat color in Siamese cats is controlled by a temperature-sensitive allele.

Q1: The gene that brings about the pale Siamese body fur is also responsible in part for the typical blue eyes of the species. What is the term for this type of inheritance?

Q2: Siamese kittens that weigh more tend to have darker fur on their bodies. Why might this be?

Q3: The Siamese cat pictured is called a "seal point" because it has seal brown extremities. Some Siamese cats show the same color pattern, but the dark areas are of a lighter color or even a different shade— for example, lilac point, red point, blue point. What results would you predict if the experiments described in the text (shaving the cat and then increasing or decreasing temperature) were conducted on cats with these color patterns?

The New Family Pet?

The silver fox is the same species as the more familiar red fox. Because of its soft, silver coat, it has been bred in captivity for over 100 years to provide fur coats, stoles, and hats for the wealthy.

In 1959 a Russian geneticist, Dmitry Belyaev, began to conduct breeding experiments on silver foxes he had purchased from a fur breeder, pairing only the tamest individuals of each generation. He determined how tame a fox was by observing its response when approached and offered food. As the foxes became tamer in each generation, they did not show a "fear response" until they were older—nine weeks instead of six weeks. (Domestic dogs develop a fear response at about 8–12 weeks.) In addition, in the tame foxes, the hormones associated with a fear response did not increase until later. These traits were all clearly influenced by the same gene or genes—an example of pleiotropy.

Another surprising result was that the foxes' appearance began to change along with the changes in behavior. They began to develop shorter tails, wider faces, and floppier ears. All of these made the adult foxes look more puppy-like, and are similar to the differences that can be observed when comparing the domestic dog to its ancestor, the wolf. Scientists conjecture that in both cases, tameness and associated changes in development, physiology, and anatomy were brought about by breeding for juvenile features.

Only 50 years and 35 generations separate this tame silver fox from its wild relative

team identified a gene involved in an eye disease that causes blindness in both humans and dogs. Her lab also identified a gene involved in kidney cancer in dogs that causes a similar syndrome in humans.

Today, Lark and Ostrander both continue their canine research, and the Georgie Project lives on. "I'm really happy with the growth of the field," says Ostrander, "and there's a lot of research coming down the pike."

Does Bigger Mean Better?

Genome size is the total amount of DNA in one copy of an organism's genome, typically measured in millions of base pairs (Mb). There is a huge range in genome sizes across plants and animals, from living organisms of 150,000 Mb to the *E. coli* genome of 4.6 Mb. Dogs and humans have smaller genomes than many species, making them easier to study genetically than, say, a lungfish.

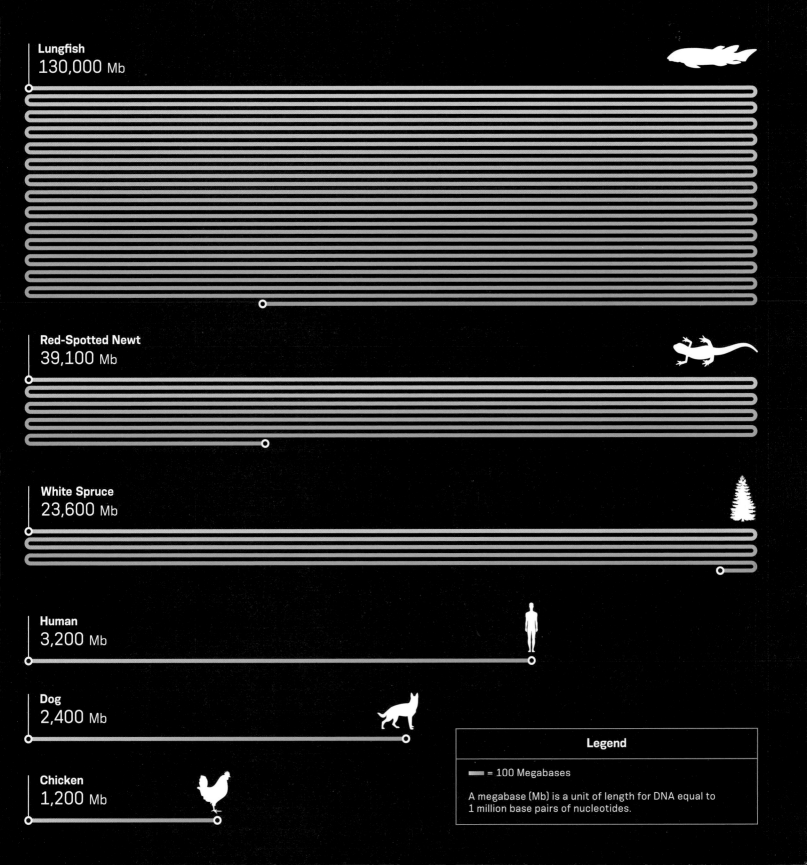

Lungfish
130,000 Mb

Red-Spotted Newt
39,100 Mb

White Spruce
23,600 Mb

Human
3,200 Mb

Dog
2,400 Mb

Chicken
1,200 Mb

Legend

━━━ = 100 Megabases

A megabase (Mb) is a unit of length for DNA equal to 1 million base pairs of nucleotides.

Most Chronic Diseases Are Complex Traits

Adisease is a condition that impairs health. It may be caused by external factors, such as infection by viruses, bacteria, or parasites, or injury produced by harmful chemicals or high-energy radiation. Nutrient deficiency can also lead to disease. Inadequate vitamin C consumption, for example, produces scurvy, once common among sailors and pirates. Disease may also be caused by the malfunction of one or more genes. Diseases caused exclusively by gene malfunction are described as genetic disorders, distinguishing them from infections and other types of diseases.

But many of the diseases that are most common in industrialized countries—heart disease, cancer, stroke, diabetes, asthma, and arthritis, for example—are caused by multiple genes interacting in complex ways with each other and with external factors. They are complex traits: malfunctions in key genes make a person susceptible to developing these diseases, but environmental factors affect whether the disease will actually appear and how severe the symptoms will be. As shown in the graph below, a large part of the estimated risk of developing colon cancer, stroke, coronary heart disease, and type 2 diabetes is avoidable. Lifestyle choices such as maintaining good nutrition, exercising regularly, and avoiding tobacco have a significant impact on our risk of developing such chronic diseases. (The word "chronic" means "unceasing," a reference to the fact that once we develop one of these diseases, we have it for the rest of our lives.)

A major goal of modern genetics is to identify genes that contribute to human disease when they fail to function normally. Researchers have identified alleles associated with increased risk of a number of common ailments, including high blood pressure, heart disease, diabetes, Alzheimer's disease, several types of cancer, and schizophrenia. The hope is that one day soon, genetic tests will tell us whether we are predisposed to a disease before we become ill with it. Then, a person carrying a risky allele could take preventive measures to reduce the chance of actually developing the condition, and treatment could be customized to fit the particular allele involved. This tailored approach to treatment, called "personalized medicine," is already being used to treat breast cancer and other chronic diseases.

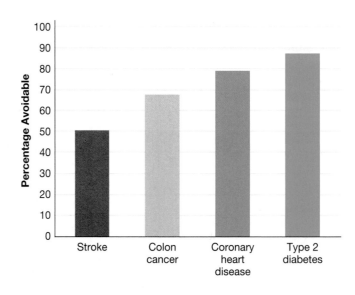

Mopsa, the puppy that Karen Miller gave Lark in 1996 in return for studying PWDs, died in April 2012, just a week short of her sixteenth birthday (**Figure 6.11**). But Lark has a new best friend, a PWD puppy he named Chou (pronounced "shoo"), for the French *petit chou*, meaning "little cabbage."

"We often make a mistake and call Chou Mopsa, because Chou looks so similar," says Lark. After all, he adds, PWDs share similar genotypes and thus similar phenotypes. And these genetic traits make them the cuddly, devoted pets that they are.

Figure 6.11

Gordon Lark and Mopsa

GEORGIE

Gordon Lark's first Portuguese water dog, a fiercely loyal and playful friend.

REVIEWING THE SCIENCE

- A **gene** is a stretch of DNA that affects one or more genetic traits. Genes are formed on **chromosomes**, thread-like molecules made of DNA and proteins.

- The **genotype** is an individual's genetic makeup, or more specifically the pair of different versions of a given gene, or **alleles**, that determine a given trait. The **phenotype** is the physical expression of an individual's genetic makeup, or more specifically the expression of a version of the given trait.

- Mendel's experiments enabled him to deduce two laws of inheritance: The **law of segregation** states that two alleles of a gene are separated during meiosis and end up in different gametes. The **law of independent assortment** states that during meiosis, the two alleles of any given gene segregate independently of any two alleles of any other gene.

- The **chromosome theory of inheritance** explains how Mendel's laws arise: genes occupy specific locations on chromosomes, and those chromosomes are randomly shuffled and recombined during meiosis.

- A grid-like diagram called a **Punnett square** can help predict the results of a **genetic cross**.

- **Mendelian traits** are genetic traits controlled by a single gene and unaffected by environmental conditions.

- **Complex traits** are those whose pattern of inheritance cannot be predicted by Mendel's laws of inheritance. A complex inheritance pattern may be due to **pleiotropy**, in which a single gene influences a number of different traits; or it may be due to **epistasis**, in which the phenotypic effect of the alleles of one gene depends on the presence of alleles for another, independently inherited gene.

THE QUESTIONS

The Basics

1 Link each of the following terms with the correct definition.

GENOTYPE

PHENOTYPE

HETEROZYGOTE

HOMOZYGOTE

DOMINANT

RECESSIVE

1. An individual that carries one copy each of two different alleles (for example, an *Aa* individual or an *I*B individual).

2. An individual that carries two copies of the same allele (for example, an *AA*, *aa*, or *II* individual).

3. An allele that does not affect the phenotype when paired with a dominant allele in a heterozygote.

4. The genetic makeup of an individual; more specifically, the two alleles of a given gene that affect a specific genetic trait in a given individual.

5. The specific version of a genetic trait that is displayed by a given individual.

6. The allele that controls the phenotype when paired with a different allele in a heterozygote individual.

2 Circle the correct terms in the following sentence:

The (**gene, allele**) for coat color has two (**genes, alleles**), one for brown coloring and one for black.

3 Circle the correct terms in the following sentences:

Cells undergo (**mitosis, meiosis**) to become gametes. This process sorts the alleles of a gene into separate gametes, which is the cause of Mendel's law of (**segregation, independent assortment**). Genes on different chromosomes also sort into separate gametes during this process, which is the cause of Mendel's law of (**segregation, independent assortment**).

4 Which of the following is a cause of variation in dogs (and other living organisms)?

(a) mutation

(b) natural selection

(c) meiosis

(d) mitosis

(e) fertilization

5 For each of the following cases, identify whether the described trait is an example of Mendelian inheritance (M) or a more complex form of inheritance (C).

_____ a. brown versus black coat color in dogs

_____ b. body size in dogs

_____ c. coat color in Siamese cats

_____ d. skin color in humans

_____ e. flower color in pea plants

6 People with blue eyes often have difficulty seeing in bright light. Is this observation an example of pleiotropy or epistasis?

7 Before Mendel conducted his experiments with pea plants, people believed that offspring were a "blend" of their parents and would show intermediate levels of their parents' traits. What would Mendel's F_1 peas have looked like if this were true?

Try Something New

8 The silver fox (see "The New Family Pet?" on page 102) belongs to the same species as the red fox: *Vulpes vulpes*. Two silver foxes will always breed true for silver offspring. A silver fox bred to a red fox will produce either all red offspring or, occasionally, half red and half silver offspring. Red foxes bred together will usually produce all red offspring but will occasionally produce silver offspring in the ratio of 3 red to 1 silver.

(a) What do these observations suggest to you about inheritance of coat color in *Vulpes vulpes*?

(b) What breeding experiment might you conduct to determine whether your hypothesis is correct?

(c) Draw a Punnett square to show your predicted results.

9 In your garden you grow Big Boy (round) and Roma (oval) tomatoes. You love the taste of Big Boys, but you think it's easier to slice Roma tomatoes. You decide to cross-pollinate a Big Boy and a Roma to see whether you can create a new strain of "Long Boys." In the first generation, all of the tomatoes are round. How would you explain this result? What would your next cross be? Write out the cross in a Punnett square, using parental genotypes. What proportion of the next generation, if any, would be oval?

10 For several hundred years, goldfish have been selectively bred in China and Japan for body color and shape, tail shape, bulging eyes, and even fleshy head growths.

Wild goldfish   Pet-shop goldfish   Black moor goldfish

Imagine that you have a tank of pet-shop goldfish and have just added a couple of black moor goldfish, hoping that they will breed. When the eggs laid by the black moor female (P generation) hatch and the young fish (F_1 generation) begin to develop, you are shocked to see that they are going to be orange. How would you explain this result in terms of the inheritance of body color in goldfish? What breeding experiment could you conduct to test your hypothesis?

11 In 2009, a large team of researchers including Elaine Ostrander and Gordon Lark published the results of its research on coat inheritance in dogs. The study began by focusing on dachshunds and Portuguese water dogs, but then widened to more than 80 breeds. The scientists were able to explain 95 percent of the variation in dog coat types with just two alleles at each of three genes, each inherited independently of the other. These genes coded for coat length, wave or curl in the coat, and the presence of "furnishings" (the moustache and eyebrows often seen in wire-haired dogs; see photo). Long-haired dogs carry two copies of the long-hair allele, which is recessive to the short-hair allele. Dogs with furnishings can be either homozygous or heterozygous for the furnishings allele; dogs without furnishings are homozygous for the no-furnishings allele.

(a) What is the genotype of a long-haired dog without furnishings?

(b) What are the possible genotypes of a short-haired dog with furnishings? List all possibilities.

(c) Create a Punnett square of two dogs heterozygous for hair length and furnishings, and calculate the phenotype ratios of their offspring.

Leveling Up

12 **Doing science** Want to get involved in dog research? If you have a purebred as a pet, you can. Find out whether the Dog Genome Project at the National Institutes of Health is doing research on your pet's breed. If they are, you can send in a swab of your dog's saliva and contribute to science. Visit the NIH website (http://research.nhgri.nih.gov/dog_genome/) for more information.

13 **Is it science?** The November 18, 2003, issue of *Weekly World News* printed a story about a woman who, after repeatedly watching the movie *Shrek* while taking fertility drugs, gave birth to a baby who looked like the main character, an ogre named Shrek. Like Shrek, the newborn had dull green skin, a large flat nose, and ears protruding from stems. From what you know about genetics, do you think it's possible for a developing fetus to change so drastically (from a normal-looking baby to a "Shrek" baby) because its mother was obsessed with a movie? Why or why not? How would you explain your answer to someone who believed this news report?

14 **What do *you* think?** Many people are critical of those who breed or purchase purebred dogs, arguing that there are many mixed-breed dogs waiting to be adopted from shelters. They also point out that mixed-breed dogs are less likely than purebred dogs to suffer from genetic diseases. Those who prefer a particular breed argue that there is a strong genetic influence on dog personality and behavior, and that they don't want any surprises when they add a new member to their family. What do you think?

A Deadly Inheritance

How researchers identified a mysterious genetic disorder, and their risky effort to develop a cure.

After reading this chapter you should be able to:

- Diagram a chromosome, identifying genes, alleles, and loci.
- Interpret a human pedigree to determine whether a given condition is recessive, dominant, or sex-linked.
- Review a human karyotype to identify the sex chromosomes and any abnormalities in chromosome number.
- Explain how sex is genetically determined in humans, and how sex determination relates to the inheritance of sex-linked traits.
- Compare and contrast the inheritance of recessive, dominant, and sex-linked disorders.
- Use a Punnett square to calculate the probability of inheriting a particular disorder.

GENETICS

CHAPTER
07

CHROMOSOMES
AND HUMAN
GENETICS

Felix clutches his mother's side. Her arm is wrapped tightly around him. Lying in a rumpled white hospital bed, Felix looks away from the two red tubes protruding from his body. Blood pumps out one of his sides into a humming machine next to the bed and is pumped back into the other side. Dr. Christoph Klein steps into the room, clothed head to toe in blue scrubs, and offers a smile and a reassuring word (**Figure 7.1**).

Born in 2005, Felix was still a newborn when he began to bleed. His parents rushed him to an intensive care unit. The bleeding eventually stopped, but the hospital visits didn't. Three years later, Felix was diagnosed with a rare and deadly disease: Wiskott-Aldrich syndrome (WAS). "Wiskott-Aldrich was the very last diagnosis I would ever want to receive," Felix's mother later said. "Every day I prayed, 'Please, Lord, let this chalice be removed from us.'"

Felix's pediatrician feared the worst. WAS patients suffer from recurring infections, pneumonia, bleeding, and rashes; they often develop leukemia or lymphoma and die of complications due to infections. Some patients can be treated with a bone marrow transplant. However, if the donor is not a matched sibling or a close unrelated match, survival rates are low.

So, in 2009, Felix's family appeared in Klein's office at the Hannover Medical School in Germany, holding tightly to the hope that Klein might be able to save Felix's life. Klein, a pediatrician who has made a career of studying rare diseases, was running a clinical trial testing a new treatment for WAS. The treatment was risky and the results unknown, but it was Felix's only hope to be cured of his deadly disease—one that had perplexed scientists for decades.

A Mysterious Malady

More than 70 years earlier, in 1937, three young brothers had come to see Dr. Alfred Wiskott, also a German pediatrician. At first, Wiskott had no idea what was wrong. The boys bled abnormally: their blood was unable to clot, and they had bloody diarrhea. They also had recurring ear infections and blistering, weeping rashes on their skin. Wiskott recorded their symptoms, but he could not help the boys. All three died at a young age.

What had killed the brothers? Their parents also had four daughters, all of whom were healthy, so it was unlikely that an infection, a toxin, or an environmental factor had caused the illness. Instead, Wiskott suspected, the boys might have inherited a disease from their parents.

Thanks to the chromosome theory of inheritance, Wiskott knew that the boys had inherited hereditary material, in the form of chromosomes, from their parents. Recall from Chapter 6 that offspring inherit one chromosome from the mother and one from the father. Wiskott suspected that a gene on one of the inherited chromosomes was causing the mysterious illness.

A disease caused by an inherited mutation, passed down from a parent to a child, is a **genetic disorder**. Wiskott recognized the importance

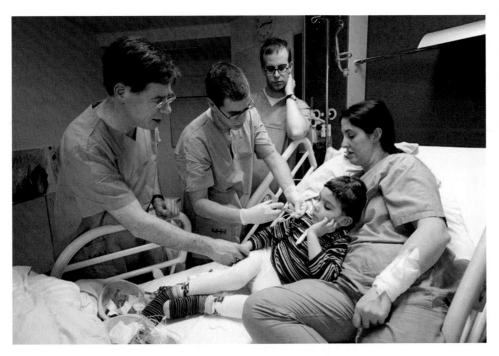

Figure 7.1

Felix and Dr. Klein

Dr. Christoph Klein examines Felix. Felix suffers from a rare genetic disease.

of studying genetic disorders, since such studies could lead to the prevention of or cure for a disease. But daunting problems have long plagued the study of human genetic disorders. From a biological point of view, humans have a long generation time, select their own mates, and decide whether and when to have children. In addition, human families tend to be much smaller than would be ideal for a scientific study. From an ethical point of view, geneticists and physicians cannot intervene and perform experiments directly on humans to determine how genetic disorders are inherited.

Painful Pedigree

Twenty years after Wiskott described the first cases of the mysterious disease and determined that it was inherited, an American pediatrician, Dr. Robert Anderson Aldrich, solved the next piece of the puzzle. Aldrich met a six-month-old boy with anemia, bloody diarrhea, and general weakness. After several emergency room visits, the baby died.

Aldrich sat down with the boy's mother to review possible causes. After an hour of asking questions, he still had no idea what might have brought on the illness. Finally, he asked about other relatives who might have had a similar illness. The boy's grandmother, who had tagged along to the meeting, exclaimed sadly, "Just like all the rest of them." Other male infants in the family, it turned out, had died under similar circumstances.

Given that information, Aldrich worked with the mother and grandmother to trace the family's history back six generations by drawing a **pedigree**, a chart similar to a family tree that shows genetic relationships among family members over two or more generations of a family's medical history (**Figure 7.2**, top). Pedigrees provide scientists with a way to analyze information in order to learn about the inheritance of a particular disorder. Aldrich found that 16 male infants in the family, but no females, had died of the syndrome (**Figure 7.2**, bottom).

Because of the "remarkable family history," as he described it, Aldrich concluded, like Wiskott, that the illness was a genetic disorder, caused by a mutation passed down from parent to child.

CHRISTOPH KLEIN

A medical doctor and cancer researcher, Christoph Klein is now a professor at the University of Munich. In 2010, Klein began testing a new gene therapy to treat young children with Wiskott-Aldrich syndrome, a rare and life-threatening disease. The therapy, though still experimental, has been very successful.

Genetic disorders can be caused by mutations in individual genes or by abnormalities in chromosome number or structure.

Every species has a characteristic number of chromosomes—humans have 23 pairs of homologous chromosomes, for a total of 46, while mosquitoes have 3 pairs, or 6 in total. Each chromosome has a particular structure, with genes arranged on it in a precise sequence. Any change in the chromosome number or structure, compared to what is typical for a species, is considered a **chromosomal abnormality**.

The two most common types of chromosomal abnormalities in humans are changes in the overall number of chromosomes and changes in chromosome structure, such as a change in the length of an individual chromosome (**Figure 7.3**). Changes in the number of chromosomes in humans are usually lethal. Down syndrome, in which individuals receive three copies of chromosome 21, is an exception to this rule, as individuals with Down syndrome can live long, healthy lives. Changes in chromosome structure can also have dramatic effects. For example, cri du chat syndrome, caused by a deletion on chromosome 5, results in slowed growth, a small head, and mental retardation.

ALFRED WISKOTT

A German pediatrician who lived from 1898 to 1978, Alfred Wiskott described the cases of three young brothers with a serious bleeding disease—what would later be called Wiskott-Aldrich syndrome. Wiskott deduced that the boys had inherited a gene from their parents that was causing the illness.

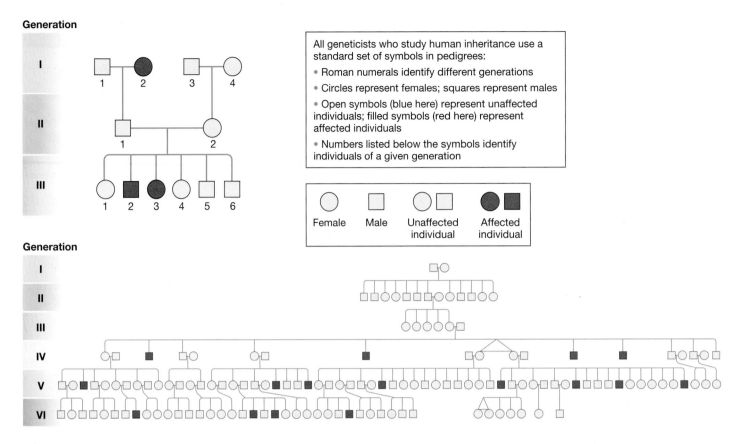

Figure 7.2

Patterns of inheritance can be analyzed in family pedigrees

The cystic fibrosis pedigree at the top of this figure shows six children (generation III), two of whom are affected with the disease. The bottom pedigree is of a family with a history of Wiskott-Aldrich Syndrome.

Q1: How many males and how many females in total does Aldrich's pedigree contain?

Q2: What proportion of males and what proportion of females were affected by the disorder?

Q3: Why did Aldrich hypothesize that the disease was X-linked? (You will need to read ahead to answer this question.)

The Special Case of the Sex Chromosomes

Aldrich continued to wonder whether the disease was caused by a chromosomal abnormality or by a single inherited mutation—a change in the DNA of a gene (see Chapter 8 for more on mutation).

In his 1954 paper describing the disease, Aldrich suggested that since primarily males inherit the syndrome, it is caused by a mutation on a **sex chromosome**—one of the two

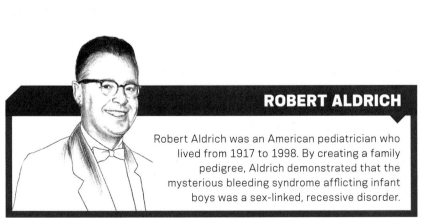

ROBERT ALDRICH

Robert Aldrich was an American pediatrician who lived from 1917 to 1998. By creating a family pedigree, Aldrich demonstrated that the mysterious bleeding syndrome afflicting infant boys was a sex-linked, recessive disorder.

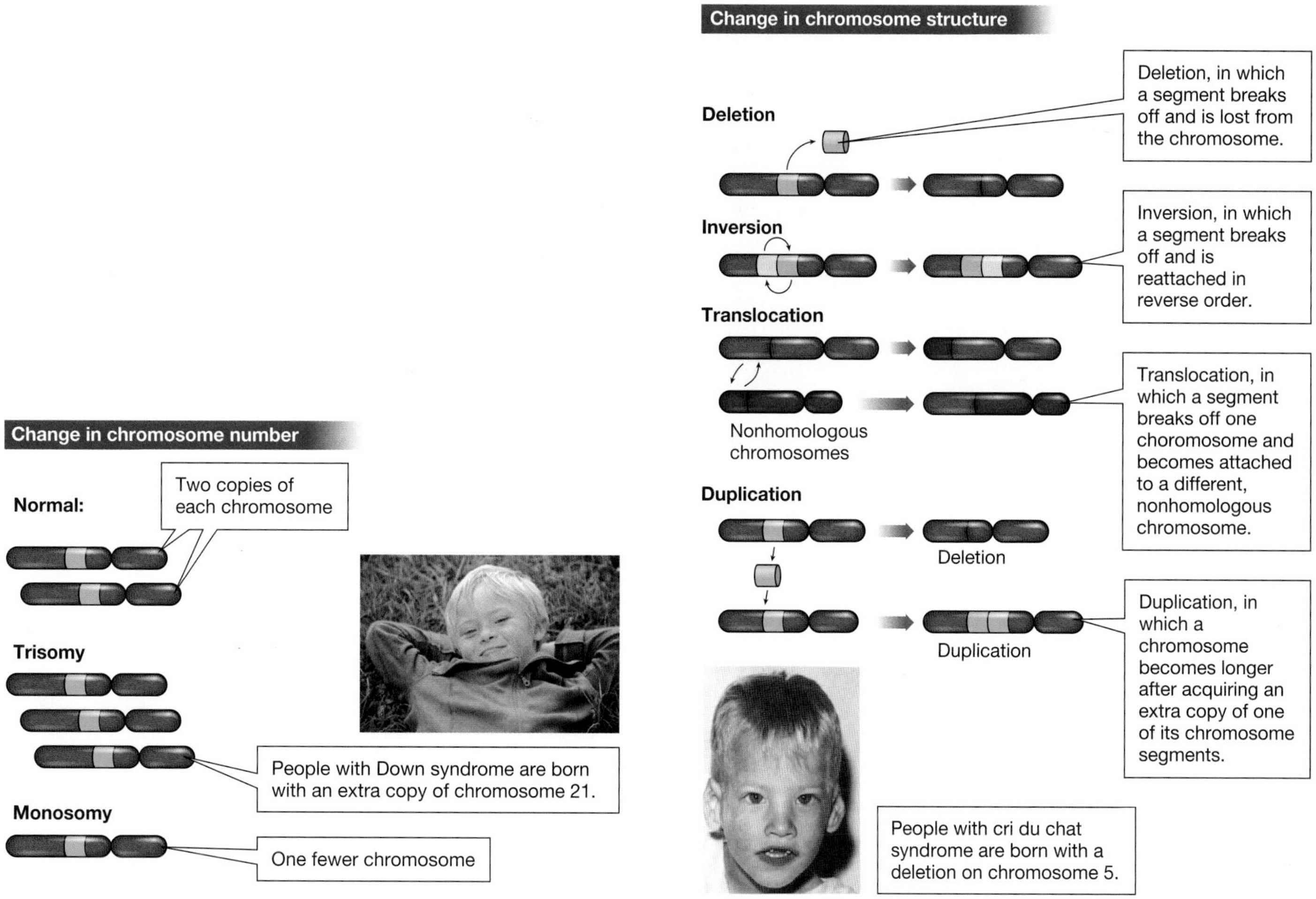

Deletion

Deletion, in which a segment breaks off and is lost from the chromosome.

Inversion

Inversion, in which a segment breaks off and is reattached in reverse order.

Translocation

Nonhomologous chromosomes

Translocation, in which a segment breaks off one chromosome and becomes attached to a different, nonhomologous chromosome.

Duplication

Deletion

Duplication

Duplication, in which a chromosome becomes longer after acquiring an extra copy of one of its chromosome segments.

People with cri du chat syndrome are born with a deletion on chromosome 5.

Change in chromosome number

Normal:

Two copies of each chromosome

Trisomy

People with Down syndrome are born with an extra copy of chromosome 21.

Monosomy

One fewer chromosome

Figure 7.3

Chromosomal abnormalities can cause serious genetic disorders

Any increase or decrease in the number of chromosomes almost invariably results in spontaneous abortion of the fetus, which is estimated to occur in up to 20 percent of all pregnancies—Down syndrome is an exception. Changes in chromosome structure may have relatively minor or more severe effects, depending on the size and location of the change.

Q1: Why are changes in chromosome *number* almost always more severe than changes in chromosome *structure*?

Q2: In which part of meiosis would you predict that chromosomal abnormalities are produced? (Refer back to Chapter 5 if necessary.)

Q3: Create a mnemonic to help remember the four kinds of structural changes (e.g., Doctors Improve Treatment Daily).

chromosomes that determine sex. All other chromosomes are called **autosomes**. The majority of chromosomes in the human genome are autosomes—homologous chromosomes exactly alike in terms of length, shape, and the genes they carry (**Figure 7.4**). Autosomes are labeled with the numbers 1 through 22 (for example, chromosome 4), and sex chromosomes are assigned letter names. In humans, males have one X chromosome and one Y chromosome, whereas females

have two X chromosomes. The Y chromosome in humans is much smaller than the X chromosome.

Because human females have two copies of the X chromosome, all the gametes (eggs) they produce contain one X chromosome, passed on to their offspring. Males, however, have one X chromosome and one Y chromosome, so odds are that half their gametes (sperm) will contain an X chromosome and half will contain a Y chromosome (**Figure 7.5**).

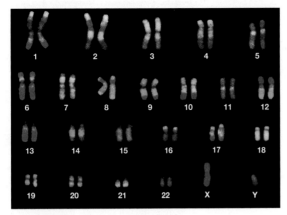

Figure 7.4

A human karyotype

To research chromosomal abnormalities, a scientist will take a photograph of a cell's chromosomes during mitosis and then pair up each set of homologous chromosomes to create a **karyotype**. In humans, the autosomes are numbered 1 through 22, and the sex chromosomes are designated X or Y.

> **Q1:** Is this the karyotype of a male or a female?
>
> **Q2:** How would the karyotype of a person with Down syndrome differ from this karyotype?
>
> **Q3:** Chromosome size correlates roughly with the number of genes residing on it. Why are an extra copy of chromosome 21 and a missing Y chromosome two of the least damaging chromosomal abnormalities?

Aldrich recognized that the family's disease was tightly linked to sex and concluded that the disease was caused by sex-linked inheritance—an inherited mutation on the X or Y chromosome. But where exactly was the mutation located?

The physical location of a gene on a chromosome is called its **locus** (plural "loci"). Because a gene can occur in different versions, or alleles, a diploid cell can have two different alleles at a given locus on a pair of homologous chromosomes. If the two alleles at a locus are different, the cell is heterozygous for the gene. If the two alleles at a locus are identical, the cell is homozygous for the gene (**Figure 7.6**).

But the sex chromosomes are different. Roughly 1,240 of the estimated 20,000 human genes are found solely on the X or Y chromosome, and approximately 1,180 of those 1,240

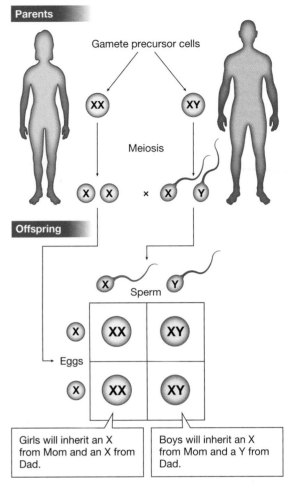

Figure 7.5

Dad's chromosomes determine baby's sex

> **Q1:** What are the odds that a given *egg* cell will contain an X chromosome? A Y chromosome?
>
> **Q2:** If a couple has two daughters, does that mean that their next two children are more likely to be sons? Explain your reasoning. *Hint:* Refer back to "What Are the Odds?" on page 98.
>
> **Q3:** Sisters share the same X chromosome inherited from their father, but they may inherit different X chromosomes from their mother. What is the probability that brothers share the same Y chromosome? What is the probability that brothers share the same X chromosome?

genes are located on the X chromosome, while only about 60 are located on the much smaller Y chromosome.

Figure 7.6

Genetic loci on homologous chromosomes

The genes shown here take up a larger portion of the chromosome than they would if they were drawn to scale. The average human chromosome has more than a thousand different genes interspersed with large stretches of noncoding DNA.

Q1: How do we know whether two chromosomes are homologous?

Q2: In one sentence, explain how the terms "gene," "locus," and "chromosome" are related.

Q3: If hair color were determined by a single gene, what would be an example of an allele for this gene?

In a pair of homologous chromosomes, one is inherited from the male parent, and the other from the female parent.

Paternal homologue

Maternal homologue

A genetic locus is the location of a particular gene on a chromosome.

At each genetic locus, an individual has two alleles, one on each homologous chromosome.

AA = Homozygous dominant

bb = Homozygous recessive

Cc = Heterozygous

Three gene pairs at three different loci

These 1,240 genes are said to be **sex-linked**. Sex-linked genes on the X chromosome are **X-linked**. Sex-linked genes on the Y chromosome are **Y-linked**. One of these is the *SRY* gene (short for "*s*ex-determining *r*egion of *Y*"). *SRY* functions as the "master sex switch," committing the sex of the developing embryo to male. In the absence of this gene, a human embryo develops as a female.

The *SRY* gene does not act alone; in both males and females, other genes on the autosomes and sex chromosomes directly influence the development of the many sexual characteristics that distinguish men and women. The *SRY* gene plays a crucial role because when present, it causes other genes to produce male sexual characteristics, but when absent, those genes produce female sexual characteristics. For example, individuals with androgen insensitivity syndrome (AIS) are female in appearance, and some say they "feel female," even though they have XY chromosomal makeup and their ovaries fail to develop normally. **Figure 7.7** shows a group of people with AIS, who are genetically male but physically female.

Besides this syndrome, there are no well-documented cases of disease-causing Y-linked genes. X chromosomes, however, contain genes

Figure 7.7

Androgen insensitivity syndrome

AIS is usually due to a genetic mutation that makes people with the condition unable to respond normally to male hormones such as testosterone. It can also be caused by a deletion of the *SRY* gene from the Y chromosome.

known to be involved in many human genetic disorders. Aldrich correctly concluded that the gene leading to the family's bleeding disease was located on the X chromosome, so it was X-linked.

X Marks the Spot

Thanks to advances in molecular biology tools, in 1994 researchers determined that the gene causing WAS is on the X chromosome, and called it, unsurprisingly, *WAS* (gene names are typically italicized, while the name of the disorder, WAS, is not). *WAS*, they discovered, is the genetic code for a protein crucial for the correct formation and function of blood cells and immune system cells. Without a healthy copy of this gene, individuals acquire blood and immune system disorders, are susceptible to infections, and have increased risk of lymphoma, cancer of the lymph nodes.

Felix inherited a mutated version of *WAS* from his mother. We can use a Punnett square to illustrate how the X-linked recessive mutation for WAS is inherited. We label the recessive mutated *WAS* allele *a*, and in the Punnett square we write this allele as X^a to emphasize the fact that it is on the X chromosome. Then we label the dominant, healthy allele *A* and write this allele as X^A in the Punnett square (**Figure 7.8**). Individuals like Felix's mother, who have only one copy of a recessive allele, are said to be **genetic carriers** of the disorder. They can pass on the disorder allele, but they do not have the disease.

If a carrier female like Felix's mother, with genotype $X^A X^a$, has children with a normal male (with genotype $X^A Y$), each of their sons will have a 50 percent chance of getting the disorder. Felix had a 50 percent chance of getting WAS, and he did.

Males of genotype $X^a Y$ suffer from the condition because the Y chromosome does not have a copy of that gene. In other words, because males cannot be heterozygous for any X-linked genes, the effects of an *a* allele cannot be masked. In general, males are more likely than females to get recessive X-linked disorders, because they need to inherit only a single copy of the disorder allele to exhibit the disorder. Females,

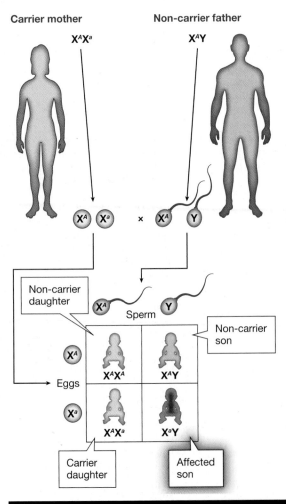

Carrier mother

$X^A X^a$

Non-carrier father

$X^A Y$

X^A X^a × X^A Y

Non-carrier daughter

Sperm

X^A Y

Non-carrier son

Eggs

X^A

$X^A X^A$ $X^A Y$

X^a

$X^A X^a$ $X^a Y$

Carrier daughter

Affected son

Figure 7.8

X-linked recessive conditions are more common in males

The recessive disorder allele (*a*) is located on the X chromosome and is denoted by X^a. The dominant normal allele (*A*) on the X chromosome is denoted by X^A.

Q1: Which of the children specified in this Punnett square represents Felix? What is his genotype?

Q2: Explain why Felix is neither homozygous nor heterozygous for the *WAS* gene.

Q3: Create a Punnett square to illustrate the offspring if Felix were to have children with a non-carrier woman. What is the probability that a son would have WAS? What is the probability that a daughter would be a carrier of WAS?

Genetic Diseases Affecting Americans

Wiskott-Aldrich Syndrome is only one of many genetic conditions seen in humans. Here are some of the most common genetic diseases in the United States, most of which can be identified in newborns using genetic testing.

U.S. births per year: 4,000,000

• = 1 birth

6,037 births
Down Syndrome

1,140 births
Cystic Fibrosis

800 births
Marfan Syndrome

Male U.S. births per year: 2,050,000

• = 1 male birth

590 births
Duchenne Muscular Dystrophy

400 births
Hemophilia

400 births
Fragile X Syndrome

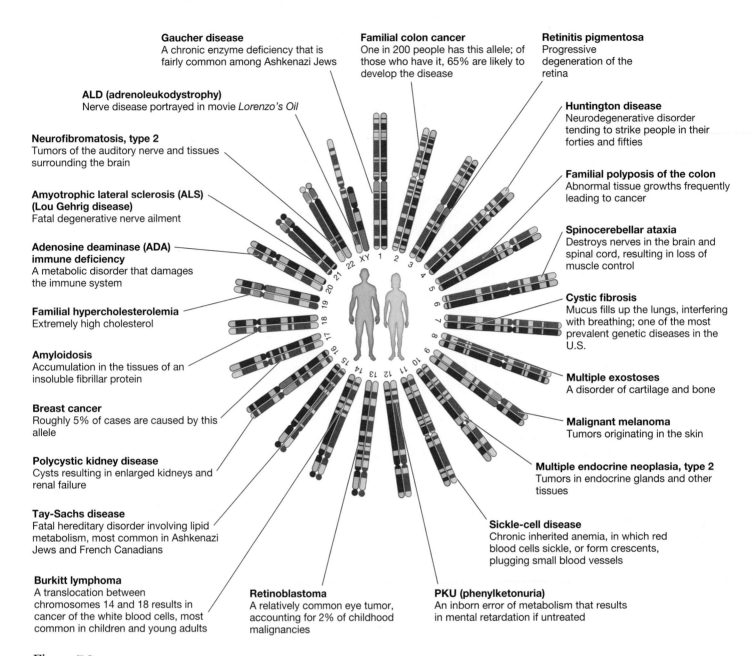

Gaucher disease
A chronic enzyme deficiency that is fairly common among Ashkenazi Jews

ALD (adrenoleukodystrophy)
Nerve disease portrayed in movie *Lorenzo's Oil*

Neurofibromatosis, type 2
Tumors of the auditory nerve and tissues surrounding the brain

Amyotrophic lateral sclerosis (ALS) (Lou Gehrig disease)
Fatal degenerative nerve ailment

Adenosine deaminase (ADA) immune deficiency
A metabolic disorder that damages the immune system

Familial hypercholesterolemia
Extremely high cholesterol

Amyloidosis
Accumulation in the tissues of an insoluble fibrillar protein

Breast cancer
Roughly 5% of cases are caused by this allele

Polycystic kidney disease
Cysts resulting in enlarged kidneys and renal failure

Tay-Sachs disease
Fatal hereditary disorder involving lipid metabolism, most common in Ashkenazi Jews and French Canadians

Burkitt lymphoma
A translocation between chromosomes 14 and 18 results in cancer of the white blood cells, most common in children and young adults

Familial colon cancer
One in 200 people has this allele; of those who have it, 65% are likely to develop the disease

Retinoblastoma
A relatively common eye tumor, accounting for 2% of childhood malignancies

Retinitis pigmentosa
Progressive degeneration of the retina

Huntington disease
Neurodegenerative disorder tending to strike people in their forties and fifties

Familial polyposis of the colon
Abnormal tissue growths frequently leading to cancer

Spinocerebellar ataxia
Destroys nerves in the brain and spinal cord, resulting in loss of muscle control

Cystic fibrosis
Mucus fills up the lungs, interfering with breathing; one of the most prevalent genetic diseases in the U.S.

Multiple exostoses
A disorder of cartilage and bone

Malignant melanoma
Tumors originating in the skin

Multiple endocrine neoplasia, type 2
Tumors in endocrine glands and other tissues

Sickle-cell disease
Chronic inherited anemia, in which red blood cells sickle, or form crescents, plugging small blood vessels

PKU (phenylketonuria)
An inborn error of metabolism that results in mental retardation if untreated

Figure 7.9

Single-gene disorders

Mutations of single genes that lead to genetic disorders are found on the X chromosome and on each of the 22 autosomes in humans. In each of these mutations, the healthy allele at that locus codes for an important function; for example, the sickle-cell allele is a mutation in the gene that codes for the hemoglobin protein, critical for carrying oxygen in the blood. For clarity, only one such genetic disorder per chromosome is shown.

Q1: Which chromosome contains the gene for cystic fibrosis? For Tay-Sachs disease? For sickle-cell disease?

Q2: No known genetic disorders are encoded on the Y chromosome. Why do you think this is?

Q3: In your own words, explain why most single-gene disorders are recessive rather than dominant.

Figure 7.10

Zoe is diagnosed with cystic fibrosis
Zoe's parents were both carriers for cystic fibrosis, a recessive genetic disorder, and Zoe inherited the disease allele from both parents.

on the other hand, must inherit two copies to be affected. X-linked recessive inheritance explains why boys are more likely than girls to get WAS.

Other X-linked genetic disorders in humans include hemophilia, a serious condition in which minor cuts and bruises can cause a person to bleed to death, and Duchenne muscular dystrophy, a lethal disorder that causes muscles to waste away, often leading to death at a young age. Both of these X-linked disorders are caused by recessive alleles.

More Common, but No Less Deadly: Zoe's Story

X-linked recessive disorders like WAS are rare compared to autosomal recessive disorders. Both sexes are equally likely to be affected by these **recessive genetic disorders**, since both males and females have two copies of autosomal chromosomes and identical odds of being homozygous or heterozygous for a disorder allele. Several thousand human genetic disorders are inherited as recessive traits on autosomes. These include sickle-cell disease, Tay-Sachs disease, and the most common, fatal genetic disease in the United States: cystic fibrosis (**Figure 7.9**).

Scott and Jada first began to suspect something was wrong when their newborn daughter, Zoe, didn't put on any weight. Every time Zoe ate, her belly became hard and bloated. She screamed in pain.

"I was beside myself," recalls Jada. "I knew something wasn't right." Like Felix, Zoe spent the first year of her life in and out of the pediatrician's office, until the eve of her first birthday, April 6, 2005. On that day, Scott and Jada sat with Zoe in a children's hospital in Florida, waiting for a second opinion. The doctor came in and asked them to sit down. The diagnosis was cystic fibrosis (**Figure 7.10**).

Cystic fibrosis (CF) is a lethal recessive genetic disorder caused by one or more mutations in the cystic fibrosis transmembrane regulator gene (*CFTR*). A *CFTR* mutation causes the body to produce abnormally thick, sticky mucus, which clogs the airways and leads to lung infections. The thick mucus also obstructs the pancreas, preventing enzymes from reaching the intestines, where they are needed to break down and digest food. The average life span of people with CF who live to adulthood is approximately 35 years. There is no cure.

Recessive genetic disorders vary in severity; some, like cystic fibrosis, are lethal, whereas others have relatively mild effects. Adult-onset lactose intolerance, for example, is caused by a single recessive allele that leads to a shutdown in the production of lactase, the enzyme that digests milk sugar.

The only individuals who get a disorder caused by an autosomal recessive allele (*a*) are those who have two copies of that allele (*aa*). Usually, when a child inherits a recessive genetic disorder, both parents are heterozygous; that is, they both have the genotype *Aa* (**Figure 7.11**). It is also possible for one or both parents to have the genotype *aa* and thus the disease. Because the *A* allele is dominant and does not cause the disorder, heterozygous individuals (*Aa*) like Scott and Jada, Zoe's parents, are genetic carriers of the disorder; that is, they carry the disorder allele (*a*) but do not have the disease.

If two carriers of a recessive genetic disorder have children, the patterns of inheritance are the same as for any other recessive trait. Each child has a 25 percent chance of not carrying the disorder allele (genotype *AA*), a 50 percent chance of being a carrier (genotype *Aa*), and a 25 percent

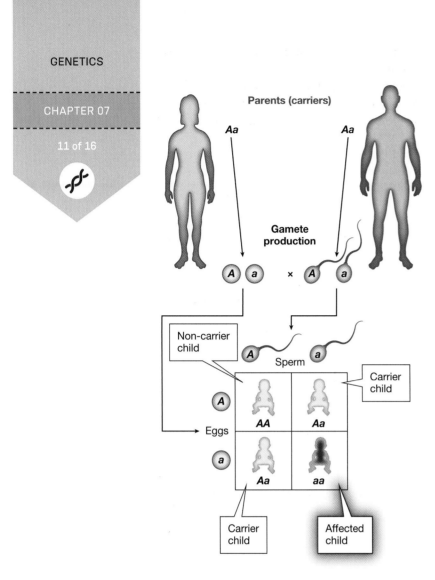

Parents (carriers)

Aa   *Aa*

Gamete production

(A) (a) × (A) (a)

Non-carrier child

(A) (a)
Sperm

Carrier child

Eggs

AA   Aa

Aa   aa

Carrier child

Affected child

Figure 7.11

Inheritance of cystic fibrosis, an autosomal recessive disorder

The patterns of inheritance for a human autosomal recessive genetic disorder are the same as for any recessive trait (compare this figure with the pattern shown by Mendel's pea plants in Figure 6.5). Recessive disorder alleles are denoted *a*. Dominant, normal alleles are denoted *A*. Here, the parents are a carrier female (genotype *Aa*) and a carrier male (genotype *Aa*).

Q1: Which of the children in this Punnett square represents Zoe? What is her genotype?

Q2: If Zoe's parents had another child, what is the probability that the child would have cystic fibrosis? That the child would be a CF carrier?

Q3: If Zoe is able to have a child of her own someday, and the other parent is not a carrier of cystic fibrosis (he would likely be tested before they chose to have children), what is the probability that the child would have cystic fibrosis? That the child would be a carrier?

chance of actually getting the disorder (genotype *aa*). Zoe did not beat the odds.

These percentages reveal one way in which lethal recessive disorders such as cystic fibrosis can persist in the human population. Although homozygous recessive individuals (with genotype *aa*) often die before they are old enough to have children, carriers (with genotype *Aa*) are not harmed by the disorder. In a sense, the *a* alleles can hide in heterozygous carriers, and those carriers are likely to pass the disorder allele to half of their children. An estimated one in 29 European Americans has a mutated *CFTR* gene. Recessive genetic disorders can also arise in the human population because new mutations produce new copies of the disorder alleles.

Deadly with One Allele

Cystic fibrosis is an example of a recessive genetic disorder, in which a child, like Zoe, inherits two recessive copies of a disorder allele. A more rare type of inherited disease is a **dominant genetic disorder**, caused by an autosomal dominant allele (*A*). In this case, the allele that causes a disorder cannot "hide" in the same way that a recessive allele can: *AA* and *Aa* individuals get the disorder; only *aa* individuals are symptom-free (**Figure 7.12**). These disorders are more rare than recessive disorders because a dominant genetic disorder often produces serious negative effects immediately upon birth, and individuals with the *A* allele may not live long enough to reproduce. Hence, few people with a dominant genetic disorder pass the allele on to their children.

For this reason, most cases of a dominant genetic disorder are produced by a new mutation in a generation. For example, achondroplasia, a form of dwarfism, is caused by a mutation in a gene involved in bone growth. People with achondroplasia have a decreased life span, so very few of them live long enough to pass the mutation on to their offspring. Instead, infants with achondroplasia are born to unaffected parents at a rate of between one in 10,000 and one in 100,000. Almost all of these births are to fathers older than 35 years, who produce this mutation during sperm production.

Huntington disease, a dominant genetic disorder, is an exception to that rule because symptoms of the disease—uncontrolled movements and loss of intellectual faculties caused by dying brain cells—arise later in life, in one's forties, after the person carrying that allele has had the opportunity to reproduce. In this way the allele is readily passed from one generation to the next. For this reason, some couples whose families have a history of Huntington's may choose to screen their developing fetus for the gene that causes the disorder.

Replacing Deadly Genes: A Work in Progress

Most inherited genetic disorders, including cystic fibrosis and Huntington disease, have no cure. Patients and their families do everything they can simply to manage the disease. At the age of 7, Zoe takes 25 pills each day just to digest food. She also takes antibiotics and mucus thinners, uses nasal sprays, and spends two hours a day doing breathing therapy. "We don't talk about the future," says Jada, Zoe's mother. "She doesn't know that this is life-shortening, that it's a progressive disease." Jada's voice cracks. "But she is aware that she's not like everybody else."

Scientists have not given up the race to find effective treatments, even cures, for genetic disorders. Thanks to Felix's doctor, Christoph Klein, WAS is one of the few such disorders for which an effective treatment has been identified.

It was in 2003 that Klein, then at Boston Children's Hospital, had the first glimmer of hope that there might be a therapy for boys afflicted with WAS. He and colleagues believed that they might be able to treat WAS using **gene therapy**, a technique for correcting defective genes responsible for disease development. Gene therapy is a type of **genetic engineering**, the permanent introduction of one or more genes into a cell, tissue, or organism. Klein hoped that by correcting the defective *WAS* gene in young affected boys, he could offer a short-term treatment,

and potentially even a permanent cure, for the disorder.

First, Klein and his team tested their plan in mice. They bred mice that lacked the entire *WAS* gene and thus could not produce the important

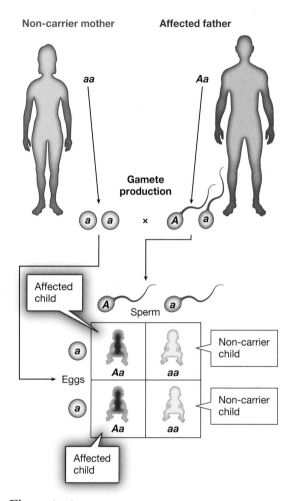

Figure 7.12

Inheritance of an autosomal dominant disorder
The pattern of inheritance for a human autosomal dominant genetic disorder is the same as for any other dominant trait. This Punnett square shows the possible children of a normal female (genotype *aa*) and an affected male (genotype *Aa*).

Q1: What is the probability that a child with one parent who has an autosomal dominant disorder will inherit the disease?

Q2: Why are there no carriers with a dominant genetic disorder?

Q3: Because dominant genetic disorders are rare, it is extremely rare for both parents to have the condition (genotype *Aa*). Draw a Punnett square with two *Aa* parents. What proportion of the offspring would have the disorder? What proportion would be normal?

Prenatal Genetic Screening

How is the baby? This is one of the first questions we ask after a child is born. Usually everything is fine, but sometimes, as with Felix and Zoe, the answer can be devastating. Today, some parents choose to have one of several prenatal genetic screening tests performed to check their baby's health before it is born.

In **amniocentesis**, a needle is inserted through the abdomen into the uterus to extract a small amount of amniotic fluid from the pregnancy sac that surrounds the fetus. This fluid contains fetal cells (often sloughed-off skin cells) that can be tested for genetic disorders. Another method is **chorionic** (kohr-ee-*AH*-nik) **villus sampling** (**CVS**), in which a physician uses ultrasound to guide a narrow, flexible tube through a woman's vagina and into her uterus, where the tip of the tube is placed next to the villi, a cluster of cells that attaches the pregnancy sac to the wall of the uterus. Cells are removed from the villi by gentle suction and then tested for genetic disorders.

Risks associated with amniocentesis and CVS, including vaginal cramping, miscarriage, and premature birth, have declined quite dramatically in recent years because of advances in technology and more extensive training. Recent studies suggest that the risk of miscarriage after CVS and amniocentesis is essentially the same: about 0.06 percent. The tests are widely used by parents who know they face an increased chance of giving birth to a baby with a genetic disorder. Older parents, for example, might want to test for Down syndrome, since the risk of that condition increases with the age of the mother. A couple in which one parent carries an allele for a dominant genetic disorder (such as Huntington disease), or both parents are carriers for a recessive genetic disorder (such as cystic fibrosis), might also choose prenatal genetic screening.

Couples who elect to have such tests performed have only two choices if their fears are confirmed: they can abort the fetus, or they can give birth to a child with a genetic disorder. Prior to conception, however, couples at risk of having a child with a genetic disorder have options to minimize that risk. If they are willing and can afford the procedure, a couple can choose to have a child by **in vitro fertilization** (**IVF**), in which an egg is fertilized by a sperm in a petri dish, after which one or more embryos are implanted into the mother's uterus. In **preimplantation genetic diagnosis** (**PGD**), one or two cells are removed from the developing embryo in the dish, usually three days after fertilization. The cell or cells removed from the embryo are then tested for genetic disorders. Finally, one or more embryos that are free of disorders are implanted into the mother's uterus, and the rest of the embryos, including those with genetic disorders, are frozen. PGD is typically used by parents who either have a serious genetic disorder or carry alleles for one.

Like all other genetic screening methods, the use of PGD raises ethical issues. People who support PGD feel that amniocentesis and CVS provide parents with a bleak set of moral choices: if the fetus has a serious genetic disorder, the parents can either abort the fetus or allow a child to be born, who will live a life that may be short and full of suffering. In their view, discarding an embryo at the 4- to 12-cell stage is morally preferable to aborting a well-developed fetus, or to giving birth to a child that will suffer the devastating effects of a serious genetic disorder. Those opposed to PGD agree that the moral choices are bleak, but they argue that once fertilization has occurred, a new life has formed and it is immoral to end that life, even at the 4- to 12-cell stage. What do you think?

WAS protein. These mice had some of the same symptoms as Felix, including a reduction in the number of blood and immune system cells. Then the researchers used a virus to insert a healthy copy of the *WAS* gene into the mice's blood cells, where they hoped it would produce the WAS protein. Months later, the researchers examined the mice. Klein was thrilled to find that the mouse cells were expressing healthy WAS protein, and that mature blood and stem cells were being produced in normal quantities.

Bolstered by these results, the scientists tested the technique in human cells and finally, in 2005, began a human clinical trial in Germany. The first two boys, both three years old, were admitted in 2006. Between 2006 and 2009, ten boys were admitted to the clinical trial. Felix was admitted in 2009.

During the collection of cells from his body, Felix had to lie still for nine hours as a machine pumped blood out of his body in order to extract cells. These cells were then taken to a laboratory where scientists used a genetically modified virus to insert a healthy copy of *WAS* into them (**Figure 7.13**). The doctors then pumped the genetically engineered cells back into Felix's body. And the waiting began.

Klein published the early results in 2010. Gene therapy was successful in nine of the boys. After gene therapy, all nine are doing much better with regard to bleeding and infections. They are able to play soccer like their healthy peers, they respond to vaccinations, and they have not developed severe infectious diseases. They are in "excellent condition" and happily attending school, says Klein. He and colleagues proved that gene therapy for the fatal WAS diagnosis is feasible, and that WAS can be corrected.

Yet a new therapy does not come without risks. As in other gene therapy trials, several participants developed blood diseases, such as leukemia, as a side effect of the type of virus used to insert the healthy gene into the cells. For this reason the trial has been put on hold.

But Klein is optimistic. Together with other physicians and researchers, he is investigating the causes of blood diseases after gene therapy. "Everything in medicine needs to be judged on the basis of rational decisions that weigh the risks and benefits," says Klein. The pressure is on, he admits, to create safer ways to insert the gene into human cells. "We're hoping to be ready for a follow-up study starting in a year or two," he adds.

A Happy Ending for Felix

A year after the gene therapy, Felix returned to the hospital for tests. The treatment had worked. Today, his body is producing healthy, functional blood cells, and most of his symptoms are completely resolved. Felix now enjoys a normal life. In this rare case, an inherited genetic disorder was cured by gene therapy. Numerous gene therapy trials continue around the world for other genetic diseases, but few have had the same success as the WAS trial. Stimulated by Klein's results, there are now WAS gene therapy trials being held in Italy and the United States.

"The most beautiful moment is when my son smiles, hugs me, and says, 'Mum, I love you so much,'" said Felix's mother after the treatment. "We are grateful that there was a treatment that helped our son."

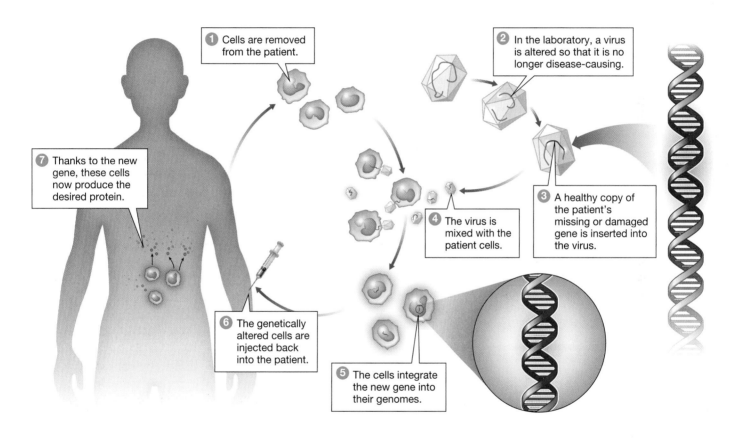

Figure 7.13

Gene therapy

In gene therapy, genetic information is transferred into cells to achieve a desired effect. Gene therapy may be used to compensate for a genetic mutation in a cell that causes the cell to malfunction.

Q1: What would be the missing or damaged gene in Felix's case? From what chromosome would a healthy copy be taken?

Q2: Why did Dr. Klein's group first conduct gene therapy on mice rather than humans? What are the advantages and limitations of this approach?

Q3: If Felix were to have children of his own someday, would they run the risk of inheriting his disorder, or has gene therapy removed that possibility? Explain your reasoning.

REVIEWING THE SCIENCE

- A **genetic disorder** is a disease caused by an inherited mutation in a gene, passed down from a parent to a child. The physical location of a gene on a chromosome is called its **locus**.

- Genetic disorders can be caused by mutations in individual genes or by **chromosomal abnormalities** (changes in chromosome number or structure).

- Every person has two **sex chromosomes**: males have one X and one Y chromosome, and females have two X chromosomes. The *SRY* gene on the Y chromosome is required for human embryos to develop as males.

- Genes found solely on the X or Y chromosome are said to be **sex-linked**. A family **pedigree** can be used to determine whether a given condition is recessive, dominant, or sex-linked.

- A **dominant genetic disorder** is caused by a dominant allele on an **autosome**. Dominant genetic disorders are more rare than **recessive genetic disorders**, which are caused by two autosomal recessive alleles. Several thousand human genetic disorders are inherited as recessive traits on autosomes.

- **Gene therapy** is a **genetic engineering** technique for correcting defective genes responsible for disease development.

THE QUESTIONS

The Basics

1 Which of the following are true of a recessive genetic disorder? Circle all that apply.

(a) They are less common than dominant genetic disorders.

(b) They are more common than sex-linked genetic disorders because there are more genes on autosomes than on sex chromosomes.

(c) Carriers inherit an allele for the disorder but do not display symptoms of the disorder themselves.

(d) Only individuals with the disorder can pass it on to their offspring.

2 Identify whether each of the following is a recessive (R), dominant (D), or sex-linked (X) genetic disorder.

_____ a. Tay-Sachs disease

_____ b. Huntington disease

_____ c. Wiskott-Aldrich syndrome

_____ d. cystic fibrosis

3 Which of the following are caused by a chromosomal abnormality? Circle all that apply.

(a) cri du chat syndrome

(b) androgen insensitivity syndrome

(c) sickle-cell disease

(d) Down syndrome

4 Use the following terms correctly in the following sentence: allele(s), chromosome(s), gene(s), locus (loci).

Two homologous _____ contain the same _____, found at the same _____, but may have the same or different copies of _____.

5 What is a genetic carrier?

6 Sometimes a segment of DNA breaks off from a chromosome and then returns to the correct place on the original chromosome, but in reverse order. This type of chromosomal structural change is called

(a) crossing-over.

(b) a translocation.

(c) an inversion.

(d) a deletion.

7 In the karyotype shown here, identify the sex chromosomes. Is this individual a male or a female?

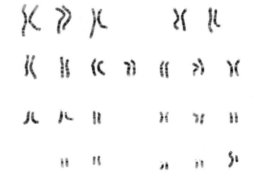

8 Link each of the following terms with the correct definition.

GENE THERAPY

IN VITRO FERTILIZATION

1. A procedure in which cells are gently suctioned from a pregnant woman's uterus to test for genetic disorders in the fetus.

2. A procedure in which a small amount of fluid (and fetal cells within it) is carefully extracted from a pregnant woman's uterus to test for genetic disorders in the fetus.

PREIMPLANTATION GENETIC DIAGNOSIS (PGD)	3. A treatment approach that seeks to correct a genetic disorder by inserting healthy copies of the mutated genes responsible for the disorder.
CHORIONIC VILLUS SAMPLING (CVS)	4. A procedure in which one or two cells are removed from a developing embryo and tested for genetic disorders; embryos that are free of genetic disorders may then be implanted into a woman's uterus.
AMNIOCENTESIS	5. A procedure in which an egg is fertilized in a petri dish, after which one or more embryos are implanted into a woman's uterus.

Try Something New

9 Which of the following sex chromosome abnormalities result in a male phenotype? Explain your reasoning.

XO, XXY, XXX, XYY, XXXY, XY missing the *SRY* gene

10 Sickle-cell disease is inherited as a recessive genetic disorder in humans; the normal hemoglobin allele (*S*) is dominant to the sickle-cell allele (*s*). For two parents of genotype *Ss* (carriers), construct a Punnett square to predict the possible genotypes and phenotypes of their children. Also list the genotype and phenotype ratios. Each time two *Ss* individuals have a child together, what is the chance that the child will have sickle-cell disease?

11 Recall that human females have two X chromosomes and human males have one X chromosome and one Y chromosome.

(a) Do males inherit their X chromosome from their mother or from their father?

(b) If a female has one copy of an X-linked recessive allele for a genetic disorder, does she have the disorder?

(c) If a male has one copy of an X-linked recessive allele for a genetic disorder, does he have the disorder?

(d) Assume that a female is a carrier of an X-linked recessive disorder. With respect to the disorder allele, how many types of gametes can she produce?

(e) Assume that a male with an X-linked recessive genetic disorder has children with a female who does not carry the disorder allele. Could any of their sons have the genetic disorder? How about their daughters? Could any of their children be carriers for the disorder? If so, which sex(es) could they be?

12 Study the pedigrees shown here. In each case, is the disorder allele dominant (*D*) or recessive (*d*)? Is the disorder allele located on an autosome or on the X chromosome?

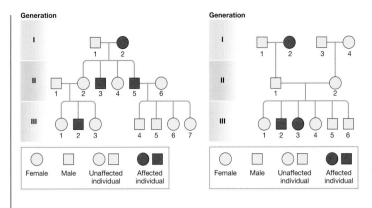

Leveling Up

13 *Write Now* biology: debating preimplantation genetic diagnosis (PGD) This assignment is designed to give you the experience of applying your knowledge of biology to a current controversy or topic of interest. You will apply the same sort of reasoning that you should be able to use as an informed citizen and consumer when making decisions that involve biology.

The scenario: The U.S. Senate Committee on Commerce, Science and Technology (CCST) is considering proposing legislation on preimplantation genetic diagnosis (PGD; see "Prenatal Genetic Screening" on page 122). The chair of the committee has invited special-interest groups to present testimony on the pros and cons of PGD. You will contribute a position paper defining PGD, describing your group's position on the technology, and making a recommendation for legislation.

Choose one of the following special-interest groups to represent (or your instructor will assign one to you):

(a) **Reproductive Specialist Group (medical doctors).** You will describe how PGD and in vitro fertilization (IVF) can be used to screen for genetic disorders or for sex selection and "family balancing," and to increase fertility in women of advanced maternal age. You will argue that legislation is not necessary, since medical association guidelines are already in place, and that it is not the place of government to judge individuals' reasons for undergoing PGD.

(b) **Genetics and Public Policy Group.** You will present data on the beliefs of the American public in this matter. You will also present the status of current legislation in the United States and how it compares to PGD legislation internationally. You will propose limited legislation based on these findings.

(c) **Parents of Down Syndrome Children.** You will argue that people with disabilities already suffer from prejudice, and that widespread use of such testing will cause even more prejudice. You will propose legislation prohibiting PGD for almost all conditions, with the exception of deadly, infant-onset disease.

(d) **Americans with Cystic Fibrosis.** You will present the case of a couple, both of whom are carriers for CF, and discuss their options. You will propose limited legislation that disallows PGD for sex selection and nondisease conditions.

Killer Spinach, Deadly Sprouts

Scientists struggle to stay abreast of dangerous strains of E. coli sweeping across the United States and Europe.

After reading this chapter you should be able to:

- Describe the structure of DNA, using appropriate terminology.

- Use the base-pairing rules to determine a complementary strand of DNA based on a given template strand.

- Label a diagram of replication, identifying the location of each step in the process.

- Describe how PCR and gene sequencing technology are used to identify and track *E. coli* outbreaks.

- Explain the cause of DNA replication errors and describe how they are repaired.

- Give an example of a mutation and its potential effects on an organism.

- Compare and contrast the genomes of prokaryotes and eukaryotes.

CHAPTER
08

WHAT
GENES ARE

On Monday, August 21, 2006, Polly Costello bought a bag of Dole baby spinach at a supermarket in Bellevue, Nebraska. Over the next several days Costello's family, including her 81-year-old mother, Ruby Trautz, ate from the bag. On Saturday, Trautz started having stomach cramps and diarrhea. On Sunday, she began passing blood, and Costello rushed her to the hospital. By Thursday, Trautz was dead.

Exactly a week later and almost 400 miles away in Wisconsin, 77-year-old Marion Graff, an avid salad consumer, passed away. Next was an 86-year-old woman in Maryland. Then a 2-year-old boy in Idaho. For each patient, doctors uncovered the same cause: *E. coli* poisoning.

Most types of *Escherichia coli*, a common bacterium (**Figure 8.1**), are not dangerous; in fact, billions of harmless *E. coli* live in your gut right now. But the type isolated from the food-poisoning victims in 2006 was no run-of-the-mill *E. coli*; it was O157:H7. *E. coli* O157:H7 attaches to cells lining the large intestine and releases toxins, causing severe cramping and diarrhea. This dangerous type of *E. coli* first burst onto the scene in a food-poisoning outbreak in Oregon in 1982 and then resurfaced again and again, peaking in 1993, when 732 people across Washington State became sick and 4 died from O157:H7-contaminated hamburgers. Now, in 2006, it appeared to have reared its ugly head once again. Americans panicked.

On September 13, Bob Brackett, then the director of the Center for Food Safety and Applied Nutrition at the U.S. Food and Drug Administration (FDA), got the first e-mail about the outbreak. Then another. And another. Within 24 hours, thanks to a national database called PulseNet that enables public health officials to track food-borne illnesses, eight states were reporting 50 cases, and most victims recalled having eaten spinach recently.

The 2006 O157:H7 outbreak was more extreme than previous outbreaks. First, its scope was nationwide; no one was safe (but there *are* ways of decreasing your risk of contracting a food-borne illness; see **Figure 8.2**). Second, victims landed in the hospital more frequently than ever before, and many of them developed kidney damage and disease.

"At that time, all we knew was that it had to do with spinach, and that it was nationwide in

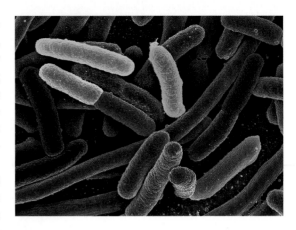

Figure 8.1

Escherichia coli

E. coli is a common and usually harmless bacterium, but some variants may cause illness and even death.

scope," says Brackett, who is now the director of the Institute for Food Safety and Health at the Illinois Institute of Technology. Then the FDA did something it had never before done: it issued a nationwide warning advising consumers not to eat a specific food product—spinach (**Figure 8.3**).

Public health officials and scientists worked feverishly to identify where the *E. coli* had originated, testing patient samples, bags of spinach, and fields that grew spinach. On September 20, researchers in New Mexico finally matched the *E. coli* DNA to a field at Paicines Ranch in central California, farmed by Mission Organics, a grower that sells to Dole. No one ever identified how the spinach was tainted, though it has been suggested that the produce was contaminated by *E. coli* in feces from wild pigs or nearby cattle.

Yet scientists faced a concern even more pressing than how the spinach had become contaminated. Before the 2006 outbreak was over, at least 5 people were dead and more than 205 others across 26 states had become very sick, with 15 percent of them suffering kidney failure. So why was this O157:H7 outbreak more harmful and infectious than prior outbreaks?

The bacteria appeared to have changed—to have evolved to become more dangerous. But how? A pair of scientists in Michigan was soon to find the answer: several small but important changes in the organism's DNA.

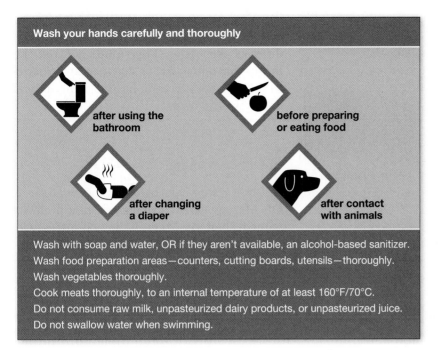

Wash your hands carefully and thoroughly

after using the bathroom

before preparing or eating food

after changing a diaper

after contact with animals

Wash with soap and water, OR if they aren't available, an alcohol-based sanitizer.
Wash food preparation areas—counters, cutting boards, utensils—thoroughly.
Wash vegetables thoroughly.
Cook meats thoroughly, to an internal temperature of at least 160°F/70°C.
Do not consume raw milk, unpasteurized dairy products, or unpasteurized juice.
Do not swallow water when swimming.

Figure 8.2

Staying healthy
Follow these simple steps from the Centers for Disease Control and Prevention to minimize your chance of an infection.

Defining DNA

DNA (**deoxyribonucleic acid**), the genetic code of life, is built from two parallel strands of repeating units called **nucleotides**. Each nucleotide is composed of the sugar deoxyribose, a phosphate group, and one of four bases: adenine, cytosine, guanine, or thymine. The bases of a single strand are connected by bonds between the phosphate group of one nucleotide and the sugar of the next nucleotide. The two strands are connected by bonds linking the bases on one strand to the bases on the other, like the rungs that connect the two sides of a ladder (**Figure 8.4**). The term **base pair** (or nucleotide pair) refers to two bases held together by one of these bonds; that is, a base pair is two nucleotides that form one rung of the ladder. The DNA ladder twists into a spiral called a **double helix** (**Figure 8.5**).

But nucleotides do not form base pairs willy-nilly. Adenine (A) on one strand can pair only with thymine (T) on the other strand (see Figure 8.5); similarly, cytosine (C) on one strand can pair only with guanine (G) on the other strand. These **base-pairing rules** have an important consequence:

when the sequence of bases on one strand of the DNA molecule is known, the sequence of bases on the other, complementary strand of the molecule is automatically known as well. The fact that A can pair only with T and that C can pair only with G allows the original strands to serve as "template strands" on which new strands can be built (see Chapter 9 for more on building new DNA strands).

Still, the four bases can be arranged in any order along a single strand of DNA, and each DNA strand is composed of millions of these bases, so a tremendous amount of information can be stored in a DNA sequence and in a genome. The O157:H7 genome has about 4.6 million base pairs, for example, while the human genome has about 3.2 *billion* base pairs. The sequence of bases in DNA differs between

Figure 8.3

The 2006 outbreak of *E. coli* was linked to spinach

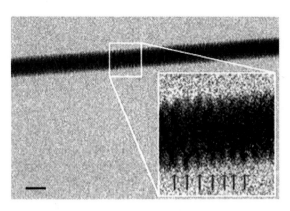

Figure 8.4

What DNA actually looks like

In November 2012, Italian researchers used an electron microscope to directly visualize DNA for the first time. This is the single thread of double-stranded DNA that they saw.

species and between individuals within a species, and these differences in genotype can result in different phenotypes (**Figure 8.6**). In *E. coli*, this means that one strain, or subtype, can be harmless while another strain has the ability to produce powerful toxins.

Tracking *E. coli*

In 2004, two years before the new strain of O157:H7 appeared, a young assistant professor named Shannon Manning joined forces with the late Thomas Whittam, a distinguished evolutionary biologist at Michigan State University (MSU), to use *E. coli* DNA to track bacterial infections around the state. The duo worked with the Michigan Department of Community Health to collect samples from hospitals across Michigan and made note of the different strains of *E. coli* that surfaced in different areas—creating a crime map of sorts for the bacteria. The samples steadily filled the lab. "Before we knew it, we had this huge collection of *E. coli*," says Manning.

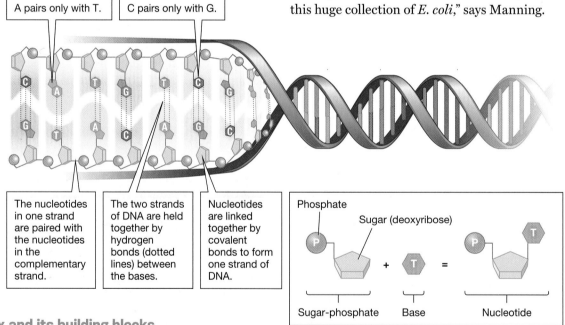

A pairs only with T.

C pairs only with G.

The nucleotides in one strand are paired with the nucleotides in the complementary strand.

The two strands of DNA are held together by hydrogen bonds (dotted lines) between the bases.

Nucleotides are linked together by covalent bonds to form one strand of DNA.

Phosphate

Sugar (deoxyribose)

P + T = P T

Sugar-phosphate | Base | Nucleotide

Figure 8.5

The DNA double helix and its building blocks

A molecule of DNA consists of two complementary strands of nucleotides that are twisted into a spiral around an imaginary axis, rather like the winding of a spiral staircase.

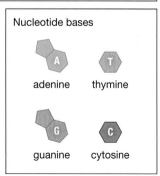

Nucleotide bases

A adenine T thymine

G guanine C cytosine

Q1: Name two base pairs.

Q2: Why is the DNA structure referred to as a ladder? What part of the DNA represents the rungs of the ladder? What part represents the sides of the ladder?

Q3: Is the hydrogen bond that holds the base pairs together a strong or weak chemical bond? Refer to Chapter 2 to review chemical bonds, if needed.

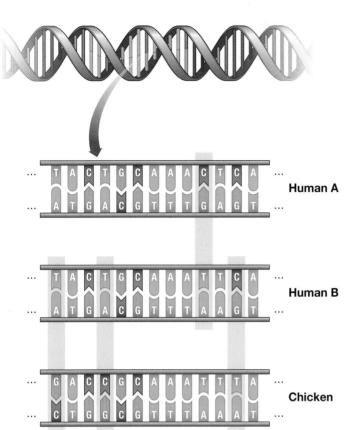

Figure 8.6

The sequence of bases in DNA differs among species and among individuals within a species

The sequence of bases in a hypothetical gene is compared for two humans (A and B) and a chicken. Base pairs highlighted in blue are variant; that is, they differ between the genes of persons A and B, and between the same gene in humans and chickens.

Q1: If all genes are composed of just four nucleotides, how can different genes carry different types of information?

Q2: Do you expect to see more variation in the sequence of DNA bases between two members of the same species (such as humans) or between two individuals of different species (for example, humans and chickens)? Explain your reasoning.

Q3: Do different alleles of a gene have the same DNA sequence or different DNA sequences?

So in 2006, when an individual came to a Michigan hospital with the new, virulent form of O157:H7, doctors there immediately sent a sample to MSU. "When the outbreak hit, it was scary," says Manning. "We thought, maybe this strain has evolved and acquired new characteristics we have never seen before."

To find out how the new strain was different, Manning compared it to their large collection of *E. coli*. First, her team sequenced its DNA, literally reading the genetic code base by base. It is hard to read the DNA sequence of a small amount of genetic material, so they copied, or "replicated," the DNA, generating thousands to millions of copies of a particular sequence, to better "read" that piece of DNA.

DNA replication is the duplication of a DNA molecule. DNA replication is ongoing in our bodies: it occurs right before a cell enters mitosis, so that there is a copy of the DNA to pass along to the new cell. Cells replicate DNA in three steps:

1. The DNA molecule unwinds, and special proteins break the hydrogen bonds connecting the two strands of DNA.

2. Each strand is then used as a template for the construction of a new strand of DNA. **DNA polymerase**—a key enzyme in the replication of DNA—builds the two new strands of DNA.

3. When construction is completed, there are two identical copies of the original DNA molecule. Each copy is composed of a template strand of DNA (from the original DNA molecule) and a newly synthesized strand of DNA.

SHANNON MANNING

A professor at Michigan State University, Shannon Manning studies the evolutionary genetics of infectious diseases. In 2008 she was part of a team that sequenced the DNA from many strains of *E. coli*, including the 2006 outbreak strain, and determined why some strains of *E. coli* are more dangerous than others.

This mode of replication is known as **semiconservative replication** because one "old" strand (the template strand) is retained, or "conserved," in each new double helix (**Figure 8.7**).

The mechanics of copying DNA are far from simple. More than a dozen enzymes and proteins are needed to unwind the DNA, to stabilize the separated strands, to start the replication process, to attach nucleotides to the correct positions on the template strand, to "proofread" the results, and to join partly replicated fragments of DNA to one another.

Despite the complexity of this task, cells can copy DNA molecules containing billions of nucleotides in a matter of hours—about eight hours in humans (over 100,000 nucleotides per second). This speed is achieved in part by starting the replication of the DNA molecule at thousands of different places at once.

In the 1980s, scientists developed a way to mimic natural DNA replication in a test tube, and just as fast. The **polymerase chain reaction**, or **PCR**, is a technique that makes it possible to produce millions of copies of a DNA sequence—to "amplify" the DNA—in just a few hours, even with a small initial amount of DNA (**Figure 8.8**). Instead of special proteins, PCR relies on heat to cause the DNA to unwind and its strands to separate. Then DNA polymerase uses loose nucleotides in a solution to build two new strands of DNA. Those steps are repeated 25–40 times to produce millions of exact copies of the target region of DNA.

Manning and Whittam used PCR to amplify the O157:H7 DNA in the sample they had received. Then they put that amplified DNA into a DNA sequencer, an instrument that produces a data file listing the complete sequence of nucleotides (**Figure 8.9**). The team compared the outbreak's

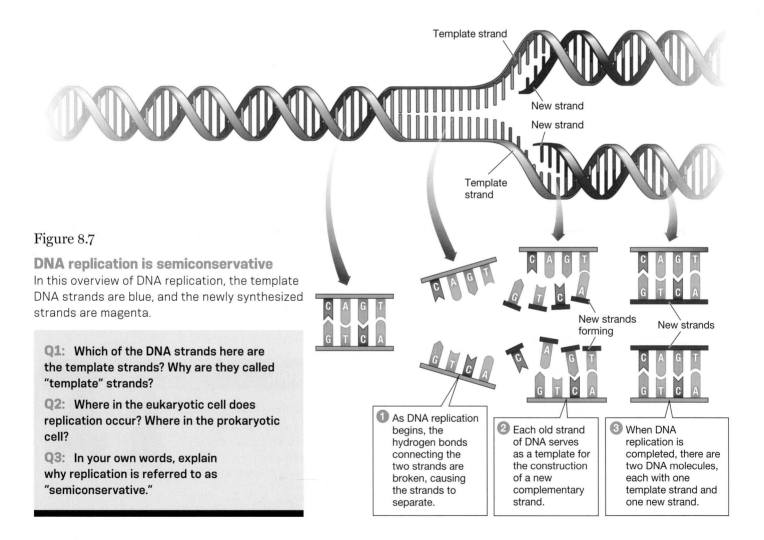

Figure 8.7

DNA replication is semiconservative

In this overview of DNA replication, the template DNA strands are blue, and the newly synthesized strands are magenta.

Q1: Which of the DNA strands here are the template strands? Why are they called "template" strands?

Q2: Where in the eukaryotic cell does replication occur? Where in the prokaryotic cell?

Q3: In your own words, explain why replication is referred to as "semiconservative."

Template strand

New strand

New strand

Template strand

New strands forming

New strands

❶ As DNA replication begins, the hydrogen bonds connecting the two strands are broken, causing the strands to separate.

❷ Each old strand of DNA serves as a template for the construction of a new complementary strand.

❸ When DNA replication is completed, there are two DNA molecules, each with one template strand and one new strand.

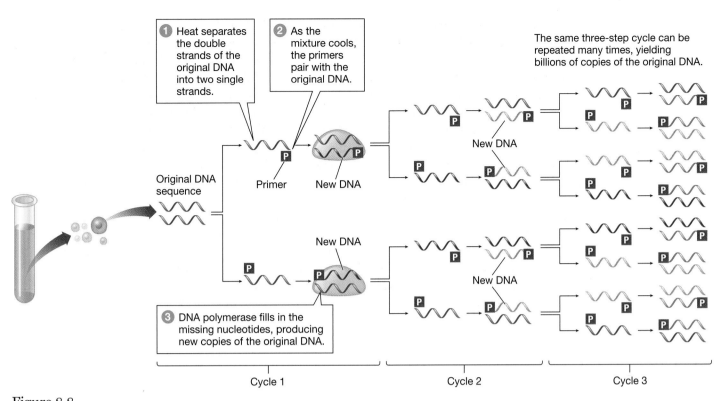

Figure 8.8

PCR can amplify small amounts of DNA more than a millionfold

Short primers consisting of synthetic DNA segments are mixed in a test tube with a sample of the target DNA, the enzyme DNA polymerase, and all four nucleotides (A, C, G, and T). The primers pair with the two ends of a gene of interest. A machine then processes the mixture and doubles the number of double-stranded versions of the template sequence. The doubling process can be repeated many times (only three cycles are shown here).

Q1: PCR replicates DNA many times to increase the amount available for analysis. Why is this process called "amplification"?

Q2: Why are DNA primers necessary for this process?

Q3: In your own words, explain how PCR is used to identify a strain of *E. coli*.

O157:H7 DNA sequence to numerous other O157:H7 genome sequences. "We looked to see what might be present in this strain that we hadn't seen in prior outbreak strains," says Manning.

Instead of looking at every base in the 4.6-million-base-pair *E. coli* genome, which would take a lot of time and effort, Manning and Whittam analyzed just 83 genes to develop a family tree of relationships among *E. coli* strains. *E. coli*, they found, can be separated into nine "clades," or closely related groups. The researchers linked those genetic data with patient symptom data to assess whether this new strain and

genetically similar strains were more likely to cause severe disease.

E. coli makes people sick when it produces toxins called Shiga toxins, which devilishly block protein synthesis in cells, causing cell death. Whittam and Manning found that different clades produce different kinds of Shiga toxins in different amounts, depending on their DNA composition. "For the first time, we know why some outbreaks cause serious infections and diseases and others don't," Whittam said in 2008.

One of those groups, clade 8, harbored particularly dangerous strains of *E. coli*, including

Figure 8.9

Machines sequence DNA

DNA sequencing machines can rapidly determine the nucleotide sequence of a DNA fragment. Here, a scientist examines a computer display showing DNA sequencing results. Each of the four nucleotide bases in DNA is represented by a different color (red, green, blue, or yellow).

Q1: Which of the computer screens in this photograph—the one in front of or the one behind the scientist—is displaying a more variable portion of DNA?

Q2: Why is it important to know the particular strain of *E. coli* in an outbreak?

Q3: Manning and Whittam did not analyze every base in the O157:H7 *E. coli* genome. Why not?

those most likely to cause kidney disease. "That suggested that there is perhaps something unique about this subgroup of O157 strains," says Manning. When Manning and Whittam examined the DNA from the spinach outbreak strain, they found, unsurprisingly, that it fit squarely into clade 8.

But how did *E. coli* branch into all these different clades with different genotypes and phenotypes, evolving to become more virulent? It mutated.

Mutant *E. coli*

When DNA is copied right before mitosis occurs in a cell, there are many opportunities for mistakes to be made. The enzymes that copy DNA sometimes insert an incorrect base in the newly synthesized strand. In addition, DNA in cells is constantly being damaged by chemical, physical, and biological agents, including energy from radiation or heat, collisions with other molecules in the cell, attacks by viruses, and random chemical accidents (some of which are caused by environmental pollutants, but most of which result from normal metabolic processes).

Replication errors and damage to DNA—especially to essential genes—disrupt normal cell functions. If not repaired, DNA damage leads to malfunctioning proteins, such as Felix's WAS protein in Chapter 7. DNA damage can also cause the death of cells and, ultimately, the death of an organism. Thankfully, cells have a way to recover: DNA polymerase immediately corrects almost all mistakes during DNA replication, "proofreading" pair bonds as they form.

DNA polymerase is not infallible. When an incorrect base is added but escapes proofreading by DNA polymerase, a mismatch error has occurred. This happens about once in every 10 million bases. But cells have another backup safety program: repair proteins that correct 99 percent of mismatch errors, reducing the overall chance of an error to one mistake in every *billion* bases (**Figure 8.10**).

On the rare occasions when a mismatch error is not corrected by repair proteins, the DNA sequence is changed, and the new sequence is reproduced the next time the DNA is replicated. A change to the sequence of bases in an organism's DNA is called a **mutation**. The extent of a mutation can range from the change of just one base in a single base pair (known as a **point mutation**) to the addition or deletion of one or more whole chromosomes (a chromosomal abnormality, covered in more detail in Chapter 7). Some change must have occurred in the O157:H7 genome to make it more dangerous to humans.

Three types of mismatch mutations can alter a gene's DNA sequence: substitutions, insertions, and deletions. In a **substitution** point mutation, one base is substituted for another in the DNA sequence of the gene. **Insertion** or **deletion** point mutations occur when a base is inserted into or deleted from a DNA sequence. Sickle-cell disease, a human genetic blood disorder, is caused by a substitution point mutation (**Figure 8.11**).

Insertions and deletions can be point mutations or can involve more than one base; sometimes thousands of bases may be added or deleted. Large insertions and deletions almost always result in the synthesis of a protein that cannot function properly. Sometimes, however, changing a few bases in a gene's DNA sequence has little or no effect. In such cases, a mutation is said to be "silent" because it produces no change

Figure 8.10

Repair proteins fix DNA damage

Large complexes of DNA repair proteins work together to fix damaged DNA.

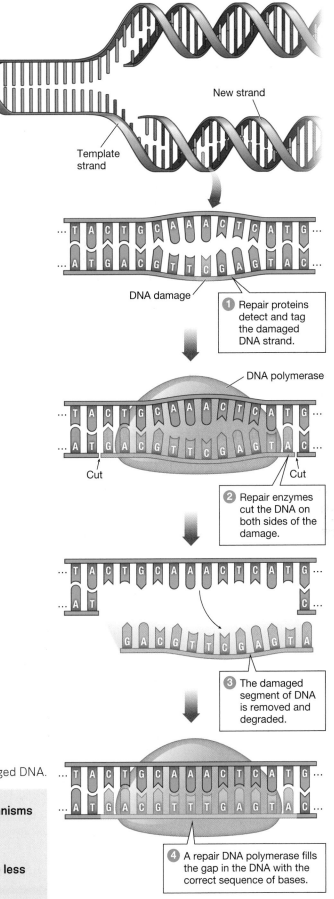

New strand

Template strand

DNA damage

1 Repair proteins detect and tag the damaged DNA strand.

DNA polymerase

Cut   Cut

2 Repair enzymes cut the DNA on both sides of the damage.

3 The damaged segment of DNA is removed and degraded.

4 A repair DNA polymerase fills the gap in the DNA with the correct sequence of bases.

Q1: Summarize how DNA repair works and why the repair mechanisms are essential for cells and whole organisms to function normally.

Q2: Is DNA repair 100 percent effective?

Q3: What would happen to an organism if its DNA repair became less effective?

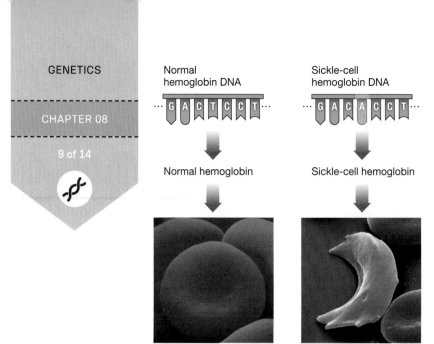

Normal hemoglobin DNA

...G A C T C C T...

Normal hemoglobin

Normal red blood cells

Sickle-cell hemoglobin DNA

...G A C A C C T...

Sickle-cell hemoglobin

A sickled red blood cell

Figure 8.11

A point mutation in the hemoglobin gene leads to sickle-cell disease

In people with the genetic disorder sickle-cell disease, a single base in the gene that makes hemoglobin, an important protein involved in oxygen transport in red blood cells, is altered. The red blood cells of people with sickle-cell disease become curved and distorted under low-oxygen conditions and can clog blood vessels, leading to serious effects, including heart and kidney failure.

Q1: What are the three types of point mutations?

Q2: Sickle-cell disease is an autosomal recessive genetic disorder. How many mutated hemoglobin alleles do people with sickle-cell disease have?

Q3: Because of improved treatments, individuals with sickle-cell disease are now living into their forties, fifties, or longer. How might this extension of life span affect the prevalence of sickle-cell disease in the population?

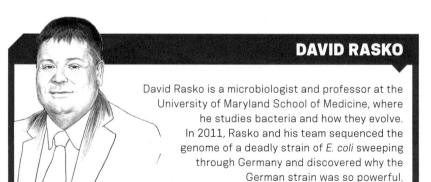

DAVID RASKO

David Rasko is a microbiologist and professor at the University of Maryland School of Medicine, where he studies bacteria and how they evolve. In 2011, Rasko and his team sequenced the genome of a deadly strain of *E. coli* sweeping through Germany and discovered why the German strain was so powerful.

in the function of the protein, and therefore no change in the phenotype of the organism.

If the mutation occurs within a gene, it will result in the formation of a new allele. Most new alleles are either neutral or harmful, but occasionally a mutation may be beneficial, as in the case of O157:H7 (beneficial to the *E. coli*, that is—not to its victims). *E. coli*, like other prokaryotes, have a genome that is quite different from that of eukaryotes (**Figure 8.12**). This allows them to share genes in two ways. First, they can acquire genes from other types of bacteria via small viruses called bacteriophages, a name often shortened to "phages." At some point during the evolution of O157:H7, it was infected by a phage carrying a Shiga toxin gene with a beneficial mutation, and incorporated that gene into its genome.

Second, *E. coli* share new mutations with nearby *E. coli* using a trick called **horizontal gene transfer** to rapidly spread new alleles to each other (see the photo in Figure 8.12). In horizontal gene transfer, the bacteria transfer genes on tiny DNA loops called plasmids from one bacterium to another, like a Frisbee tossed back and forth. Once a single bacterium has a new allele of a gene, it can rapidly transfer that allele to many other bacteria. That's how *E. coli* rapidly evolves to get better at its job—infecting other organisms.

Sprout Sickness

In May 2011, five years after the spinach outbreak in the United States, the first reports of a massive *E. coli* outbreak in Germany began to surface. Hundreds of individuals began to develop kidney failure. Many types of food were blamed before the source was identified. Surprisingly, this time the culprit wasn't O157:H7. It was a new strain of the bacterium, dubbed O104:H4, and it followed an unusual pattern. The bacterium affected adults of all ages and mainly women, rather than primarily young children and the elderly, as in previous *E. coli* outbreaks.

At the University of Maryland School of Medicine, microbiologist David Rasko and his colleagues were one of the first teams to sequence the genome of the German outbreak. Like Manning and Whittam, Rasko recognized that the

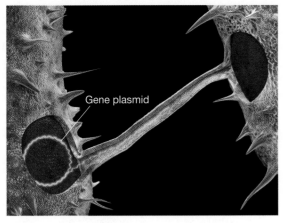

Gene plasmid

Prokaryotic genomes

Typically have several million base pairs of DNA.
Hold DNA in a single chromosome.
Share DNA via horizontal gene transfer (pictured above).
Genes tend to be organized by function and are turned on or off as one unit.
Most of the DNA codes for proteins.

Eukaryotic genomes

Most have hundreds of millions to billions of base pairs of DNA.
Distribute DNA among multiple chromosomes (pictured above).
Horizontal gene transfer is uncommon.
Most genes with related functions are *not* grouped near one another on chromosomes, or even on the same chromosome.
Much of the DNA does not code for proteins.

Figure 8.12

Prokaryotic and eukaryotic genomes differ in many ways

Q1: Why does horizontal gene transfer enable *E. coli* to evolve more quickly than it could by cell division alone?

Q2: In your own words, summarize the major differences between the genomes of prokaryotes and eukaryotes.

Q3: What is the advantage of having a large genome, with DNA spread out among many chromosomes, as in eukaryotes? What is the advantage of having a small genome on a single chromosome, as in prokaryotes?

best way to understand the genome sequence was to compare it to others. "I told my collaborators we weren't sequencing a single genome," says Rasko. "Genomes by themselves are interesting, but they provide no context."

Rasko's team sequenced the genome from the new strain of *E. coli* in a week, running three different DNA sequencing machines 24 hours a day, and then compared it to seven other O104 strains and numerous other *E. coli* strains, including O157:H7. The new German strain, they saw, was

unique because it combined two powerful traits of *E. coli* that had never previously been found together: the abilities to produce Shiga toxins *and* to form brick-like walls of bacteria in the intestine. The strain also had genes that made it antibiotic resistant, again acquired through mutation and horizontal gene transfer from other bacteria. All told, the new strain of *E. coli* marauding across Germany was a stew of dangerous traits.

The 2011 outbreak, which was ultimately traced back to sprout seeds in Egypt, lasted 3

Grocery Cart Outbreaks

Foodborne illness outbreaks are notoriously hard to solve and can occur in a wide range of foods, including everyday staples like peppers and peanut butter. Here are some of the biggest culprits of recent years.

Jalapeños — 2008
Salmonella enterica

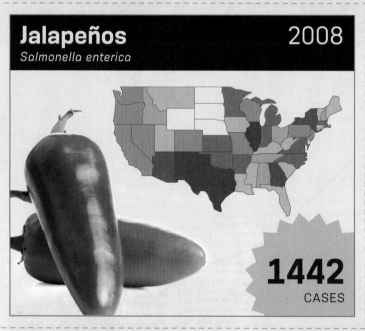

1442 CASES

Peanut Butter — 2009
Salmonella Typhimurium

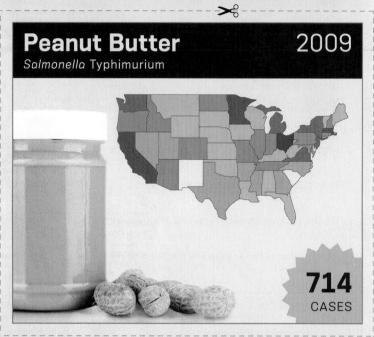

714 CASES

Peppered Salami — 2009
Salmonella Montevideo

272 CASES

Cantaloupe — 2012
Salmonella Typhimurium & *Salmonella* Newport

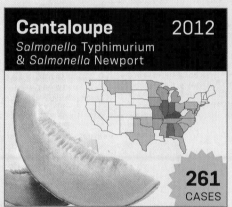

261 CASES

Spinach — 2006
Escherichia coli

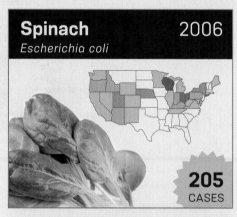

205 CASES

Cantaloupe — 2011
Listeria monocytogenes

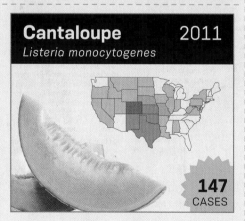

147 CASES

Ground Turkey — 2011
Salmonella Heidelberg

136 CASES

Legend

1-5 cases	6-10 cases	11-20 cases	21-40 cases	41-100 cases	100+ cases

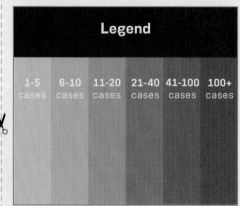

months and infected more than 3,900 people. Of these victims, 855 developed kidney failure, and 53 died. Since then, researchers have been on the lookout for other outbreaks of O104 (**Figure 8.13**). "Is this the emergence of another really nasty bug?" asks Manning. "Are we back to 1982? We'll have to wait and see."

With their shorter generation time and ability to share new genes through horizontal gene transfer, bacteria are able to evolve new ways to infect humans more quickly than we are able to evolve defenses against these infections. That is why scientists are constantly researching new ways to detect and classify emerging strains of *E. coli* and other food-borne pathogens, including a way to identify bacteria present in quantities too small to be found by current methods.

We need all the help we can get, says Rasko. "Bacteria are six steps ahead of us," he explains. "I expect outbreaks of different [strains] to continue to occur."

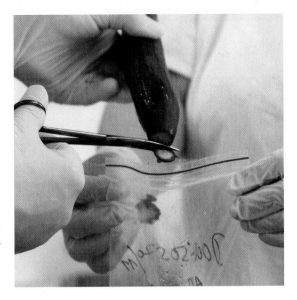

Figure 8.13

Scientists continue to research ways to monitor and control food-borne pathogens
Researchers collect tissue for DNA testing from a suspect cucumber during a recent *E. coli* breakout.

REVIEWING THE SCIENCE

- Genes are composed of **DNA**, which consists of two parallel strands of repeating units called **nucleotides** twisted into a **double helix**.

- The four nucleic bases are adenine (A), cytosine (C), guanine (G), and thymine (T). These bases pair together according to **base-pairing rules**: A can pair only with T, and C can pair only with G.

- **DNA replication** occurs in all living organisms prior to mitosis. The double helix unwinds, and the two strands break apart. Each strand of DNA serves as a template from which a new strand is copied. **DNA polymerase** builds each new strand of DNA.

- The **polymerase chain reaction**, or **PCR**, is a laboratory technique to amplify the DNA from a small initial amount to millions of copies. Amplified DNA can then be sequenced to examine specific genes or mutations.

- DNA is subject to damage by physical, chemical, and biological agents, and errors in DNA replication are common. DNA polymerase

"proofreads" the DNA during replication and corrects most mistakes. Repair proteins are a backup repair mechanism and correct any errors that DNA polymerase misses.

- A change to the sequence of bases in an organism's DNA is called a **mutation**. Three types of mismatch mutations can alter a gene's DNA sequence: **substitutions**, **insertions**, and **deletions**. If only a single base is altered, it is a **point mutation**.

- Prokaryotic and eukaryotic genomes differ in significant ways. Prokaryotes typically have several million base pairs of DNA, all contained in a single chromosome, and individuals share that DNA via horizontal gene transfer. Eukaryotes have hundreds of millions to billions of base pairs of DNA distributed among multiple chromosomes.

THE QUESTIONS

The Basics

1 DNA replication results in

(a) two DNA molecules—one with two old strands, and one with two new strands.

(b) two DNA molecules, each of which has two new strands.

(c) two DNA molecules, each of which has one old strand and one new strand.

(d) none of the above

2 The DNA of cells is damaged

(a) thousands of times per day.

(b) by collisions with other molecules, chemical accidents, and radiation.

(c) not very often and only by radiation.

(d) both a and b

3 The DNA of different species differs in the

(a) sequence of bases.

(b) base-pairing rules.

(c) number of nucleotide strands.

(d) location of the sugar-phosphate portion of the DNA molecule.

4 If a strand of DNA has the sequence CGGTATATC, then the complementary strand of DNA has the sequence

(a) ATTCGCGCA.

(b) GCCCGCGCT.

(c) GCCATATAG.

(d) TAACGCGCT.

5 Mutation

(a) can produce new alleles.

(b) can be harmful, beneficial, or neutral.

(c) is a change in an organism's DNA sequence.

(d) all of the above

6 Link each of the following terms with the correct definition.

NUCLEOTIDE   1. Two complementary bases joined by hydrogen bonds.

BASE PAIR   2. The nitrogen-containing component of a nucleotide; there are four variants of this component.

DNA MOLECULE   3. A strand of nucleotides linked together by covalent bonds between a sugar and a phosphate; two strands are linked by hydrogen bonds between complementary bases.

BASE   4. A phosphate, a sugar, and a nitrogen-containing base.

7 Label the diagram of replication shown here with the appropriate terms: (a) base pair, (b) base, (c) nucleotide, (d) template strand, (e) newly synthesized strand, (f) separating strands.

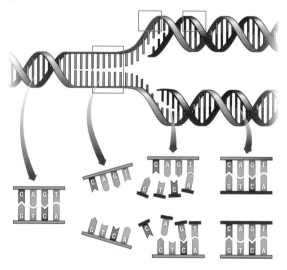

8 Circle the correct terms in the following sentences: To identify a new strain of *E. coli*, researchers first replicate the DNA many times using (**PCR, gene sequencing**). This process increases the amount of DNA so that they are able to determine the order of bases in the *E. coli* using (**PCR, gene sequencing**).

9 Place the following steps of DNA replication and repair in the correct order by numbering them from 1 to 5.

_____ a. A template strand begins to be replicated.

_____ b. If the incorrect base is not identified and replaced, it remains as a point mutation in the DNA.

_____ c. DNA polymerase identifies and replaces most incorrect bases with the correct base, complementary to the base on the template strand.

_____ d. An incorrect base is added to the growing strand of DNA.

_____ e. Proteins identify and replace any incorrect bases missed by DNA polymerase.

10 Given the template DNA sequence GCAGCATGTT, identify each of the following mutations to the complementary strand as an insertion (I), a deletion (D), or a substitution (S).

_____ a. CGTCGTACA

_____ b. CGTGGTACAA

_____ c. CGTCGTACTAA

11 Indicate whether each the following describes a prokaryote genome (P) or a eukaryote genome (E).

_____ a. Contains several million base pairs of DNA.

_____ b. DNA is distributed among several chromosomes.

_____ c. DNA is shared between individuals via horizontal gene transfer.

_____ d. Most of the DNA consists of genes for proteins.

_____ e. A large amount of the DNA is noncoding.

_____ f. Within a gene, there is little noncoding DNA.

_____ g. Genes with related functions are separated across the genome.

_____ h. Genes with related functions are turned on and off as a unit.

Try Something New

12 Using base-pairing rules to guide you, for a DNA double helix that contains 20 percent adenine (A), specify the percentage of (a) thymine, T; (b) guanine, G; and (c) cytosine, C.

13 Each human cell contains over 1,000 times as much DNA as *E. coli* bacteria have. Do we have over 1,000 times as many genes as *E. coli* has? Explain.

14 Scientists estimate that genes encoding proteins make up less than 1.5 percent of the human genome. Other genes in our cells encode different types of nonprotein RNA molecules. The rest of our genome consists of various types of **noncoding DNA**, defined as DNA that does not code for any kind of functional RNA. Some of the remaining DNA has regulatory functions—for example, controlling gene expression. Some of it has architectural functions, such as giving structure to chromosomes or positioning them at precise locations within the nucleus. Since these sections of DNA are noncoding, does it matter if replication errors occur within them? Explain your reasoning.

Leveling Up

15 *Write Now* **biology: pink slime or lean finely textured beef?** Your aunt has e-mailed you a link to a 2009 *The New York Times* article on ammonia-treated beef being used in school lunches. She tells you that her friends refer to it as "pink slime" and the beef industry calls it LFTB, short for "lean finely textured beef." Her friends are worried that it will increase the risk of their children becoming infected with pathogenic *E. coli*, *Salmonella*, or another food-borne pathogen. But the industry websites she visited say that LFTB is safer than regular ground beef because it is treated with ammonia to kill any pathogens. She asks you to research the matter for her and recommend whether she should begin sending your young cousins to school with home-packed lunches rather than allowing them to eat cafeteria food.

Write a letter to your aunt addressing the following points (using about a paragraph for each one).

(a) Explain exactly how the beef (LFTB, or "pink slime") is produced and processed.

(b) Describe how food-borne pathogens end up in ground meats. Is LFTB more or less likely to harbor these pathogens? Explain why.

(c) If one of your cousins were to be sickened by pathogenic *E. coli*, what would be the potential risks?

(d) In your final paragraph, advise your aunt as to whether she should begin packing lunches for your cousins or allow them to continue eating what the school cafeteria serves. You may also provide an alternative recommendation. Justify your opinion with data and logic.

Tobacco's New Leaf

One company's quest to produce tomorrow's drugs in today's plants.

After reading this chapter you should be able to:

- Explain how an organism uses gene expression to create a phenotype from a genotype.
- Caption a diagram of transcription, identifying each step in the process and the relevant molecules.
- Caption a diagram of translation, identifying each step in the process and the role of each type of RNA.
- Determine the correct amino acid from a given codon.
- Explain why the genetic code is universal.
- Describe how a cell can increase or decrease its expression of particular genes, and why this ability is important for an organism.

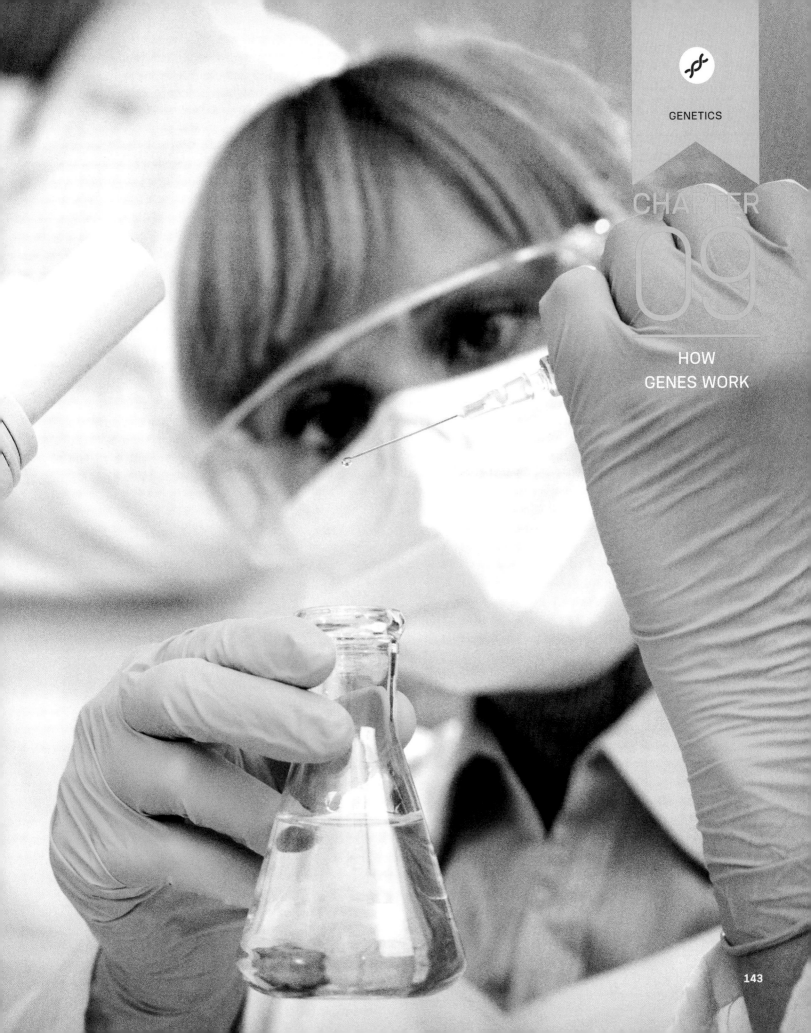

CHAPTER
09

HOW
GENES WORK

The greenhouse is vast, the size of half a football field. A loud, steady thrum reverberates in the room as massive metal fans push air through the hot, humid space. Mike Wanner, a tall, serious man with gray hair and piercing brown eyes, walks through rows and rows of leafy plants, each a foot and a half tall. He stoops to rub a leaf between his fingers, then raises his hand to his nose and sniffs. "That's what tobacco smells like," Wanner shouts, straining to be heard over the noise of the fans.

Wanner is vice president of operations at Medicago, a Canadian biotechnology company growing tobacco at a facility outside Durham, North Carolina, once home to Lucky Strike cigarettes (**Figure 9.1**). But these plants are not being grown to smoke, chew, or dip. Instead, they serve a dramatically different purpose: this tobacco will be used to make flu vaccines.

Influenza vaccines are normally grown in chicken eggs—a process that can take months. As an alternative, companies and researchers have begun experimenting with "biopharming"—manufacturing vaccine proteins in plants. Since most genes contain instructions for building proteins, scientists insert a gene that codes for the protein that interests them. This production of a protein from a gene occurs via **gene expression**—the process by which genes are transcribed into RNA and then translated to make proteins (**Figure 9.2**).

Proteins participate in virtually every process inside and between cells, so gene expression is the fundamental way by which genes influence the structure and function of a cell or organism. It is the process through which an organism's genotype gives rise to its phenotype (see Chapter 6). All prokaryotes, eukaryotes, and viruses utilize gene expression.

In addition to being involved in every process in a cell, proteins are the key components of some vaccines. Injected pieces of a virus's proteins activate the human immune system to defend against that virus in the future (**Figure 9.3**). When the body's immune system is exposed to such a vaccine, it recognizes the viral protein as an invader and mounts an attack against it. Upon later exposure to the actual flu virus, a vaccinated individual's immune system is already armed and bristling. The protein itself can't cause the flu, because it does not contain genetic material.

Figure 9.1

Growing tobacco to treat the flu
Tobacco plants grow in a Medicago greenhouse in Durham, North Carolina.

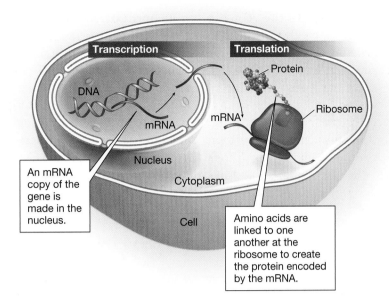

Transcription

DNA

mRNA

An mRNA copy of the gene is made in the nucleus.

Nucleus

Cytoplasm

Cell

Translation

Protein

mRNA

Ribosome

Amino acids are linked to one another at the ribosome to create the protein encoded by the mRNA.

Figure 9.2

An overview of gene expression

Genetic information flows from DNA to RNA to protein during gene expression, which occurs in two steps: first *transcription*, and then *translation*. The transcription of a protein-coding gene produces an mRNA molecule, which is then transported to the cytoplasm, where translation occurs and the protein is made with the help of ribosomes.

To make vaccines, scientists must produce large quantities of viral proteins. The traditional method is to inject chicken eggs with a virus, let the virus multiply in the chicken cells, and then extract the virus, remove its genetic material, and prepare a vaccine from the leftover viral proteins. Unlike chicken eggs, however, plants can be grown in vast quantities, and they grow rapidly—often in just days or weeks. "The big advantage of plant systems is that they can produce massive amounts of proteins very inexpensively," says James Roth, director of the Center for Food Security and Public Health at Iowa State University, who studies biopharming in plants and animals and is not associated with Medicago. In the event of a pandemic flu outbreak, Medicago could produce vaccines 6 times faster and 12 times cheaper than traditional egg manufacturing can, the company states (**Figure 9.4**).

Figure 9.3

How vaccines work

A vaccine trains the body's immune system to fight infection.

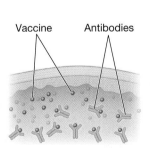

Vaccine

A vaccine with a harmless form of a virus (or other organism) is injected under the skin.

Vaccine   Antibodies

The vaccine stimulates the immune system to produce antibodies (in green) that recognize the virus.

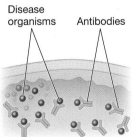

Disease organisms   Antibodies

When the individual is exposed to the virus after vaccination, the new antibodies are primed to attack and destroy the invader.

Q1: Describe in one sentence how a vaccine creates immunity to a virus.

Q2: Why is it impossible to become infected with a virus from a vaccine composed of viral proteins?

Q3: Natural immunity occurs without a vaccine, just by being exposed to a particular stimulus, like the chicken pox virus. Explain why people don't get chicken pox twice.

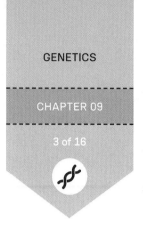

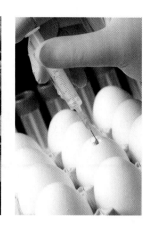

	Tobacco	Egg
Speed from outbreak to vaccine production:	1 month	6 months
Cost to produce 50 million flu vaccines:	$36 million	$400 million
Allergy risks:	Minimal	Individuals with egg allergies cannot receive the vaccine
Availability:	Not yet approved by the Food and Drug Administration	Approved and currently the most widely used technology to produce flu vaccines

Figure 9.4

Tobacco or egg?

Using tobacco plants to produce proteins for influenza vaccines has several advantages over the traditional approach using chicken eggs.

Q1: Why do you think biopharming with plants is faster than biopharming with eggs?

Q2: Why is speed of production so important for vaccines?

Q3: Why must tobacco-derived vaccines, or any new medications for that matter, be approved by the FDA?

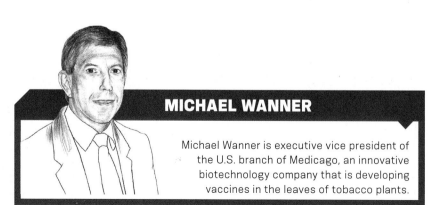

MICHAEL WANNER

Michael Wanner is executive vice president of the U.S. branch of Medicago, an innovative biotechnology company that is developing vaccines in the leaves of tobacco plants.

In April 2012, Medicago put its tobacco plants to the test, running the North Carolina manufacturing facility at full tilt for 30 days to see how many doses of vaccine it could produce from the tobacco in a month. The U.S. Department of Defense had given the company millions of dollars to test whether it could quickly produce enough pandemic flu vaccine to stem an outbreak. The pressure is on, says Wanner, standing inside the facility. He looks out over his crop. "We'll see what happens."

Fighting the Flu with Tobacco

In 1997, Louis-Phillippe Vézina, then a research scientist at Agriculture and Agri-Food Canada and an associate professor at Université Laval in Quebec City, Canada, decided to start a company. A plant biotechnologist, Vézina wanted to explore the possibility of manufacturing proteins in alfalfa plants, and thus he called his company Medicago, the name of the genus to which alfalfa belongs. Later, as the company grew, Vézina and his team discovered that tobacco produces higher yields of proteins in a shorter time frame than alfalfa does, so the company switched plants.

By 2005, Medicago had begun to receive calls asking about its product and when it would begin clinical trials, the first step toward getting a drug approved by the U.S. Food and Drug Administration (FDA). "I was surprised. Even the FDA was asking, 'When are we going to see your vaccine?'" recalls Vézina. "They were keen on technologies like this, because they know how much vaccines cost, and they're interested in anything that can decrease that cost."

One of the first vaccines that Medicago produced was a vaccine for influenza virus H1N1, or swine flu, the most common cause of the flu in 2009 (**Figure 9.5**). In the previous swine flu pandemic, it had taken months for vaccines grown in chicken eggs to reach the market. In contrast, Medicago produced its vaccine lots, ready for testing, in just 19 days after receiving the H1N1 genetic sequence from government officials.

To make a flu vaccine in plants, Vézina and his team use genetic engineering in much the same way that it was used to create a gene therapy for Wiskott-Aldrich syndrome, as described in Chapter 7. They identify and synthesize a single viral gene that codes for hemagglutinin, a protein found on the surface of flu viruses, including H1N1. To make large quantities of the protein, the scientists at Medicago first insert the hemagglutinin gene into small, rod-shaped bacteria called *Agrobacterium*, which infect plants. Loaded with the hemagglutinin gene, these "agrobacteria" are then exposed to the tobacco plants: a robotic arm lifts a tray containing 5-week-old tobacco plants secured to the surface, flips the tray upside down, and dips the plants into a liquid solution swimming with the bacteria. Once the plants are immersed in the liquid, the technicians turn on a vacuum, sucking air out of the leaves and pulling bacteria into the leaves—like dipping a sponge into water, squeezing it, and then releasing it to soak up the liquid.

When the plants are returned to the greenhouse, their leaves are floppy and almost translucent, like wet tissue paper. Now things are cooking: once the bacteria are inside the plant leaf, the bacterial cells release the hemagglutinin gene into the plant cell, where it is transported into the nucleus to begin the process of gene expression—making a protein from DNA via the two steps of transcription and translation.

"When you get the *Agrobacterium* solution into the leaf, it will invade the cells and there will be a burst of gene expression in the plant cell," said Vézina. "The plant takes over and uses its machinery to produce the protein." (**Figure 9.6.**)

Two-Step Dance, DNA to Protein: *Transcription*

The first step in gene expression—in this case, for a flu gene in a plant cell—is **transcription**, the synthesis of RNA based on a DNA template. In the nucleus, an enzyme called **RNA polymerase** binds to a segment of DNA near the beginning of the gene, called a **promoter**. The promoter contains a specific sequence of DNA bases that the RNA polymerase recognizes and binds. At

Figure 9.5

Swine flu

In 2009, people around the world donned masks as a precaution against influenza virus H1N1, nicknamed "swine flu" because it contained DNA from bird, swine, and human flu viruses. The virus caused a pandemic, killing an estimated 284,500 people.

Medicago, scientists attach specific promoters to the hemagglutinin gene so that the plant cell's RNA polymerase can identify the gene and actively transcribe it, maximizing the rate of transcription, says Vézina.

Once bound to the promoter, the RNA polymerase unzips the DNA double helix at the beginning of the gene, separating a short portion of the two strands. Only one of the two DNA strands is used as a template and thus is called the **template strand**. The RNA polymerase begins to move down the DNA template strand, constructing a **messenger RNA (mRNA)** molecule, a strand of nucleotides complementary to the DNA, from

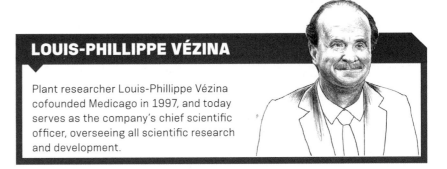

LOUIS-PHILLIPPE VÉZINA

Plant researcher Louis-Phillippe Vézina cofounded Medicago in 1997, and today serves as the company's chief scientific officer, overseeing all scientific research and development.

free nucleotides floating around in the nucleus (**Figure 9.7**). RNA does not have all the same bases as DNA: its four bases are adenine (A), cytosine (C), guanine (G), and uracil (U). Those bases pair with the four DNA bases according to the following rules: A pairs with T, C pairs with G, G pairs with C, and U pairs with A.

Part of the reason that tobacco cells produce so much hemagglutinin so quickly is that the hemagglutinin gene inserted into the cells contains a special DNA sequence that triggers multiple RNA polymerases to transcribe a hemagglutinin gene at a single time. As for any gene, as an RNA polymerase moves away from the promoter and travels

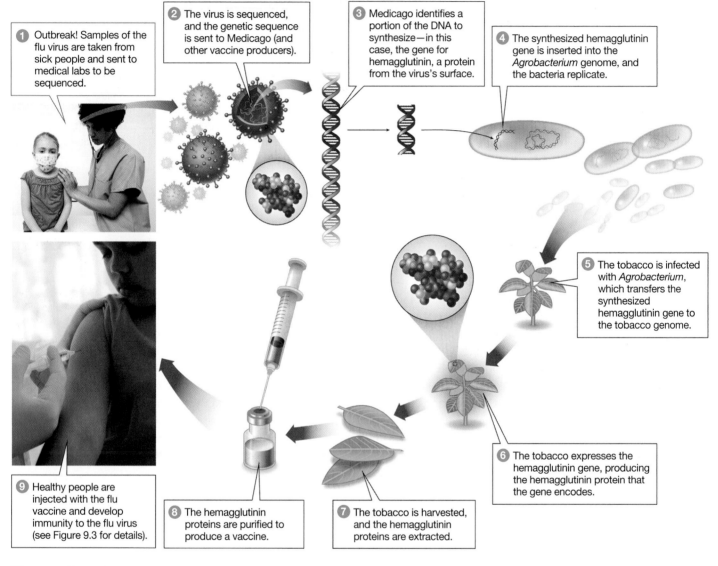

Figure 9.6

From outbreak to vaccine

During flu season or any other viral outbreak, a large network of medical professionals and scientists is activated to create a vaccine quickly and effectively.

Q1: In which of the steps illustrated here does DNA replication occur? In which steps does gene expression occur?

Q2: Why do vaccine producers not simply replicate the entire viral genome? Why do they instead isolate the gene for one protein and replicate only that gene?

Q3: What role do the bacteria play in this process? Why are they needed?

down the template strand, another RNA polymerase can bind at the promoter and start synthesizing a second mRNA on the heels of the first. At any given time, therefore, many RNA polymerases can be traveling down a DNA template simultaneously, each synthesizing an mRNA.

Transcription stops when the RNA polymerase reaches a special sequence of bases called a **terminator**. In eukaryotic cells, the mRNA then undergoes an elaborate sequence of modifications that prepare it to leave the nucleus. These steps include chemical modification of both ends of the mRNA, as well as a process called RNA splicing.

Most eukaryotic genes (and many viral genes) are embedded with stretches of sequences that don't code for anything, called **introns**. The stretches of DNA in a gene that carry instructions for building the protein are called **exons**. Because of this patchwork construction, with genes made of introns and exons, newly transcribed mRNA (pre-mRNA) is also a patchwork of coding sequences intermixed within noncoding sequences. During **RNA splicing**, the introns are snipped out of a pre-mRNA and the remaining pieces of mRNA—the exons—are joined to generate the mature mRNA (**Figure 9.8**). This mRNA is then ready to leave the nucleus.

To review, transcription occurs when RNA polymerase binds to a promoter, unzips the DNA helix, and constructs a strand of mRNA based on the DNA template strand. Transcription ends at the terminator sequence, and the mRNA is then processed, at which time noncoding introns are spliced out of the sequence.

Two-Step Dance, DNA to Protein: *Translation*

The microscopic molecular dance inside the tobacco cells continues with translation. Once the hemagglutinin gene has been transcribed into mRNA in the nucleus, it is time to make the protein—the actual product that will be extracted from the tobacco leaves. First the mRNA is transported from the nucleus, where it was made, to the site of protein synthesis: the **ribosomes** in the cytoplasm. To escape the nucleus, the long

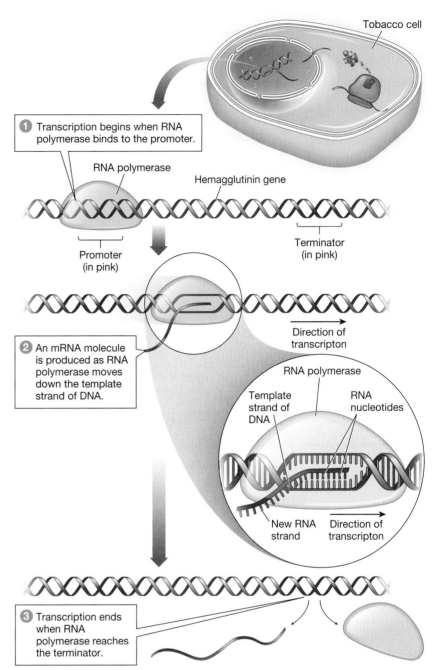

Figure 9.7

Plants making proteins. I: Transcription
RNA polymerase transcribes the hemagglutinin gene into a molecule of RNA.

Q1: Why is only one strand of DNA used as a template?

Q2: If a mutation occurred within the promoter or terminator region, do you think it would affect the mRNA transcribed? Why or why not?

Q3: The template strand of a gene has the base sequence TGAGAAGACCAGGGTTGT. What is the sequence of RNA transcribed from this DNA, assuming that RNA polymerase travels from left to right on this strand?

strand of mRNA passes through a nuclear pore, like a noodle slipping through the hole of a colander. Once the mRNA molecule arrives in the cytoplasm, the information it contains must be translated, with the help of ribosomes, from the language of mRNA (nitrogenous bases) to the language of proteins (amino acids). **Translation** is the process by which ribosomes convert the information in mRNA into proteins.

During translation, ribosomes "read" the mRNA code like a grocery list, and collect the corresponding amino acids (bread, milk, etc.), linking them in the precise sequence dictated by mRNA (**Figure 9.9**). Ribosomes read the mRNA information in sets of three bases at a time, and each unique sequence of three mRNA bases is called a **codon**. The average gene for hemagglutinin has about 1,770 bases, of which 1,695 code for the protein. That makes 565 codons (1,695 divided by 3), and therefore the hemagglutinin protein is composed of 565 amino acids.

There are 64 different ways that four bases (A, C, G, U) can be arranged to create a three-base sequence (because $4^3 = 64$). Therefore there are 64 possible codons (**Figure 9.10**). Most of the 64 codons specify a particular amino acid. A couple of amino acids are specified by only one codon, while other amino acids are specified by anywhere from two to six different codons. Some codons do not code for any amino acid and instead act as signposts that communicate to the ribosomes where they should start or stop reading the mRNA. The **start codon** (the codon AUG) is the ribosome's starting point on the mRNA strand, and there are three possible **stop codons** (UAA, UAG, or UGA). By beginning and ending at fixed points, the cell ensures that the mRNA message is read precisely the same way every time.

The information specified by all 64 possible codons is the **genetic code**. The genetic code has several significant characteristics. First, it is *unambiguous*: each codon specifies only one amino acid. It is also *redundant*: since there are a total of 64 codons but only 20 amino acids, several different codons call for the same amino acid, as already mentioned. Finally, the genetic code is virtually *universal*: nearly every organism on Earth uses the same code, from agrobacteria to tobacco cells to human cells—a feature that illustrates the common descent of all organisms.

Making a protein from an mRNA strand requires two additional types of RNA: The first is **ribosomal RNA (rRNA)**, which is an important component of ribosomes. The second is **transfer RNA (tRNA)**, which is the caddy for the process, delivering specific amino acids to the ribosomes as the codons are read off the mRNA "list".

There are many types of RNA. Each tRNA specializes in binding to a specific amino acid and recognizes and pairs with a specific codon in the mRNA, like a puzzle piece that fits one amino acid on one end and one codon on the other. At one end of a tRNA molecule, a special sequence of three

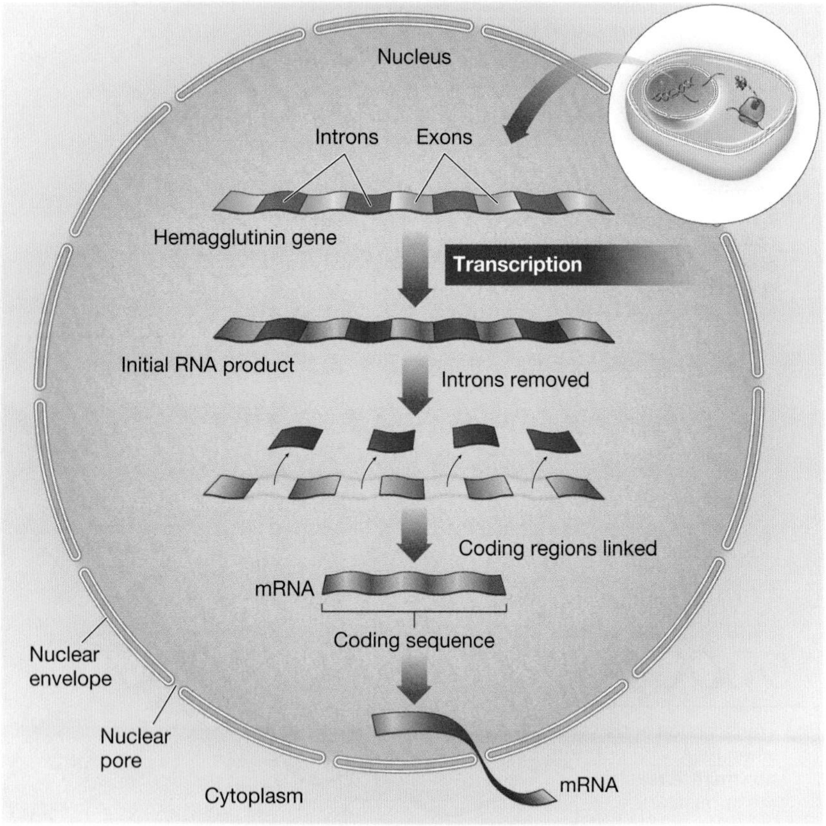

Figure 9.8

Processing mRNA for export to the cytoplasm
In eukaryotes, introns must be removed before an mRNA leaves the nucleus.

Q1: In your own words, define RNA splicing. When during gene expression does it occur?

Q2: What do you predict would happen if the introns were not removed from RNA before translation? Why would it be a problem if the introns were not removed?

Q3: Where is the mRNA destined to go once it has been transported out of the nucleus?

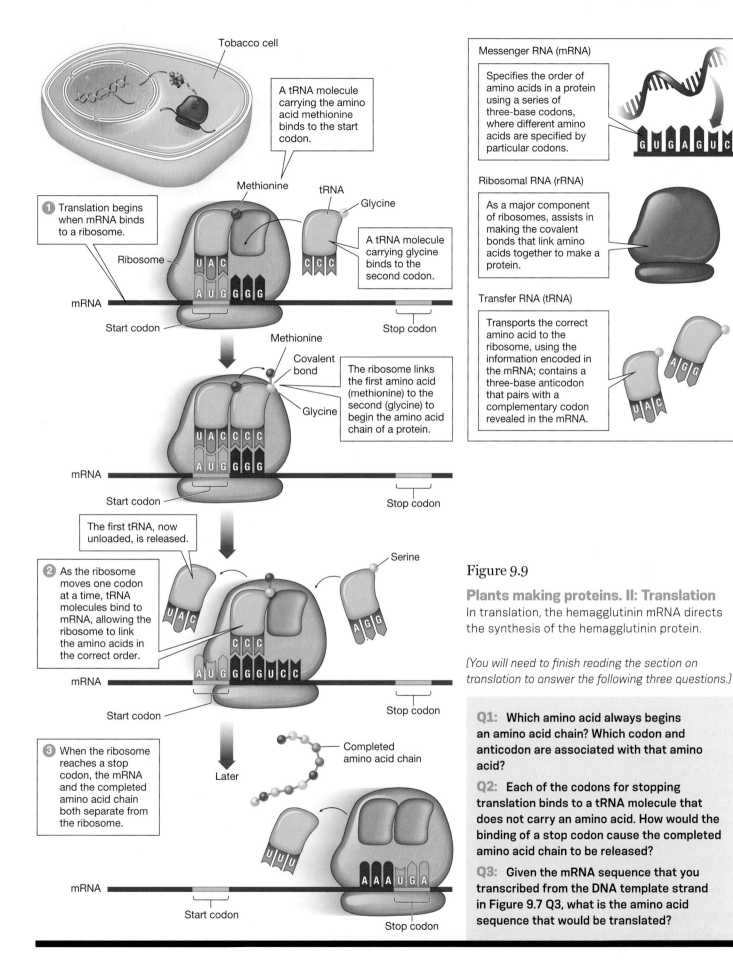

Tobacco cell

A tRNA molecule carrying the amino acid methionine binds to the start codon.

Messenger RNA (mRNA)

Specifies the order of amino acids in a protein using a series of three-base codons, where different amino acids are specified by particular codons.

GUGAGUC

Ribosomal RNA (rRNA)

As a major component of ribosomes, assists in making the covalent bonds that link amino acids together to make a protein.

Transfer RNA (tRNA)

Transports the correct amino acid to the ribosome, using the information encoded in the mRNA; contains a three-base anticodon that pairs with a complementary codon revealed in the mRNA.

Methionine

tRNA

Glycine

1 Translation begins when mRNA binds to a ribosome.

A tRNA molecule carrying glycine binds to the second codon.

Ribosome

U A C

C C C

mRNA

A U G G G G

Start codon

Stop codon

Methionine

Covalent bond

The ribosome links the first amino acid (methionine) to the second (glycine) to begin the amino acid chain of a protein.

Glycine

U A C C C C

mRNA

A U G G G G

Start codon

Stop codon

The first tRNA, now unloaded, is released.

Serine

2 As the ribosome moves one codon at a time, tRNA molecules bind to mRNA, allowing the ribosome to link the amino acids in the correct order.

U A C

A G G

C C C

mRNA

A U G G G G U C C

Start codon

Stop codon

3 When the ribosome reaches a stop codon, the mRNA and the completed amino acid chain both separate from the ribosome.

Completed amino acid chain

Later

U U U

mRNA

A A A U G A

Start codon

Stop codon

Figure 9.9

Plants making proteins. II: Translation
In translation, the hemagglutinin mRNA directs the synthesis of the hemagglutinin protein.

(You will need to finish reading the section on translation to answer the following three questions.)

Q1: Which amino acid always begins an amino acid chain? Which codon and anticodon are associated with that amino acid?

Q2: Each of the codons for stopping translation binds to a tRNA molecule that does not carry an amino acid. How would the binding of a stop codon cause the completed amino acid chain to be released?

Q3: Given the mRNA sequence that you transcribed from the DNA template strand in Figure 9.7 Q3, what is the amino acid sequence that would be translated?

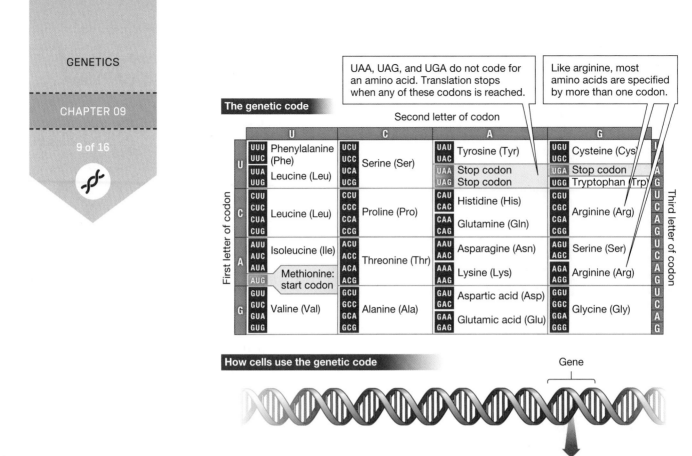

Figure 9.10

The genetic code

The genetic code at the top of the figure is composed of the 64 possible codons found in the mRNA. Each codon specifies an amino acid or is a signal that starts or stops translation. The genetic code is used during the translation of mRNA to protein.

Q1: How many codons code for isoleucine? For tryptophan? For leucine?

Q2: What codons are associated with asparagine? With serine?

Q3: From the mRNA sequence that you transcribed from the DNA template strand in **Figure 9.7** (Q3), remove only the first A. What amino acid sequence would be translated as a result of this change? How does that sequence compare to the amino acid sequence you translated from the original mRNA sequence? *Bonus:* What kind of mutation is this? (See Chapter 8.)

nitrogenous bases, called an **anticodon**, binds the correct codon on the mRNA. At the other end, the specific amino acid attaches (see Figure 9.9).

Let's recap. For translation to occur, an mRNA molecule must first bind to a ribosome. The ribosomal machinery then "scans" the mRNA until it finds a start codon (AUG). Next the ribosome recruits the appropriate tRNAs one by one, as determined by the codons read in the mRNA sequence. A special site on the ribosome facilitates the linking of one amino acid to another, like beads on a string. Finally, the ribosome reaches a stop codon. The amino acid chain cannot be extended further, because none of the tRNAs will recognize and pair with any of the three stop codons. At this point the mRNA molecule and the completed amino acid chain separate from the ribosome. The new protein then folds into its compact, specific three-dimensional shape and is ready to go to work in the cell.

But this process does not always go as planned. In Chapter 8 we learned that a mutation is a change in the sequence of bases in DNA. Mutations affect an organism by disrupting or preventing the healthy formation of a protein. For example, a mutation can cause a DNA sequence not to be translated or transcribed, prompt the amino acid chain to end prematurely, or make the final protein fold incorrectly, among other possibilities. While single base substitutions are not always a problem, single-base insertions and deletions cause a genetic "frameshift," shifting all subsequent codons "downstream" by one base (**Figure 9.11**). This shift scrambles the entire downstream DNA message, causing the ribosomes to assemble a very different sequence of amino acids from the mutation point onward—as if every letter in this phrase were shifted to the right once while retaining the word length and spaces between words (a si fever ylette ri nthi sphras ewer eshifte dt oth erigh tonc ewhil eretainin gth ewor dlengt han dspace sbetwee nword s).

Tweaking Gene Expression

Gene regulation enables organisms to change which genes they express in response to internal signals (from inside the body) or external cues in the environment. In this way, gene regulation

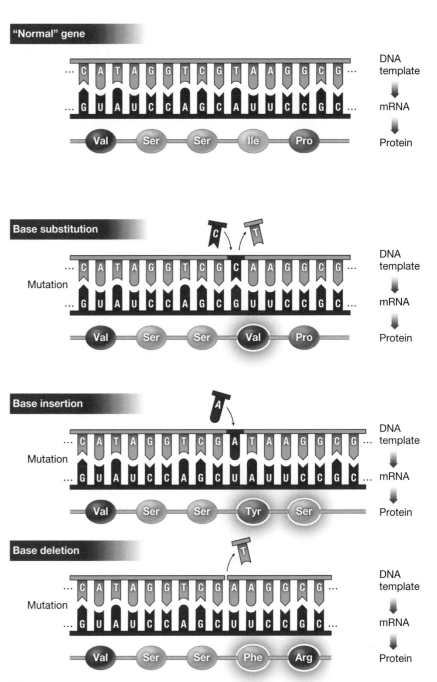

Figure 9.11

Effects of point mutations

Q1: Why is an insertion or a deletion in a gene more likely to alter the protein product than a substitution, such as A for C, would?

Q2: Which would you expect to have more impact on an organism—a point mutation as shown here, or the insertion or deletion of a whole chromosome, as discussed in Chapter 8?

Q3: Which mechanisms in a cell prevent mutations? (*Hint:* Refer back to Chapter 8 if needed.)

The Deadly Price of a Pandemic

The Spanish flu of 1918 devastated populations across Europe, killing not only the young and old but also healthy adults. Since then, new strains of flu have reared their ugly heads, and doctors and vaccine manufacturers struggle to anticipate the next outbreak.

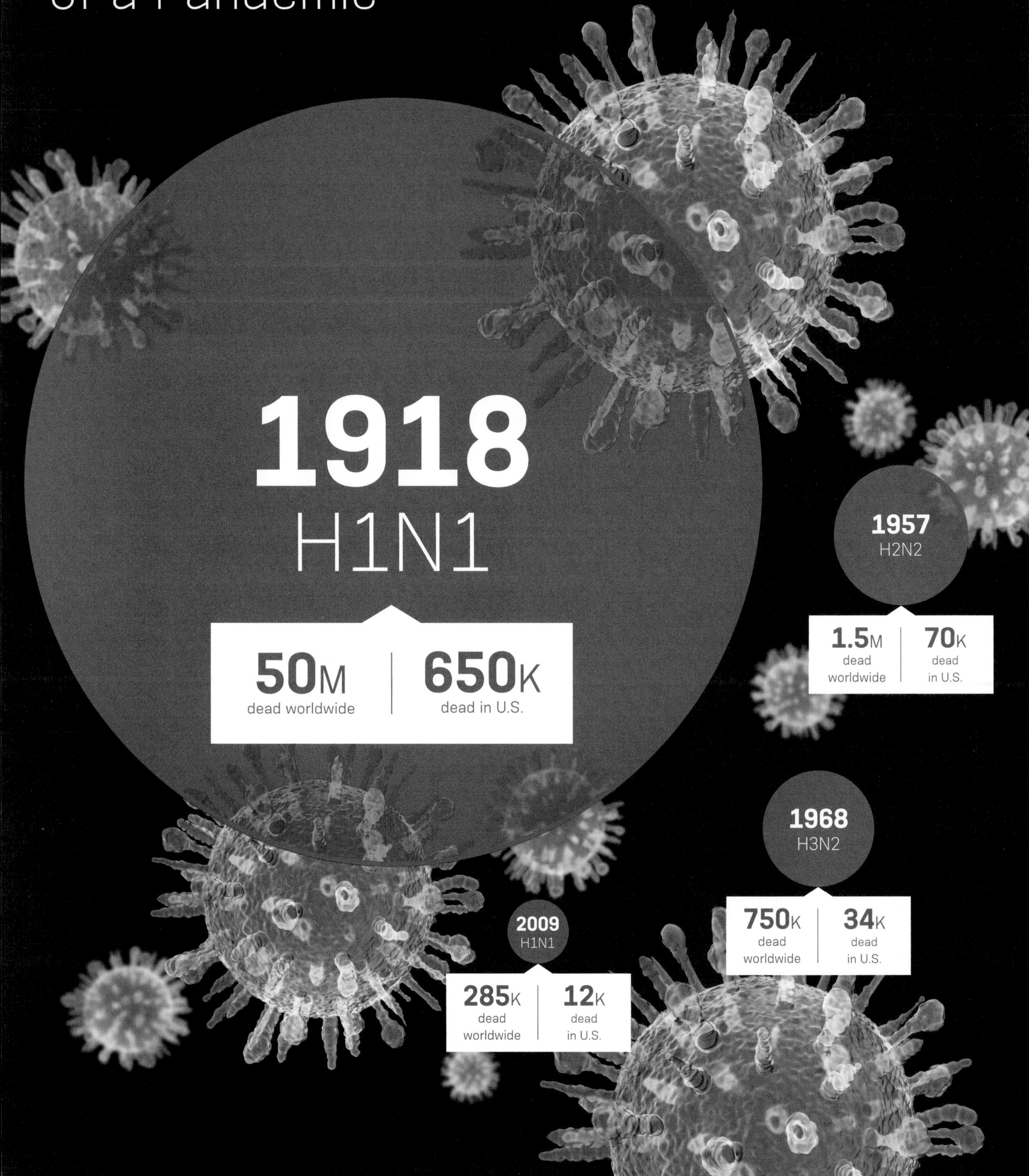

1918
H1N1

50M
dead worldwide

650K
dead in U.S.

1957
H2N2

1.5M
dead worldwide

70K
dead in U.S.

1968
H3N2

750K
dead worldwide

34K
dead in U.S.

2009
H1N1

285K
dead worldwide

12K
dead in U.S.

enables organisms to adapt to their surroundings by producing different proteins as needed. All cells in a multicellular individual have essentially the same DNA-based information, yet different cells express different sets of genes, and within a given cell the pattern of gene expression can change over time. Single-celled organisms, such as bacteria, face a more difficult challenge: they are directly exposed to their environment, and they have no specialized cells to help them deal with changes in that environment. One way they meet this challenge is to express different genes at different times.

The expression of most genes in prokaryotes and eukaryotes is regulated by both internal and external signals. Many genes are also developmentally regulated, meaning that their expression can change, sometimes dramatically, as an organism grows and develops. But for all living cells, a few genes are always expressed at a low level; their transcription is not regulated, because these genes are needed at all times.

Inside a tobacco cell, as in most living cells, the expression of many genes can be turned on or off, slowed down (**down-regulated**), or sped up (**up-regulated**). Gene expression is regulated at many different points in the cell, including DNA packing (the way DNA is compressed or

unwound in the genome), transcription, mRNA processing, and several points during translation (**Figure 9.12**).

At Medicago, the company takes advantage of gene regulation in tobacco cells to produce as

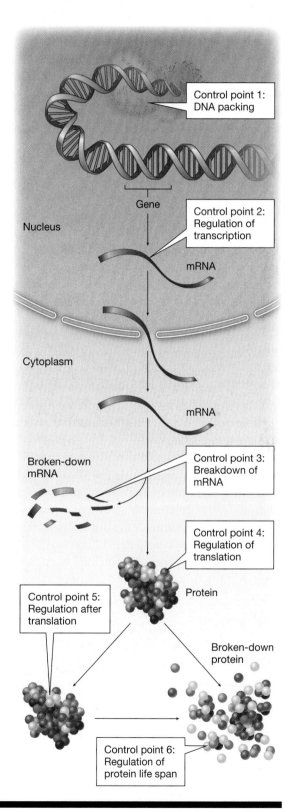

Figure 9.12

How gene expression is regulated

In eukaryotes, gene expression can be controlled at several points along the pathway from gene to protein to phenotype: before transcription, at transcription, during RNA processing, or at translation. Gene expression can also be regulated after translation, by control of the activity or life span of the protein.

Q1: As illustrated here, at what control point is transcription regulated?

Q2: What is a possible advantage of regulating gene expression before versus after transcription?

Q3: If you wanted to up-regulate the production of the hemagglutinin protein in a tobacco plant carrying the hemagglutinin gene, at which control point(s) would that be possible? Justify your reasoning.

much hemagglutinin protein as possible: after the agrobacteria are vacuum-sucked into the leaves and transcription and translation begin to actively occur in the tobacco cells, the plants are sent to grow in an incubation room, where technicians can alter the humidity, temperature, and amount of light to maximize the amount of protein expressed by the plants.

"We've been able to tweak environmental conditions of the plants to boost gene expression," says Vézina. "You name it, we've tried it." It's an important step in the process, adds Wanner. "We determined what the best conditions for protein expression are," he says.

To the Market

It's the end of April, and the final steps are being taken to isolate and purify the hemagglutinin protein from the tobacco plants, to see how much vaccine can be made in a month. After the plants have incubated for several days, the leaves are stripped and diced into green confetti, and then digested with enzymes to break up the leaf material so that the desired proteins are released into solution. The resulting solution, which resembles green-pea soup, is filtered several times to isolate clusters of hemagglutinin, which will then be processed into a vaccine product that is safe to inject into people.

Medicago is not the first company to produce a human drug using a plant. The first human-like enzyme was produced from tobacco back in 1992 at Virginia Polytechnic Institute. Numerous other plant biopharming companies have sprung up since then, experimenting with various plant species, including corn, soybean, duckweed, and more. But the field is not without risks and controversy. "The main concern and risk is spread—that the gene will get out into nature and spread," says Roth. "But there are techniques to make sure that doesn't happen, and it is closely regulated."

One of those techniques is to grow the plants in contained environments, where there is no risk of contaminating food crops. Israel-based biotech company Protalix Biotherapeutics, for example, grows carrot cells inside in large hanging bags of fluid and cells. In May 2012, the FDA approved the first biopharmed drug for humans, produced by Protalix—a therapy for Gaucher disease, a rare genetic disorder—grown in the company's carrot cells (**Figure 9.13**). "This approval demonstrates a proof of concept for the power of this technology," the CEO of Protalix told *Nature*.

Today, Medicago has completed safety trials for its pandemic flu vaccine and has had positive results in the first of two large clinical trials in humans testing their seasonal flu vaccine's efficacy. When the April manufacturing test was completed, the company had produced an astounding 10 million doses of flu vaccine in a single month. It would have taken five to six months to produce the same amount using the traditional method of growing vaccines in chicken eggs.

Medicago continues to develop flu vaccines, as well as novel vaccines against rotavirus and rabies virus. "It might have taken a bit longer than we thought for biopharming to be accepted, but it deserves the visibility and attraction it has now," says Vézina. "Biopharming is here to stay. I'm convinced of this."

Figure 9.13

Producing human enzymes in carrot cells

REVIEWING THE SCIENCE

- Most genes contain instructions for building proteins. **Gene expression** is the process by which genes are transcribed into RNA and then translated into a protein.

- During **transcription**, which occurs in the nucleus, RNA polymerase binds the **promoter** of a gene and produces a **messenger RNA (mRNA)** version of the gene sequence from free nucleotides.

- Next, during **RNA splicing**, **introns** are snipped out of the pre-mRNA sequence, and the remaining **exons** are joined. The mRNA is transported out of the nucleus.

- During **translation**, which occurs in the cytoplasm, ribosomes convert the sequences of bases in an mRNA molecule to the sequence of amino acids in a protein, with the help of **ribosomes** composed of **ribosomal RNA (rRNA)**, and **transfer RNA (tRNA)**.

- Ribosomes read the mRNA information in sets of three bases at a time, called **codons**. There are 64 possible codons, including a **start codon** (AUG) and three possible **stop codons** (UAA, UAG, or UGA).

- The **genetic code** is universal because nearly all organisms on Earth use the same code.

- Gene expression is regulated at many different points in the pathway from gene to protein. Organisms rely on **gene regulation** to respond to signals inside the body and external cues in the environment. Many genes are developmentally regulated; their expression changes as an organism grows and develops.

THE QUESTIONS

The Basics

1 Link each of the following terms with the correct definition.

GENE EXPRESSION   1. RNA is made using the information in the DNA sequence of a gene.

GENE REGULATION   2. The flow of information from gene to protein.

TRANSCRIPTION   3. The control of gene expression in response to environmental or developmental needs.

TRANSLATION   4. Amino acids are linked in the precise sequence dictated by an mRNA base sequence.

2 For each of the following, identify the type of RNA involved (mRNA, rRNA, tRNA).

_____ a. Transports the correct amino acid to the ribosome, using the information encoded in the mRNA.

_____ b. Is a major component of ribosomes.

_____ c. Specifies the order of amino acids in a protein using a series of three-base codons, where different amino acids are specified by particular codons.

_____ d. Contains a three-base anticodon that pairs with a complementary codon revealed in the mRNA.

_____ e. Assists in making the bonds that link amino acids together to make a protein.

3 Using the genetic code shown in Figure 9.10, find the amino acid coded by each of the following codons.

(a) AAU

(b) UAA

(c) AUA

(d) GGG

(e) CCC

4 Using the genetic code shown in Figure 9.10, find one of the codon(s) that code for each of the following amino acids.

(a) arginine

(b) alanine

(c) methionine

(d) glycine

5 Circle the correct terms in the following sentences:
The genetic code demonstrates (**ambiguity, redundancy**) because some amino acids are coded by more than one codon. The lack of (**ambiguity, redundancy**) in the genetic code is evidenced by the fact that each codon codes for one, and only one, amino acid.

6 Which of the following are possible reasons that a cell would regulate its expression of a gene? (Circle all that apply.)

(a) an increased need for a particular enzyme

(b) a decreased need for a particular enzyme

(c) increasing temperature in the external environment

(d) changing needs as an organism ages

(e) death

7 In the diagram of transcription shown here, fill in the blanks with the appropriate term: (a) gene; (b) promoter; (c) terminator; (d) RNA polymerase; (e) mRNA.

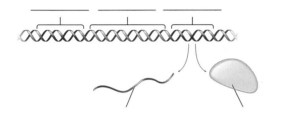

8 Place the following steps of translation in the correct order by numbering them from 1 to 9.

_____ a. A tRNA molecule carrying the amino acid methionine binds at its anticodon site to the mRNA start codon.

_____ b. The ribosome links the first amino acid to the second amino acid to begin the amino acid chain.

_____ c. The ribosome continues to link each amino acid to the growing amino acid chain.

_____ d. The ribosome reaches a stop codon.

_____ e. An mRNA binds to a ribosome.

_____ f. The mRNA and the completed amino acid chain separate from the ribosome.

_____ g. The first tRNA, separated from its amino acid, releases from the mRNA.

_____ h. A tRNA molecule carrying the second amino acid binds to the second mRNA codon.

_____ i. Each tRNA releases from the mRNA after it is separated from its amino acid.

9 How is gene expression similar to DNA replication, and how is it different? Give at least one similarity and one difference.

Try Something New

10 Your roommate, who is also taking a biology class, has become a little confused. He informs you that the genetic code is known to be ambiguous because a given genotype may give rise to a variety of phenotypes during gene expression (for example, his twin brother is an inch taller and more tan than he is). You like your roommate and would like him to pass his next biology exam, so write him a brief note explaining (a) why the genetic code is not, in fact, ambiguous and (b) how gene expression derives a phenotype from a genotype.

11 Some diseases (for example, Huntington's and Parkinson's) appear to be related to increasing protein levels in brain cells, which lead eventually to cell death. At which of the checkpoints shown in Figure 9.13 might a gene regulation error be occurring with these diseases? Identify one checkpoint at which the error would be of increased up-regulation, and one checkpoint at which the error would be of increased down-regulation.

Leveling Up

12 **What do *you* think?** Most people carry two copies of a normal gene that codes for an enzyme, glucosylceramidase, that is involved in breaking down lipids that are no longer needed in cells. (Enzymes are proteins that cause specific chemical changes; they are biological catalysts.) One in 100 people in the United States carries a recessive mutation that codes for a defective glucosylceramidase enzyme. And about one in 40,000 people carries two copies of the mutation and displays the symptoms of Gaucher disease. These symptoms, caused by the accumulation of lipids in cells, include anemia, enlarged organs, swollen glands and joints, and, in severe cases, neurological problems and early death.

Enzyme replacement therapy is effective but is very expensive—about $200,000 annually—and must be continued, every 2 weeks, for life. Protalix Biotherapeutics, an Israeli biotech company working with the U.S.-based Pfizer Pharmaceuticals, has developed a process to genetically modify carrots to produce a replacement enzyme. The biopharmed enzyme will cost about 25 percent less than the standard enzyme therapies, which are grown in mammalian cell lines. Protalix is now working on treatments for other enzyme deficiency diseases.

The FDA's May 2012 approval of the drug developed by Protalix alarmed some environmental activists and health advocates, who fear that Protalix's genetically modified carrot is the thin edge of a wedge that will lead to an underregulated and potentially dangerous industry. There is some legitimacy to their concerns: the USDA does not require an environmental impact assessment for biopharmed crops; nor does it require biotech companies to share the location of their test fields or the identity of the biopharmed molecules being produced. Furthermore, the USDA is not sufficiently staffed to effectively monitor companies involved in biopharming.

What do you think? Should biopharming be allowed in the United States? If so, under what conditions and with what limits? For example, should it be allowed to produce drugs for only life-threatening illnesses, or only under highly controlled conditions? Be prepared to discuss your observations and reflections in class.

13 **Life choices.** Go to the Centers for Disease Control and Prevention (CDC) influenza website and read the pages "Key Facts about Seasonal Flu" (under "Flu Basics") and "Prevention—Flu Vaccine." You can also go to the Mayo Clinic's influenza website. Then answer the following questions.

(a) What is the flu? How is it passed on?

(b) What are symptoms and complications of the flu?

(c) How can you decrease your chance of getting the flu, and what treatments are available if you become infected?

(d) What are the benefits and risks of the flu vaccine?

(e) Why is there a new flu vaccine every year?

(f) Why is the flu vaccine more effective in some years than others?

(g) Who would you recommend should get the flu vaccine? Explain your reasoning.

(h) Do you get a flu vaccine every year? Why or why not?

Whale Hunting

Fossil hunters discover Moby-Dick's earliest ancestor—a furry, four-legged land-lover.

After reading this chapter you should be able to:

- Define evolution and list the six types of evidence for evolution.
- Compare and contrast artificial selection and natural selection.
- Explain how the theory of evolution is supported by the fossil record.
- Describe how homologous and vestigial traits support the theory of common descent.
- Describe why even distantly related species have similar genes.
- Use knowledge about evolution and continental drift to make predictions about geographic locations of fossils.
- Describe how similarities in embryonic development among species provide evidence of their evolutionary past.

CHAPTER

10

EVIDENCE FOR EVOLUTION

Fossils break all the time. This time, the 50-million-year-old ear bone of a small, deerlike mammal called *Indohyus* snapped clean off the skull. Sheepishly, the young laboratory technician cleaning the fossil handed the broken piece to his boss, paleontologist and embryologist J.G.M. 'Hans' Thewissen at Northeast Ohio Medical University. Thewissen tenderly turned the preserved animal remains over in his hand. Then, as the tech reached for the fossil to glue it back onto the animal's skull, Thewissen went rigid.

"Wow, that is weird," said Thewissen. The *Indohyus* ear bone, which should have looked like the ear bone of every other land-living mammal—like half a hollow walnut shell, but smaller—was instead razor thin on one side and very thick on the other (**Figure 10.1**). "Wow," repeated Thewissen. This wasn't the ear of a deer, or any other land mammal. Thewissen squinted closer. "It looks just like a whale," he said.

Although they live in the ocean like fish, whales are mammals like us. So are dolphins and porpoises (**Figure 10.2**). Like all other mammals, whales are warm-blooded, have backbones, breathe air, and nurse their young from mammary glands. Numerous fossils have been found documenting whales' unique transition from land-living mammals to the mammoths of the sea, during which whale populations developed shorter and shorter legs and longer tails. But one crucial link in the fossil record was missing: the closest land-living relatives of whales. What did the ancestors of whales look like before they entered the water? Staring at the strange fossil in his hand, Thewissen realized he could be holding the ear of that missing link.

Whales are but one of the many organisms that share our planet. Every species is exquisitely fit for life in its particular environment: whales in the open ocean, hawks streaking through the sky, tree frogs camouflaged in the green leaves of a rainforest. There is a great diversity of life on Earth—animals, plants, fungi, and more—with each species well-matched to its surroundings. This diversity of life is due to evolution.

"Evolution," in everyday language, means "change over time." In science, biological

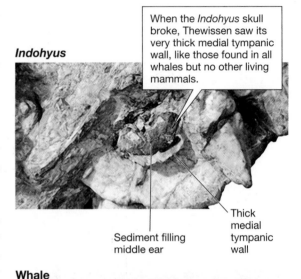

Indohyus

When the *Indohyus* skull broke, Thewissen saw its very thick medial tympanic wall, like those found in all whales but no other living mammals.

Sediment filling middle ear

Thick medial tympanic wall

Whale

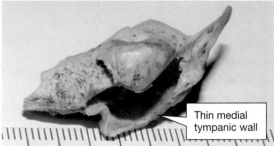
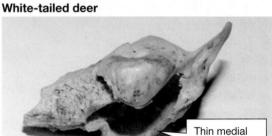

Thick medial tympanic wall

White-tailed deer

Thin medial tympanic wall

Figure 10.1

The mysterious ear bone
The inside of the middle ear, normally full of air, is filled with sediment in the *Indohyus* fossil, above. (*Source*: *Indohyus* and whale photos courtesy of J.G.M. 'Hans' Thewissen.)

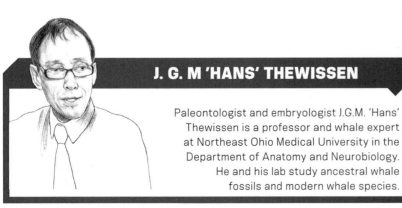

J. G. M 'HANS' THEWISSEN

Paleontologist and embryologist J.G.M. 'Hans' Thewissen is a professor and whale expert at Northeast Ohio Medical University in the Department of Anatomy and Neurobiology. He and his lab study ancestral whale fossils and modern whale species.

evolution is a change in the overall inherited characteristics of a group of organisms over multiple generations. Whales, for example, evolved from four-legged land-living animals into sleek emperors of the ocean, slowly becoming suited for the water over tens of millions of years. Whales changed as a population. Populations evolve; individuals do not.

You may wonder how we can be sure of evolutionary change, especially for a transformation as extreme as that of a furry, four-legged beast into a whale. There is strong evidence for evolution not only from fossils like those that Thewissen studies, but also from features of existing organisms, common patterns of how embryos develop, DNA evidence, geographic evidence, and even direct observation of organisms evolving today—including man's best friend.

Artificial to Natural

As we saw in Chapter 6, all dogs are a single species. That species is *Canis lupus familiaris*, a subspecies of the gray wolf. Scientists estimate that domestication of the gray wolf began about 16,000 years ago as the animals were habituated to human groups. From there, people bred the wolf for desired qualities, such as friendliness and the ability to follow commands. Millennia later, dogs were selectively bred for specific traits like the long legs of a greyhound or the short snout of a bulldog. This selective process has led to incredible variation in the size and shape of dog breeds, from 6-pound Chihuahuas to 200-pound Great Danes (**Figure 10.3**). Dogs are a clear example of evolution that we can directly observe, and they evolved via artificial selection. **Artificial selection** is brought about by **selective breeding**, in which humans allow only individuals with certain inherited characteristics to mate.

Through selective breeding, humans have crafted enormous evolutionary changes within not only dogs, but many other domesticated organisms, including ornamental flowers, pet birds, and food crops. This is a fact, not conjecture. We can observe evolution happening through artificial selection via selective breeding.

Artificial selection happens when humans choose which individuals of a particular species are allowed to breed. Without the intervention of humans, can the environment itself "choose" who survives and breeds? It does. In nature, evolution occurs mainly via *natural* selection. **Natural**

Figure 10.2

A montage of mammals
Earth boasts an amazing diversity of life, as evidenced here. These mammals are only a few of the almost 5,500 species of mammals currently living on the planet.

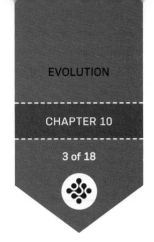

selection is the process by which individuals with advantageous inherited characteristics for a particular environment survive and reproduce at a higher rate than do individuals with other, less useful characteristics. In other words, whoever has the most kids wins!

After the environment "chooses" the winners—those who successfully breed the most—the characteristics of those individuals become more common in successive generations because they have produced more offspring. For example, in 1977, a terrible drought struck the Galápagos Islands off the coast of Ecuador. One species of small ground finches—petite birds with sharp, pointy beaks—starved as the small, tender seeds they ate became scarce. But some heat-loving, drought-resistant plants still produced large, hard seeds. The finches with larger beaks could eat those seeds. They survived and reproduced, and by 1978, in just one generation, the average beak size in the population had increased (**Figure 10.4**).

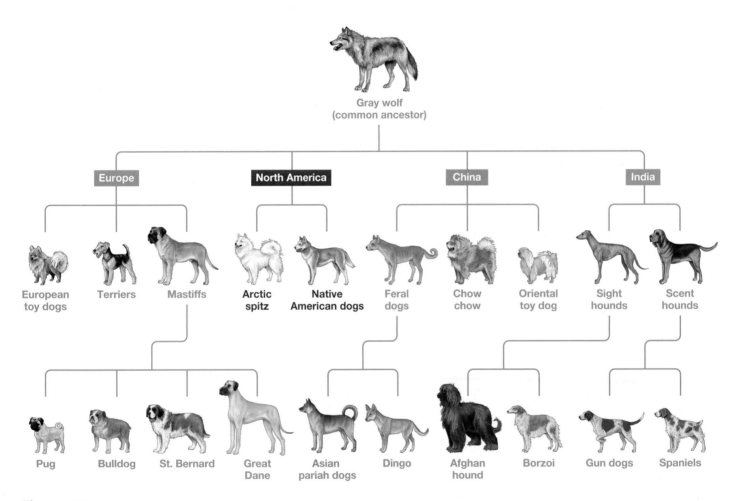

Figure 10.3

Selective breeding of dogs produces myriad traits

Dogs were domesticated only a few times, and always from gray wolves. Thus, the remarkable diversity of dogs represents the effects of selective breeding on a small number of lineages of domesticated wolves.

Q1: What is selective breeding, and how does it work?

Q2: Describe how selective breeding leads to artificial selection.

Q3: Name as many organisms as you can whose current phenotype is due to artificial selection.

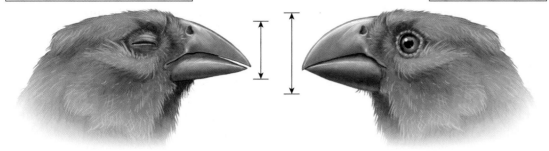

Birds with smaller beaks had nothing to eat and died, thus failing to have sex, produce offspring, and pass on their genetic traits.

Birds with large beaks passed on their large beak size to their offspring.

Figure 10.4

Natural selection results in larger beak size in finches

After a drought, only birds with larger beaks were able to eat the available food: large hard seeds. In the span of just one generation, the average size of the species' beak was visibly larger.

Q1: What is natural selection?

Q2: If humans are the selective force in artificial selection, what is the selective force in natural selection?

Q3: Compare and contrast artificial selection and natural selection. Name two ways in which they are similar. How are they different?

This is one of many examples of how a population can evolve via natural selection so that more and more individuals have beneficial traits, and fewer and fewer have disadvantageous traits. This is called **adaptation**—an evolutionary process by which a population becomes better matched to its environment over time. The finch population quickly adapted to its new, drier environment. Over time, the small-beaked finches died off and the large-beaked birds survived and reproduced; the finch population had adapted to its environment in just a few years. Other adaptations take millions of years, such as whale ancestors adapting to aquatic life.

It is important to realize that biological evolution includes human evolution. Surveys taken during the past 10 years reveal that almost half the adults in the United States do not believe that humans evolved from earlier species of animals. This statistic is startling because evolution has been a settled issue in science for nearly 150 years. Scientists like Thewissen go to work every day and see evolution in action. In fact, the vast majority of scientists of all nations agree that the evidence for evolution is overwhelming.

Six lines of evidence provide compelling support for biological evolution:

1. Direct observation of evolution through artificial selection
2. Fossil evidence
3. Shared characteristics among living organisms
4. Similarities and differences in DNA
5. Biogeographic evidence
6. Common patterns of embryo development

Nowhere is all this evidence more present and intriguing than in one of the most dramatic transitions to occur on Earth: the evolution of small, land-living mammals into dolphins, porpoises, and mighty whales.

Fossil Secrets

Laying his hands on the *Indohyus* fossils was no easy task for Thewissen. Beginning in 2003, he made an annual pilgrimage to Dehradun, India, a city nestled in the foothills of the Himalayas. There, he visited the widow of Anne Ranga Rao, an Indian geologist who had hoarded piles of fossils excavated from Kashmir, a disputed border area between India and Pakistan. Most early

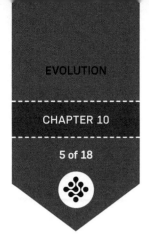
whale fossils have come from the India-Pakistan region, where whales first evolved. But because of political tensions, it is too dangerous to travel to Kashmir, much less dig for **fossils**. Fossils are the mineralized remains of formerly living organisms or the impressions of formerly living organisms (**Figure 10.5**).

Thewissen was frustrated by his inability to travel to Kashmir and collect fossils. The fossil record enables biologists to reconstruct the history of life on Earth, and it provides some of the strongest evidence that species have evolved over time. The relative depth or distance from the surface of Earth at which a fossil is found is referred to as its *order* in the fossil record. The ages of fossils correspond to their order: older fossils are found in deeper, older rock layers.

The fossil record contains excellent examples of how major new groups of organisms arose from previously existing organisms. The record includes numerous **intermediate fossils**, evidence of species with some similarities to the ancestral group (land-living mammals) and some similarities to the descendant species (whales). Thewissen spent decades studying these intermediates—from the first known whale, the wolflike *Pakicetus* that waded in shallow freshwater; to the larger crocodile-like *Ambulocetus* that stalked its prey underwater; to the fully aquatic *Dorudon*, with its blowhole, flippers, and tail (**Figure 10.6**). Yet Thewissen and others had long been searching for the animal that preceded them all—the ancestor of whales that lived on land. If Thewissen had to guess where

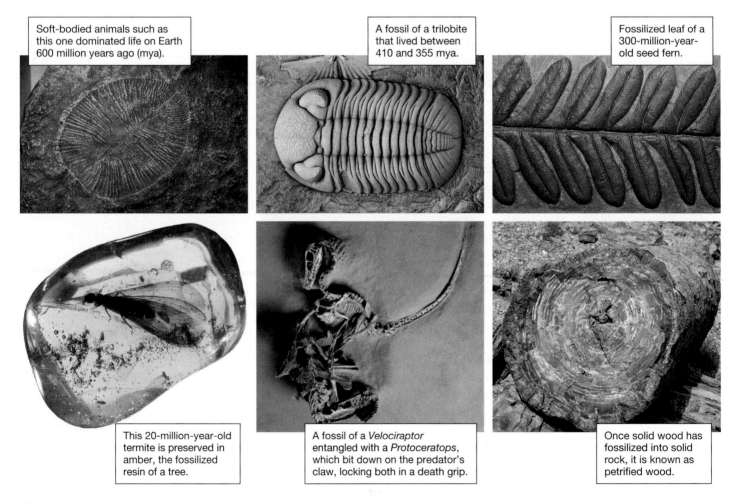

Soft-bodied animals such as this one dominated life on Earth 600 million years ago (mya).

A fossil of a trilobite that lived between 410 and 355 mya.

Fossilized leaf of a 300-million-year-old seed fern.

This 20-million-year-old termite is preserved in amber, the fossilized resin of a tree.

A fossil of a *Velociraptor* entangled with a *Protoceratops*, which bit down on the predator's claw, locking both in a death grip.

Once solid wood has fossilized into solid rock, it is known as petrified wood.

Figure 10.5

Fossils through the ages

Myriad fossils exist, ranging from imprints of organisms, to preserved organisms, to completely mineralized bone and wood. Each fossil can be dated, and when compiled, they can tell the life history of Earth.

Watching Evolution Happen

In the laboratory, scientists can manipulate populations of organisms to watch evolution in real time via artificial selection. In 2012, researchers at the University of California, Irvine, manipulated the growing environment of *Escherichia coli* bacteria, exposing them to far hotter temperatures than normal, to see if they would adapt. It turned out most of the bacteria adapted via one of just two primary pathways.

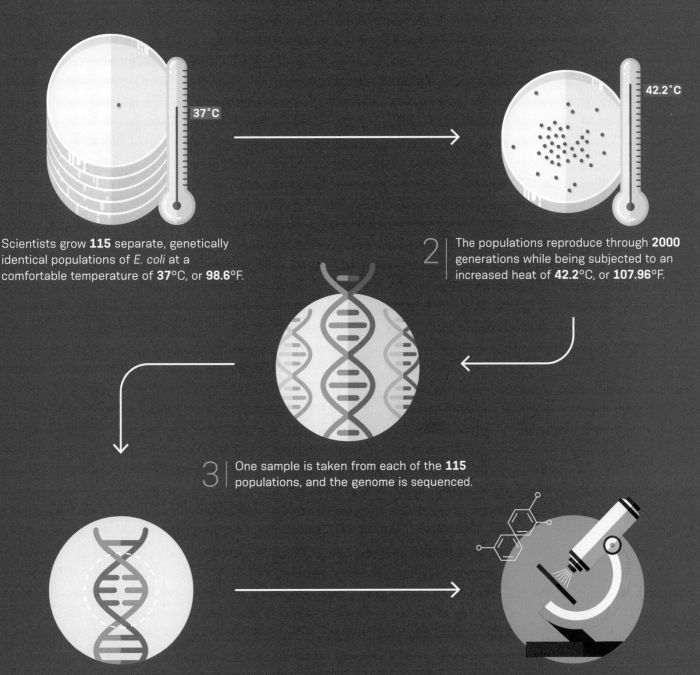

1 | Scientists grow **115** separate, genetically identical populations of *E. coli* at a comfortable temperature of **37**°C, or **98.6**°F.

37°C

2 | The populations reproduce through **2000** generations while being subjected to an increased heat of **42.2**°C, or **107.96**°F.

42.2°C

3 | One sample is taken from each of the **115** populations, and the genome is sequenced.

4 | **1258** molecular changes, averaging **11** genetic mutations per clone, are detected. To help survive the heat, the *E. coli* tend to mutate along two different trajectories:

- Mutations in the RNA polymerase complex, an enzyme that transcribes RNA.

- Mutations in the rho gene, which encodes a protein that stops RNA transcription.

5 | The research continues: the next step is to figure out how the mutations in the RNA polymerase complex and in rho helped the *E. coli* survive the heat.

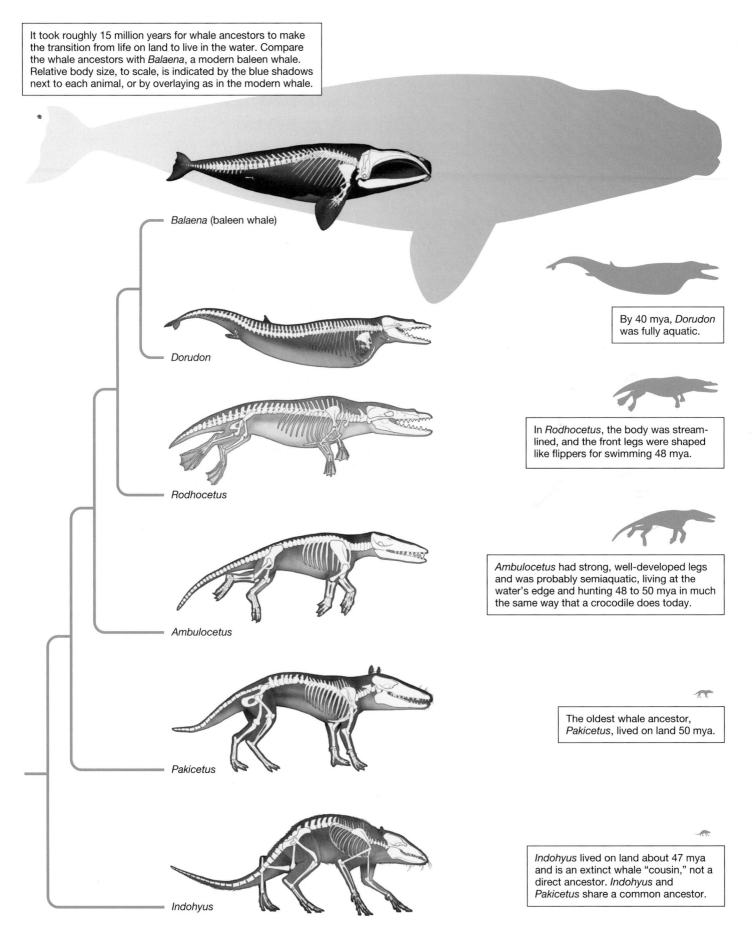

It took roughly 15 million years for whale ancestors to make the transition from life on land to live in the water. Compare the whale ancestors with *Balaena*, a modern baleen whale. Relative body size, to scale, is indicated by the blue shadows next to each animal, or by overlaying as in the modern whale.

Balaena (baleen whale)

Dorudon

By 40 mya, *Dorudon* was fully aquatic.

Rodhocetus

In *Rodhocetus*, the body was stream-lined, and the front legs were shaped like flippers for swimming 48 mya.

Ambulocetus

Ambulocetus had strong, well-developed legs and was probably semiaquatic, living at the water's edge and hunting 48 to 50 mya in much the same way that a crocodile does today.

Pakicetus

The oldest whale ancestor, *Pakicetus*, lived on land 50 mya.

Indohyus

Indohyus lived on land about 47 mya and is an extinct whale "cousin," not a direct ancestor. *Indohyus* and *Pakicetus* share a common ancestor.

Figure 10.6

Skeletons and body sizes of modern whales and fossil ancestors

These reconstructed skeletons from modern whales (top) and various ancestors are in chronological order.

Q1: What is the general definition of a fossil?

Q2: Describe how the fossil record provides strong evidence for evolution.

Q3: What is meant by the term "intermediate fossil" when referring to the fossil record?

those fossils might be, it was in Kashmir, and potentially in Ranga Rao's basement.

Unfortunately, Ranga Rao's widow was protective of the fossils, paranoid that someone might steal her husband's property and legacy. But each year Thewissen visited her, chatted with her, and gained her trust. When she passed away in 2007, she made Thewissen cotrustee of her estate, and suddenly the fossils, which had sat in dusty piles for 30 years, were available for study.

"I focused on taking the rocks back to the U.S., and having my fossil preparers remove the fossils from the rocks, which is very difficult," says Thewissen. From Ranga Rao's collection, Thewissen identified more than 400 bones that belonged to *Indohyus*. By collecting a thighbone here and a jawbone there, his team compiled a Frankenstein-like skeleton of a single *Indohyus*

individual (**Figure 10.7**). After the discovery of the whalelike ear, the researchers looked even more carefully at the features of the fossils and found additional evidence that *Indohyus* was a relative of whales. This unassuming little animal, with a pointy snout and slender legs tipped with hooves, lived close to and had an affinity for the water.

Thewissen and his team found their first clues about *Indohyus*'s lifestyle from its teeth. Oxygen in the molecules that make up teeth comes from the water and food that an animal ingests. Levels of oxygen isotopes in *Indohyus*'s teeth match those of water-going mammals today, suggesting that *Indohyus* lived near and potentially spent a significant amount of time in the water. It also had large, crushing molars with levels of carbon isotopes that suggest it grazed on plants, as do

Figure 10.7

Fossilized skeleton of *Indohyus*, oldest cousin of the whales

A reconstructed fossilized skeleton of *Indohyus* was compiled from multiple sources and locations. The illustration is an artist's depiction of the living animal about 47 million years ago. (*Source*: Photo courtesy of J.G.M. 'Hans' Thewissen.)

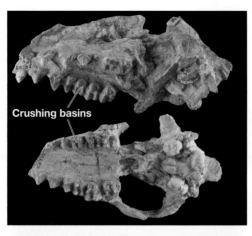

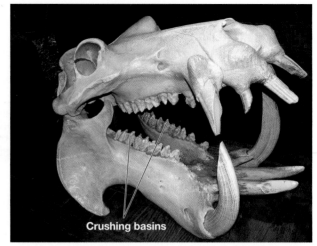

Crushing basins

Crushing basins

The molars of *Indohyus* (top left) are similar to the shape of molars in contemporary aquatic plant-eating animals like hippos (top right and bottom left) and muskrats (not pictured). These molars have crushing basins for grinding up tough plant fibers.

Figure 10.8

Comparing the skulls and jaws of fossilized *Indohyus* and a modern hippopotamus
These organisms' teeth indicate their ability to eat plant material.

hippopotamuses or muskrats that graze near and in water (**Figure 10.8**).

Lisa Cooper, a graduate student in Thewissen's lab at the time, identified another adaptation to the water: *Indohyus*'s leg bones. From the outside, the limbs of *Indohyus* look like those of any other mammal walking around on land. But on

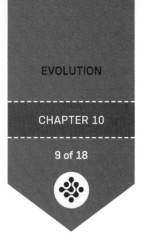

LISA COOPER

Lisa Cooper is an assistant professor at Northeast Ohio Medical University in the Department of Anatomy and Neurobiology. She earned her PhD in Thewissen's lab.

the inside, it's another story. Cooper cut out a section of bone from a limb, ground it down until she could see light through it, and then looked at the bone under a microscope. She saw that a thick layer of bone was wrapped around the bone marrow.

"Hans already had lots of bones of the earliest whales, and they all had extraordinarily thick bones," says Cooper, now an assistant professor at Northeast Ohio Medical University. Modern animals that live in shallow water, such as manatees and hippos, also have thick bones, which help prevent them from floating and enable them to dive quickly (**Figure 10.9**). "It isn't just isolated to whales," says Cooper. "Bones have thickened again and again as different groups of vertebrates entered the water. When you trace back through the fossil record, there is a pretty good correlation between thickness of bone and whether something was living in the water."

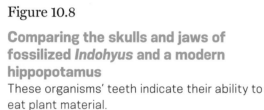

Pakicetus, an extinct water-dwelling whale ancestor	Manatee, a water-dwelling mammal	*Indohyus*, an extinct water-dwelling whale cousin

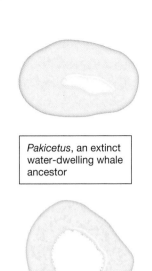

Hippopotamus, a land- and water-dwelling animal	Polar bear, a land-dwelling mammal	Rat, a land-dwelling mammal

Figure 10.9

Cross sections of the femurs of *Indohyus*, water-dwelling animals, and land-dwelling animals

Aquatic organisms have thick, dense bone around a narrow marrow space, while terrestrial organisms have thin bone and a large marrow space. *Indohyus* bone structure is an example of a homologous trait that is shared with other water-dwelling animals.

Q1: Why do water-dwelling animals have thicker bones than land-dwelling animals?

Q2: Why does this thick-bone adaptation suggest a water-dwelling lifestyle?

Q3: How did this adaptation likely increase survival or reproduction in *Indohyus*?

Indohyus's thick bones are an example of an **adaptive trait**, a feature that gives an individual improved function in a competitive environment. By being able to easily wade and dive in water, *Indohyus* had an advantage over other organisms in escaping predators and accessing plants to eat on the river floor. Adaptive traits take many forms, from an anatomical feature like *Indohyus*'s bones, to behaviors, to the functions of individual proteins. Echolocation in bats, for example, is an adaptation for catching insects in the dark. Stick insects have adaptations that help them avoid detection by predators; they physically and behaviorally mimic the plants they live on (**Figure 10.10**).

The Ultimate Family Tree

Thick bones are not restricted to just the water-loving ancestors of whales, as we noted, but can be seen in many other animals such as hippos. This similarity across organisms is another type of evidence for evolution—shared characteristics between species. Many shared characteristics—such as thick bones for animals that take to shallow water, or sexual reproduction via egg and sperm, or eukaryotic cells—result from

organisms sharing traits that evolved from a **common ancestor**, an organism from which many species have evolved. A group of organisms have **common descent** if they share a common ancestor.

Figure 10.10

Stick insects avoid detection by predators

Stick insects are well adapted to their environment; they move slowly and look just like the branches on which they live.

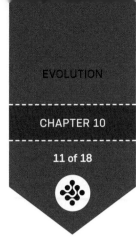

When one species splits into two, the two resulting species share similar features, or **homologous traits**, because they have common descent—though these features may begin to look different from one another over time (**Figure 10.11**). For example, whales are so different from humans that it can be difficult to find similarities, but we evolved from a common mammalian ancestor and do share homologous traits. Humans and whales both nurse their young and have a single lower jawbone because our common ancestor had those traits.

Vestigial traits are another type of trait that many organisms have because of a common ancestor. These characteristics are a piece of the evolutionary past, inherited from a common ancestor but no longer used. Vestigial traits may appear as reduced or degenerated parts whose function is hard to discern (**Figure 10.12**). For example, many modern whales have vestiges of thighbones, also called femurs, embedded in the skin next to the pelvis. In land mammals, birds and other tetrapod vertebrates, these bones are critical for walking, running, and jumping. Aquatic whales have no need of this bone.

Whales also have small muscles devoted to nonexistent external ears, apparently from a time when they were able to move their ears, as land animals such as dogs do for directional hearing. Vestigial traits are not adaptations. In fact, they can be detrimental. Most humans no longer need wisdom teeth to replace lost teeth during adolescence and yet, most people still have them. They tend to erupt around the 20th year and can cause severe pain and displace other teeth, and they usually require removal.

Clues in the Code

Within every organism is one of the strongest pieces of evidence for evolution: DNA. Living things universally use DNA as hereditary, or genetic, material (see Chapter 8 for review). The fact that all organisms on Earth—even those as different as bacteria, redwood trees, and humans—use the same genetic code is further evidence that the great diversity of living things evolved from a common ancestor.

Researchers have analyzed the DNA sequences of whales and other animals and shown that, of all animals, whales are most closely related to

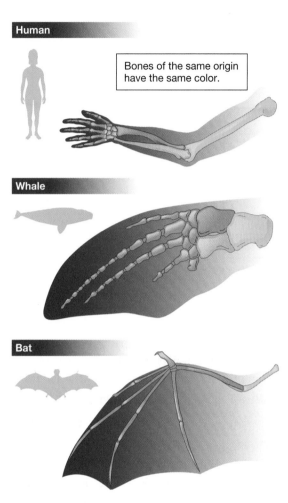

Figure 10.11

Homologous traits are shared characteristics inherited from a common ancestor

The human arm, the whale flipper, and the bat wing are homologous structures due to common descent. All three structures have a matching set of five digits and a matching set of arm bones that have been altered by evolution for different functions.

Q1: What is meant by the term "common ancestor"? Give an example.

Q2: How are homologous structures among organisms evidence for evolution?

Q3: Aside from skeletal structural similarities, what other commonalities among organisms are considered homologous?

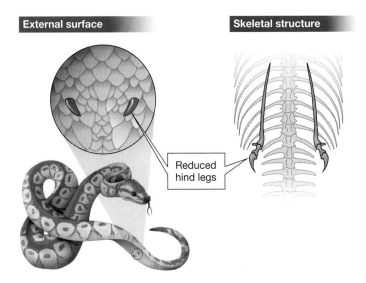

External surface

Skeletal structure

Reduced hind legs

Figure 10.12

Vestigial traits are reduced or degenerated remnants with no apparent function
Snakes are limbless reptiles with no apparent use for the degenerated remnants of hind legs that they still have. The python shown here has extremely reduced hind legs that are only barely visible externally.

Q1: How are vestigial structures among organisms evidence for evolution? Give examples.

Q2: Are vestigial structures also homologous structures? Explain.

Q3: Why do vestigial structures still exist if they are no longer useful?

even-toed ungulates—a group of hoofed mammals that includes modern deer, giraffes, camels, pigs, and hippos. *Indohyus* is an example of an extinct even-toed ungulate. Molecular studies therefore confirm the prediction that whales and *Indohyus* share a common ancestry.

The DNA sequence similarity between whales and hippos shows that hippos are whales' closest living relatives. Whale DNA is more similar to hippo DNA than to the DNA of other marine mammals, such as seals and sea lions. **DNA sequence similarity** is a measure of how closely related two DNA molecules are to each other. For example, in two DNA molecules of the same gene from two different species, 83 percent of the nucleotides might be identical and in the same order (**Figure 10.13**). These two species would share a more recent common ancestor than two species for whom the DNA sequence similarity of the same gene was only 72 percent identical. In fact, the insulin gene of our closest living relative on Earth, the chimpanzee, has 98 percent similarity to the human insulin gene. That

comparability implies that humans and chimpanzees share a very recent common ancestor.

The fact that these separate lines of evidence—anatomical features and DNA—yield the same result over and over again for diverse groups of organisms is strong evidence for evolution.

Whale evolution is "one of the best case studies documenting how a vertebrate can go from a terrestrial to an aquatic environment," says Lisa Cooper. Another type of evidence supporting whale evolution comes from the locations where whale fossils have been found on Earth.

Birthplace of Whales

Earth's continents are on massive tectonic plates, which slowly move over time in a process called continental drift or plate tectonics. For example, each year South America and Africa drift farther apart by about an inch. Although they are

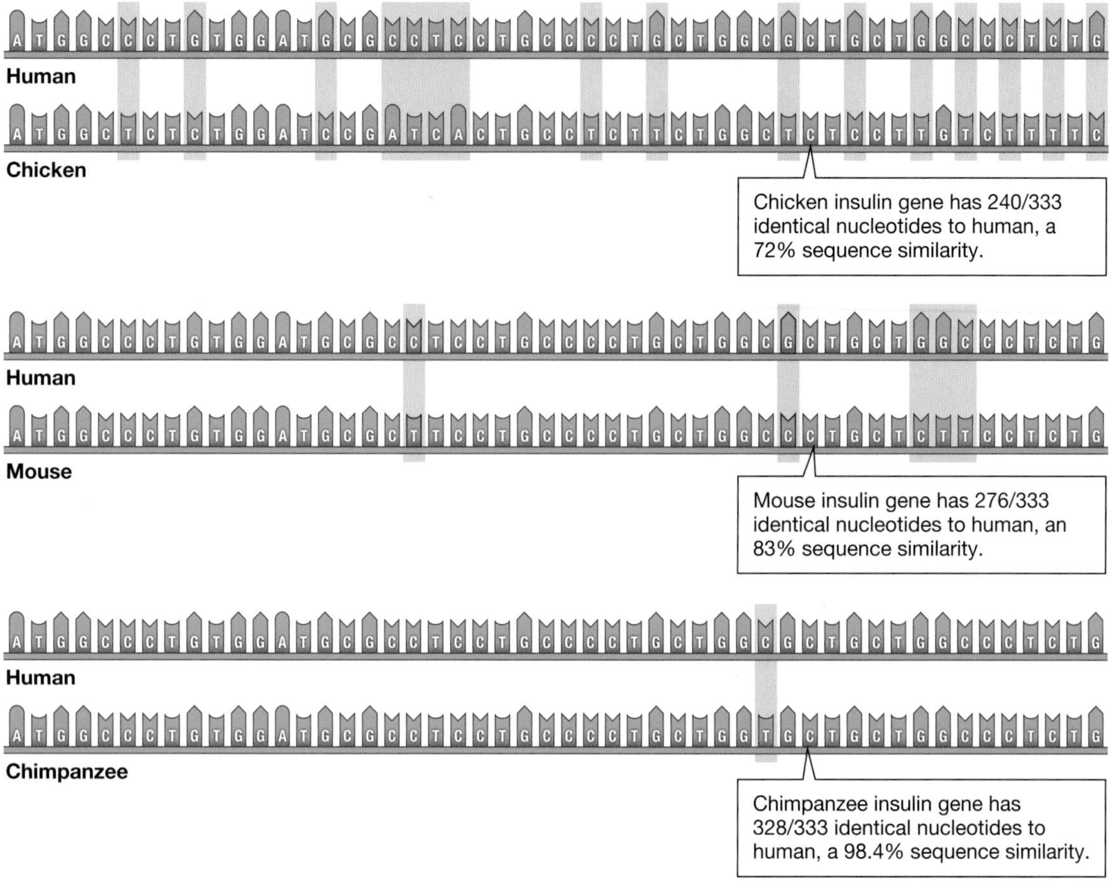

Chicken insulin gene has 240/333 identical nucleotides to human, a 72% sequence similarity.

Mouse insulin gene has 276/333 identical nucleotides to human, an 83% sequence similarity.

Chimpanzee insulin gene has 328/333 identical nucleotides to human, a 98.4% sequence similarity.

Figure 10.13

DNA sequence similarities of the insulin gene

The complete coding sequence of the human insulin gene is 333 nucleotides. Only the first 50 nucleotides are shown here. Unshaded paired sequences are identical nucleotides at that position; those shaded in yellow are different.

Q1: If a sequence from another species were compared and showed a 96 percent sequence similarity to humans, would that species be more closely related to humans than chimpanzees are?

Q2: Are similarities in the DNA sequences of genes considered evolutionary homology? Explain.

Q3: How is the increased similarity in the DNA sequences of genes between more-related organisms—and the decreased similarity between less-related organisms—evidence for evolution? Use the examples in this figure to support your answer.

separating from one another now, about 250 million years ago South America, Africa, and all of the other landmasses of Earth had drifted together to form one giant continent called Pangaea. About 200 million years ago, Pangaea slowly began to split up, ultimately forming the continents as we know them today.

We can use knowledge of evolution and continental drift to make predictions about the biogeography of a species—the geographic locations where its fossils will be found. For example, today the lungfish *Neoceratodus fosteri* is found only in northeastern Australia, but its ancestors lived during the time of Pangaea, and, as predicted, fossils of those ancestors are found on all continents except Antarctica (**Figure 10.14**).

The biogeography of whale fossils matches the pattern predicted by evolution: all early species of

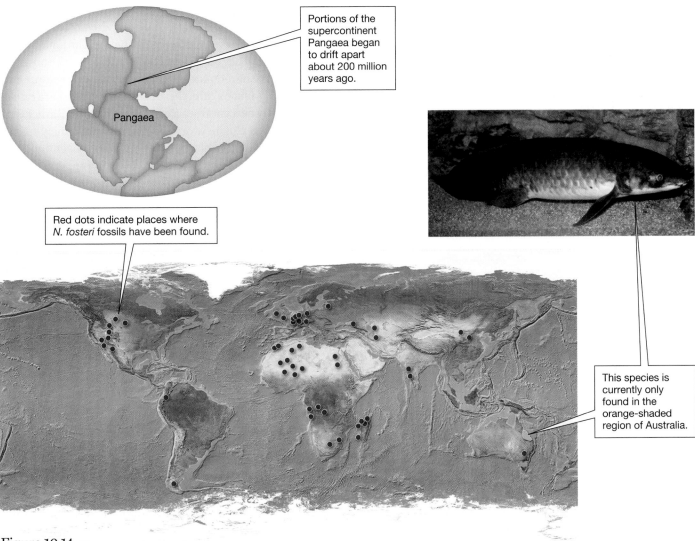

Figure 10.14

The biogeography of a lungfish reflects its evolutionary past
Ancestors of the freshwater lungfish *Neoceratodus fosteri* lived during the time of Pangaea. *N. fosteri* fossils have been found on all continents except Antarctica.

Portions of the supercontinent Pangaea began to drift apart about 200 million years ago.

Pangaea

Red dots indicate places where *N. fosteri* fossils have been found.

This species is currently only found in the orange-shaded region of Australia.

Q1: Why should we expect to find *N. fosteri* fossils all over the world, given that it first evolved in Pangaea?

Q2: Can we use biogeographic evidence to support evolution without using fossil evidence? Explain and give examples.

Q3: Can we use DNA sequence similarities together with biogeography as evidence for evolution? Explain, using examples.

whales, such as the crocodile-like *Pakicetus* that lived in rivers and lakes but did not swim in the ocean, are found near India and Pakistan. "It makes sense," says Thewissen. "You don't have crocodiles crossing the Atlantic." But fossils of fully aquatic species that emerged about 40 million years ago— the protocetids were adept swimmers—are geographically much more widespread, found as far away from Pakistan as Canada. "Protocetids are

good swimmers, so we find their fossils all around the world," says Thewissen.

Growing Together

Though Thewissen has built a career on finding and describing whale fossils, he recently became enamored with another vein of evolutionary

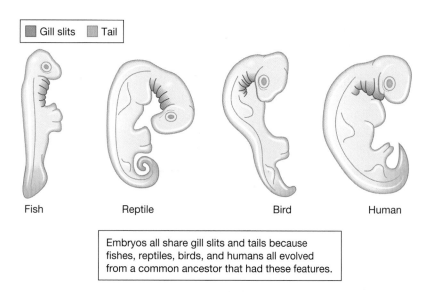

Embryos all share gill slits and tails because fishes, reptiles, birds, and humans all evolved from a common ancestor that had these features.

Figure 10.15

Evolutionary history can be extrapolated from similarities in embryo development

Complex structures in descendant species are generally elaborations of structures that existed in their common ancestor.

Q1: How are the similarities among organisms during early development evidence for evolution? Give examples.

Q2: Are the similar structures among vertebrate species during embryogenesis homologous structures? Explain.

Q3: Why do embryonic structures still exist at points during embryogenesis if they are not used after birth?

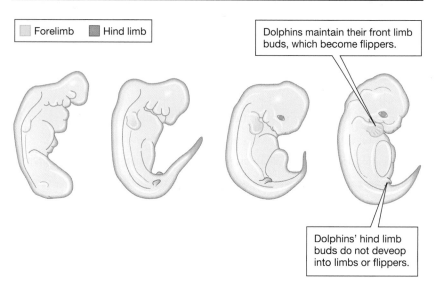

Dolphins maintain their front limb buds, which become flippers.

Dolphins' hind limb buds do not deveop into limbs or flippers.

Figure 10.16

Dolphin embryonic development

The embryos of dolphins from weeks 4–9 of development show the formation and then subsequent loss of the hind limb buds. (*Source:* Based on a photo by J.G.M. 'Hans' Thewissen, permission granted.)

evidence: embryology. A major prediction of evolution is that organisms should carry within themselves evidence of their evolutionary past, and they do. Evidence of evolution can be observed in shared patterns of **embryonic development**.

Once again, these common patterns are caused by descent from a common ancestor. Rather than evolving new organs "from scratch," new species inherit structures that may have been modified in form and sometimes even in function.

Upon fusion of sperm and egg, an animal embryo begins to grow and develop. The manner in which an embryo develops, especially at the early stages, may mirror early developmental stages of ancestral forms. For example, anteaters and some baleen whales do not have teeth as adults, but as fetuses they do. And the embryos of fish, amphibians, reptiles, birds, and mammals (including humans) all develop pharyngeal pouches or gill slits (**Figure 10.15**). In fish, the pouches develop into gills that adult fish use to absorb oxygen underwater. In human embryos, these same features become parts of the ear and throat.

"I was interested to get embryos to look at some of these processes that we see happen in evolution, to see if they happen in development," says Thewissen. The first trait he examined was hind limbs. "We know, from fossil evidence, that early whales lose their hind limbs," says Thewissen. So he wondered if hind limbs exist in whale or dolphin embryos (dolphins are also mammals and are closely related to whales). And if so, what makes them subsequently disappear before the animal is born.

Examining spotted dolphin embryos, Thewissen saw that when the embryos are the size of a pea, they do develop hind limb buds, but by the time they grow into the size of a bean, the limb buds are gone. In 2006, he and researchers at several other universities studied the genes that are active in whale and dolphin embryos, and concluded that whales' hind limbs regressed over millions of years through small changes in a number of genes relatively late in embryonic development (**Figure 10.16**). The loss of hind limbs in the embryo corresponds to the disappearance of hind limbs in the fossil record. "It was an awesome way to combine embryology with fossils," says Thewissen.

With their discovery of *Indohyus*, Thewissen, Cooper, and their team bridged a 10-million-year gap in the fossil record, identifying a transition species that is the closest relative to whales. It is another rock in the mountain of evidence surrounding whale evolution.

In the end, evolution is supported by mutually reinforcing, independent lines of evidence: direct observation, fossils, shared characteristics among living organisms, similarities and divergences in DNA, biogeographic evidence, and common patterns of embryo development. Just as the theory of gravity forms the foundation of physics, so evolution is the central tenet of biology.

There is no question that evolution happens. The intriguing, fascinating question is *how* does it happen? Through scientific research, we know that whales descended from a group of land-living mammals including the petite *Indohyus*, munching on freshwater plants and splashing through the shallows. But how did a population of small, furry animals become the mammoths of the sea? Next we investigate the mechanisms of evolution—how and why it works.

REVIEWING THE SCIENCE

- Biological **evolution** is the overall change in characteristics of a population of organisms over time.

- **Artificial selection** results in biological evolution. Humans choose which organisms survive and reproduce—a process known as **selective breeding**.

- **Natural selection** is the process by which individuals with advantageous genetic characteristics for a particular environment survive and reproduce at a higher rate than competing individuals with other, less useful characteristics.

- **Fossils** are the preserved remains (or their impressions) of formerly living organisms. The fossil record enables biologists to reconstruct the history of life on Earth, and it provides some of the strongest evidence that species have evolved over time.

- Many similarities among organisms are due to the fact that the organisms evolved via **common**

descent from a **common ancestor**. When one species splits into two, the two resulting species share similar features, called **homologous traits**. If a homologous trait is no longer useful, it is called a **vestigial trait**.

- The fact that organisms as different as bacteria, redwood trees, and humans show **DNA sequence similarity** is evidence that the great diversity of living things descended or evolved from a common ancestor.

- **Biogeography** uses knowledge about evolution and plate tectonics to make predictions about the geographic locations where fossils will be found.

- Similarities in **embryonic development** of different organisms suggest that the complexities of modern organisms arose through evolutionary modifications of traits inherited from common ancestors.

THE QUESTIONS

The Basics

1 If two different organisms are closely related evolutionarily, then they will

(a) be similar in size.

(b) share a recent common ancestor.

(c) have very different DNA sequences in their genes.

(d) be randomly located throughout the world.

2 All mammals have tailbones and muscles for moving a tail. Even humans have a reduced tailbone and remnant tail-twitching muscles, though these features have no apparent usefulness. These traits in humans would best be described as

(a) convergent structures.

(b) fossil evidence.

(c) evidence of biogeography.

(d) vestigial traits.

3 Reduced tailbones and the associated remnant muscles in humans are an example of what type of evidence for common descent?

(a) artificial selection

(b) homologous traits

(c) biogeography

(d) fossil evidence

4 When two organisms are very *distantly* related in an evolutionary sense,

(a) they should have extremely similar embryonic development.

(b) they must share a very recent common ancestor.

(c) the sequences of DNA in their genes should be less similar (more different) than those of two more closely related organisms.

(d) they should share more homologous traits than two more closely related organisms share.

5 Link each of the following terms with the correct definition.

BIOGEOGRAPHY	A reconstruction of the history of life on Earth.
FOSSIL RECORD	The similarities in the nucleotide sequences among related organisms.
DNA SEQUENCE SIMILARITY	The similarities among organisms that are due to the fact that the organisms evolved from a common ancestor.
EMBRYONIC SIMILARITY	The geographic locations where related organisms and fossils are found.
HOMOLOGOUS TRAITS	Specifically during development, complex structures in descendant species that are generally elaborations of structures that existed in the common ancestor.

Try Something New

6 Cat DNA is much more similar to dog DNA than to tortoise DNA. Why is that?

(a) Cats and dogs are both carnivores and take in similar nutrients.

(b) Cats and dogs have lived together with humans for a long period of time, so they have grown more similar.

(c) Cats and dogs have more offspring during their lifetime than tortoises have, so their DNA changes less rapidly.

(d) Cats and dogs have a common ancestor that is more recent than the common ancestor of cats and tortoises.

7 DNA sequences were analyzed from humans and three other mammals: species X, Y, and Z. Which of these mammals is most closely related to humans? (*Note:* Regions identical to human DNA are shown in bold type.)

Human:
AATGCTTTGGGGGATCGCGAGCGCAGCGC

Species X:
GGGTTTTT**ATCGC**TATATATATATA

Species Y:
AATGCTTTGGGGGATCGCGAGCGCATATA

Species Z:
AATGCGGGTTTTT**ATC**TATATATATATATA

(a) species X

(b) species Y

(c) species Z

(d) two of the above

8 Which of the following is *not* an example of artificial selection?

(a) Your younger sibling got a hamster as a birthday present, and it turned out to be pregnant. Several of the offspring had long, silky hair, and your sibling put them together in an enclosure to try to produce more baby hamsters with long, silky hair. Your sibling continued to breed long- and silky-haired hamsters to each other and now, several years later, your sibling's bedroom is full of cages with long- and silky-haired hamsters.

(b) Your mother has been saving the best seeds from her lima bean plants every summer and replanting them the next year. She likes seeds that are plump and bright green and saves only these each year. Within 10 years, almost all of her lima beans are plump and bright green.

(c) Farmer Brown has a duck that can type. He sets up an online dating profile for his duck to find a female duck that can also type. Seven years after the arranged wedding of the two typing ducks, Farmer Brown has an entire flock of ducks, most of which can type.

(d) Female fish in a natural pond have variable skill at depositing their eggs near the murky shore where the eggs are better hidden from predators. Eggs that are not deposited near the murky shore are quickly devoured by fish. Female hatchlings from the eggs deposited near the murky shore grow up to be good at depositing eggs near the shore and therefore have a survival advantage.

9 The fossil record shows that the first mammals evolved 220 million years ago. The supercontinent Pangaea began to break apart 200 million years ago. Therefore, fossils of the first mammals should be found

(a) on most or all of the current continents.

(b) only in Antarctica.

(c) on only one or a few continents.

(d) only in Africa.

Leveling Up

10 **What do *you* think?** The prerequisites for medical school application always include courses in cell biology, genetics, and biochemistry but rarely include a formal course in evolution. Do you think medical schools should require a formal course in evolution as a prerequisite for admission? Why or why not? Research the issue and support your case using information you find.

11 ***Write Now* biology: evidence for evolution** This assignment is designed to expand your knowledge of the evidence for evolution. View the following videos, found at http://www.pbs.org/wgbh/evolution/educators/teachstuds/svideos.html, and answer the questions accompanying each one.

Video 1: "Isn't Evolution Just a Theory?"
Why is evolution not *just* a theory? Use specifics from the video to defend your answer.

Video 2: "Who Was Charles Darwin?"
Why do you think Charles Darwin's ideas and book *On the Origin of Species* were so groundbreaking and "revolutionary"? Use specifics from the video to defend your answer.

Video 3: "How Do We Know Evolution Happens?"
Describe how the video portrays whale evolution. Include specific examples of transitional fossils described in the video, and explain why the scientists at the time the video was made considered the fossils to be whale ancestors.

Video 5: "Did Humans Evolve?"
DNA sequences of different species can be used to provide evidence of common descent. Using examples from the video, explain why DNA sequence similarity is the best evidence of evolution on Earth.

Video 6: "Why Does Evolution Matter Now?"
Describe why the theory of evolution matters to the field of medicine and to individual doctors. Use the example of tuberculosis from the video to support your answer. How does this video affect your answer to question 10?

Battling Resistance

An antibiotic-resistant superbug is evolving to overcome our drugs of last defense. How do we stop the perfect pathogen?

- -

After reading this chapter you should be able to:

- Understand that evolution occurs only in populations, not in individuals.
- Describe how natural selection improves reproductive success of a population in its environment and how it differs from sexual selection.
- Compare convergent evolution and evolution by common descent.
- Describe how DNA mutations create new alleles at random.
- Understand how gene flow works and give examples of when this process inhibits evolution.
- Define genetic drift and give examples of genetic bottlenecks and the founder effect.

EVOLUTION

CHAPTER

11

MECHANISMS
OF EVOLUTION

mycı
on, USP

awn Sievert vividly remembers June 14, 2002—the day disaster struck. Sievert is an epidemiologist, or "disease detective": someone who studies the patterns and causes of human disease. At the time, she was working at the Michigan Department of Community Health, monitoring reported cases of antibiotic-resistant bacteria, a major healthcare concern. Sievert spent a significant amount of her time investigating outbreaks of an increasingly common and worrisome microbe called MRSA, short for methicillin-resistant *Staphylococcus aureus*.

S. aureus, commonly known as "staph," is a small, round bacterium that usually lives benignly in our nostrils and on our skin. But on rare occasions, staph slips beneath the surface of a burn or cut and causes an infection, which can be dangerous—especially for individuals with suppressed immune systems, such as the elderly and patients on chemotherapy.

Staph is one of the most common causes of hospital infections today, and it is treated with antibiotics, drugs that kill bacteria but not human cells. Penicillin was the first antibiotic used against staph (**Figure 11.1**), but the wily microbe evolved resistance to penicillin even before the drug became commercially available to the public in the 1940s.

When penicillin stopped working against staph, doctors switched to an antibiotic called methicillin. Methicillin worked for about 20 years, until populations of staph evolved widespread resistance to that antibiotic as well. To the chagrin of doctors and patients everywhere, bacteria adapt rapidly to new threats and share antibiotic resistance genes among themselves. These tiny microbes are the Navy SEALs of evolution—the best of the best at evolving.

As we saw in Chapter 10, biological *evolution* is a change in the frequencies of inherited traits in a population over several generations. Staph adapted to the presence of methicillin. Only bacteria with traits protecting the microbe from the antibiotic survived. These traits were passed within populations and down from one generation to the next until methicillin resistance was frequent across staph populations (**Figure 11.2**).

Today, MRSA is rampant in hospitals, so doctors have been forced to turn to one of medicine's last lines of defense against the "superbug." Vancomycin, a strong, blunt antibiotic that was first isolated from the mud of the Borneo jungle, is considered one of the "drugs of last resort" for fighting these serious infections—which brings us back to June 14, 2002, and Dawn Sievert.

On that day, a lab technician at a dialysis center in Detroit, Michigan, took two swabs of an infected foot ulcer of a 40-year-old diabetic woman (**Figure 11.3**). The patient previously suffered from numerous foot infections, including MRSA, for which she had been treated with vancomycin for 6½ weeks. The swabs of this latest infection were sent to a local laboratory, where technicians grew the bacteria in a dish to test its susceptibility to antibiotics.

When the results of the first test came in, the laboratory staff immediately picked up the phone and called Sievert's office. The bacteria, they told Sievert and the health department, appeared to be resistant to vancomycin. "First, we needed laboratory confirmation and had to control any potential for panic," says Sievert. Her team asked the local lab to run its own test again and to send the health department a sample to independently test. Both teams waited.

The tests came back from each lab. They were both positive; the woman's foot was infected with the first reported vancomycin-resistant strain of *S. aureus*. VRSA had arrived.

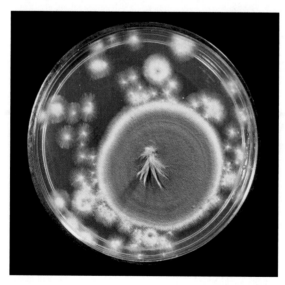

Figure 11.1

Penicillin, produced by mold, kills bacteria

The mold *Penicillium*, the fuzzy white growth with blue spores growing on this petri dish, secretes the antibiotic penicillin into the agar medium surrounding it.

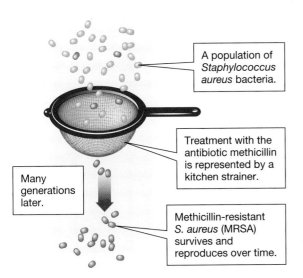

A population of *Staphylococcus aureus* bacteria.

Treatment with the antibiotic methicillin is represented by a kitchen strainer.

Many generations later.

Methicillin-resistant *S. aureus* (MRSA) survives and reproduces over time.

Figure 11.2

Natural selection results in resistance to antibiotics

The use of antibiotics allows any bacteria that are randomly resistant to the antibiotic to survive and reproduce. Over time, the frequency of resistant bacteria increases in the surviving populations.

Q1: What is natural selection selecting for here?

Q2: Why do bacteria that are not randomly resistant to antibiotics die out when exposed to antibiotics?

Q3: Why is the antibiotic represented by a kitchen strainer in this figure?

Figure 11.3

MRSA infections in the foot of a patient with diabetes

Birth of a Superbug

"At that point, it was the first ever VRSA in the world," says Sievert. Unfortunately, it was not the last. Through evolution, staph first survived penicillin, then methicillin, and finally vancomycin (**Figure 11.4**). The evolution of *S. aureus* is a profound example of a species changing over time.

In the mid-1800s, two English biologists, Charles Darwin and Alfred Wallace, studied the diversity of life and concluded that species were

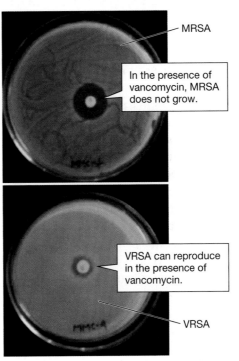

MRSA

In the presence of vancomycin, MRSA does not grow.

VRSA can reproduce in the presence of vancomycin.

VRSA

Figure 11.4

MRSA versus VRSA in a vancomycin-resistance test

In the presence of vancomycin, MRSA does not grow, while VRSA grows relatively well.

Q1: What is the difference between MRSA and VRSA?

Q2: Why is there a clear zone around the paper disk in the top dish and not the bottom dish?

Q3: Why is the lack of a clear zone around the paper disk in the bottom dish so alarming?

not, as was generally thought at that time, the unchanging result of separate acts of creation. Instead, both men came to the bold new conclusion that species "descend with modification" from ancestor species; that is, new species arise from previous species.

Descent with modification, we know today, occurs not only in populations of large organisms like whales and finches, but also in the tiniest single-celled bacteria and viruses. We generally think of evolution happening over millions of years, but some evolutionary changes, such as adaptations for antibiotic resistance, take place over very short time spans as particular alleles spread rapidly through a population. Recall that *alleles* are different versions of the same gene (sequences of DNA) produced by random mutation, and therefore *allele frequencies* are percentages of specific alleles in a population (**Figure 11.5**). Evolution corresponds to changes in the proportions of alleles in a population over time.

When allele frequencies in a population change, becoming more or less common, the attributes or phenotypes of the population change as well; that is, the population evolves. As more and more staph containing the allele for methicillin resistance survived and reproduced, the whole population of staph evolved, becoming new, more powerful bacteria. But how exactly does this happen? Where do new alleles come from, and how do the frequencies of alleles in a population change?

Sievert and a host of other researchers and doctors experienced the emergence of a new allele firsthand. When the results of the foot ulcer test came in positive for VRSA, Sievert's team immediately called the Centers for Disease Control and Prevention (CDC), a government agency that investigates disease outbreaks and makes public health recommendations. The Michigan

DAWN SIEVERT

Infectious diseases epidemiologist Dawn Sievert works for the Centers for Disease Control and Prevention. While at the Michigan Department of Community Health, Sievert investigated the first-ever VRSA infection.

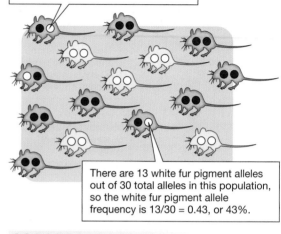

Because there are 15 mice depicted in this population, there are 30 alleles possible in this gene pool (15 mice multiplied by 2 alleles per mouse).

There are 13 white fur pigment alleles out of 30 total alleles in this population, so the white fur pigment allele frequency is 13/30 = 0.43, or 43%.

●●●●●●●●●●●●●●●●● 17/30 = 57%
○○○○○○○○○○○○○ 13/30 = 43%

Figure 11.5

Allele frequencies are calculated as percentages in a population

These mice have two white fur pigment alleles and appear white, have two black fur pigment alleles and appear gray, or have one black and one white allele and appear gray. To calculate the white fur pigment allele frequency in the population, the number of white alleles is counted and divided by the total number of alleles.

Q1: What would the white fur pigment allele frequency be if three of the homozygous black allele mice (having two black alleles) were heterozygous (having one white and one black allele) instead?

Q2: What would the white fur pigment allele frequency be if all of the white mice died and were therefore removed from the population? Would the black fur pigment allele frequency be affected? If so, how?

Q3: What would the white fur pigment allele frequency be if all of the gray mice died and were therefore removed from the population?

Health Department and the CDC converged on the dialysis center where VRSA had been found. They pored over the medical history of the patient, examined her wound, took swabs from the nostrils and wounds of anyone who had come in contact with her, and then waited anxiously to see whether the dangerous microbe had spread.

Thankfully, they found that the vancomycin-resistant bug had not spread. "We're lucky. If that highly resistant organism had the ability to spread rapidly, we'd have some very sick people at risk in hospitals and other healthcare settings," says Sievert.

Yet it was not the end of VRSA. Since 2002, there have been 12 additional cases of vancomycin-resistant staph infections in the United States: in the urine of a woman in New York with multiple sclerosis, in the toe wound of a diabetic man in Michigan, in the triceps wound of a woman in Michigan, and more. So far, each infection has been isolated; the microbe has never been transmitted from person to person, as MRSA has. VRSA, while dangerous, doesn't appear to spread through the human population.

But that's not to say it won't evolve. When MRSA first emerged, it seemed restricted to hospital settings. Today, however, clinicians have been horrified to see cases of MRSA pop up from simple scrapes on a playground, proving that the microbe is out in the community, where it can do widespread damage. In theory, the same is possible for VRSA, raising the bone-chilling specter of the superbug evolving into an "apocalyptic bug," as one reporter called it.

So how did staph acquire vancomycin resistance 13 separate times? How likely is vancomycin resistance to become more widespread? We can find answers to these questions by understanding four mechanisms by which evolution occurs:

1. Natural selection

2. Mutation

3. Gene flow

4. Genetic drift

Bacteria are perfect organisms for examining these evolutionary mechanisms for the same reason that they are so dangerous—because they evolve incredibly fast.

Rising Resistance

Harvard Medical School microbiologist Michael Gilmore has long tracked the ways that bacteria evolve antibiotic resistance. After staph evolved widespread resistance to methicillin in the 1980s and vancomycin began to be used in hospitals, "we waited and waited and waited" for vancomycin resistance to emerge, says Gilmore. He knew that once vancomycin was widely used to kill staph, the microbe would evolve a way to avoid the poison and continue to reproduce. The process by which a population gains one or more alleles that enable it to survive better than other populations is called *natural selection*, as we saw in Chapter 10. Darwin and Wallace were the first to propose natural selection, and today we know that it is the central driver of evolution.

During natural selection, individuals with particular inherited characteristics survive and reproduce at a higher rate than other individuals in a population. Natural selection acts by favoring some phenotypes over others (**Figure 11.6**). For example, in an environment where bacteria are exposed to vancomycin, the bacteria that can resist the antibiotic will live on and reproduce, while those that cannot will perish. Although natural selection acts directly on the phenotype, not on the genotype, of a population, the alleles that code for a trait favored by natural selection tend to become increasingly common in future generations. Bacteria that survive an antibiotic attack, for example, pass on alleles that confer that resistance to their offspring.

Natural selection acts on adaptive traits, and therefore a population can become better suited to survive and reproduce in its environment. The overall reproductive success of the population within an environment is consistently improved over time through natural selection. And unfortunately for us, natural selection is the mechanism by which staph is adapting to our use of vancomycin.

MICHAEL GILMORE

Michael Gilmore is a microbiologist at Harvard Medical School. He and his laboratory uncovered the genetic basis for the recent emergence of VRSA.

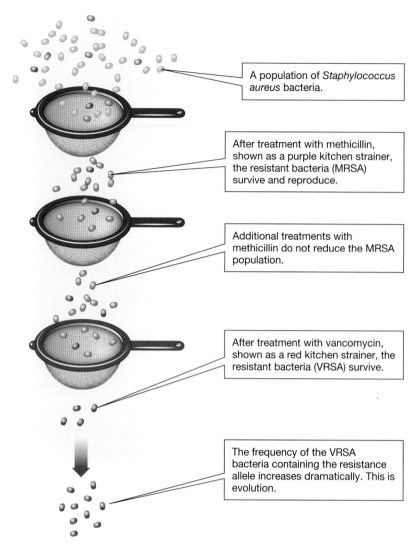

A population of *Staphylococcus aureus* bacteria.

After treatment with methicillin, shown as a purple kitchen strainer, the resistant bacteria (MRSA) survive and reproduce.

Additional treatments with methicillin do not reduce the MRSA population.

After treatment with vancomycin, shown as a red kitchen strainer, the resistant bacteria (VRSA) survive.

The frequency of the VRSA bacteria containing the resistance allele increases dramatically. This is evolution.

Figure 11.6

Evolution happens

Imagine there is a population of *Staphylococcus aureus* bacteria living on your skin. Most of them are susceptible to the antibiotic methicillin (purple strainer). A few, however, are randomly resistant to methicillin (purple), like the ones in Figure 11.2. A few of these bacteria, in turn, could also be resistant to vancomycin (red).

Q1: Why does the population of *S. aureus* bacteria *not* pose a life-or-death health threat outright?

Q2: Why do the vancomycin-resistant bacteria have a higher frequency in the population after treatment with vancomycin?

Q3: If this figure used the mouse example of allele frequency from Figure 11.5 and the white mice increased in numbers like the vancomycin-resistant bacteria here did, what would happen to the allele frequency of the white fur pigment allele? What would happen to the black fur pigment allele frequency?

Because staph's vancomycin resistance adaptation is very recent, scientists are studying the patterns of how natural selection gives rise to VRSA. There are three common patterns of natural selection that we observe in nature: *directional selection, stabilizing selection,* and *disruptive selection.* Whatever the pattern, all types of natural selection operate by the same principle: individuals with certain forms of an inherited trait have better survival rates and produce more offspring than do individuals with other forms of that trait.

Directional selection is the most common pattern of natural selection, in which individuals at *one extreme* of an inherited phenotypic trait have an advantage over other individuals in the population. The peppered moth provides a vivid example.

Before 1959, dark-colored moths had risen in frequency in both England and the United States after industrial pollution blackened the bark of trees, causing dark-colored moths to be harder for bird predators to find than light-colored moths. A reduction in air pollution following clean-air legislation, enacted in 1956 in England and in 1963 in the United States, caused the bark of trees to become lighter, and light-colored moths became harder for predators to find than dark-colored moths. As a result, the proportion of dark-colored moths plummeted because they were easily seen and eaten by predators (**Figure 11.7**). Similarly, when methicillin became widely used to fight staph, MRSA evolved via directional selection: the bacteria in hospital settings that were resistant to the antibiotic survived, while those that were not perished.

In cases of **stabilizing selection**, individuals with *intermediate values* of an inherited phenotypic trait have an advantage over other individuals in the population. Birth weight in humans provides a classic example of this pattern of natural selection (**Figure 11.8**). Historically, light or heavy babies did not survive as well as babies of average weight, and as a result there was stabilizing selection for intermediate birth weights. Today, however, this stabilizing trend is not as strong, because advances in the care of premature babies and an increase in the use of cesarean deliveries for large babies have allowed babies of all weights to thrive.

Finally, **disruptive selection** occurs when individuals with *either extreme of an inherited trait* have an advantage over individuals with an intermediate phenotype. This type of selection is the

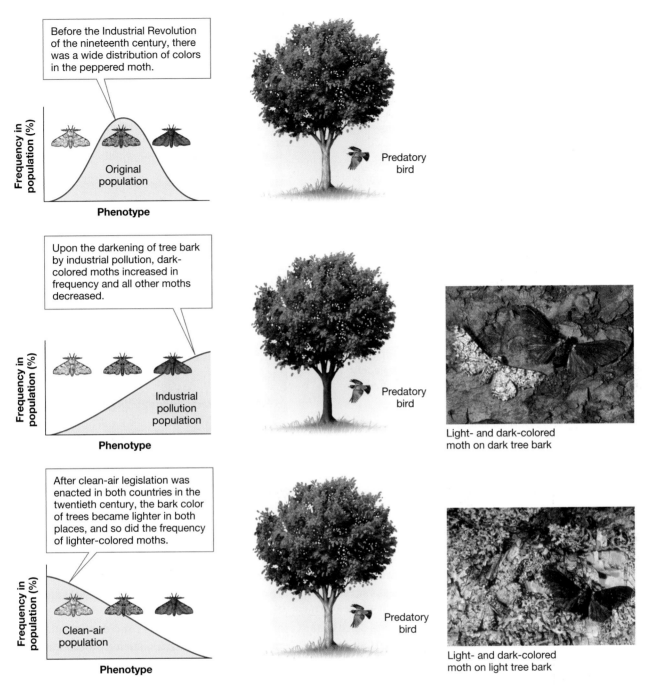

Before the Industrial Revolution of the nineteenth century, there was a wide distribution of colors in the peppered moth.

Frequency in population (%)

Original population

Phenotype

Upon the darkening of tree bark by industrial pollution, dark-colored moths increased in frequency and all other moths decreased.

Frequency in population (%)

Industrial pollution population

Phenotype

Predatory bird

Light- and dark-colored moth on dark tree bark

After clean-air legislation was enacted in both countries in the twentieth century, the bark color of trees became lighter in both places, and so did the frequency of lighter-colored moths.

Frequency in population (%)

Clean-air population

Phenotype

Predatory bird

Light- and dark-colored moth on light tree bark

Figure 11.7

The peppered moth has undergone directional selection two different times in the last 200 years

The Industrial Revolution of the nineteenth century in both England and the United States caused extreme air pollution due mostly to soot from the mass burning of coal.

Q1: If one extreme phenotype makes up most of a population after directional selection, what happened to the individuals with the other phenotypes?

Q2: What do you think would happen to the phenotypes of the peppered moth if the tree bark was significantly darkened again by disease or pollution?

Q3: What do you think would happen to the phenotypes of the peppered moth if the tree bark became a medium color, neither light nor dark?

Figure 11.8

Stabilizing selection and human birth weight from 1935 to 1946

This graph is based on data for 13,700 babies born between 1935 and 1946 in a hospital in London. In countries that can afford intensive medical care for newborns, the strength of stabilizing selection has been greatly reduced in recent years. However, even with improved care of premature babies and cesarean deliveries of large babies, babies at the extremes of newborn weight still survive at a lower rate than those closer to the median weight.

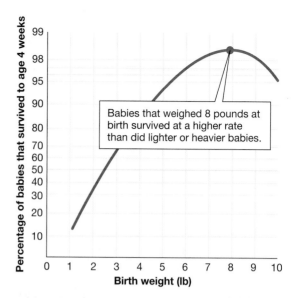

Babies that weighed 8 pounds at birth survived at a higher rate than did lighter or heavier babies.

Q1: Think of another example of stabilizing selection in human biology. Has modern technology or medicine changed its impact on the resulting phenotypes?

Q2: How do you think a graph of birth weight versus survival for a developing country with little health care would compare to the graph shown here?

Q3: How do you think a graph of birth weight versus survival for an affluent city in the United States today would compare to the graph shown here?

least commonly observed in nature, but one example is the beak size within a population of the birds called African seed crackers (**Figure 11.9**). During one dry season, birds with large beaks survived on hard seeds and birds with small beaks survived on soft seeds, but birds with intermediate-sized beaks fed inefficiently on both types of seeds. As a result, the birds with intermediate-sized beaks died, and the birds with large and small beaks lived. Therefore, natural selection favored both large-beaked and small-beaked birds over birds with intermediate beak sizes.

Any of these patterns of natural selection can cause distantly related organisms to evolve similar structures because they survive and reproduce under similar environmental pressures. This type of evolution, called **convergent evolution**, results in organisms that have different genetics but appear very much alike. Cacti found in North American deserts and distantly related plants found in African and Asian deserts offer an excellent example of convergent evolution. These two types of desert plants have very different genetics, but they look similar and function in a similar manner. In another example, sharks and dolphins are only distantly related (sharks are fish, and dolphins are marine mammals), yet they both evolved for success as predators in the ocean and share common characteristics, such as a streamlined body (**Figure 11.10**). When species share characteristics because of convergent evolution and not because of modification by descent from a recent common ancestor, those characteristics are called **analogous traits** (instead of homologous traits).

Enter Enterococcus

Intent on determining how VRSA was evolving after the first Michigan infection, Gilmore began to track the appearance of the bug. He and others wanted to know where the allele for

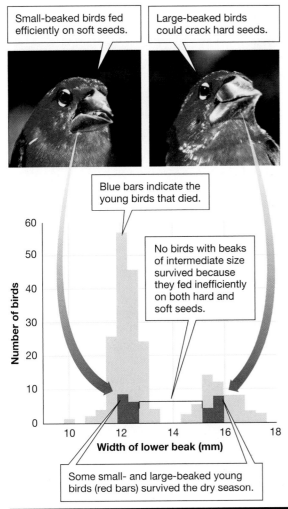

Small-beaked birds fed efficiently on soft seeds.

Large-beaked birds could crack hard seeds.

Blue bars indicate the young birds that died.

No birds with beaks of intermediate size survived because they fed inefficiently on both hard and soft seeds.

Some small- and large-beaked young birds (red bars) survived the dry season.

Figure 11.9

Disruptive selection for beak size

Among a group of young African seed crackers hatched in one year, only those with small or large beaks survived the dry season, when seeds were scarce. Although many of the small- and large-beaked birds did not survive, none of the intermediate-beaked birds survived.

Q1: Almost all birds starved during the dry season depicted here. What type of selection would have been present if only the intermediate-beaked birds had survived (instead of the small- and large-beaked birds)?

Q2: Think of another example of disruptive selection. Now change the parameters so that your example illustrates directional selection instead. Which individuals survive? Which individuals die?

Q3: Of the three patterns of natural selection presented in this discussion, which one always results in two different phenotypes left standing?

vancomycin resistance came from, and how staph had acquired it 13 different times.

As Gilmore gathered information about each subsequent infection of VRSA, searching for patterns, he noticed some peculiar things. First, most of the infections were turning up in people with diabetes, typically in bad foot wounds. Second, in most cases, when scientists looked closely at the samples, they saw not only staph but also a small spherical bacterium called *Enterococcus*, which had evolved resistance to vancomycin years earlier (**Figure 11.11**). Upon close observation, Gilmore and others noted that the enterococci, growing cozily side by side with staph, contained a vancomycin-resistance gene identical to the one present in the staph. This suggested that staph had acquired vancomycin resistance directly from *Enterococcus*, rather than via random mutations in its own genome.

New alleles in a species emerge via mutation. A **mutation** is a change in the sequence of any segment of DNA in an organism, and it is the only means by which new alleles are generated. DNA mutations create new alleles at random, thereby providing the raw material for evolution. In this sense, all evolutionary change depends ultimately on mutation. Mutations can stimulate the rapid evolution of populations by providing new genetic variation—differences in genotypes between individuals within a population. Then, natural selection and other mechanisms of evolution act on the resulting phenotypes.

In sexually reproducing species, genetic mutations that occur in an organism's germ line cells—the cell lineage that produces gametes such as eggs and sperm—can contribute to evolution. Mutations in other cells of the body, such as skin or blood cells, can affect the individual by causing cancer or other problems, but those mutations are not passed to that individual's offspring. If mutations are not passed to offspring, they cannot contribute to evolution.

These three plants evolved from very distantly related plants via natural selection to adapt to life in the desert.

Natural selection has caused two distantly related animals—sharks (a type of fish, left) and dolphins (a type of mammal, right)—to look very similar.

Figure 11.10

Natural selection can result in convergent evolution

Q1: How is convergent evolution different from evolution by common descent?

Q2: What is the main difference between a homologous structure and an analogous structure?

Q3: Why is convergent evolution considered evidence for evolution (see Chapter 10)?

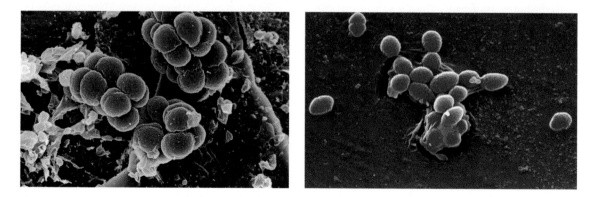

Figure 11.11

Staphylococcus aureus (left) and *Enterococcus* (right)

The same does not hold true for bacteria, which are single cells and reproduce via asexual reproduction. All genetic mutations in a bacterial cell are passed on to the offspring. (See Chapter 5 for a review of cellular replication.) When a mutation is passed to an offspring, if that mutation increases the individual's ability to reproduce, it is favored by natural selection. These favored mutated genes are passed from parent to offspring, spreading through future generations in a way that alters the population as a whole.

But bacteria don't always wait for the right random mutations to pop up in their genomes. Sometimes they simply borrow new alleles from one another. That appeared to be the case with VRSA, researchers found. "In most cases, VRSA has developed in a perfect storm of a very bad wound that's not healing and is a soup mix of organisms, *Enterococcus* and staph coming together and sharing genes," says Sievert.

Horizontal gene transfer is the process by which bacteria pass genes to one another (**Figure 11.12**). Bacteria store these genes on small, circular pieces of DNA called plasmids. They send these plasmids to each other through small tunnels, like mailing the instructions for new traits. The physical process of transferring plasmids is called *conjugation*, and the physical connection created between two bacteria involved in horizontal gene transfer is known as a *conjugation tube*. Some bacteria also participate in horizontal gene transfer without using a connecting tunnel. Such direct uptake of plasmid DNA is known as *natural transformation*.

Horizontal gene transfer is one example of another mechanism by which evolution occurs: gene flow. **Gene flow** is the exchange of alleles between populations. Gene flow can occur between two different species—in this case, between a population of staph and a population of *Enterococcus*—or between two populations of the same species, such as strains of staph passing methicillin resistance among themselves in the community. An individual that migrates between two otherwise isolated populations of a species may facilitate gene flow as well (**Figure 11.13**). Gene flow can also occur when only gametes move from one population to another, as happens when wind or pollinators like insects transport pollen from one population of plants to another.

The introduction of new alleles via gene flow can have dramatic effects. Two-way gene flow consists of an exchange of alleles between one

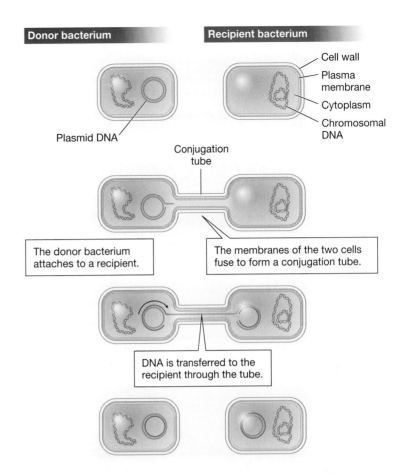

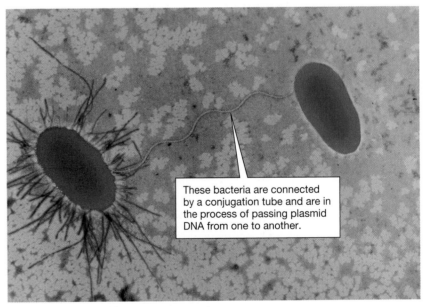

Figure 11.12

Horizontal gene transfer accelerates the rate of evolution in prokaryotes

The diagram depicts the horizontal gene transfer of plasmid DNA through conjugation. In the case of VRSA, *Staphylococcus aureus* acquired the vancomycin resistance gene through horizontal gene transfer from *Enterococcus*.

population and another, so it tends to make the genetic composition of different populations more similar. If one strain of staph shares the methicillin resistance allele with another strain of staph, for example, the two populations become more alike. A mutual exchange of alleles through gene flow can *counteract* the effects of the other mechanisms, such as mutation, that tend to make populations more different from one another.

It appeared that gene flow was responsible for the emergence of VRSA—staph picked up an allele for vancomycin resistance from *Enterococcus*. But Gilmore made another observation that suggested why this horizontal gene transfer was happening now, and not back in the 1980s when vancomycin had first begun to be widely used.

Gilmore noted that all the VRSA samples taken from patients were clonal cluster 5, or CC5, strains of *S. aureus*. "There were implications that there was something special about clonal cluster 5 strains that was leading to this vancomycin resistance bubbling up [exclusively] in these strains," says Gilmore. CC5 strains appeared to have somehow evolved to readily take up and use vancomycin resistance alleles from *Enterococcus*.

Primed for Pickup

In 2012, Gilmore and his team at Harvard analyzed the DNA of 11 of the 12 known cases of VRSA at the time (samples from 1 of the 12 weren't available, and the 13th case had yet to occur). In the genome of every sample of VRSA, they found three traits that demonstrate how the

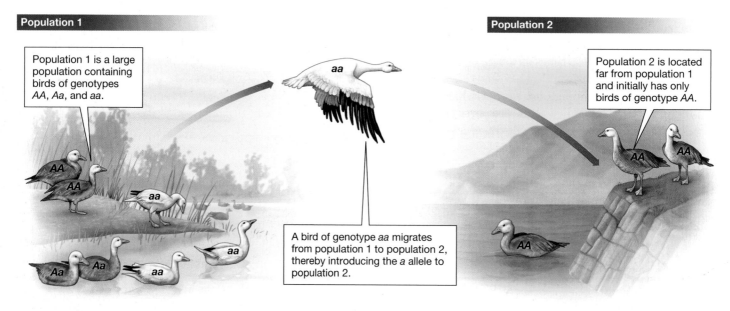

Figure 11.13

Migrants can move alleles from one population to another

As a result of gene flow between these two genetically disparate populations of geese, the populations will eventually become more similar to each other. Mating between the migrant *aa* genotype goose and a population 2 goose of *AA* genotype will result in *Aa* genotype offspring. Continued mating of the *Aa* genotypes will result in all three possible genotypes (*AA*, *Aa*, and *aa*). The resulting population 2 gene pool now looks much more similar to the gene pool of population 1.

Q1: If a goose with genotype *AA* had migrated instead of the goose with genotype *aa*, would this still be considered gene flow? Why or why not?

Q2: If a goose with genotype *Aa* had migrated instead of the goose with genotype *aa*, would this still be considered gene flow? Why or why not?

Q3: If the goose with genotype *aa* migrated to population 2 as shown but failed to mate with any of the *AA* individuals, would this still be considered gene flow? Why or why not?

Race Against Resistance

Each year in the U.S., over two million people become infected with antibiotic-resistant bacteria, including methicillin-resistant *Staphylococcus aureus* (MRSA), resulting in at least 23,000 deaths. To combat antibiotic resistance, doctors need novel antibiotics, yet there are fewer and fewer new antibiotics coming to pharmacies each year.

Antibacterial drugs approved by the FDA

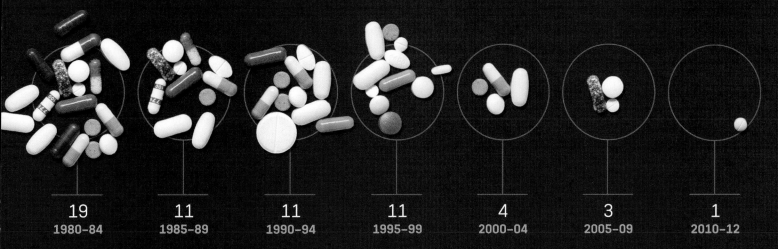

19	11	11	11	4	3	1
1980–84	1985–89	1990–94	1995–99	2000–04	2005–09	2010–12

Incidence of strains of bacteria resistant to three common antibiotics

Ciprofloxacin-resistant *Campylobacter*

Ciprofloxacin-resistant *Salmonella* Typhi

/ / / Erythromycin-resistant Group B *Streptococcus*

/////// Clindamycin-resistant Group B *Streptococcus*

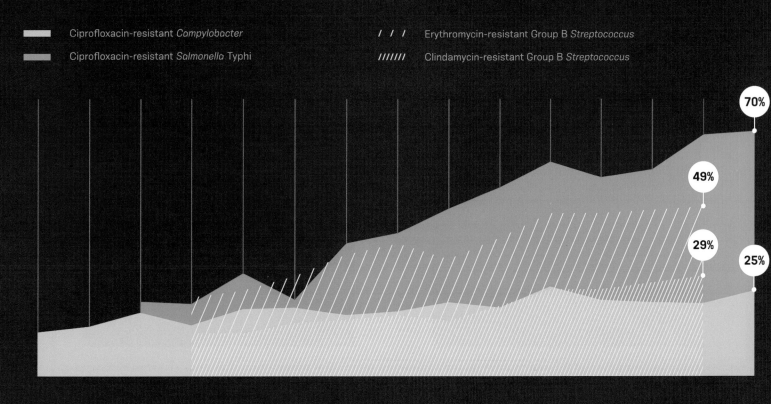

70%

49%

29%

25%

1997 1998 1999 2000 2001 2002 2003 2004 2005 2006 2007 2008 2009 2010 2011

CC5 strains of staph evolved to effectively pick up the allele for vancomycin resistance.

First, all the vancomycin-resistant CC5 staph bacteria have the same mutation in a gene called *DprA*. *DprA* appears to be involved in preventing horizontal gene transfer from occurring. A mutation in this gene might make it easier for the staph to take up DNA from other bacteria, such as *Enterococcus*.

Second, the CC5 strains lack a set of genes that encode an antibiotic that kills other bacteria. Perhaps this antibiotic normally kills *Enterococcus* near the staph, which would explain why horizontal transfer between the two species does not typically occur.

Finally, Gilmore and his team found that in place of that missing set of antibiotic genes, the vancomycin-resistant CC5 staph have a unique cluster of genes encoding proteins that confuse the human immune system. These proteins could make it easier for staph and other bacteria to grow in a wound because the host immune system would be less able to fight them off.

Both the lack of the antibiotic genes and the presence of new genes create a perfect storm for a mixed infection, in which different species of pathogens mingle in a festering soup of contamination. Mixed infections are breeding grounds for antibiotic resistance because they are sites of gene flow among very different organisms. In this explanation, the CC5 staph evolved via the three mechanisms—mutation, natural selection, and gene flow (via horizontal transfer)—to be more susceptible to take up the vancomycin-resistance allele from *Enterococcus*.

In addition to natural selection (and its strange permutation called *sexual selection*; see "Sex and Selection"), mutation, and gene flow, there is one

Sex and Selection

Staphylococcus and other bacteria replicate asexually by copying their DNA and dividing in two. But reproduction gets more complicated when sex is added to the equation. Another mechanism by which species evolve is called sexual selection. In **sexual selection**, nature selects a trait that increases an individual's chance of mating—even if that trait decreases the individual's chance of survival.

Sexual selection favors individuals that are good at getting mates, and it often helps explain differences between males and females in size, courtship behavior, and other traits. Species whose males and females are distinctly different in appearance, as seen in peacocks, lions, and ducks, are said to exhibit **sexual dimorphism**. In many species, the members of one sex—often females—are choosy about whether to mate. In birds, for example, brightly colored males may perform elaborate displays in their attempts to woo a mate. In other species, males may attract attention by other means, such as calling vigorously; females then select as their mates the males with the loudest calls.

Yet some characteristics that increase an individual's chance of mating can *decrease* its chance of survival. For example, male túngara frogs perform a complex mating call that may or may not end in one or more "chucks." Females prefer to mate with males that emit chucks, but frog-eating bats use that same sound to help them locate their prey. As a result, a frog's attempt to locate a mate can end in disaster.

Male túngara frogs face an ecological tradeoff: the same type of call that is most successful at attracting females for mating also attracts predatory bats to dinner.

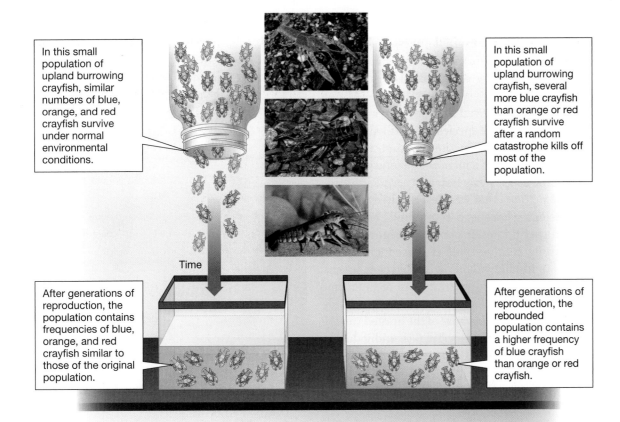

In this small population of upland burrowing crayfish, similar numbers of blue, orange, and red crayfish survive under normal environmental conditions.

In this small population of upland burrowing crayfish, several more blue crayfish than orange or red crayfish survive after a random catastrophe kills off most of the population.

Time

After generations of reproduction, the population contains frequencies of blue, orange, and red crayfish similar to those of the original population.

After generations of reproduction, the rebounded population contains a higher frequency of blue crayfish than orange or red crayfish.

Figure 11.14

A genetic bottleneck is a type of genetic drift

Two small populations of upland burrowing crayfish are compared. One, but not the other, experiences a genetic bottleneck event. This is genetic drift—a change in the frequency of a trait that is not associated with natural selection. In fact, the blue crayfish could be less well adapted to the environment than the other crayfish are.

Q1: Why do you think a genetic bottleneck is more likely to occur in a small population than in a large population?

Q2: Genetic drift is often described as a "chance event." Name several chance events that could cause a genetic bottleneck.

Q3: Which resulting population has the most genetic diversity?

other mechanism by which organisms evolve. Beneficial alleles like antibiotic resistance for bacteria are usually selected for and maintained in a population by the nonrandom action of natural selection. But in some cases, chance events can cause alleles from a parent generation to be selected for inclusion in the next generation.

Genetic drift is a change in allele frequencies produced by random differences in survival and reproduction among the individuals in a population. In a generation, some individuals may, just by chance, leave behind more descendants than other individuals do. In this case, the genes

of the next generation will be the genes of the "lucky" individuals, not necessarily the "better" individuals. Genetic drift occurs in all populations, including mammals and bacteria, but it is more likely to cause evolution in a small population than in a large one.

There are two mechanisms by which genetic drift occurs: *genetic bottlenecks* and the *founder effect*. A **genetic bottleneck** is a drop in the size of a population, for at least one generation, that causes a loss of genetic variation (**Figure 11.14**). A genetic bottleneck can threaten the survival of a population.

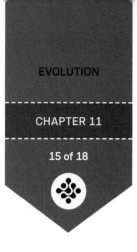

In the 1970s, for example, the population of the endangered Florida panther plummeted because of hunting and habitat destruction. The species barely escaped extinction. At one point, experts believed that only six wild individuals in the whole species were still alive. This rapid population reduction created a genetic bottleneck in which a lot of the genetic variation within the species was lost, and severe inbreeding among those that were left resulted in maladies including low sperm counts and abnormally shaped sperm in male panthers (**Figure 11.15**). Thankfully, panther numbers have increased to about 80–100 individuals in recent years, in part because of breeding programs. All of the members of this rebounded population are incredibly similar genetically, because of the limited variety of alleles present in the original six individuals.

Figure 11.16

Early Dutch colonists in South Africa

The **founder effect** occurs when a small group of individuals establishes a new population isolated from its original, larger population. For example, in South Africa a population of people called Afrikaners descended primarily from a few Dutch colonists (**Figure 11.16**). Today, the Afrikaner population has an unusually high frequency of the allele that causes Huntington disease because those original colonists by chance carried the allele when they settled in the area.

After Vancomycin

When the initial case of VRSA popped up back in 2002, the patient's foot was carefully treated with an old topical drug that is rarely used today. The only other options were one or two new drugs that hadn't yet been released on the market. "There are still some antibiotics we can use for treatment, but we recommend those be used very responsibly in order to maintain their effectiveness," says Sievert. "You don't want to give them out when not necessary because you don't want to see resistance to them. You save them as the end of the line, so something is available if the commonly used drugs end up failing due to resistance." After nine months of careful treatment,

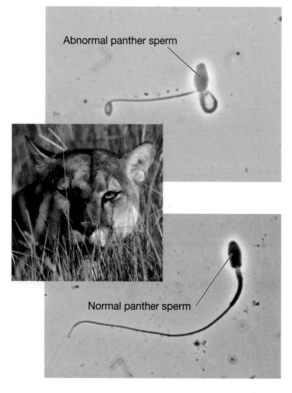

Abnormal panther sperm

Normal panther sperm

Figure 11.15

Abnormally shaped sperm in the rare Florida panther

Florida panthers have more abnormal sperm than do panthers from other populations—a possible effect of a genetic bottleneck promoting the increase of harmful alleles in the Florida panther population.

the patient healed and was able to keep her foot. Other cases of VRSA have similarly been resolved using the newer antibiotics and the delicate care of doctors.

But what happens when even these drugs no longer work against staph? Few new antibiotics are currently being developed by pharmaceutical companies, says Frank DeLeo, acting chief of the Laboratory of Human Bacterial Pathogenesis at the National Institute of Allergy and Infectious Diseases, and bacteria evolve more quickly than we can develop new antibiotics. It's a vicious cycle, he says. "As we continue to use antibiotics, populations of microbes will develop resistance to antibiotics."

VRSA has not yet become widespread, as MRSA has. Scientists believe the reason may be that staph do not handle the resistance allele for vancomycin very well; the gene from *Enterococcus* is bulky and difficult to manage, so the staph that receive the gene are often outcompeted by other staph in the environment. "Though it allows them to survive, they don't do well with it. There's a fitness cost to having this resistance," says Gilmore. In the absence of the drug, the staph are better off without the resistance allele, so perhaps they lose it or don't pick it up when not in the presence of vancomycin. VRSA therefore remains "a rare, unstable organism in the environment that isn't very hardy, so it doesn't last and spread," says Sievert. "That's the good news."

But there's still trepidation that staph may adapt and get comfortable with this resistance gene, just as staph adapted to methicillin resistance, says Gilmore. If staph were to survive and reproduce more easily, with the resistance gene, the gene would likely spread throughout the population. "We don't know if that's a possibility," says Gilmore. "My suspicion is that it is." (See "How Can *You* Make a Difference? Help Prevent Antibiotic Resistance!")

In fact, in July 2012 the CDC reported that a man in Delaware was the thirteenth individual to develop a VRSA infection. Initial reports showed that this strain of VRSA was not a CC5 strain,

How Can *You* Make a Difference? Help Prevent Antibiotic Resistance!

The U.S. Centers for Disease Control and Prevention (CDC) recommends the following guidelines:

- Take antibiotics exactly as the doctor prescribes. Do not skip doses. Complete the prescribed course of treatment, even when you start feeling better.
- Only take antibiotics prescribed for you; do not share or use leftover antibiotics. Antibiotics treat specific types of infections. Taking the wrong medicine may delay treatment and allow bacteria to multiply.

- Do not save antibiotics for the next illness. Discard any leftover medication once the prescribed course of treatment is completed.
- Do not ask for antibiotics when your doctor thinks you do not need them. Remember that antibiotics have side effects. When your doctor says you don't need an antibiotic, taking one might do more harm than good.
- Prevent infections by practicing good hand hygiene and getting recommended vaccines.

SOURCE: http://www.cdc.gov/Features/AntibioticResistance.

says Gilmore, suggesting that vancomycin resistance is spreading to other strains of staph. "It's disconcerting," he adds. It seems other populations of staph may now also be evolving to pick up vancomycin resistance. "The main worry is that this would move into a strain that was highly transmissible in the community," adds DeLeo. "The potential is there."

REVIEWING THE SCIENCE

- Natural selection for inherited traits occurs in three common patterns: directional, stabilizing, and disruptive.

- In **directional selection**, individuals at one phenotypic extreme of a given genetic trait have an advantage over all others in the population.

- In **stabilizing selection**, individuals with intermediate phenotypes have an advantage over all others in the population.

- During **disruptive selection**, individuals with either extreme phenotype have an advantage over those with an intermediate phenotype.

- In **convergent evolution**, distantly related organisms (those without a recent common ancestor) evolve similar structures in response to similar environmental challenges.

- All mechanisms of evolution depend on the genetic variation provided by new **alleles** created by **mutation**.

- **Sexual selection** occurs when a trait increases an individual's chance of mating even if it *decreases* that individual's chance of survival.

- **Gene flow** is the exchange of alleles between separate populations.

- **Genetic drift** is the change in allele frequencies produced by *random* differences in survival and reproduction in a small population and generally occurs through one of two processes: a genetic bottleneck or the founder effect.

- A **genetic bottleneck** occurs when a drop in the size of a population leads to a loss of genetic variation in the new, rebounded population.

- The **founder effect** occurs when a few individuals from a large population establish a new population, leading to a loss of genetic variation in the new, isolated population.

THE QUESTIONS

The Basics

1 The founder effect is a type of _____ in which individuals in one small group of a large population _____.

(a) genetic drift; establish a new population at a distinct location and reproduce

(b) gene flow; establish a new population at a distinct location and reproduce

(c) genetic drift; are the only ones to survive and then subsequently reproduce

(d) gene flow; are the only ones to survive and then subsequently reproduce

2 Unlike natural selection, _____ is not related to an individual's ability to survive and may result in offspring that are less well adapted to survive in a particular environment.

(a) genetic drift

(b) sexual selection

(c) directional selection

(d) convergent evolution

3 In a population, which individuals are most likely to survive and reproduce?

(a) The individuals that are the most different from the others in the population.

(b) The individuals that are best adapted to the environment.

(c) The largest individuals in the group.

(d) The individuals that can catch the most prey.

4 Which of the following statements about convergent evolution is true?

(a) It demonstrates how similar environments can lead to different physical structures.

(b) It demonstrates how similar environments can lead to the same physical structures.

(c) It demonstrates that similarity of structures is due to descent from a common ancestor.

(d) It demonstrates that similarity of structures is due to random chance.

5 Evolution is most accurately described as

(a) a change in allele frequencies in an individual over time.

(b) a change in allele frequencies in a species over time.

(c) a change in allele frequencies in a population over time.

(d) a change in allele frequencies in a community over time.

Try Something New

6 A study of a population of the goldenrod wildflower finds that large individuals consistently survive and reproduce at a higher rate than small or medium-sized individuals. Assuming size is an inherited trait, the most likely evolutionary mechanism at work here is

(a) disruptive selection.

(b) directional selection.

(c) stabilizing selection.

(d) natural selection, but it is not possible to tell whether it is disruptive, directional, or stabilizing.

7 Two large populations of the same species found in neighboring locations that have very different environments are observed to become genetically more similar over time. Which evolutionary mechanism is the most likely cause of this trend?

(a) gene flow

(b) mutation

(c) natural selection

(d) genetic drift

8 The Tasmanian devil, a marsupial indigenous to the island of Tasmania (and formerly mainland Australia as well), experienced a population bottleneck in the late 1800s when farmers did their best to eradicate it. After it became a protected species, the population rebounded, but it is now experiencing a health crisis putting it at risk for disappearing again. Many current Tasmanian devil populations are plagued by a type of cancer called devil facial tumor disease, which occurs inside individual animals' mouths. Afflicted Tasmanian devils can actually pass their cancer cells from one animal to another during mating rituals that include vicious biting around the mouth.

Unlike the immune systems of other species, including humans, the immune system of the Tasmanian devil does not reject the passed cells as foreign or nonself (just as we reject a liver transplant from an unmatched donor), but accepts them as if they were their own cells. Why would a population bottleneck result in the inability of one devil's immune system to recognize another devil's cells as foreign?

(a) Population bottlenecks cause individuals in the resulting population to be more genetically similar to each other. In this case, they are so similar that they do not distinguish another devil's cells as different from their own cells.

(b) Population bottlenecks result in individuals that are always highly susceptible to cancer, regardless of the type.

(c) Population bottlenecks result in individuals that are more genetically diverse, making it more likely that mutations will occur that cause cancer.

(d) Population bottlenecks result in individuals that appear more similar to each other, even though they do not share a recent common ancestor.

9 Global warming is causing more and more ice to melt each year at far northern latitudes, exposing more bare ground than ever before. These vast areas of brown ground coloration make polar bears (which are white) much more conspicuous to their prey. Recently, an infant polar bear was born with brown fur. This polar bear survived to adulthood and has sired several offspring with brown fur. Which of the following is a plausible explanation of how the brown fur trait appeared in these polar bears?

(a) A polar bear realized it would be better to be brown in order to hide more effectively. It induced mutations to occur in its fur pigment gene, which resulted in a change in pigment from white to brown fur.

(b) One or more random mutations occurred in the fur pigment gene in an individual polar bear embryo, which resulted in a change in pigment from white to brown fur.

(c) Increased temperatures due to global warming caused targeted mutations in the fur pigment gene in an individual polar bear embryo, which resulted in a change in pigment from white to brown fur.

(d) A female polar bear realized it would be better for her offspring to be brown and therefore mated with a grizzly bear to achieve this result.

10 In the garden shed belonging to one of this text's authors, stabilizing selection has occurred over the past 10 years in the house mouse, *Mus musculus*. Which of the following scenarios is an example of stabilizing selection?

(a) Small and medium-sized mice cannot reach the seed shelf in the shed and therefore are at a disadvantage for finding food, so they do not survive and reproduce as well as large mice.

(b) Small mice cannot reach the seed shelf, and large mice are easily seen by hawks circling above. Medium-sized mice therefore survive and reproduce better than both small and large mice.

(c) Small mice can easily cross the yard to the vegetable garden, and large mice can easily reach the seed shelf. Medium-sized mice have trouble with the seed shelf and are seen by hawks in the yard. Small and large mice survive and reproduce much better than medium-sized mice.

(d) All of these are examples of stabilizing selection.

(e) None of these are examples of stabilizing selection.

Leveling Up

11 **What do *you* think?** One way to prevent a small population of a plant or animal species from going extinct is to deliberately introduce some individuals from a large population of the same species into the smaller population. In terms of the evolutionary mechanisms discussed in this chapter, what are the potential benefits and drawbacks of transferring individuals from one population to another? Do you think biologists and concerned citizens should take such actions?

12 *Write Now* biology: mechanisms of evolution
This assignment explores the mechanisms of evolution through five selected short stories from *Welcome to the Monkey House* by Kurt Vonnegut Jr. Answer the questions associated with each story.

"Harrison Bergeron"
What message is this story trying to send? Cite examples from the story and relate them to the mechanisms of evolution from this chapter.

"Welcome to the Monkey House"
Is this story an example of sexual selection? Why or why not? Cite examples from the story and from this chapter to support your thinking.

"The Euphio Question"
If technology could produce such an instrument, how would it affect the evolution of humans? What about the evolution of other species on Earth?

"Unready to Wear"
Relate this story to as many of the mechanisms of evolution from this chapter as you can. Cite examples from the story and the chapter to support your thinking.

"Tomorrow and Tomorrow and Tomorrow"
Do you think these types of drugs are a good or bad thing? Where would you draw the line on technology's ability to extend life? How would drugs like these affect the natural selection and evolution of humans? What about the evolution of other species on Earth?

Fast Lizards, Slow Corals

The rapid evolution of lizards and the slow growth of corals offer clues about how species are born.

After reading this chapter you should be able to:

- Describe how adaptive traits lead to greater reproductive fitness and how differences in fitness lead to evolution by natural selection.

- Compare and contrast the biological and the morphological species concepts.

- Define speciation and understand that genetic divergence between populations must occur for new species to emerge.

- Explain how geographic isolation can lead to allopatric speciation.

- Explain how the different forms of sympatric speciation occur.

- Understand the role of coevolution in speciation.

201

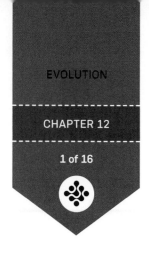

In 1971, the United States was at war in Vietnam, a gallon of gas cost 40 cents, a computer engineer sent the first e-mail—and an Israeli biologist named Eviatar Nevo captured 10 lizards on a small, rocky island off the coast of Croatia. Each lizard was about the length of a pinky finger and as heavy as a nickel. There was nothing too remarkable about them.

Nevo and his team released the captured lizards, a species called *Podarcis sicula*, on a nearby island just three miles away, within sight of the original island but separated by a deep ocean gulf (**Figure 12.1**). Pod Mrčaru was a smaller, plant-covered island already inhabited by two other lizard species. Nevo was curious to find out what would happen when the three species began to compete for resources on Pod Mrčaru. But Nevo never got the chance to return to the island. Shortly after he and his team departed,

unrest broke out across Croatia, and war held the region in a stranglehold throughout the 1980s and 1990s.

Scientists did not return to Pod Mrčaru until 2004. While visiting the University of Antwerp in Belgium, biologist Duncan Irschick from the University of Massachusetts, Amherst, and his Belgian colleagues decided to go investigate the mysterious island. One of the researchers was familiar with Nevo's obscure paper published in 1972 and thought it would be interesting to see what, if anything, had happened to the lizards. Further, Irschick would be able to complete some research on lizard behavior that he had planned to do in Europe anyway. "We didn't know what we'd find," recalls Irschick. "We decided, 'Let's just go check.'" This time, the lizards they found *were* remarkable.

Leaping Lizards

Back in the 1970s, Nevo's ten *P. sicula* lizards had been well adapted to the rocky, sparsely vegetated environment of the island Pod Kopište. They had specific **adaptive traits**, inherited characteristics that enabled them to survive and reproduce successfully on the island. Adaptive traits can be structural features, biochemical traits, or behaviors. In this case, the lizards were fast, with long legs that may have helped them catch insects, which made up most of their diet on the island. They were also territorial, fighting with other lizards over space and mating partners. These adaptive traits, among others, enabled the lizards to survive and reproduce better than competitors lacking those traits on Pod Kopište.

When transplanted to Pod Mrčaru, however, the lizards faced a new environment. Either they would adapt to it, or they would die. The term **adaptation** is commonly applied to adaptive traits *or* the process of evolution through natural selection, as discussed in Chapter 11. Therefore, an adaptation can be a trait that is advantageous to an individual or a population, or in broader terms, it can be the evolutionary *process* of natural selection that enables a good match between a population of organisms and their environment. If the lizards were going to live, they would need to undergo the process of adapting to their new environment.

When Irschick and his team landed on the island of Pod Mrčaru in 2004, they set to work

Figure 12.1

View of Pod Kopište from Pod Mrčaru

The island Pod Kopište (top), the original home of the *Podarcis sicula* lizards (bottom), is only three miles from their new home on the island Pod Mrčaru.

catching lizards. It wasn't hard; the island was swarming with them. Irschick could sit on a rock and simply pick up lizards as they ran by. There were several thousand lizards on the seven-acre island. When Irschick and his team looked closely, they realized that *all* the lizards were *P. sicula*. The ten transplants had wiped out the other two species of lizards. What's more, the adapted lizards looked strange (**Figure 12.2**). "They were these really big, chunky lizards," recalls Anthony Herrel, then a postdoctoral fellow in Irschick's lab and now a researcher at the French National Centre for Scientific Research in Paris. "They were unlike lizards on any of the other islands."

Twice a year for three years, the group returned to Pod Mrčaru to collect, weigh, and measure the lizards. They also took small pieces of the tails (which grow back) to test the lizards' DNA and compare it to the original population of *P. sicula* on Pod Kopište. Those DNA tests confirmed that the lizards on Pod Mrčaru were indeed descended from the original ten that Nevo had transported to the island. In addition to size and weight measurements, the team tested the strength of the lizards' bite force. They also dissected two samples of dead lizards.

What the researchers found was unprecedented. In about 33 years, from the time Nevo stepped off the island to the time Irschick stepped on, the species had evolved dramatically. The descendant lizards' heads were larger and shaped differently, making the lizard's bite much stronger than that of the original lizards brought to the

Pod Kopište, the island from which the original ten *Podarcis sicula* lizards were taken, is rocky and sparsely vegetated.

Pod Mrčaru is a smaller island covered with lush vegetation.

Figure 12.3

Pod Kopište's sparse vegetation and Pod Mrčaru's lush vegetation

island. The descendant lizards also had a unique digestive-tract structure called a cecal valve: a set of muscles between the large and small intestine that slow down food digestion, enabling the lizards to better process the cellulose of plants. It's really rare for lizards to have these structures, says Herrel. Only a few plant-eating lizard species, like iguanas, have cecal valves.

The new adaptive traits were the result of the lizards adapting to a different food source, says Irschick: On Pod Kopište, the lizards ate primarily insects, but Pod Mrčaru has an abundant supply of plants, including leaves and stems of local shrubs and grasses (**Figure 12.3**). Flushing the stomachs of a few lizards revealed that the lizards'

Longer, wider head

Thicker body

New digestive structure

Figure 12.2

Podarcis sicula lizard on Pod Mrčaru, 33 years after introduction of the species

DUNCAN IRSCHICK

Duncan Irschick is a biologist at the University of Massachusetts, Amherst, who studies animal function and evolution, specializing in research on animal movement and gecko adhesion.

diet was made up of two-thirds plant material. Over time, the lizards that by chance evolved a new head shape and digestive structures survived and reproduced better than those that did not, because these traits enabled them to take advantage of the new food sources on Pod Mrčaru.

The lizards evolved not only physical adaptive traits, but behavioral ones too. They became less territorial and less aggressive and were mating more often, probably because there was more food available. Adaptations, whether behavioral, physical, or biochemical, have three important characteristics. First, they show a close match between organism and environment; in this case, the lizards evolved to match the ecosystem on Pod Mrčaru. Second, adaptations are often complex, such as the lizards' new gut structure. Finally, adaptations help the organism accomplish important functions, such as feeding and mating.

The lizards on Pod Mrčaru show that evolution by natural selection can improve the adaptive traits of organisms over not only long periods of time, as with the evolution of whales described in Chapter 10, but also over surprisingly short periods of time. In just 33 years, the *P. sicula* lizards adapted both physical and behavioral traits that helped them flourish on the new island. "It was really, truly rapid evolution across multiple facets," says Irschick.

What Makes a Species?

Despite how impressive the lizards' quick adaptation may be, natural selection does not always result in a perfect match between an organism and its environment. In many cases, animals fail to adapt successfully. In fact, scientists estimate that 99 percent of all species that have ever lived are now extinct. Every extinct species is a silent testament to a failure to adapt in the face of adversity.

Today, Irschick is hesitant to call the Pod Mrčaru lizards a new species until he has done more testing. Most commonly, the term **species** is used to refer to members of a group that can mate with one another to produce fertile offspring. According to the **biological species concept**, a species is a group of natural populations that can interbreed to produce fertile offspring and cannot breed with other such groups; that is, they are **reproductively isolated** from other populations (**Figure 12.4**). Irschick and Herrel have yet to test whether the newly adapted lizards can still mate with their cousins back on Pod Kopište. According to the biological species concept, if the two populations of lizards can still readily breed, they are not different species. This same idea holds true for all sexually reproducing organisms. (It's not obvious why most organisms reproduce sexually! See "Why Sex?" on the following page.)

But the definition of "species" is not a black-and-white issue; it is a frequently discussed, multifaceted topic in evolution, and the biological species concept doesn't always apply. For example, not all species can be defined by their ability to interbreed, such as prokaryotes like bacteria, which reproduce asexually. To handle

Figure 12.4

Two South Pacific rattlesnakes confirm that they are the same species
These snakes are in the middle of a mating ritual and will successfully mate.

Q1: What are the three requisite parts of the biological species concept?

Q2: How would you design an experiment to determine whether two populations are distinct species according to the biological species concept?

Q3: For which types of populations does the biological species concept *not* work as a way of determining how they're related?

Figure 12.5

These tree frogs would likely be categorized as two different species under the morphological species concept

Genetically, these frogs are similar enough to be considered differently colored variations of the same species. Breeding these frogs with each other will determine their classification under the biological species concept.

Q1: What is the definition of a morphological species?

Q2: Using the morphological species concept, how would you determine whether two populations are distinct?

Q3: How is genetic divergence between populations determined?

these cases and others, scientists may use biogeographic information, DNA sequence similarity, and **morphology**—the organisms' physical characteristics—to identify and distinguish species. The **morphological species concept** is based on the notion that most species can be identified as separate and distinct groups of organisms solely on the basis of their physical characteristics.

Yet the morphological species concept does not always agree with the biological species concept (**Figure 12.5**). The lizards adapted to the island of Pod Mrčaru are physically different enough from their cousins on Pod Kopište to be considered a different species, says Herrel, yet their DNA sequences are close to identical, so he believes they might still be able to interbreed. **Genetic divergence**, or the presence of differences in the DNA sequences of genes, is absolutely required for **speciation**, the process by which one species splits to form two species or more.

Why Sex?

Sex is ubiquitous in the animal kingdom. An estimated 99 percent of multicellular eukaryotes are capable of sexual reproduction, which involves the joining of two haploid gametes produced through meiosis. Yet sex is very costly for individuals, so scientists have struggled to explain why it is so prevalent compared to asexual reproduction. And no, it's not because sex feels good—the first eukaryotes to engage in sex were single-celled protists some two billion years ago, long before animals developed neurons capable of giving an individual a sense of pleasure.

Costs of Sex:

1. Time and energy must be invested to find and woo a mate.

2. A parent passes on only 50% of their genetic material to an offspring, as opposed to 100% that is passed on through asexual reproduction.

3. During sexual reproduction, gene combinations that may have benefited the parents are shuffled and broken apart during meiosis and recombination.

Possible Benefits of Sex:

1. The genetic diversity created by sexual reproduction is critical for adaptation to new environments.

2. Sexual reproduction can help a population get rid of bad gene alleles and generate new beneficial alleles.

3. Rapid genetic change that occurs through sexual recombination can help species evolve resistance to parasitic infections.

The Kaibab squirrel is confined to the North Rim of the Grand Canyon.

Abert's squirrel lives on the South Rim and other southern locations, all the way into Mexico

Figure 12.6

The Grand Canyon is a geographic barrier for squirrels

The Kaibab squirrel population became isolated from the Abert's squirrel population when the Colorado River cut the Grand Canyon—which is as deep as 6,000 feet in some places. With gene flow between them blocked, probably beginning about five million years ago, the two populations diverged.

Q1: What is the definition of gene flow? How is gene flow blocked by geographic barriers?

Q2: Name as many types of geographic barriers as you can. Which do you think would be the best at blocking gene flow?

Q3: Are geographic barriers universal for all species? If not, name a geographic barrier that would block gene flow for one species but not another.

These two populations of lizards may continue to diverge, becoming more and more different because of their **geographic isolation** from one another, and may eventually become two unarguably distinct species. This is one of the most common ways that new species form: individuals of a single population become geographically separated from one another. This process can begin when a newly formed geographic barrier, such as a river, a canyon, or a mountain chain, isolates two populations of a single species (**Figure 12.6**). Such geographic isolation can also occur when a few members of a species colonize a region that is difficult to reach, such as an island located far outside the usual geographic range of the species.

Geographically isolated populations are disconnected genetically; there is little or no gene flow between them. Gene flow usually occurs via mating events between individuals of different populations. Without gene flow, the other mechanisms of evolution we saw in Chapter 11—mutation, genetic drift, and natural selection—can more easily cause populations to diverge from one another. If populations remain isolated long enough, they can evolve into new species. The formation of new species from geographically isolated populations is called **allopatric speciation** (from *allo*, "other"; *patric*, "country"), as shown in **Figure 12.7**.

ANTHONY HERREL

Anthony Herrel worked as a research fellow in Duncan Irschick's lab before starting his own lab at the French National Centre for Scientific Research in Paris, where he now studies the evolution of feeding and locomotion in vertebrates.

But what happens when there isn't a physical barrier between two populations? In the ocean, for example, plants and animals can drift around or swim extremely long distances. There are no physical barriers between them, yet there are many separate, unique species. How do new species form when populations are free to mix and mingle?

Caribbean Corals

Carlos Prada did not spend much of his time at the University of Puerto Rico in a classroom. Instead, he spent it in the ocean, studying corals, a type of marine invertebrate. On regular dives onto the Caribbean reef, the graduate student documented the morphology—the physical shapes and sizes—of a species of coral named *Eunicea flexuosa*, commonly called "sea fans." At 12 different locations along the Puerto Rican coast, Prada observed different morphologies of the coral at different depths. In shallow areas, less than 15 feet from the surface, the sea fans were wide, with a broad network of thick branches like a bush. In deeper areas, between 15 and 55 feet from the surface, the sea fans were much taller and spindly, resembling trees with networks of thin branches.

Figure 12.7

Physical barriers can produce allopatric speciation by blocking gene flow
Allopatric speciation can occur when populations are separated by a geographic barrier, such as a rising sea.

Q1: What factors must be present for allopatric speciation to occur?

Q2: If a geographic barrier is removed and the two reunited populations intermingle and breed, what attributes must the offspring have in order to conclude by the biological species concept that the two populations are still the same species?

Q3: If the two populations in question 2 are determined to still be the same species, did allopatric speciation occur?

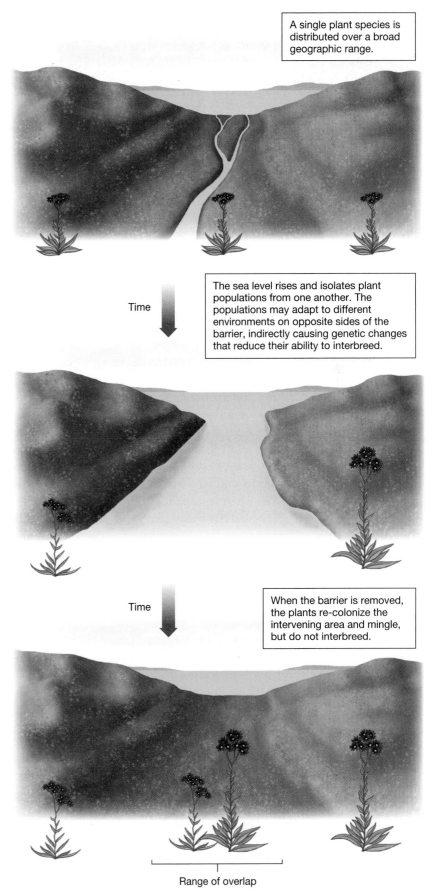

A single plant species is distributed over a broad geographic range.

Time

The sea level rises and isolates plant populations from one another. The populations may adapt to different environments on opposite sides of the barrier, indirectly causing genetic changes that reduce their ability to interbreed.

Time

When the barrier is removed, the plants re-colonize the intervening area and mingle, but do not interbreed.

Range of overlap

Shallow-water sea fan

Deep-water sea fan

Carlos Prada transplants coral to waters of different depths.

Figure 12.8

Carlos Prada transplants corals to waters of different depths
These two corals were once considered the same species, *Eunicea flexuosa*, commonly called a "sea fan." (*Source: Photos courtesy of Carlos Prada.*)

"They're supposed to be the same species, but they really looked different at different depths," says Prada. Curious about whether the morphologies were due to genetics or just a result of the surrounding environment shaping the animals as they grew, Prada began carefully transplanting the corals, moving deep-water sea fans to shallow depths and vice versa (**Figure 12.8**). He found that when transplanted, the corals did change. The shallow-water sea fans became taller and more spindly when planted in deep water, and the deep-water sea fans became wider in shallow waters, but—critically—neither made a complete transition to the alternate shape. The lack of a total transformation by either form to the other suggested to Prada that the corals, while they likely share a common ancestor, are actually two species adapted genetically to their respective water depths.

When Prada finished his graduate work in Puerto Rico, he e-mailed a professor at Louisiana State University who studied speciation in ocean animals. With wavy, bleached-blond hair, Michael Hellberg looks more like a California surfer than a professor, but this young evolutionist, who calls his lab the "Hellhole Crew," has long been fascinated with how one species splits to become two, especially in the ocean. "Say you have a new lake forming, and a species becomes isolated in the lake. Then it's pretty obvious there's not going to be a lot of interbreeding to fight against, and the species just adapts. To me, there's no mystery in that," says Hellberg. "I've always tried to target groups where species look closely related and where ranges of the species overlap. That makes things a lot harder."

Hellberg welcomed Prada into his crew, and the two set out to extend Prada's work to find out whether his idea—that coral adapt to different depths in the ocean, causing the formation of new species—was unique to coral reefs in Puerto Rico or could be observed in other areas around the Caribbean. With Hellberg's support, Prada traveled to the Bahamas, Panama, and Curaçao to observe and take samples from sea fan colonies.

As he waited for Prada to return home with the data, Hellberg remained skeptical of the idea of **ecological isolation**, that two closely related species in the same territory could be reproductively isolated by slight differences in habitat. But

MICHAEL HELLBERG AND CARLOS PRADA

Michael Hellberg (right) is an evolutionary biologist at Louisiana State University who studies how species evolve in marine environments. Carlos Prada (left) is a graduate student in Hellberg's lab at LSU. Prada initiated and conducted a study about how corals form new species in the ocean.

as Prada sampled more and more locations, the evidence convinced him. Indeed, everywhere he looked, Prada saw the same thing he had observed in Puerto Rico: broad, leaflike coral at shallow depths, and tall, sticklike coral in deeper waters. Distance between the populations didn't matter; some shallow-water and deep-water fans were close enough together that Prada could reach out and bend them to touch each other. But depth did matter. The coral were physically close enough to easily interbreed in the water, yet the species had somehow become specialized to two different depths (**Figure 12.9**).

Prada brought his coral samples back to the lab and performed tests to see how genetically similar the two groups of corals were. Prada and Hellberg found that all the shallow-water sea fans across the Caribbean were more closely related to each other than they were to any of the deep-water sea fans. The same held true for the deep-water sea fans: they were more closely related to each other than to any of the shallow-water fans. In the DNA data, Hellberg saw some genetic exchange between the two populations in the past, yet that exchange of genetic materials had been limited, and each group had bred primarily with its own population.

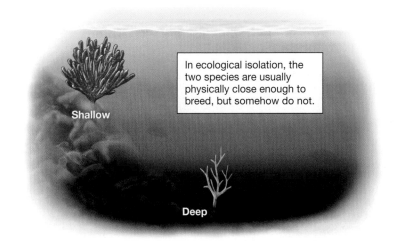

In ecological isolation, the two species are usually physically close enough to breed, but somehow do not.

Figure 12.9

Depth of water provides ecological isolation for sea fans
The two different depths shown here differ slightly in their quantity and quality of light, the force of the waves or current, the amount of sediment deposited on them, the number and type of predators, and the availability and type of food.

organisms only 15 feet below the surface receive a lot of sunlight, while those at 30 feet receive less of it, and at different wavelengths. Prada found different types of algae living on the two coral populations (**Figure 12.10**), so it might be that the sea fans adapted to be better hosts for the most successful algae at their depth.

Different Depths, Different Habitats

How did a few yards of depth produce such different habitats that the *E. flexuosa* corals evolved different adaptations to each depth? The scientists are still investigating the differences between the shallow and deep habitats, but they have some hypotheses. One is that the coral have adapted their morphologies, and possibly their biochemistry, to suit different symbiotic algae that grow on them at different depths.

Symbiotic algae live on coral and, through photosynthesis, use sunlight to produce energy and organic compounds that coral use to maintain and grow calcium carbonate skeletons. In turn, the coral provide the algae with a sheltered place to live and produce carbon dioxide that the algae use during photosynthesis. (This cooperative relationship between species is known as mutualism; see Chapter 17.) But different species of algae have different light and nutrient requirements, and

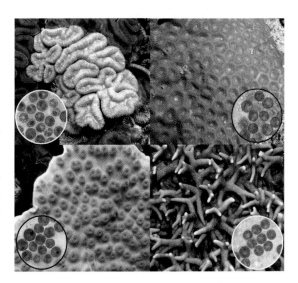

Figure 12.10

Different species of corals and their resident algae
Although all these algae (single-celled eukaryotic organisms) look the same under a microscope, they are actually different species with different light requirements and photosynthetic capacities.

Flowers that coevolved with hummingbirds produce nectar specific to the birds' diet, are colored to attract the birds, and are shaped specifically to fit the hummingbirds' bills.

Figure 12.11

Coevolution at its finest
This hummingbird's bill fits perfectly into this flower for easy access to the hummingbird's favorite food, nectar. The hummingbird feasts and then distributes the flower's pollen by carrying the pollen along to its next meal. It's a win-win situation!

Q1: Describe how coevolution is distinct from evolution as described in Chapters 10 and 11.

Q2: Is coevolution the same thing as convergent evolution, described in Chapter 11?

Q3: Do you think one species' adapting over time to feed specifically and extremely successfully on another species is an example of coevolution? Why or why not?

So Many Chromosomes

New plant species can form in a single generation as a result of **polyploidy**, a condition in which an individual gains an extra full set or two (or three) of chromosomes. Humans and most other eukaryotes are *diploid* (having two sets of chromosomes), but some organisms are *triploid* (three sets) or *tetraploid* (four sets), or have an even higher number of chromosomes. Polyploidy is invariably fatal in people, but in many plant species it is not lethal.

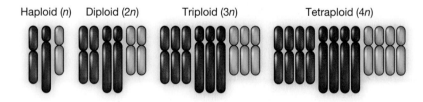

Haploid (*n*)   Diploid (2*n*)   Triploid (3*n*)   Tetraploid (4*n*)

Polyploidy can occur when two different species hybridize to produce an offspring with an odd number of chromosomes, resulting in sterility. The sterile hybrid may then spontaneously double its chromosome number, resulting in a fertile plant with an even number of chromosomes. This doubling of the chromosomes can lead to reproductive isolation because the chromosome number in the gametes of the new polyploid no longer matches the number in the gametes of either of its parents, so the plant cannot breed with others of its species.

Polyploidy has had a large effect on life on Earth: more than half of all plant species alive today are descended from species that originated by polyploidy. A few animal species also appear to have originated by polyploidy, including several species of lizards, salamanders, and fish.

In many cases in the ocean, as with coral and algae, the interaction between two species so strongly influences their survival that the two species have evolved in tandem—a phenomenon known as **coevolution**. The term "coevolution" encompasses a wide variety of ways in which the evolution of an adaptation in one species causes a reciprocal adaptation in another species. Another example of coevolution is the relationship between hummingbirds and certain species of flowers (**Figure 12.11**).

Alternatively, or in addition to hosting specific algae, Hellberg hypothesizes that the shallow-water and deep-water sea fans may similarly have adapted to the type and density of sedimentation found at different ocean depths. Or perhaps different predators are more abundant at different depths. Whatever the major habitat factors driving the corals' adaptation, natural selection led to the formation of two different coral populations tailored to the ocean depths at which they grow.

Today, Hellberg considers the two populations different species. "What makes this interesting is that you have populations that could potentially interbreed, yet they maintain their differences in the face of each other," says Hellberg. "For just about any biological question you want to ask, they are independent entities." The formation of new species in the *absence* of geographic

On the Diversity of Species

There are an estimated 8.74 million eukaryotic species on Earth, yet we are only familiar with a fraction of those: scientists estimate that 86% of land species and 91% of aquatic species have not been discovered. Of the known species, insects top the chart with an estimated one million different species. (Prokaryotes, not included here, are so plentiful and diverse that scientists have yet to even estimate their quantity or diversity.)

Numbers of known species

Fish
31,200

Crustaceans
47,000

Mollusks
85,000

Fungi
99,000

Arachnids
102,200

Insects
1,000,000

Plants
310,100

Mammals
5,500

Sponges
6,000

Amphibians
6,500

Reptiles
8,700

Jellyfish and Polyps
9,800

Birds
10,000

Millipedes and centipedes
16,100

Segmented worms
16,800

Flatworms
20,000

Roundworms
25,000

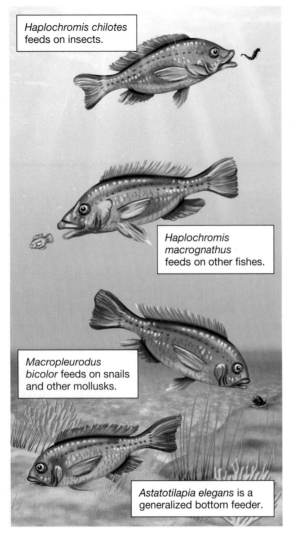

Haplochromis chilotes feeds on insects.

Haplochromis macrognathus feeds on other fishes.

Macropleurodus bicolor feeds on snails and other mollusks.

Astatotilapia elegans is a generalized bottom feeder.

Figure 12.12

Sympatric speciation drives diversity among Lake Victoria cichlid species
Scientists have described some 500 species of cichlid fish in Lake Victoria. Genetic analyses indicate that they all descended from just two ancestor species over the past 100,000 years. These four species show some of the differences in feeding behavior and morphology.

Q1: What is the main difference between allopatric and sympatric speciation?

Q2: Name two events that must happen for both allopatric speciation and sympatric speciation to occur.

Q3: Do you think all of the 500 species in Lake Victoria arose through sympatric speciation? Why or why not?

isolation is called **sympatric speciation** (from *sym*, "together"), shown in **Figure 12.12**. Sympatric speciation is a particularly important process in plants (see "So Many Chromosomes" on page 210).

Still, a question lingers: Why don't the two coral species interbreed? They appear to be capable of it, because of evidence of some genetic mixing in the past, says Hellberg, yet there is no evidence that interbreeding is common—only a few rare hybrids grow on the ocean floor between them. When two species are reproductively isolated from each other, we say that **reproductive barriers** exist between those species. Reproductive barriers are often divided into two categories: *prezygotic* and *postzygotic*.

Barriers that prevent a male gamete (such as a human sperm) and a female gamete (such as a human egg) from fusing to form a zygote are **prezygotic barriers**. Prezygotic barriers act *before* the zygote exists. By definition, all allopatric speciation events are caused by prezygotic geographic barriers. No zygote can form if a mountain lies between the male and female in question. Some sympatric speciation events can also be caused by prezygotic barriers, such as the failure of one bird species to recognize the appropriate ritual mating dance of another species (**Figure 12.13**).

Barriers that prevent zygotes from developing into healthy and fertile offspring are called **postzygotic barriers**, which act *after* the zygote is formed. By definition, postzygotic barriers cannot be geographic, so they always result in sympatric speciation events.

A wide variety of cellular, anatomical, physiological, and behavioral mechanisms generate pre- and postzygotic reproductive barriers, but they all have the same overall effect: little or no mating occurs and therefore few or no alleles are exchanged between species (**Table 12.1**). Reproductive barriers ensure that the members of a species share a unique genetic heritage: a set of genes and alleles that is typical of the species but different from that of all other species. Hellberg suspects that some prezygotic barriers exist between the two sea fan species—perhaps the gametes of the two species don't fuse successfully, as in some sea urchins (**Figure 12.14**)—but there may also be a postzygotic barrier such as generation time.

Blue-footed boobies point their beaks, wings, and tails upward in a mating dance called "sky pointing."

Figure 12.13

The blue-footed booby courtship dance is a prezygotic, behavioral reproductive barrier

This species of booby has a unique ritual dance that must be accurately completed before mating. Other booby species do not perform exactly the same dance and therefore do not mate with the blue-footed booby.

Q1: What does "prezygotic" mean?

Q2: How is the ritual dance a prezygotic reproductive barrier?

Q3: What are some other prezygotic reproductive barriers besides a ritualistic mating dance?

It seems likely that the reason hybrids between the two species do not succeed is the uniquely long generation time for these corals, says Prada. Sea fans don't reach reproductive age until they are 15–20 years old, and they continue reproducing until they are 60 years old or more. So, while coral gametes and larvae can and do disperse far from their parents and may interbreed or take root at incorrect depths, natural selection then has 15–20 years to winnow out the less successful offspring. Therefore, if there are small differences in survival rates between the two species

Table 12.1

Reproductive Barriers that Isolate Two Species in the Same Geographic Region

Type of Barrier	Description	Effect
Prezygotic		
Ecological isolation	The two species breed in different portions of their habitat, in different seasons, or at different times of day.	Mating is prevented.
Behavioral isolation	The two species respond poorly to each other's courtship displays or other mating behaviors.	Mating is prevented.
Mechanical isolation	The two species are physically unable to mate.	Mating is prevented.
Gametic isolation	The gametes of the two species cannot fuse, or they survive poorly in the reproductive tract of the other species.	Fertilization is prevented.
Postzygotic		
Zygote death	Zygotes fail to develop properly, and they die before birth.	No offspring are produced.
Hybrid performance	Hybrids survive poorly or reproduce poorly.	Hybrids are not successful.

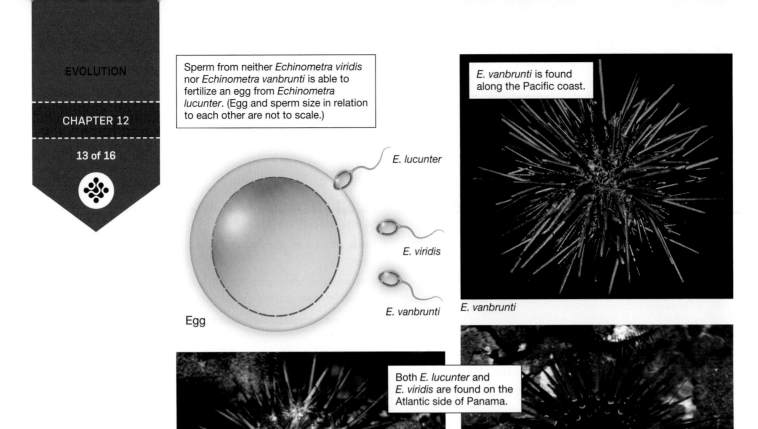

Sperm from neither *Echinometra viridis* nor *Echinometra vanbrunti* is able to fertilize an egg from *Echinometra lucunter*. (Egg and sperm size in relation to each other are not to scale.)

E. lucunter

E. viridis

E. vanbrunti

Egg

E. vanbrunti is found along the Pacific coast.

E. vanbrunti

Both *E. lucunter* and *E. viridis* are found on the Atlantic side of Panama.

E. lucunter

E. viridis

Figure 12.14

Gametic isolation between sea urchin species

The sea urchin species found along both coasts of Panama illustrate gametic isolation as a reproductive barrier. At one point long ago, the Panama landmass was not continuous and did not separate the two oceans. These three species likely evolved through both allopatric and sympatric speciation from a recent common ancestor after the oceans were separated.

Q1: Which species is/are sympatric with *E. lucunter*?

Q2: Which species is/are allopatric with *E. lucunter*?

Q3: If gametes are incompatible, what will be the result of a mating event between them?

at a particular depth—if a deep-water coral fares slightly worse in a shallow area than a shallow-water coral does—those differences will become amplified over 15 years, and by the time the corals reach reproductive age it is likely that only the shallow-water species will still be alive in the shallow area.

New species form and adapt to their environments in many different ways, from island-bound lizards to depth-dependent corals. And from those varied beginnings comes the vast diversity of life that exists on our planet today, the result of billions of years of evolution. In the next two chapters, we explore the dramatic transitions in life-forms that have walked, crawled, and swum around our planet, from small dinosaurs with wings to the most successful, most cunning species ever to traverse the planet: humans.

REVIEWING THE SCIENCE

- An **adaptive trait** is an inherited characteristic that improves an individual's chances of surviving and reproducing in a specific environment.

- Natural selection leads to **adaptation**, the process that improves the reproductive success of organisms in their environment over time. Adaptation does not produce "perfection."

- According to the **biological species concept**, a **species** is a group of populations that interbreed and can produce live and fertile offspring.

- The **morphological species concept** defines a species as any group of organisms with extraordinarily similar physical characteristics.

- The process of **speciation**, one species splitting into two or more species, is the by-product of **reproductive isolation** and **genetic divergence** between populations.

- Although the biological and morphological species concepts do not always agree, speciation can occur only when there is genetic divergence between populations.

- **Allopatric speciation** occurs when populations of a species become **geographically isolated**, limiting gene flow and making genetic divergence more likely.

- Speciation that occurs between populations lacking geographic isolation is called **sympatric speciation**. Sympatric isolating mechanisms can be **prezygotic barriers** (before zygote formation) or **postzygotic barriers** (after zygote formation).

- When speciation and changes in adaptive traits of one species strongly affect the adaptive traits in another species, **coevolution** has occurred.

THE QUESTIONS

The Basics

1 When two populations are reproductively isolated, what else must occur for speciation to happen?

(a) gene flow

(b) genetic divergence

(c) coevolution

(d) convergent evolution

(e) none of the above

2 Traits that are inherited and improve an individual's ability to survive and reproduce are called _____ traits.

(a) adaptive

(b) polymorphic

(c) biological

(d) sympatric

(e) allopatric

3 Examples of allopatric isolating mechanisms include all of the following *except*

(a) mountain ranges.

(b) oceans.

(c) mating dances.

(d) wide rivers.

(e) All of these are allopatric isolating mechanisms.

4 Prezygotic isolating mechanisms prevent hybrid offspring from occurring between species because

(a) the resulting offspring are not fertile and cannot reproduce.

(b) egg and sperm fuse and form a zygote, but it does not survive.

(c) egg and sperm do not ever meet or, if they do, cannot fuse to form a zygote.

(d) all of the above

5 Defining a species by whether two populations can mate and produce viable and fertile offspring is called the

(a) biological species concept.

(b) morphological species concept.

(c) biogeographic species concept.

(d) none of the above

6 The morphological species concept is based on determining species by

(a) their physical characteristics.

(b) the DNA sequences in their genes.

(c) their ability to mate with each other.

(d) where they live or have lived.

(e) none of the above

7 Which of the following sympatric isolating mechanisms prevents mating because the two species are physically unable to mate?

(a) ecological isolation

(b) behavioral isolation

(c) mechanical isolation

(d) gametic isolation

(e) all of the above

Try Something New

8 Which of the following is an adaptive trait?

(a) A faster-growing individual reaches reproductive age earlier.

(b) An individual is more successful at attracting mates.

(c) An individual is better camouflaged in its environment.

(d) all of the above

(e) none of the above

9 Researchers from the Smithsonian Institution were startled to discover that the 3 species of *Starksia* blennies they had been studying in the Caribbean islands were really 10 different species. How could these researchers have thought that these 10 different species of reef fish were only 3 species?

(a) They were using the morphological species concept to determine the number of species they were studying, and there were only 3 different color morphs (differently colored variations) among the fish.

(b) They were using the biological species concept to determine the number of species they were studying, and they saw all 10 different species interbreeding.

(c) They were using the morphological species concept to determine the number of species they were studying, and there were 10 different color morphs among the fish.

(d) none of the above

10 Distinct species that are able to interbreed in nature are said to *hybridize*, and their offspring are called *hybrids*. The gray oak and the Gambel oak can mate to produce fertile hybrids in regions where they co-occur. However, the gene flow in nature is sufficiently limited that, overall, the two species remain phenotypically distinct. If the hybrid offspring survive well and reproduce to the extent that there is a large population of hybrid individuals that breed between themselves but do not interbreed with either original parent species (the gray and the Gambel), which of the following would you say led to the new hybrid species?

(a) prezygotic reproductive barriers

(b) sympatric speciation

(c) allopatric speciation

(d) postzygotic reproductive barriers

(e) none of the above

11 The eastern and western meadowlarks are different species, although they are very similar in morphology. The two species do not interbreed, even though their range overlaps in parts of the upper Midwest. These grassland species are reproductively isolated because each sings a distinctly different song, and a female will mate only with the male that sings the melody unique to her species. What types of reproductive barriers are at work in this example?

(a) allopatric speciation and prezygotic barriers

(b) sympatric speciation and postzygotic barriers

(c) allopatric speciation and postzygotic barriers

(d) sympatric speciation and prezygotic barriers

(e) none of the above

12 The four-eyed fish, *Anableps anableps*, really has only two eyes that function as four, enabling the fish to see clearly through both air and water. *A. anableps* is a surface feeder, so the ability to see above water helps it locate prey such as insects. Its unique eyes also enable it to scan simultaneously for predators attacking from above (such as birds) or below (such as other fish). This species of fish _____ crafted by natural selection.

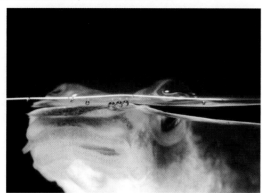

Anableps anableps

(a) is an example of coevolution

(b) has an evolutionary adaptation

(c) is an example of sympatric speciation

(d) is an example of allopatric speciation

Leveling Up

13 *Write Now* **biology: adaptations**

Select an organism (other than humans) that you find interesting. Research two adaptations of your organism and describe them in detail. Explain carefully why each of these features is considered an important adaptation for its species. Discuss how a change in the environment might change the usefulness of these adaptations.

14 **What do *you* think?** Apply your understanding of adaptive evolution to organisms that cause infectious human diseases, such as HIV, the virus that causes AIDS. How do our efforts to kill or prevent infection with such organisms affect their evolution? Are the evolutionary changes we promote usually beneficial or harmful for us? Explain your answer.

15 **What do *you* think?** According to the Defenders of Wildlife organization, there were 20–30 million American range bison in the old West. During the late 1800s the American range bison were hunted to the brink of extinction, leaving behind a bottleneck population of only 1,091 individuals. The population has since rebounded to about 500,000 bison. Unfortunately, almost all of these bison are the descendants of these few individuals crossbred with domestic cattle by ranchers. Scientists and conservationists want to genetically test bison to find those of pure bison origin to preserve the species. Only these, they argue, should be called American range bison and be allowed to roam free in the national parks as bison. They think hybrids should be confined to farms and ranches, should be called "beefalo" rather than bison, and should not be afforded the protection that pure bison currently have. What do you think? Should bison tainted with cattle genes be removed from free-range parks? Should the government spend scarce conservation monies on genetic testing and breeding efforts to preserve the pure bison population? Investigate conservation efforts and the costs of genetically testing and relocating bison to help you with your decision. Is speciation at the hands of human beings now part of evolution as we know it?

The First Bird

A fresh wave of fossils challenges the identity of the first dino-bird.

Identify critical events on a time line of life on Earth and explain their evolutionary importance.

Create an accurate evolutionary tree, given a group of organisms and their traits.

Explain how new scientific data may change an evolutionary tree without challenging our understanding of evolution.

For each kingdom, identify key characteristics and give an example organism.

Explain the impact of mass extinctions on biodiversity.

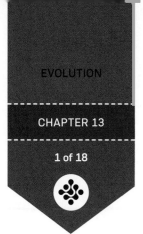

In 1861, quarry workers in Germany unearthed the fossil of a crow-sized bird. The 150-million-year-old preserved skeleton looked bizarre: it had feathers and clawed hands like a bird, but it had teeth and a long, bony tail like a reptile.

After close examination by paleontologists, the fossil was hailed as a transitional form between birds and reptiles and labeled the oldest known bird. Named *Archaeopteryx*, the feathered dinosaur became famous around the world as the first solid evidence that birds descended from dinosaurs (**Figure 13.1**). Charles Darwin, who proposed the theory of descent with modification, called the discovery "a grand case for me."

Over 100 years later, a young fossil hunter named Xu Xing bent over the dirt at a dig site in Liaoning province in northern China. It was Xu's first time in the field since finishing college. An excellent student, Xu had wanted to study economics at Peking University in Beijing, but at the time, students in China did not pick their majors, and Xu was assigned paleontology. Luckily, it was promising work: farmers in Liaoning had recently begun to uncover huge numbers of fossils, and research was booming.

So, as a grudging paleontologist, Xu began searching for dinosaurs in Liaoning. He had a particular interest in feathered dinosaurs. By the late twentieth century, the idea that birds descended from dinosaurs was no longer a hypothesis, but a well-established scientific theory backed by a mountain of data. And atop that mountain of data sat *Archaeopteryx*, the earliest known bird.

Little did Xu know, as he dug through the dirt that day, that he would soon be the one to knock *Archaeopteryx* from its perch.

Dinosaurs and Domains

Although the *Archaeopteryx* fossil is old—150 million years old, in fact—much older fossils have been found.

Our solar system and Earth formed 4.6 billion years ago. Some of the oldest known rocks on Earth are 3.8 billion years old and contain carbon deposits that hint at life. Cell-like structures have been found in layered mounds of sedimentary

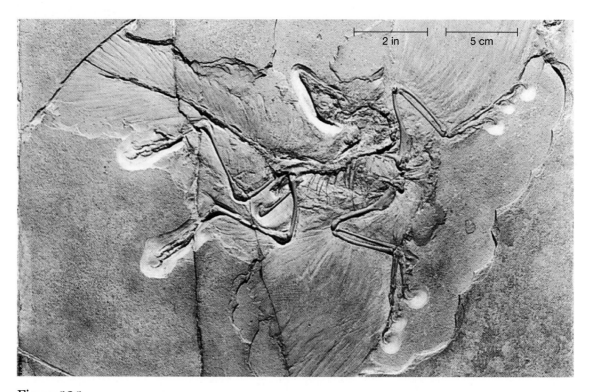

Figure 13.1

The original *Archaeopteryx* fossil
There are 11 complete *Archaeopteryx* fossils, but this one from Germany was the first to be found. Note the faint impressions of feathers on the wings and tail.

rock called stromatolites that formed 3.5 billion years ago, and projections based on DNA analysis also support the idea that life appeared on Earth at that time.

The question of how life arose from nonlife is one of the greatest riddles in biology, but scientists have little doubt that all life on Earth is related. As noted in Chapter 1, all living organisms are united by a basic set of characteristics. Life shares this set of common properties because all living organisms descended from a common ancestor, known as the *universal ancestor*. This hypothetical ancestral cell is placed at the base of the tree of life. From that cell, all life emerged, and life throughout history and today is so diverse that biologists have created a classification system to organize it into categories.

The **domains** form the highest hierarchical level in the organization of life, describing the most basic and ancient divisions among living organisms. There are three domains of life (**Figure 13.2**):

- **Bacteria**, which includes familiar disease-causing bacteria such as *E. coli*

- **Archaea**, which consists of single-celled organisms best known for living in extremely harsh environments

- **Eukarya**, which includes all other living organisms, from amoebas to plants to fungi to animals

Humans, dinosaurs, and birds are all part of the Eukarya domain. They are *eukaryotes*. Bacteria

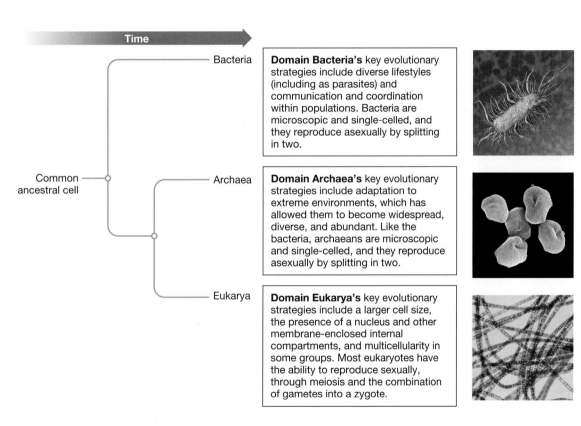

Figure 13.2

Three domains of life
This "tree of life" shows the relationships of the three domains of life.

Q1: Why is there a shared line from the universal ancestor for Archaea and Eukarya?

Q2: Where would birds be found within this figure? What about humans?

Q3: To which domain would you expect a disease-causing organism to belong? What if the organism were multicellular?

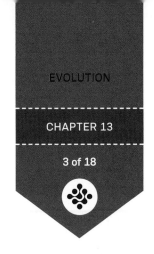

and Archaea are two different domains—Archaea are more closely related and in some ways more similar to Eukarya than to Bacteria—yet because neither Bacteria nor Archaea are eukaryotes, the two have traditionally been lumped under a common label: *prokaryotes.*

Prokaryotes first appear in the fossil record at about 3.5 billion years ago (**Figure 13.3**), but the first eukaryotes did not evolve until a billion years later. Luckily for us, and all other eukaryotes, roughly 2.8 billion years ago a group of bacteria evolved a type of photosynthesis that releases oxygen as a by-product. As a result, the oxygen concentration in the atmosphere increased over time, and about 2.1 billion years ago the first single-celled eukaryotes evolved. When the oxygen concentration reached its current level, by about 650 million years ago (mya), the evolution of larger, more complex multicellular organisms became possible, including fish, then land plants, then

insects, amphibians, and reptiles. One group of reptiles, which would eventually dominate most other species, was the dinosaurs. Dinosaurs first appeared during the Triassic period, about 230 mya, and they took over the planet.

Feathered Friends

Xu may not have wanted to be a paleontologist when he left for college, but by the time he graduated, he was hooked on dinosaurs. Over the next 20 years, Xu became one of the most productive researchers in his field. To date, he has discovered and named more than 60 extinct species—mostly dinosaurs, but also a reptile and a salamander. And the majority of those dinosaur fossils have feathers.

As scientists traced back the **lineage**, or line of descent, from birds to dinosaurs, it became

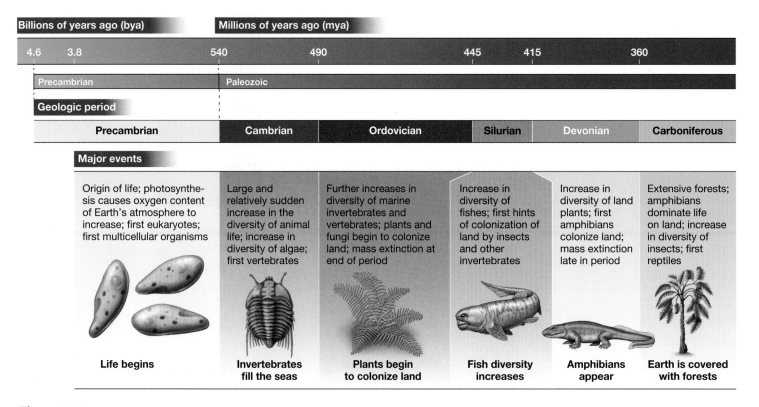

Figure 13.3

The geologic timescale and major events in the history of life

The history of life can be divided into 12 major geologic time periods, beginning with the Precambrian (4.6 bya to 540 mya) and extending to the Quaternary (2.6 mya to the present). This timescale is not drawn to scale; to do so would require extending the diagram off the book page to the left by more than five feet (1.5 m).

clear that birds are most closely related to theropods—fast-moving dinosaurs that ran on two legs and had hollow, thin-walled bones (as birds do). Theropods were a diverse group of dinosaurs (**Figure 13.4**). Most were carnivores, though a few were herbivores, insectivores, or omnivores. There were theropods that could swim and eat fish. Some theropod species boasted enlarged scales, and of course, many theropods had feathers.

Xu and other scientists map out lineages using a diagram called an **evolutionary tree**, a model of evolutionary relationships among groups of organisms based on similarities and differences in their DNA, physical features, biochemical characteristics, or some combination of these. An evolutionary tree maps the relationships between ancestral groups and their descendants, and it clusters the most closely related groups on neighboring branches.

In an evolutionary tree, organisms under consideration are depicted as if they were leaves at the tips of the tree branches. A given ancestor and all its descendants make up a **clade**, or branch, on the evolutionary tree. *Archaeopteryx* and all subsequent animals that evolved from it are considered a clade (**Figure 13.5**).

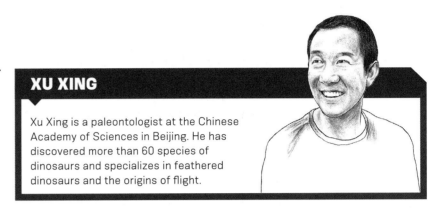

XU XING

Xu Xing is a paleontologist at the Chinese Academy of Sciences in Beijing. He has discovered more than 60 species of dinosaurs and specializes in feathered dinosaurs and the origins of flight.

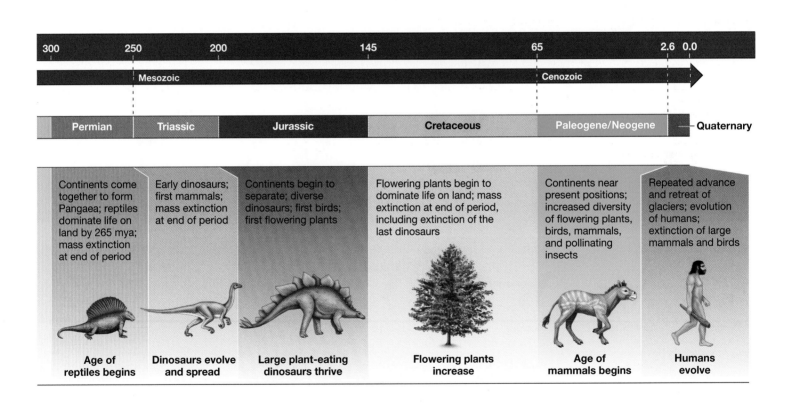

| 300 | 250 | 200 | 145 | 65 | 2.6 | 0.0 |

Mesozoic   Cenozoic

| Permian | Triassic | Jurassic | Cretaceous | Paleogene/Neogene | Quaternary |

Continents come together to form Pangaea; reptiles dominate life on land by 265 mya; mass extinction at end of period

Early dinosaurs; first mammals; mass extinction at end of period

Continents begin to separate; diverse dinosaurs; first birds; first flowering plants

Flowering plants begin to dominate life on land; mass extinction at end of period, including extinction of the last dinosaurs

Continents near present positions; increased diversity of flowering plants, birds, mammals, and pollinating insects

Repeated advance and retreat of glaciers; evolution of humans; extinction of large mammals and birds

Age of reptiles begins

Dinosaurs evolve and spread

Large plant-eating dinosaurs thrive

Flowering plants increase

Age of mammals begins

Humans evolve

Q1: During what geologic period did life on Earth begin?

Q2: How long ago did species begin to move from water to land? What period was this?

Q3: In what period would *Archaeopteryx* have been alive?

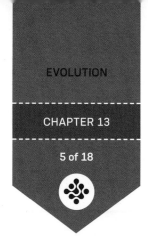

A **node** marks the moment in time when an ancestral group split, or diverged, into two separate lineages. The node represents the **most recent common ancestor** of the two lineages in question—that is, the most *immediate* ancestor that *both* lineages share. For over 100 years, researchers considered *Archaeopteryx* the most recent common ancestor of both birds and dinosaurs and thus placed it at the root of the avian clade—the first bird (**Figure 13.5**, top).

The first bird, that is, until Xu stumbled across a new fossil that threw the field into controversy and made paleontologists gasp in shock. In 2008, Xu visited the Shandong Tianyu Museum of Nature, a dinosaur museum in eastern China. There, he happened upon a unique fossil that had been collected by a farmer in Liaoning and sold through a dealer to the museum. The fossil, entombed in yellowish rock, shows a small, bird-like dinosaur seemingly craning its neck forward and spreading its short wings (**Figure 13.6**). "I saw it and said, 'Oh, this is an important species,'" recalls Xu. He asked the museum to let him study it.

As Xu examined the fossil, named *Xiaotingia zhengi*, he wondered where it belonged on the dinosaur-bird evolutionary tree. To find out, he analyzed *shared derived traits* of the fossil and

Spinosaurus   *Tyrannosaurus*   *Velociraptor*   *Deinonychus*

Archaeopteryx   *Eosinopteryx*   *Aurornis*   *Xiaotingia*

Figure 13.4

Dinosaurs large and small

Theropods ranged in size from tiny, like a chicken, to huge, like the group's most famous member, *Tyrannosaurus rex*.

Q1: In what ways were theropods the same as modern birds? Give at least two similarities.

Q2: In what ways did theropods differ from modern birds? Give at least two differences.

Q3: Birds are often referred to as "living dinosaurs." Is this accurate? Why or why not?

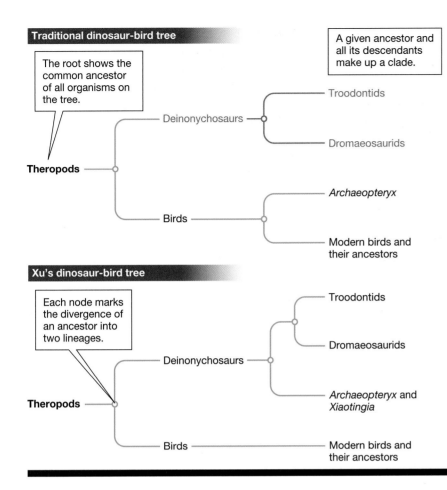

The root shows the common ancestor of all organisms on the tree.

A given ancestor and all its descendants make up a clade.

Troodontids

Deinonychosaurs

Dromaeosaurids

Theropods

Archaeopteryx

Birds

Modern birds and their ancestors

Xu's dinosaur-bird tree

Each node marks the divergence of an ancestor into two lineages.

Troodontids

Dromaeosaurids

Deinonychosaurs

Theropods

Archaeopteryx and *Xiaotingia*

Birds

Modern birds and their ancestors

Figure 13.5

The evolutionary origins of birds

(*Top*) The traditional evolutionary tree showing *Archaeopteryx* as an early bird, split off from the deinonychosaur (birdlike, carnivorous dinosaur) groups Troodontidae and Dromaeosauridae.

(*Bottom*) The evolutionary tree proposed by Xu after the discovery of *Xiaotingia*, with *Archaeopteryx* as a deinonychosaur rather than an early bird.

Q1: In the traditional tree, identify the node showing the common ancestor for early birds and dinosaurs.

Q2: What do both the traditional tree and Xu's tree suggest about troodontids and dromaeosaurids?

Q3: In both trees, identify the node for the common ancestor of *Archaeopteryx* and other birds. In what way are the nodes different in the two trees?

similar early birds. **Shared derived traits** are the unique features common to all members of a group that originated in the group's most recent common ancestor and then were passed down in the group (but not in groups that are not direct descendants of that ancestor). In this case, the original ancestor in question was assumed to be *Archaeopteryx*, and the shared traits included feathers, clawed hands, and a long, bony tail.

By comparing *Xiaotingia* to *Archaeopteryx* and other related species, Xu created a new evolutionary tree of early birds (**Figure 13.5**, bottom). Suddenly, *Archaeopteryx* wasn't in the avian clade. Instead, *Archaeopteryx* and *Xiaotingia* were in a different clade, grouped with deinonychosaurs—the small, birdlike, carnivorous dinosaurs commonly called raptors. (The term "raptor" refers to carnivorous modern birds like hawks and owls, but it is also used informally to describe this group of dinosaurs.) *Archaeopteryx*, Xu believed, was not the first bird, but a raptor. It had feathers, but its descendants did not evolve into birds.

"It was a big change," says Xu. Yet, he adds, it wasn't entirely unexpected. In the last 10–20 years, more and more fossils of early avian species have been discovered—and the more Xu compared *Archaeopteryx*, discovered in Europe, to other early birds discovered in China, the less

5 cm

Figure 13.6

Xiaotingia zhengi: A controversial new leaf on the dinosaur-bird evolutionary tree

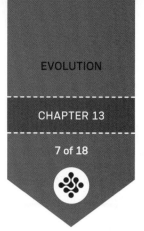

it looked like a bird and the more it looked like a raptor. Early birds have small, thick skulls and two toes on each foot; *Archaeopteryx*, on the other hand, has a long, almost pointy skull and three toes on its feet. "*Archaeopteryx* is just so different from other early birds," says Xu.

Xu published his revised evolutionary tree in 2011. It took the scientific world by storm. Some researchers embraced the idea: "Perhaps the time has come to finally accept that *Archaeopteryx* was just another small, feathered bird-like theropod fluttering around in the Jurassic," wrote one paleontologist in the journal *Nature*. Others disagreed, arguing that Xu's analysis was not convincing. For them, *Archaeopteryx* remained the first bird.

The History of Life on Earth

Xu knew that one new fossil was but a single shred of evidence and therefore unlikely, by itself, to convince the paleontology community. "The evidence is not very strong," he admits, "but it was a question we wanted to discuss and add more analyses to, including old and new fossils." The identity of the first bird is important because our understanding of how flight evolved is based on the classification of *Archaeopteryx*. If *Archaeopteryx* is not the species from which flight evolved, then our understanding of flight is wrong.

Xu's redesign of the early bird evolutionary tree raised another important question: Did flight evolve more than once? If *Archaeopteryx*, which looked capable of flight, was a bird, then flight probably evolved just once, in the avian lineage. If *Archaeopteryx* was not a bird, and was instead simply a raptor that could fly, then flight evolved at least twice—once in raptors and once in avians—but only avians went on to evolve into modern birds.

Biological classification helps us answer important evolutionary questions like these. In addition to recognizing three broad domains of life, biologists group life into six distinct **kingdoms**, the second-highest level in the hierarchical classification of life (**Figure 13.7**). **Bacteria** and **Archaea** are each kingdoms unto their own. The four kingdoms of the Eukarya domain are **Protista**, a diverse group that includes amoebas and algae; **Plantae**, which encompasses all plants; **Fungi**, which includes mushrooms, molds, and yeasts; and **Animalia**, which encompasses all animals, including dinosaurs, birds, and humans. The members of each of these kingdoms share evolutionary innovations that adapted the organisms to their environment, enabling them to live and reproduce successfully (**Figure 13.8**).

The earliest forms of life evolved in water. During the Precambrian period, about 650 mya, the number of organisms appearing in the fossil record increased. At that time, much of Earth was covered by shallow seas, which were filled with small, mostly single-celled organisms that floated freely in the water.

Then, about 540 mya, the world experienced an astonishing burst of evolutionary activity, with a dramatic increase in the diversity of life. Most of the major living animal groups first appear in the fossil record during this time, popularly known as the **Cambrian explosion**. The Cambrian explosion changed the face of life on Earth from a world of relatively simple, slow-moving, soft-bodied scavengers and herbivores to a world filled with a diversity of large, fast predators. The presence of predators sped up the evolution of

Figure 13.7

Six kingdoms of life

The domain Bacteria is equivalent to the kingdom Bacteria, and the domain Archaea is equivalent to the kingdom Archaea. The domain Eukarya encompasses four kingdoms: Protista (protists, an artificial grouping that includes organisms such as amoebas and algae), Plantae (plants), Fungi (including yeasts and mushroom-producing species), and Animalia (animals).

Q1: What group of organisms shares a common ancestor with plants?

Q2: Are fungi more closely related to plants or to animals? Does this surprise you? Why or why not?

Q3: If you were to create an evolutionary tree in which amoebas were included within the kingdom of organisms to which they were the most closely related (rather than with protists, where they are currently placed), where would you put them?

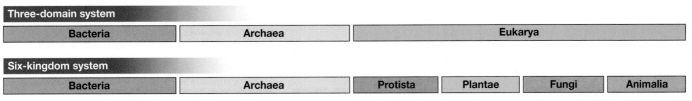

Three-domain system		
Bacteria	Archaea	Eukarya

Six-kingdom system					
Bacteria	Archaea	Protista	Plantae	Fungi	Animalia

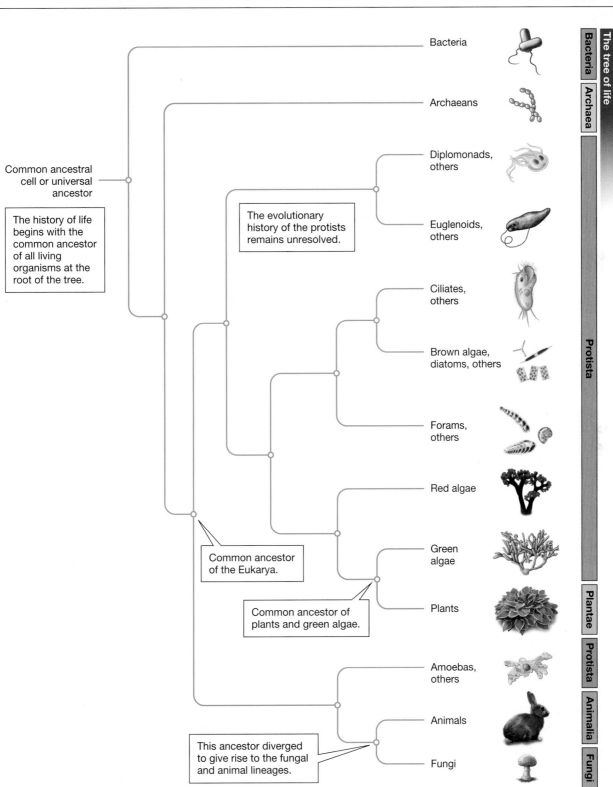

Common ancestral cell or universal ancestor

The history of life begins with the common ancestor of all living organisms at the root of the tree.

The evolutionary history of the protists remains unresolved.

Common ancestor of the Eukarya.

Common ancestor of plants and green algae.

This ancestor diverged to give rise to the fungal and animal lineages.

Bacteria

Archaeans

Diplomonads, others

Euglenoids, others

Ciliates, others

Brown algae, diatoms, others

Forams, others

Red algae

Green algae

Plants

Amoebas, others

Animals

Fungi

The tree of life

Bacteria

Archaea

Protista

Plantae

Protista

Animalia

Fungi

Protista

The kingdom Protista is an artificial grouping defined by what members of this group are not: protists are not plants, animals, fungi, bacteria, or archaeans.

Although most protists are harmless, the best-known ones are pathogenic, like *Plasmodium vivax*, the protist that causes malaria.

Animal-like protists are consumers.

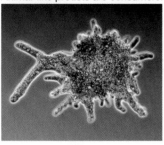

Fungus-like protists are decomposers.

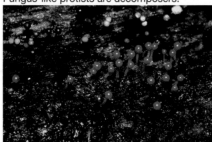

Plant-like protists are photosynthetic.

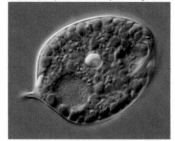

Most protists are single-celled and microscopic, and they can swim with the help of one or more flagella, or by waving a carpet of tiny hairs called cilia.

Protista

Diplomonads, others	Euglenoids, others	Ciliates, others	Brown algae, diatoms, others	Forams, others	Red algae	Green algae

Common ancestor of the Eukarya.

Plantae

Plants are multicellular autotrophs and mostly terrestrial. Because plants are producers, they form the basis of essentially all food webs on land.

Bryophytes (Mosses and liverworts)

Plants have a waxy covering, known as the *cuticle*, that covers their above-ground parts. A waxy cuticle holds in moisture—an important adaptation to life on land.

Lycophytes (Ferns and allies)

Gymnosperms (Conifers and others)

Gymnosperms were the first plants to evolve pollen, a microscopic structure that contains sperm cells, which freed them from a dependence on water for fertilization. Gymnosperms were also the first to evolve seeds, which can be disseminated so they will not compete with the mother plant for sunlight, or for water and nutrients in the soil.

Angiosperms (flowering plants)

Flowering plants, or angiosperms, are dominant and the most diverse group of plants on our planet. The keys to the success of angiosperms are the flower, a structure that evolved through modification of early plant reproductive organs, and the fruit, a fleshy ovary wall that protects and helps disperse the seeds inside it.

Figure 13.8

The four kingdoms of the domain Eukarya

Zygomycetes (molds)

Fungi play several roles in terrestrial ecosystems. Many are decomposers, acting as garbage processors and recyclers by speeding the return of the nutrients in dead and dying organisms to the ecosystem.

Fungi digest organic material outside the body and absorb the molecules released as breakdown products.

Ascomycetes (cup fungi)

Basidiomycetes (mushrooms)

Fungi are similar to animals in that they store surplus food energy in the form of glycogen. Like some animals, such as insects and lobsters, fungi produce a tough material called *chitin* that strengthens and protects the body. Unlike animals, fungal cells have a protective cell wall that wraps around the plasma membrane and encases the cells.

Plantae	Protista	Animalia	Fungi

Plants   Amoebas, others   Animals   Fungi

Animals are multicellular ingestive heterotrophs, obtaining energy and carbon by ingesting food. All animals are consumers, and some are important decomposers in the ecosystems they inhabit.

Animal cells differ from those of plants and fungi in that they lack cell walls. Instead, many cells in the animal body are enveloped in, or attached to, a felt-like layer known as the *extracellular matrix*. An important evolutionary innovation of animals is the development of true tissues. Most animals have two or three main tissue layers that give rise to a structurally complex body.

Lophotrochozoans: rotifers, flatworms, annelids, mollusks, and others

Sponges

Cnidarians, ctenophores

Ecdysozoans: nematodes, tardigrades, arthropods, and others

Echinoderms, hemichordates

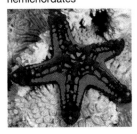

Chordates

The sponges are the most ancient animal lineage. Cnidarians, a group that includes jellyfish and corals, evolved next. The remaining animal phyla fall into two groups: the *protostomes* and *deuterostomes*, distinguished by different patterns of embryonic development. Protostomes comprise more than 20 separate subgroups, including mollusks (such as snails), annelids (segmented worms), and arthropods (including spiders and insects). Deuterostomes include echinoderms (sea stars and their relatives) and the chordates. The chordates are a large group encompassing all animals with backbones, such as fish, birds, and humans.

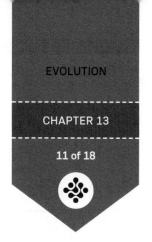
Cambrian herbivores, judging by the variety of scales and shells and other protective body coverings typical of many Cambrian, but not Precambrian, fossils (**Figure 13.9**).

The Cambrian explosion, however, occurred primarily in the oceans. Because life first evolved in water, the colonization of land by living organisms posed enormous challenges. Indeed, many of the functions basic to life, including support, movement, reproduction, and the regulation of heat, had to be handled very differently on land than in water. About 480 mya, plants were the first organisms to meet these challenges. These early terrestrial colonists were single-celled or had just a few cells.

Fungi are thought to have made their way onto land next, according to new studies. For example,

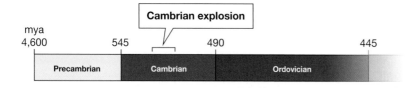

Figure 13.9

Cambrian biodiversity

During the Cambrian period, animal diversity increased dramatically. The remains of many of these species have been found in Canada (the Burgess Shale), China (the Maotianshan Shale), and fossil beds in Greenland and Sweden. Some of the fossils look familiar, resembling sponges and brachiopods, but do not appear to be related to any living groups of animals.

scientists have found fossils of terrestrial fungi that are 455–460 million years old. Fungi are absorptive heterotrophs: They digest organic material outside the body and absorb the molecules that are released as breakdown products. Because fungi do not fossilize well, their early evolutionary history is shrouded in mystery. Reconstructing the evolutionary history of eukaryotes from DNA data, scientists estimate that the common ancestor of fungi and animals diverged from all other eukaryotes about 1.5 billion years ago, and fungi diverged from their closest cousins, the animals, about 10 million years after that.

Next, land plants evolved and diversified from the original green algae that made it to land. By 360 mya, at the end of the Devonian period, Earth was covered with plants. As new groups of land plants arose, they evolved key innovations, including a waterproof cuticle, vascular systems, structural support tissues (such as wood), leaves, roots, seeds, and specialized reproductive structures. These and other important changes enabled plants to cope with life on land.

The first land animals likely emerged around 400 mya. Many of the early animal colonists on land were carnivores; others fed on living plants or decaying plant material. Insects, which are currently the most diverse group of terrestrial animals, first appeared roughly 400 mya, and they played a major role on land by 350 mya.

The first vertebrates to colonize land were amphibians, the earliest fossils of which date to about 365 mya (see Chapter 14 for more on vertebrates). Early amphibians descended from lobe-finned fish. Amphibians were the most abundant large organisms on land for about 100 million years. Then, in the late Permian period, reptiles, which evolved from a group of reptile-like amphibians, took over as the most common vertebrate group. Reptiles were the first group of vertebrates that could reproduce without returning to open water, because they lay amniotic eggs that have a built-in food source and are protected from drying out by a hardened shell, compared to the jelly-like sac that encloses the eggs of other vertebrates. The evolution of the amniotic egg was a major event in the history of life because it established a new evolutionary branch, the amniotes, which later included all reptiles, birds, and mammals.

And so, with the rise of reptiles 230 mya, the age of dinosaurs began.

Figure 13.10

Another feathered dinosaur, *Eosinopteryx brevipenna*, supports Xu's new tree

Figure 13.11

A namesake fossil, *Aurornis xui*, illuminates the dinosaur-bird transition

Tussling with Trees

In January 2013, a year and a half after Xu published his controversial paper about *Xiaotingia* and *Archaeopteryx*, his work received unexpected support. Pascal Godefroit, a paleontologist at the Royal Belgian Institute of Natural Sciences, reported the discovery of a feathered dinosaur called *Eosinopteryx brevipenna. Eosinopteryx* had been dug up by a commercial collector in northeastern China in the same area in which *Xiaotingia* was discovered. The tiny 161-million-year-old dinosaur was preserved as a virtually complete skeleton, with its legs bent and arms out, as if it were about to jump (**Figure 13.10**).

When Godefroit and his team added *Eosinopteryx* to the evolutionary tree of feathered dinosaurs, they came to the same surprising conclusion that Xu had: *Archaeopteryx* was not a bird. Instead, *Archaeopteryx* was a deinonychosaur along with *Eosinopteryx* and *Xiaotingia*. All three species share traits that early birds did not have, such as arms longer than their legs, reduced tail plumage, and primitive feather development.

But science is a continuously changing process, and only four months later, Godefroit published data about another birdlike fossil that caused him to revise his hypothesis once again. This time it was a feathered dinosaur that Godefroit found collecting dust in the archives of a Chinese museum. The 18-inch-long fossil had small, sharp teeth and long forelimbs. Godefroit believed the dinosaur, named *Aurornis xui* to honor Xu's work, probably couldn't fly, but instead used its wings to glide from tree to tree. But its other features, including the hip bones, were clearly shared by modern birds (**Figure 13.11**).

Using *Aurornis*, Godefroit constructed another evolutionary tree. This time he started from scratch, compiling data on almost a thousand characteristics of skeletons of 101 species of dinosaurs and birds. "It's very impressive," paleontologist Mike Lee at the South Australian

PASCAL GODEFROIT

Pascal Godefroit is a paleontologist at the Royal Belgian Institute of Natural Sciences in Brussels. He has discovered numerous feathered dinosaurs.

Museum told *National Geographic*. "They considered more than twice as much anatomical information as even the best previous analyses."

Contrary to his first study, Godefroit's second tree placed *Archaeopteryx* back on its roost in the bird family, although no longer as the oldest bird. That place belongs to *Aurornis*, says Godefroit (**Figure 13.12**).

Still, the debate is not over. "Of course we need more evidence and more work," says Xu. "Many of these new species are a possible candidate for the earliest bird." There are likely to be many more fossils that will shake up the bird tree, says Xu, but that's a good problem to have. "There are so many new species, it just makes it difficult for us all to agree."

Xu, Godefroit, and others are sure to continue to dig up dinosaur fossils for some time yet. Dinosaurs arose about 230 mya and dominated the planet from about 200 mya to about 65 mya. Then the majority of dinosaurs became extinct, except for those that evolved into birds.

As the fossil record shows, species have regularly gone extinct throughout the history of life. The rate at which this has happened—that is, the number of species that have gone extinct during a given period—has varied over time, from low to very high. At the upper end of this scale, the fossil record shows that there have been five **mass extinctions**, periods of time during which great numbers of species went extinct (**Figure 13.13**).

Although difficult to determine, the causes of the five mass extinctions are thought to include such factors as climate change, massive volcanic eruptions, changes in the composition of marine and atmospheric gases, and sea level changes. The Cretaceous extinction event occurred about 65 mya and wiped out three-quarters of plant and animal species on Earth, including non-avian dinosaurs. The cause of this mass extinction is suspected to be the impact of a massive comet or asteroid in the Gulf of Mexico, which choked the skies around the planet with debris

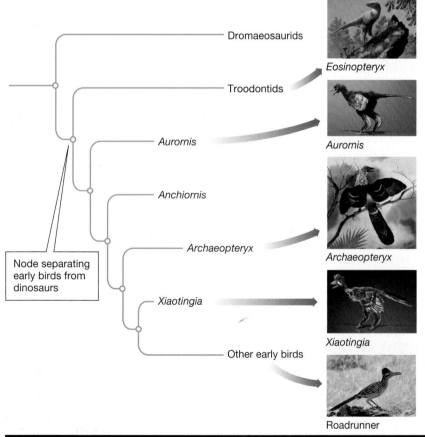

Dromaeosaurids

Eosinopteryx

Troodontids

Aurornis

Aurornis

Anchiornis

Node separating early birds from dinosaurs

Archaeopteryx

Archaeopteryx

Xiaotingia

Xiaotingia

Other early birds

Roadrunner

Figure 13.12

The early birds

Godefroit's 2013 study places *Archaeopteryx* and *Xiaotingia* with birds rather than dinosaurs, as in the traditional dinosaur-bird evolutionary tree (see Figure 13.5, top). However, it places *Aurornis* as the earliest known bird on the evolutionary tree.

Q1: Is *Xiaotingia* an earlier or later bird than *Archaeopteryx* in this tree?

Q2: If a future study, based on more fossils or new measurements, placed *Archaeopteryx* back with dinosaurs, would this suggest that birds are not related to dinosaurs? Why or why not?

Q3: If you were to create an evolutionary tree of modern birds, where would you expect to place the roadrunner (based on its appearance in the above figure) as compared to a house sparrow or pigeon?

The Sixth Extinction

On at least five occasions, mass extinctions have occurred across the globe, caused at different times by climate change, volcanic eruptions, and possible asteroids. Today, scientists agree we are in the midst of a sixth extinction, and this time, we, the human race, are the cause.

Four extinct species

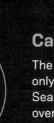

Passenger Pigeon

The passenger pigeon, the most common bird in North America 200 years ago, was hunted to extinction in the 19th century. The last wild bird was shot in 1900, and the last captive bird died in 1914.

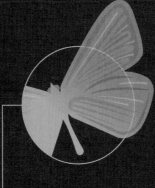

Xerces Blue Butterfly

The Xerces Blue butterfly once lived on sand dunes around San Francisco—until their habitat was destroyed by urban development. It was last seen in 1943.

Caribbean Monk Seal

The Caribbean monk seal was the only seal native to the Caribbean Sea and Gulf of Mexico. It was overhunted for oil, and its food sources were overfished. It was last sighted in 1952.

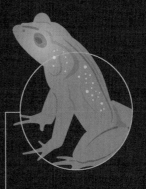

Golden Toad

Once common in the cloud forests of Monteverde, Costa Rica, the golden toad has not been seen since 1989. Pollution and global warming likely contributed to its extinction.

Number of species known to be extinct or extinct in the wild since 1500

Arachnids	9
Crustaceans	12
Reptiles	22
Amphibians	36
Insects	58
Fishes	71
Mammals	79
Plants	134
Birds	145
Mollusks	324

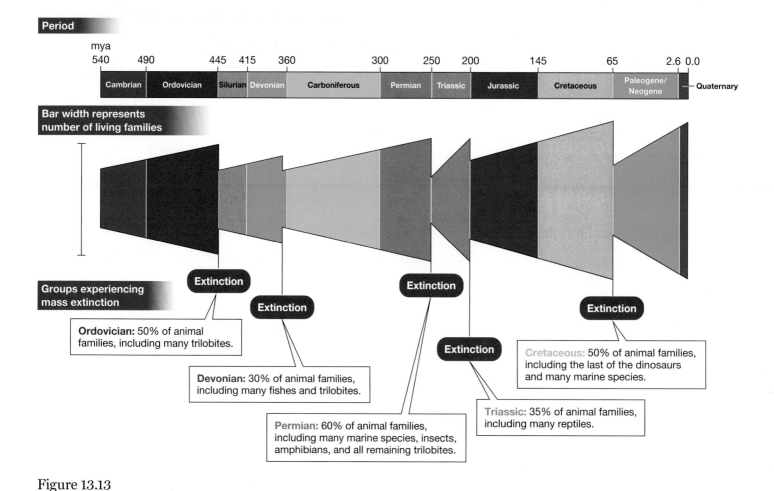

Period

Bar width represents number of living families

Groups experiencing mass extinction

Extinction

Ordovician: 50% of animal families, including many trilobites.

Extinction

Devonian: 30% of animal families, including many fishes and trilobites.

Extinction

Permian: 60% of animal families, including many marine species, insects, amphibians, and all remaining trilobites.

Extinction

Triassic: 35% of animal families, including many reptiles.

Extinction

Cretaceous: 50% of animal families, including the last of the dinosaurs and many marine species.

Figure 13.13

Mass extinctions and biodiversity

In addition to the marine and terrestrial animal groups shown here, plant groups were severely affected by the five mass extinctions that have occurred in Earth's history. After each extinction, life again diversified.

Q1: What extinction event occurred about 200 mya? What animal groups were most affected by this event?

Q2: Which of the mass extinctions appears to have removed the most animal groups? How long ago did this extinction occur?

Q3: The best studied of the mass extinctions is the Cretaceous extinction. Why do you think it has been better studied than the other extinctions?

and decreased plants' ability to photosynthesize. As the plants died, so, too, did animals further up the food chain.

The effects of mass extinctions on the diversity of life are twofold: First, entire groups of organisms perish, changing the history of life forever. Second, the extinction of one or more dominant groups of organisms can provide new opportunities for groups of organisms that previously were of relatively minor importance, thereby dramatically altering the course of evolution. The extinction of dinosaurs, for example, paved the way for mammals. Today, one species of mammal, *Homo sapiens*, dominates life on land. And our evolution has been as wonderful and dramatic as that of dinosaurs.

REVIEWING THE SCIENCE

- The first single-celled organisms resembled bacteria and probably evolved about 3.5 billion years ago.

- Scientists use **evolutionary trees** to model ancestor-descendant relationships among different organisms. The tips of branches represent existing groups of organisms, and each **node** represents the moment when an ancestor split into two descendant groups. A **clade** is an ancestral species and all its descendants.

- The most basic and ancient branches of the tree of life define three **domains**: **Bacteria**, **Archaea**, and **Eukarya**. All life-forms fall into one of these three domains. The domains are further divided into six **kingdoms**: the prokaryotic **Bacteria** and **Archaea**; and the eukaryotic **Protista**, **Fungi**, **Plantae**, and **Animalia**.

- Closely related groups of organisms share distinctive features that originated in their **most recent common ancestor**. These **shared derived traits** are used to identify the **lineages** of a species.

- The release of oxygen by photosynthetic bacteria caused oxygen concentrations in the atmosphere to increase. Rising oxygen concentrations made possible the evolution of single-celled eukaryotes about 2.1 billion years ago. Multicellular eukaryotes followed about 650 mya.

- Life in the oceans changed dramatically during the **Cambrian explosion**, when large predators and well-defended herbivores suddenly appear in the fossil record.

- The land was first colonized by plants (about 480 mya), fungi (about 460 mya), and invertebrates (insects; about 400 mya), which were followed later by vertebrates (about 365 mya).

- There have been five **mass extinctions** during the history of life on Earth. The extinction of a dominant group of organisms provides new opportunities for other groups.

THE QUESTIONS

The Basics

1 *Archaeopteryx*

(a) belongs to the domain Prokarya.

(b) belongs to the kingdom Protista.

(c) lived during the Precambrian period.

(d) was much larger than modern birds.

(e) is an early example of the evolution of birds.

2 Identify the domain and kingdom for each of the following organisms.

_____; _____a. *Escherichia coli* bacteria

_____; _____b. chanterelle mushrooms

_____; _____c. a palm tree

_____; _____d. green algae

_____; _____e. you

3 Place the following evolutionary events in order from earliest to most recent, numbering them 1 to 5.

_____a. Cambrian explosion

_____b. origin of life on Earth

_____c. plants' transition to land

_____d. oxygen-rich environment created by bacteria

_____e. evolution of birds from dinosaurs

4 Fill in the tree below to show the evolutionary relationships among the three domains and six kingdoms of life.

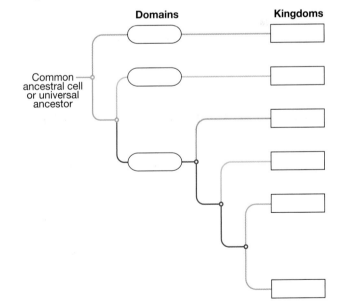

5 Link each of the following terms with the correct definition.

CLADE   1. A distinctive feature that originated in two groups' most recent common ancestor.

NODE   2. An ancestor and all its descendants.

LINEAGE   3. The point at which an ancestral group splits into two separate lineages.

EVOLUTIONARY TREE   4. A diagram showing the evolutionary relationships among a related group of organisms.

SHARED DERIVED TRAIT   5. The line of descent of a group of organisms.

6 Circle the correct terms in the following sentence:
The domains Archaea and Bacteria are referred to as (**prokaryotes, eukaryotes**). The domain (**Eukarya, Prokarya**) includes four kingdoms. The kingdom (**Plantae, Fungi**) was first to make the transition to land. The kingdom (**Animalia, Plantae**) is most closely related to the kingdom Fungi.

7 All of the following are considered possible causes of the five mass extinctions *except*

(a) climate change

(b) change in the composition of atmospheric or marine gases

(c) comet or asteroid strike

(d) worldwide thunderstorms

(e) volcanic eruptions

Try Something New

8 Create an evolutionary tree showing the relationships of the following organisms, based on their shared traits.

Species A: hairless, no tail, broad pelvis

Species B: hair-covered, long tail, narrow pelvis

Species C: hairless, short tail, broad pelvis

Species D: hair-covered, short tail, narrow pelvis

9 Which kingdom(s) might contain organisms with the following traits (you may need to review Figure 13.8 to answer):

____ a. motile, with flagella or cilia

____ b. all organisms within the kingdom are single-celled

____ c. found in extreme environments (for example, high temperature, low oxygen, high salt)

____ d. multicellular with organ systems

____ e. photosynthetic

10 The traditional dinosaur-bird tree (Figure 13.5, top) can be restated as a hypothesis: "We hypothesize that *Archaeopteryx* is an early bird, and that birds split off from the closely related dinosaur groups Troodontidae and Dromaeosauridae." Restate Xu's tree (Figure 13.5, bottom) and Godefroit's tree (Figure 13.12) as hypotheses.

11 Symbiosis is a long-term and intimate association between two different types of organisms. A symbiotic organism may live on or inside another species. For each of the following symbiotic relationships, (1) define the relationship in one to three sentences, (2) identify the domain and kingdom of each partner in the relationship, and (3) discuss whether the relationship evolved as a mutualism (both benefit), commensalism (one benefits, the other is not affected), or parasitism (one benefits to the detriment of the other).

(a) mycorrhizae

(b) lichens

(c) microbiome

(d) hermit crabs/shells

(e) malaria

Leveling Up

12 **What do *you* think?** Is a mass extinction under way? The International Union for Conservation of Nature (IUCN), also known as the World Conservation Union, maintains what it calls its Red List, which identifies the world's threatened species. To be defined as such, a species must face a high to extremely high risk of extinction in the wild. The 2013 Red List contains 21,286 species threatened with extinction, of a total of 71,576 species assessed.

Because this assessment accounts for only about 1 percent of the world's 1.7 million described species, the total number of species threatened with extinction worldwide may actually be much larger. For example, only 4,610 of 1 million described insect species have been assessed in terms of their survival risk.

The Red List is based on an easy-to-understand system for categorizing extinction risk. It is also objective, yielding consistent results when used by different people. These two attributes have earned the Red List international recognition as an effective method to assess extinction risk.

(a) If the threatened species listed by the IUCN do become extinct and the percentages of species under threat in other taxonomic groups turn out to be similar to those listed, then the percentages of species that will go extinct will approach the proportions lost in some of the previous mass extinctions. Does this mean that a mass extinction is under way? Why or why not? Explain your reasoning, using information in Figure 13.13 to support your argument.

(b) What are the causes of the current high extinction rates? Compare the causes of the current situation with those of previous mass extinction events. In what ways are they similar, and in what ways do they differ?

(c) Why are some groups in more danger of extinction than others? Choose one group that has a large proportion of species threatened with extinction, and find out more about it (you can begin at www.iucnredlist.org):

- What kinds of habitat are they found in?

- Does a particular aspect of their ecology—for example, feeding or reproduction—make them more vulnerable to extinction?

- Is anything being done to protect them and/or their habitat?

- Has there been an increase or decrease in the number of species within the group that have been identified as vulnerable to extinction?

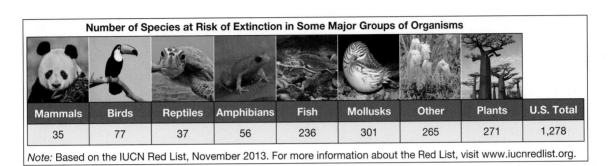

Number of Species at Risk of Extinction in Some Major Groups of Organisms

Mammals	Birds	Reptiles	Amphibians	Fish	Mollusks	Other	Plants	U.S. Total
35	77	37	56	236	301	265	271	1,278

Note: Based on the IUCN Red List, November 2013. For more information about the Red List, visit www.iucnredlist.org.

Neanderthal
Sex

The relationship between modern humans and Neanderthals just got a whole lot spicier.

After reading this chapter you should be able to:

- Describe the Linnaean hierarchy classification system.
- Understand the differences between hominids and hominins.
- Compare and contrast the inheritance of mitochondrial DNA and nuclear DNA.
- Recognize that many hominin species lived on Earth in the past.
- Describe the evidence that suggests that many of these species intermingled with our direct ancestors and made us who we are today.

CHAPTER

14

HUMAN EVOLUTION

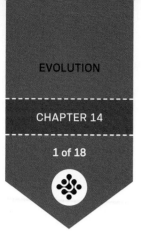

Most of us think of Neanderthals as ugly, hairy cavemen with big brains but no wits. The first Neanderthal bones were discovered in Germany in 1856; since then, we've cultivated an image of our closest extinct human relatives as hulking brutes who communicated by grunting, walked like chimps, and hit each other over the head with clubs. Yet, over the years, paleontologists have discovered fossilized vocal bones, sophisticated tools, and other evidence suggesting that Neanderthals were a fairly advanced group—and not quite as different from our own species as we'd like to believe.

Then, in 2010, scientists sequenced the Neanderthal genome, compared it to the modern human genome, and got the biggest shock of all: humans have some Neanderthal DNA. Our two species may have been a whole lot closer than we thought. "There could have been interbreeding between modern humans and Neanderthals," says Silvana Condemi, a researcher in the French National Centre of Research (CNRS) at the Aix-Marseille University in France. "We can imagine they not only exchanged culture, but exchanged genes."

That's right: Mounting evidence suggests that Neanderthals and modern humans had sex.

Blurring Lines

This story begins where many stories end—with a pile of bones. In the 2000s, Laura Longo, a curator at the Civic Natural History Museum of Verona in Italy, decided to look again at a group of fossils excavated from a rock shelter called Riparo Mezzena. Riparo Mezzena is nestled in the Lessini Mountains in northern Italy, a wide-open landscape speckled with large rocks and evergreen trees. The region is snowy and silent in the winter but green and thriving in the summer, when paleontologists come to work.

In the 1950s, paleontologists had carefully collected fossils of Neanderthals that lived around Riparo Mezzena about 35,000 years ago, late in the history of Neanderthals. But the fossils had sat at the museum, untouched, for over 50 years (**Figure 14.1**). Now, Longo believed, the fossils could help answer a hotly debated question about human history.

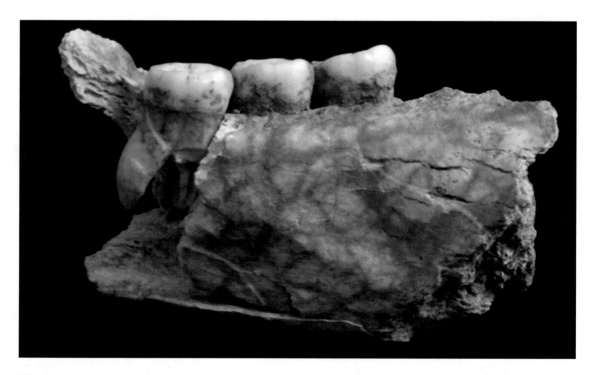

Figure 14.1

Fossil remains

This jaw was found at the Riparo Mezzena rock shelter in the Lessini Mountains in Italy. The individual it belonged to lived between 30,000 and 40,000 years ago.

Modern humans and Neanderthals lived in some of the same areas of Europe at the same times. Because of this proximity, paleontologists debate how closely the two species interacted. Some say that as modern humans expanded their territory, Neanderthals were quickly driven to extinction and therefore, the two did not live side by side. Others claim the opposite: that Neanderthals, *Homo neanderthalensis*, were slowly incorporated into the population of newly incoming humans, *Homo sapiens*.

In Chapter 13 we learned that biologists use a three-domain (Bacteria, Archaea, and Eukarya) and six-kingdom (Bacteria, Archaea, Protista, Fungi, Plantae, and Animalia) classification system for organizing life. Below the level of kingdom, we can get even more specific using the **Linnaean hierarchy**, a system of biological classification devised in the eighteenth century by a Swedish naturalist named Carolus Linnaeus.

The smallest unit of classification in the Linnaean hierarchy—the **species**—reflects individuals that are the most related to each other. Scientists do not know the exact number of species alive today; most estimates, however, fall in the range of 3 million to 30 million. So far, a total of about 1.5 million species have been collected, named, and placed in the Linnaean hierarchy.

The most closely related species are grouped together to form a **genus** (plural "genera"). Using these two categories in the hierarchy, every species is given a unique, two-word Latin name, called its **scientific name**. The first word of the name identifies the genus to which the organism belongs; the second word defines the species. Thus, *Homo sapiens* and *Homo neanderthalensis* are separate species, but both belong to the genus *Homo*.

In the Linnaean hierarchy, each species is placed in successively larger and more inclusive categories beyond genus. Closely related genera are grouped into a **family**. Closely related families are grouped into an **order**. Closely related orders are grouped into a **class**. Closely related classes are grouped into a **phylum** (plural "phyla"). Finally, closely related phyla are grouped together into a **kingdom** (**Figure 14.2**).

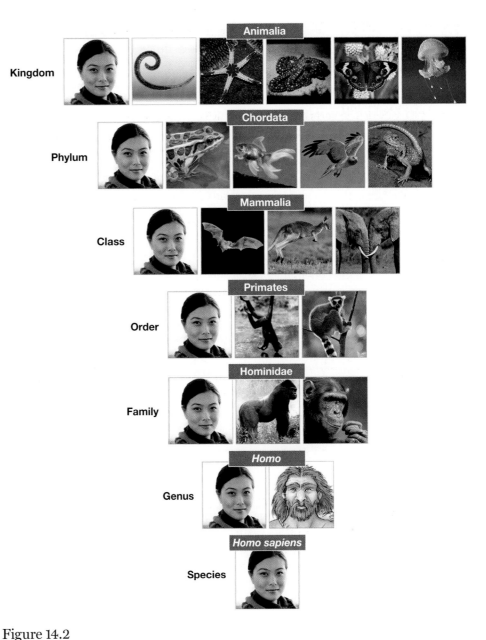

Figure 14.2

The Linnaean hierarchy of human classification
Homo sapiens is the only surviving species of the genus *Homo*.

Q1: Within which category are individuals most closely related to one another?

Q2: Within which category are individual species most distantly related?

Q3: Are individual species more closely related within the same order or within the same family?

Get a Backbone!

The phylum Chordata includes several subgroups of organisms, but we will focus on the **vertebrates**, animals with an internal *vertebral column* ("backbone") composed of a series of strong, hollow, cylindrical sections known as *vertebrae* (singular "vertebra") that enclose and protect a major nerve cord.

The **jawless fish** were the first vertebrates to evolve. Their skeletons—including the backbone—were made from a strong but flexible tissue called cartilage. Only a few groups of jawless fish have survived to the present day, most notably the lampreys. The next great leap in vertebrate evolution was hinged jaws, which enabled predators to grab and swallow prey efficiently. The evolution of teeth made jaws even more effective because teeth enabled animals to seize and tear food.

Another major step in the evolution of vertebrates was the replacement of the cartilage-based skeleton with a denser tissue strengthened by calcium salts: bone. Although the descendants of cartilaginous fish—sharks, skates, and rays—are still with us today, **bony fish** are far more diversified and widespread in both marine and freshwater environments. With more than 30,000 species, bony fish are the most diverse vertebrates today.

The advent of lungs was a crucial milestone in the transition of vertebrates onto land. **Amphibians** made this transition only partially; they can live on land but must return to the water to lay eggs and breed. The several thousand species of amphibians include frogs and salamanders.

Reptiles were the first vertebrates to head into drier environments, and they evolved a number of adaptive traits to deal with the risk of dehydration. These adaptations included skin covered in waterproof scales, a water-conserving excretory system, and the amniotic egg with its calcium-rich protective shell, which retards moisture loss while allowing the entry of life-giving oxygen and the release of waste carbon dioxide for the developing embryo. Reptiles dominated the Earth during the age of the dinosaurs, and the dinosaurs' descendants (as we saw in Chapter 13) remain with us today in the form of **birds**. Like mammals, birds are warm-blooded, but they have feathers instead of fur for insulation. At least 10,000 different species of birds are living today.

After the dinosaurs were gone, the **mammals** diversified to include more than 5,000 species, divided into three broad categories all of which feed their offspring with milk. More than 95 percent of mammals alive today are **eutherians**, a category that includes us. A unifying characteristic of eutherians is that the offspring are nourished inside the mother's body through a special organ called the placenta and are therefore born in a relatively well-developed state. The nonplacental, egg-laying mammals are classified as **monotremes**, which today consist of just one platypus species and several echidna species, all confined to Australia and New Guinea. The **marsupials** have a poorly developed placenta, resulting in offspring born early who then complete development in an external pocket or pouch. Marsupials are found mainly in Australia and New Zealand, with a few species in the Americas.

Down from the Trees

All *Homo* species, of which *H. sapiens* is the only one currently alive today, are in the Animalia kingdom, the Chordata phylum (because we have backbones; see "Get a Backbone!"), and the Mammalia class—we are mammals. As mammals, we share specific features with all other mammals, including body hair and milk produced by mammary glands. Within the mammals, we are part of the order consisting of the **primates** (**Figure 14.3**). Like all primates, we have flexible shoulder and elbow joints, five functional fingers and toes, thumbs that are **opposable** (that is, they can be placed opposite each of the other four fingers), flat nails (instead of claws), and brains that are large in relation to our body size.

Within the primates we are members of the ape family, the **hominids**. We are not just closely related to apes; we are apes. As such, we share many characteristics with other apes, especially chimpanzees, including the use of tools, a capacity for symbolic language, and the performance of deliberate acts of deception. But we are part of a distinct branch of apes called **hominins**—the "human" branch of the ape family that includes our extinct relatives, such as Neanderthals. The members of the hominin lineage have one or more humanlike features—for example, thick tooth enamel or upright posture—that set them apart from apes like gorillas and chimpanzees.

A major step in hominin evolution, and the main feature to distinguish hominins from other hominids, was the shift from moving on four legs to being **bipedal**—walking upright on two legs (**Figure 14.4**). Many skeletal changes accompanied the switch to walking upright, including the loss of opposable toes, as you will notice if you try touching your little toe with the big toe on the same foot.

The loss of opposable toes that accompanied bipedalism would have been a handicap in trees, since opposable toes help grasp branches during climbing. It is therefore likely that bipedalism was an adaptation for living on the ground. Walking on two feet freed the hands to carry food, tools, and weapons, and it also elevated the head, enabling the walker to see farther and over more things.

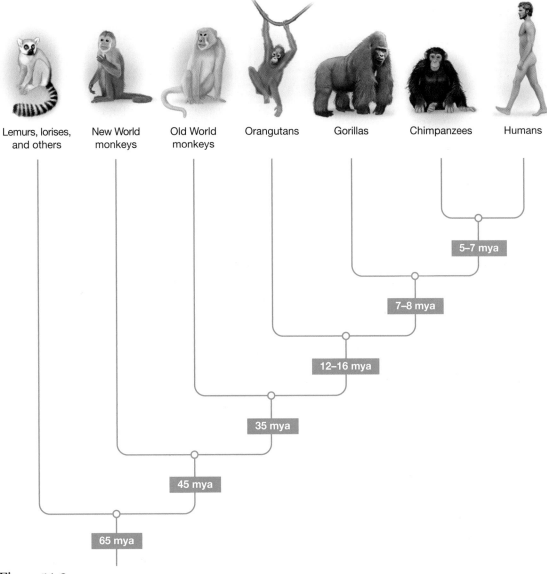

Figure 14.3

The primates include lemurs, monkeys, and apes
Genetic analyses and a series of spectacular fossil discoveries have led scientists to propose that the human lineage diverged from that of chimpanzees about 5–7 million years ago (mya). Similar evidence suggests that the evolutionary lineage leading to humans diverged from the lineage leading to gorillas about 7–8 mya, and from the lineage leading to orangutans about 12–16 mya.

Q1: According to this evolutionary tree, which primate group is most closely related to humans?

Q2: According to this evolutionary tree, which primate group is most distantly related to humans?

Q3: What are the common characteristics among all the primates, including humans?

The shift to life on the ground was probably not sudden or complete. The skeletal structure of some of the oldest fossil hominins (3–3.5 million years old) indicates that they walked upright. However, foot bones and fossilized footprints show that the hominins living at that time still had partially opposable big toes (**Figure 14.5**). Perhaps they continued to occasionally climb in trees.

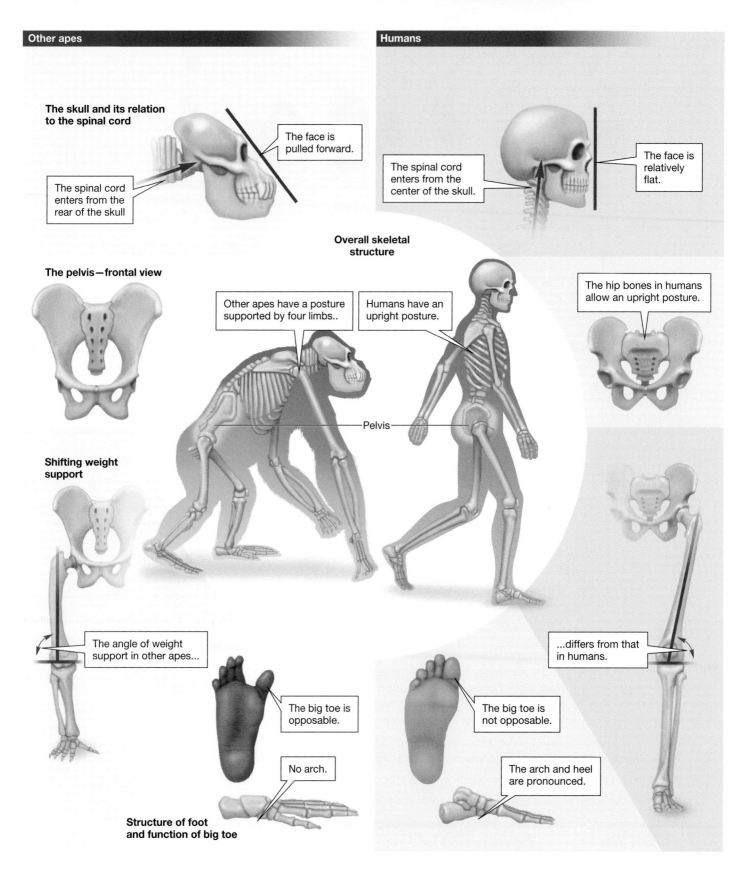

Other apes

The skull and its relation to the spinal cord

The face is pulled forward.

The spinal cord enters from the rear of the skull

The pelvis—frontal view

Overall skeletal structure

Other apes have a posture supported by four limbs..

Shifting weight support

Pelvis

The angle of weight support in other apes...

The big toe is opposable.

No arch.

Structure of foot and function of big toe

Humans

The spinal cord enters from the center of the skull.

The face is relatively flat.

Humans have an upright posture.

The hip bones in humans allow an upright posture.

...differs from that in humans.

The big toe is not opposable.

The arch and heel are pronounced.

Figure 14.4

Evolutionary differences between humans and other apes

The switch to walking upright required a drastic reorganization of primate anatomy, especially of the hip bones.

The earliest known hominin is *Sahelanthropus tchadensis*, identified from a 6- to 7-million-year-old skull discovered in 2002. Other early hominins include *Ardipithecus ramidus*, who lived 4.4 million years ago, and several *Australopithecus* species that are 3–4.2 million years old, including the first full-time walker with the first modern foot: *Australopithecus afarensis*. All of these hominins are thought to have walked upright. Their brains were still relatively small (less than 400 cubic centimeters in volume), and their skulls and teeth were more similar to those of other apes than to those of humans (**Figure 14.6**). A typical modern human has a brain volume of about 1,400 cubic centimeters. That's about the same volume as a 1.5 liter bottle of soda.

Rise of the Apes

Within the hominin branch is the *Homo* genus. The fossils identified at Riparo Mezzena were believed to be *Homo neanderthalensis* bones, but they had never been closely studied. So Laura Longo, the curator from Verona, compiled a team of researchers to analyze them, including anthropologist Silvana Condemi. Condemi had long been interested in the movement of Neanderthals across Europe and how their populations overlapped with human populations, so she joined the team to compare the Riparo Mezzena fossils to Neanderthal groups that had been dug up elsewhere around Europe, hoping to see what the fossils might say about how Neanderthals and modern humans interacted.

"For years, it was a very simple story: Neanderthals either disappeared quickly when modern humans came or they integrated with humans," says Condemi. But she suspected the dynamic was not that simple. "In some regions, Neanderthals disappeared very quickly. In other regions, we have evidence the two [species] could have

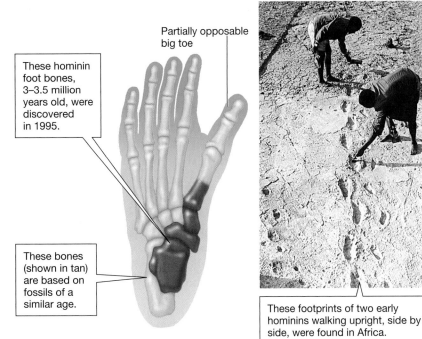

Partially opposable big toe

These hominin foot bones, 3–3.5 million years old, were discovered in 1995.

These bones (shown in tan) are based on fossils of a similar age.

These footprints of two early hominins walking upright, side by side, were found in Africa.

Figure 14.5

Early hominins had an upright stance and partially opposable big toes

Fossilized foot bones show that some hominins living between 3 and 3.5 million years ago walked upright but had partially opposable big toes.

Q1: What other reason besides continuing to use trees might explain why early hominins had partially opposable big toes?

Q2: In what way does the pattern of footprints suggest that the print makers were walking upright?

Q3: Why do you think we no longer have partially opposable big toes?

overlapped." Researchers, including Condemi, had often suggested that during the overlap, Neanderthals and humans interbred. But there was little physical evidence.

Even DNA evidence initially suggested there was no interbreeding. In 1997 and then 2004,

Q1: According to natural selection, deleterious traits disappear from a population over time. What traits must have been deleterious for ground-dwelling early hominins?

Q2: According to natural selection, advantageous traits persist in a population over time. What traits must have been advantageous for ground-dwelling early hominins?

Q3: What traits may have been neither deleterious nor advantageous for ground-dwelling early hominins?

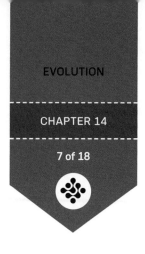
DNA from the mitochondria of Neanderthals (isolated first from a single Neanderthal fossil in 1997 and then from four Neanderthal fossils in 2004) was compared to modern human mitochondrial DNA, and the tests showed that there was no genetic overlap between the species. Mitochondrial DNA (mtDNA) is unique because it is passed down virtually unchanged from mother to child, so it can be tracked from one generation, or one species, to another (**Figure 14.7**). But modern *H. sapiens* did not have Neanderthal mitochondrial DNA, so it appeared there was no interbreeding, at least not with female Neanderthals.

That mitochondrial work was performed by Svante Pääbo and researchers at the Max Planck Institute for Evolutionary Anthropology in Leipzig, Germany. Pääbo is one of the founders of the effort to use genetics to study early humans and other ancient populations, and he has pioneered numerous techniques to extract delicate DNA from even the tiniest slivers of fossilized bones. But despite his initial findings that modern humans and Neanderthals did not share mitochondrial DNA, Pääbo still thought there might be room for some small contribution. Being a diligent scientist, he decided to look even deeper—this time at the whole Neanderthal genome.

Pääbo spent 4 years sequencing about 1.5 billion base pairs in the Neanderthal genome, using DNA extracted from the femur bones of three

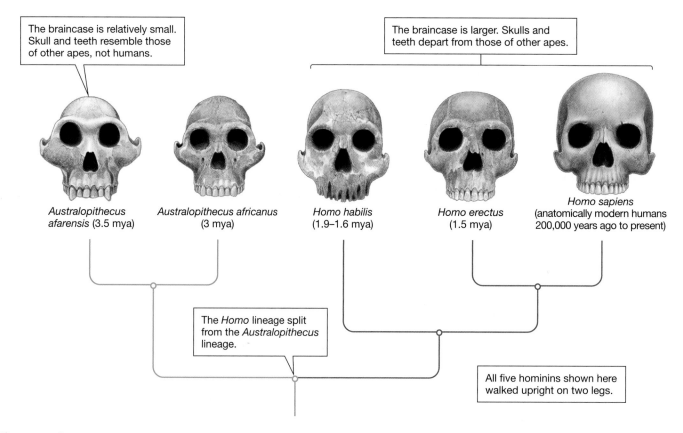

Figure 14.6

A gallery of hominin skulls

This tree shows the evolutionary relationships and the skulls of five hominin species. A complete evolutionary tree of hominins would be "bushier," with multiple side branches emerging at different times.

Q1: Where would the Neanderthal species branch be on this tree?

Q2: How would the Neanderthal skull differ from the *Homo erectus* skull?

Q3: How would the Neanderthal skull differ from the *Homo sapiens* skull?

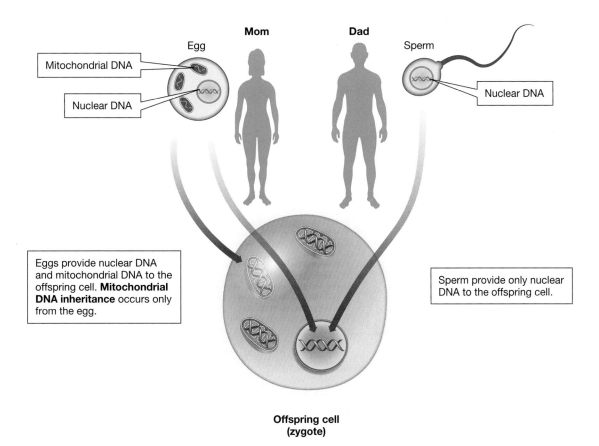

Egg Mitochondrial DNA Nuclear DNA **Mom** **Dad** **Sperm** Nuclear DNA

Eggs provide nuclear DNA and mitochondrial DNA to the offspring cell. **Mitochondrial DNA inheritance** occurs only from the egg.

Sperm provide only nuclear DNA to the offspring cell.

Offspring cell (zygote)

Figure 14.7

Mitochondrial DNA comes only from your mom (not from your dad)

In addition to the chromosomes in the nucleus (nuclear DNA), cells also have DNA in their mitochondria, called mitochondrial DNA. Mitochondria and their resident DNA all come from Mom.

Q1: Why does mitochondrial DNA come only from your mother?

Q2: If a Neanderthal-human hybrid was born to a human mother and a Neanderthal father, could you tell by mitochondrial-DNA sequencing?

Q3: If a Neanderthal-human hybrid was born to a human father and a Neanderthal mother, could you tell by mitochondrial-DNA sequencing?

38,000-year-old female Neanderthals. Then he compared that long, composite genome sequence to the genomes of five living humans from China, France, Papua New Guinea, southern Africa, and western Africa.

According to the results, all modern ethnic groups, other than Africans, carry traces of Neanderthal DNA in their genomes—between 1 and 4 percent. It turns out that most of us have a little Neanderthal in us. But whether that DNA is the result of thousands of sexual encounters between humans and Neanderthals or a few one-night stands remains unknown (**Figure 14.8**).

When Pääbo published his work in 2010, he and others admitted it was possible that the shared DNA wasn't necessarily a product of

SVANTE PÄÄBO

Svante Pääbo is a Swedish geneticist who directs the Department of Genetics at the Max Planck Institute for Evolutionary Anthropology in Leipzig, Germany. He specializes in using genetics to study early humans and other ancient populations.

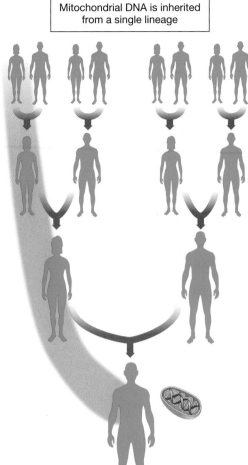

Nuclear DNA is inherited from all ancestors

Mitochondrial DNA is inherited from a single lineage

Figure 14.8

Nuclear-DNA inheritance
By sequencing the bases of nuclear DNA, scientists can determine how related an individual is to its ancestors, both male and female (see Chapter 10). Mitochondrial-DNA sequencing can determine only how related an individual is to its female ancestors on its mother's side.

Q1: If a human-Neanderthal hybrid was born to a human mother and a Neanderthal father, could you tell by whole-genome DNA sequencing?

Q2: If a human-Neanderthal hybrid was born to a Neanderthal mother and a human father, could you tell by whole-genome DNA sequencing?

Q3: Under what circumstances are scientists able to do whole-genome sequencing, and when are they restricted to mitochondrial-DNA sequencing?

interbreeding. It could have been a remnant of DNA from a common shared ancestor.

And who would that ancestor have been? The oldest *Homo* fossil fragments were found in Africa and date from 2.4 million years ago (mya), suggesting that the earliest members of the genus *Homo* originated in Africa 2–3 mya. More complete early *Homo* fossils exist from the period 1.9–1.6 mya; these fossils have been given the species name *Homo habilis*. The oldest *H. habilis* fossils resemble those of *Australopithecus africanus*, yet more recent *H. habilis* fossils

cm
198
183
168
152
137
122
107
92

Homo habilis

Homo sapiens

Homo floresiensis

Homo erectus

Paranthropus boisei

Homo heidelbergensis

Homo neanderthalensis

Figure 14.9

Meet the folks

Cartoon representations of seven hominin species depicting their average height in centimeters and their presumed features. *Homo sapiens* are represented as a 6-foot-tall male for reference.

Q1: Are you surprised by the interpretations of the hominins in this picture? If you are, explain why. If you're not surprised, explain why not.

Q2: Describe the main differences between the hominin species.

Q3: From what you've learned about these species, do you think these representations are accurate? How can you find more information about each species to help you answer this question?

show a more rounded skull and a face that isn't pulled as far forward. *H. habilis* fossils therefore provide an excellent record of the evolutionary shift from ancestral hominins (in *Australopithecus*) to more recent species, such as *Homo erectus*, the most likely candidate for a shared common ancestor of Neanderthals and modern humans (**Figure 14.9**).

Taller and more robust than *H. habilis*, *H. erectus* also had a larger brain and a skull more like that of modern humans. It is likely that by 500,000 years ago, *H. erectus* could use, but not necessarily make, fire. In addition, *H. erectus* probably hunted large animals, as suggested by a remarkable 2010 discovery in Germany of three 400,000-year-old spears, each about 2 meters long and designed for throwing with a forward center of gravity (like a modern javelin). *H. erectus*, or one of the other *Homo* ancestors, migrated from Africa about two million years ago. From there, this ancestor species spread around the Middle East and into Asia. *Homo* fossils dating from the period 1.7–1.9 mya have been found in the central Asian republic of Georgia, in China, and in Indonesia.

When a group of organisms like *H. erectus* expands to take on new ecological roles and to form new species and higher taxonomic groups, that group is said to have undergone an **adaptive radiation**. The descendants of those early *Homo* species multiplied and dispersed, presumably forming species that were able to live in a broad range of new environments (**Figure 14.10**). Some of the great adaptive radiations in the history of life occurred after mass extinctions, such as when the mammals diversified after the extinction of the dinosaurs, as described in Chapter 13. In other cases, adaptive radiations have occurred after a group of organisms has acquired a new adaptation that enables it to use its environment in new ways.

Overall, current research on *H. habilis*, *H. erectus*, and other early *Homo* species indicates that there were more species of *Homo* than was once thought, and that several of these species existed in the same places and times. More research and evidence will be necessary before general agreement is reached regarding the exact number of early *Homo* species and their evolutionary relationships.

So, was the Neanderthal DNA found in modern human genomes simply a remnant of a common ancestor? In 2012, Pääbo's team and others were able to determine the age of the pieces of Neanderthal DNA in the human genome. They found that the DNA was introduced into our genome between 40,000 and 90,000 years ago, around the same time that modern humans spread out of Africa and met the Neanderthals. A remnant of DNA from a common ancestor would have been ten times older.

Uniquely Human?

Yes, the frontal lobe of the brain is unique to human beings, and it enables us to reason like no other animal on our planet. But what about the rest of the attributes that we so commonly consider unique to us? Humans pride themselves on their intelligence and deep emotional connections to others, but are these really only human traits?

Language: Researchers once believed language was an exclusively human trait. We now know that chimpanzees in the wild use sign language, with approximately 70 different signs for distinct words. Meanwhile, other primates, birds, whales, and squid have distinct vocalizations that they use to communicate.

Memory: Others have suggested humans alone possess the ability to store memories. But dogs easily learn and remember many commands, while crows can learn and remember shapes better than human adults can, and they can use causal reasoning, not trial and error, to unlock doors and find hidden objects.

Social culture: Once thought to be strictly human, social culture is a trait that chimpanzees, Japanese macaques, and killer whales pass throughout their populations. Tool use by dolphins, elephants, and octopi varies in its specifics from population to population—a sure sign of learned behavior. Nigerian termite mounds rival human engineering with internal ventilation, including heating and cooling mechanisms, agricultural fungal gardens, storage cellars, chimneys, sanitation systems, and expressways. Some ants are known to capture and keep caterpillars to provide a constant food source by "milking" them for their sugary excretions.

Emotions: Our emotions make us human, right? Others in the animal kingdom have been documented expressing empathy (elephants), grief (dolphins, elephants), jealousy (great apes), curiosity (cats, lizards), altruism (great apes), and gratitude (whales). Great apes have been seen laughing at a clumsy fellow ape and using deception to outwit a family member.

Self-awareness: The ability to recognize oneself in the mirror, or show self-awareness, was once thought to be ours alone. As it turns out, all the great apes, some gibbons, elephants, magpies, and some whales pass the mirror test of self-awareness.

Morality: Finally, what about a sense of morality or an understanding of social norms? Monkeys and rats will not accept offered food if, in doing so, a fellow member of their species receives an electric shock.

To be sure, there *is* something unique about humans that lies at the intersection of all these abilities. However, our expanding knowledge of animal behavior can't help but make us feel more closely connected to the other species with whom we share this planet.

All in the Family

The fossil record indicates that the first *H. sapiens*, called archaic *H. sapiens*, originated between 300,000 and 400,000 years ago. Archaic *H. sapiens* bore features intermediate between those of *H. erectus* and those of "anatomically modern" *H. sapiens*—our species—which arose some 195,000 to 200,000 years ago. These ancestors

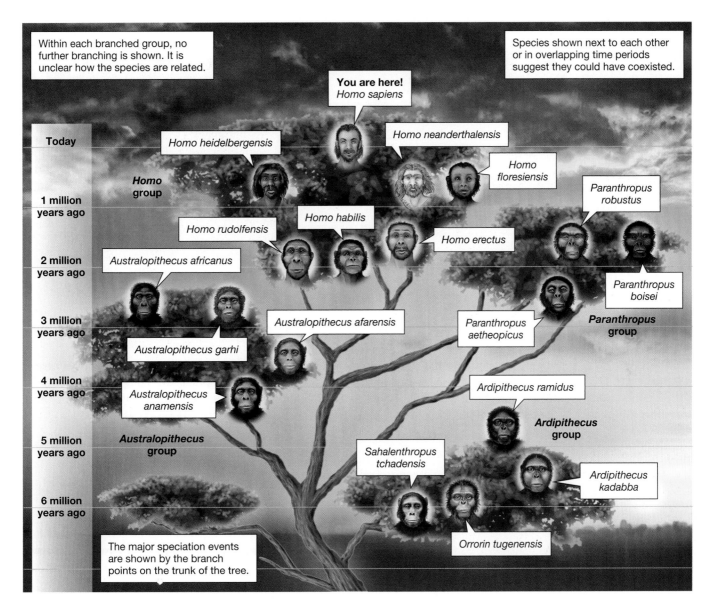

Figure 14.10

Hominin evolutionary tree

Hominin species are grouped together, with more similar species clumped on separate branches of the evolutionary tree.

Q1: Which major branch represents the oldest known group of hominins?

Q2: With what other species of *Homo* did modern humans overlap in time?

Q3: According to this tree, many hominin groups overlapped in time with other groups. Do you think these species intermingled?

of anatomically modern humans developed new tools and new ways of making tools, used new foods, and built complex shelters. (But humans are not the only organisms to do many of these things; see "Uniquely Human?" on page 250.)

What became of archaic *H. sapiens*? Early populations eventually gave rise to both the Neanderthals (who lived from 28,000 to 300,000 years ago) and us—that is, anatomically modern humans. There is some debate as to whether

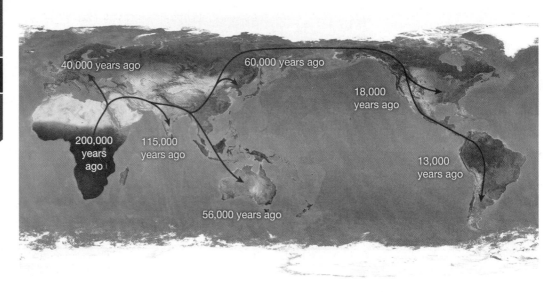

40,000 years ago

60,000 years ago

18,000 years ago

200,000 years ago

115,000 years ago

13,000 years ago

56,000 years ago

Figure 14.11

Anatomically modern humans evolved in Africa

The earliest known archaeological specimens of modern humans (*Homo sapiens*) come from Africa. The dates provided give the age of the earliest evidence that anatomically modern humans lived in different regions of the world. These dates are continually challenged by new fossil evidence that scientists must then work to confirm.

Q1: What evidence suggests that Neanderthals never lived in Africa?

Q2: How does the hypothesized origin of modern humans (*Homo sapiens*) differ from the hypothesized origin of Neanderthals (*Homo neanderthalensis*)?

Q3: What species of hominins other than the Neanderthals may have commingled with modern humans?

Neanderthals are simply an odd form of archaic *H. sapiens*, or their own distinct species. That question has yet to be resolved.

According to the **out-of-Africa hypothesis**, anatomically modern humans first evolved in Africa about 195,000 to 200,000 years ago from a unique population of archaic *H. sapiens*, and then spread into other continents to live alongside other hominins (**Figure 14.11**). Evidence from the fossil record indicates that anatomically modern humans overlapped in time with *H. erectus* and Neanderthal populations, yet remained distinct from them. Neanderthals and modern humans coexisted in western Asia for about 80,000 years, and in Europe for some 10,000 years, until modern humans completely replaced all other *Homo* populations.

But what happened in that intervening time? Were modern humans and Neanderthals friendly neighbors, or were the latter quickly wiped out by the former?

With another team, Condemi had previously used fossil evidence from southern Italy to determine that modern humans arrived on the Italian peninsula between 43,000 and 45,000 years ago, *before* the disappearance of Neanderthals.

SILVANA CONDEMI

Silvana Condemi is an anthropologist and CNRS (French National Center for Scientific Research) research director at Aix-Marseille University in France. She studies the movement of Neanderthals across Europe and how their populations overlapped with modern humans.

Hereditary Heirlooms

Because of a shared common ancestor 1.6 billion years ago, humans share DNA with all animals, plants and fungi. But how much? Take a look to see how much genetic material you have in common with other organisms.

Percentage of genes shared with humans

Humans
100%

Neanderthals
99%

Chimpanzees
90%

Mice
88%

Dogs
84%

Zebrafish
73%

Platypuses
69%

Chickens
65%

Honeybees
44%

Roundworms
38%

Grapes
24%

Baker's yeast
18%

So the two populations likely made contact in Italy. During this period, it's possible that "there was a kind of interbreeding mixture," says Condemi. But if the two species interbred and had children, what did those children look like? And why hadn't we found their bones?

Then Condemi finally got a chance to examine the Riparo Mezzena bones. One in particular caught her attention. It was a jawbone, a mandible, from a late Neanderthal living in Italy at the same time that modern humans had already made their way into Europe (see Figure 14.1). But the jawbone didn't look like a Neanderthal's, which has no chin. Instead, the face of the Riparo Mezzena individual, when reconstructed with three-dimensional imaging, had an intermediate jaw, something between no chin and a strongly projected chin. Because chins are a feature unique to modern humans (**Figure 14.12**), the jaw appeared to be something of a hybrid between a Neanderthal and a modern human. "This, in my view, could only be a sign of interbreeding," says Condemi.

To back up her hypothesis, Condemi's team analyzed the fossil's DNA. The fossil had Neanderthal mitochondrial DNA, confirming that at least the individual's mother was a Neanderthal. From the DNA and imaging evidence, Condemi and her team concluded that it was the child of a "female Neanderthal who mated with a male *Homo sapiens*" (**Figure 14.13**). What's more, she added, this evidence supports the idea of a slow transition from Neanderthals to anatomically modern humans, in which the two species intermingled in both culture and sex, rather than the abrupt extinction of Neanderthals when modern humans arrived.

It is unlikely that Condemi's finding will be the last word on Neanderthal-human interbreeding. And an even bigger question lingers for future scientists: While modern humans continued to develop their culture and radiate around the planet, Neanderthals went extinct. Why? Maybe another pile of bones will someday reveal the answer.

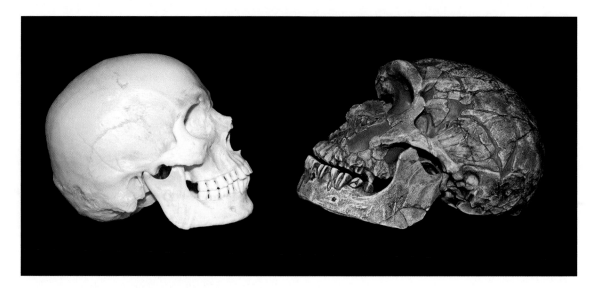

Figure 14.12

Neanderthal and human skulls
The skull on the right is from a Neanderthal; the one on the left, from a modern human. Notice the differences in the eyebrow ridge, forehead, and lower jaw.

Q1: Which skull features are distinctly modern human?

Q2: Which skull features are distinctly Neanderthal?

Q3: Why would you expect a hybrid of the two species to have intermediate features?

Figure 14.13

Neanderthal-human hybrid
A paleontological reconstruction of a Neanderthal-human hybrid.

REVIEWING THE SCIENCE

- The **Linnaean hierarchy** places each **species** in successively larger and more inclusive categories. Closely related species are grouped together into a **genus**, related genera are grouped into a **family**, related families into an **order**, related orders into a **class**, related classes into a **phylum**, and finally related phyla into a **kingdom**.

- Organisms are identified by their genus and species names, together referred to as their **scientific name**.

- Humans are **mammals** (class) and **primates** (order) with flexible shoulder and elbow joints, five functional fingers and toes, **opposable** thumbs, flat nails (instead of claws), and large brains.

- **Hominins** are characterized by **bipedalism**, the ability to walk upright on two legs. They are a branch of the great apes, the **hominids** (family).

- **Mitochondrial-DNA inheritance** occurs exclusively through the maternal egg. The sperm contributes essentially no mitochondria and none of the mitochondrial DNA.

- **Nuclear-DNA inheritance** occurs equally through both the eggs and sperm. All individuals have half maternal and half paternal nuclear DNA.

- According to the **out-of-Africa hypothesis**, anatomically modern humans first evolved in Africa and then spread to the rest of the world, possibly interbreeding with and definitely replacing other *Homo* populations.

- The descendants of early *Homo* species, such as *Homo erectus*, multiplied and dispersed. Many *Homo* species were able to live in a broad range of environments, presumably through **adaptive radiation**.

THE QUESTIONS

The Basics

1 Which of the following terms most specifically describes what occurs when a group of organisms expands to take on new ecological roles, forming new species and higher taxonomic groups in the process?

(a) speciation

(b) mass extinction

(c) evolution

(d) adaptive radiation

2 Which of the following sentences is *not* true?

(a) A single evolutionary line led from *Ardipithecus ramidus* to modern humans.

(b) Some hominid traits evolved more rapidly than others.

(c) Brain size increased greatly from early hominids to *Homo sapiens*.

(d) Toolmaking technology has improved greatly over the past 300,000 years.

3 The out-of-Africa hypothesis states that:

(a) all new species of hominins arose in Africa and then migrated to the rest of the world.

(b) many new species of hominins arose outside of Africa and then migrated back to Africa later.

(c) all hominin speciation events occurred outside of Africa.

(d) all speciation events occurred in Africa, but only *Homo sapiens* distributed themselves across the globe.

4 Which of the following features do humans lack that other primates have?

(a) forward-facing eyes

(b) short snouts

(c) flexible shoulder and elbow joints

(a) opposable big toes

5 _____ specimens have features that are intermediate between those of *Australopithecus africanus* and *Homo erectus* and provide an amazing record of the evolutionary shift from ancestral hominin characteristics seen in *Australopithecus* fossils to more recent ones seen in *H. erectus* fossils.

(a) *Homo sapiens*

(b) *Homo neanderthalensis*

(c) *Homo habilis*

(d) *Ardipithecus ramidus*

Try Something New

6 You visit your local museum of natural history and come upon an exhibit showcasing hominin fossils that date back 300,000–400,000 years ago. You notice that these fossils have features intermediate between those of *Homo erectus* and those of "anatomically modern" *Homo sapiens*. Who were these fossils?

(a) archaic *Homo erectus*

(b) archaic *Homo sapiens*

(c) *Homo neanderthalensis*

(d) *Homo habilis*

7 After years of digging, you discover fossilized remains of what appears to be an extinct hominin. When analyzed by experts, they are proclaimed to be fossils from the probable shared recent common ancestor between Neanderthals and modern humans. Who were these fossils?

(a) *Homo erectus*

(b) archaic *Homo sapiens*

(c) *Homo neanderthalensis*

(d) *Homo habilis*

8 Judging by the inability of your friend's boyfriend to use common table manners and hold an intelligent conversation, you are suspicious that he has more than the normal amount of Neanderthal DNA in his genome. What can you do to find out if your suspicions are accurate?

(a) Sequence his mitochondrial DNA and compare it to Neanderthal mitochondrial DNA to determine whether he is in the normal 1–4 percent range of overlap.

(b) Measure his brow bone and chin to determine whether he is more Neanderthal than modern human.

(c) Ask your friend about his "hairiness" and compare to all the people you know.

(d) Sequence his nuclear DNA and compare it to Neanderthal nuclear DNA to determine whether he is in the normal 1–4 percent range of overlap.

9 How does the fact that all ethnic groups *except* Africans contain some Neanderthal DNA (1–4 percent of their DNA) support the out-of-Africa hypothesis for the origin of modern humans (*Homo sapiens*)?

(a) It suggests that Neanderthals evolved and left Africa before modern humans evolved in Africa, and that modern humans intermingled with Neanderthal populations only after leaving Africa.

(b) It suggests that Neanderthals and modern humans evolved at the same time in Africa but had sex with each other only after leaving Africa and settling in other areas.

(c) It suggests that modern humans evolved and left Africa before Neanderthals evolved in Africa, and that Neanderthals intermingled with modern human populations only after leaving Africa.

(d) It suggests that Neanderthals and modern humans had sex every chance they got in all areas, but that only in Africa were the hybrid offspring shunned and forced to move out.

Leveling Up

10 *Write Now* **biology: if we were not alone** Fossil evidence indicates that in the relatively recent past (about 30,000 years ago), anatomically modern humans, or *Homo sapiens*, may have shared the planet with at least three other distinct hominins: *H. erectus*, *H. neanderthalensis*, and *H. floresiensis*. If one or more of these species were alive today, how would their existence affect the world as we know it?

11 **Is it science?** Watch the original 1968 movie *Planet of the Apes*. Document as many scientific problems with the movie as you can. Which of the apes' adaptations would be biologically possible, and which ones would be impossible, from your understanding of apes on Earth today? If this species did evolve to have the adaptive traits of *Homo*, would its members still be called apes?

12 **What do *you* think?** Read Jean M. Auel's novel *The Clan of the Cave Bear*, or watch the movie. Hypothesize about which species each of the characters in the book might have belonged to. Why do you think so? Cite examples from the story to support your ideas. The book was written in 1980; how does recent DNA sequencing evidence support or refute the story line in this book?

13 **What do *you* think?** Every year we find evidence of additional hominin species or variations of each of the *Homo* species. On the Internet, search something like "hominin species discovered" and find at least two species not mentioned in this chapter. Where do you think they fit into the evolutionary tree represented in Figure 14.10? Draw a sketch of what you think they would look like based on the descriptions you find. Indicate where they would fit in the lineup of hominins in Figure 14.9. How many more species do you think are still out there waiting to be discovered?

Amazon on Fire

Can the world's largest rainforest survive a vicious cycle of wildfires and climate change?

After reading this chapter you should be able to:

- Define the biosphere and the role that humans play in it.
- Articulate the difference between a biotic factor and an abiotic factor.
- Describe the greenhouse effect.
- Understand the water cycle and the carbon cycle.
- Explain how global warming contributes to climate change.
- List and describe the consequences of climate change.

CHAPTER

15

GENERAL PRINCIPLES OF ECOLOGY

Abiotic factors: rocks, water, air

Figure 15.3

Amazon rainforest ecology

Ecology is the study of how living org
other organisms, and how they all inte
environment.

Q1: List as many biotic and abiotic
can.

Q2: Is the forest part of the bi

part of the

burning plots
happens after-
y. **Ecology** is the
s between organ-
where the environ-
both biotic factors
biotic (nonliving)
y helps us under-
ive in, but humans
ere in ways that are
ible, to fix.

Flames rise from the forest floor, licking at Jennifer Balch's heels. Balch walks carefully ahead of the heat, stoking it, encouraging it. She tips a large metal container, dripping flaming kerosene onto another pile of dead branches on the ground (**Figure 15.1**). The forest burns behind her.

Balch steps back to look at her work—the destruction of a small patch of the Amazon rainforest. Balch, an ecologist at the University of Colorado Boulder, is in the southern part of the Amazon basin in Brazil, in the state called Mato Grosso. There, on a small square kilometer and a half of land belonging to a soybean farmer, Balch is experimenting to see what happens when the Amazon burns.

The Amazon is the largest rainforest in the world and a critical part of our planet's biosphere. The **biosphere** consists of all of Earth's organisms, plus the physical space we all inhabit. It includes inorganic chemicals like water, our nitrogen-rich atmosphere, every living organism, and more (**Figure 15.2**). Put more simply, the biosphere is the integration of all the ecosystems on Earth. It is crucial to our survival and well-being because humans depend on the biosphere for food and raw materials.

Within the biosphere, the Amazon is home to more than half of the world's millions of species of plants and animals. It also contains one-fifth of the world's fresh water. Yet the Amazon is under threat: Since 1960, the human population of the Brazilian Amazon region has increased to 25 million people from 6 million, leading to a dramatic expansion of agriculture, including major deforestation as humans cut down trees to make room for roads, cattle, and fields. The forest cover in the Amazon has declined by 20 percent during this period, and the grasslands and pastures that have taken its place, including soybean fields like the one Balch works at, are threatening the rainforest that is left.

Figure 15.1

Jennifer Balch ignites a controlled burn in an experimental plot in the rainforest

It's traditional for farmers in Brazil to clear land for agriculture by setting fires to burn away trees and brush. In addition, fire is a useful tool for clearing old grasses to encourage the growth of new grasses for cattle to feed on. Yet when these fires are not kept under control, they can spread to nearby rainforest and destroy it. Historically, this has not been a major concern because the Amazon is wet and humid, so forest fires burned slowly, extinguishing themselves before spreading too far. Yet a dramatic shift in our planet's climate is making the Amazon drier and more vulnerable to wildfires.

Our planet is warming. As a result of global climate change, the temperature in Earth's atmosphere is increasing, causing less rainfall and more drought in the Amazon, leaving the forests primed for fire. And they've begun to burn. Even worse, that loss of trees adds to the warming, resulting in a vicious cycle in which the Amazon becomes hotter and drier, burns again, and contributes to further climate change.

Balch and other scientists have been investigating whether it might be possible to intervene in that cycle to slow or stop the loss of the Amazon

JENNIFER BALCH

Jennifer Balch is an ecologist at the University of Colorado Boulder who studies how fire disturbance affects ecosystems. In 2004, she initiated a one-of-a-kind experimental burn study in the Amazon to see how wildfires affect the tropical rainforest.

Figure 15.2

The biosphere is Earth and all of its inhabitants

This is a view of Earth from space. The atmosphere, Earth's surface, and all of the organisms make up the biosphere.

Q1: Clouds cover much of Earth. Are these part of the biosphere? Explain.

Q2: Polar ice caps cover part of Earth. Are these part of the biosphere? Explain.

Q3: The photo shows Earth as it is found in our solar system surrounded by outer space. Is outer space part of the biosphere? Explain.

---

rainforest. Their latest experiment, burning plots of Amazon rainforest to see what happens afterward, is an experiment in ecology. **Ecology** is the scientific study of interactions between organisms and their environment, where the environment of an organism includes both biotic factors (other living organisms) and abiotic (nonliving) factors (**Figure 15.3**). Ecology helps us understand the natural world we live in, but humans continue to change our biosphere in ways that are difficult, perhaps even impossible, to fix.

Hot and Dry

Balch began setting fires in the Amazon in 2004 as a graduate student at the Yale School of Forestry & Environmental Studies. She is an ecologist, a scientist fascinated by how organisms and environments interact with and affect each other.

Balch has always been interested in the interaction between humans and fire. "Fire is integral to the human experience. We play with fire as kids, and we use fire in our cars, through combustion, on a daily basis," says Balch. "But humans have an imperfect relationship with fire. We are vulnerable to fire; we don't completely control this tool."

Balch decided to try to control it just enough to learn about fire's impact on the Amazon. By performing a planned burn in a restricted area, her team could take measurements that cannot be obtained by studying an accidental or escaped fire, such as inventories of the plants in the area before and after a fire, a census of the animals in the area, and more (**Figure 15.4**). With a crew of other researchers, Balch orchestrated an experiment using three 0.5-square-kilometer plots of forest. One control plot would never be burned,

Abiotic factors: rocks, water, air

Biotic factors: plants, animals, microbes

Figure 15.3

Amazon rainforest ecology

Ecology is the study of how living organisms (biotic factors) interact with other organisms, and how they all interact with their nonliving (abiotic) environment.

Q1: List as many biotic and abiotic factors in this photograph as you can.

Q2: Is the forest part of the biotic or abiotic environment? Explain.

Q3: Is the river part of the biotic or abiotic environment? Explain.

---

one experimental plot would be burned once every three years, and a second experimental plot would be burned once every year.

To prevent fires in their two experimental plots from spreading, each plot was protected by a firebreak—a perimeter around the plot, several meters wide, cleared of all plants, sticks, grasses, and debris. The team then walked about 10 kilometers of trail cutting through the plots, dripping flaming kerosene from special tanks and watching the forest burn.

The experiment began not a moment too soon. In 2005, the Amazon experienced an extreme drought. Experts called it a "hundred-year drought," meaning that such a drought is expected to occur, on average, only once in a century. But just five years later, in 2010, there was another hundred-year drought, this one even more widespread and severe. And in both cases, extensive wildfires followed the drought, destroying over 30,000 square miles of rainforest. In 2005 alone, fires increased 20-fold in the year following the drought. "It was a huge increase," says Michael Coe, a researcher at the Woods Hole Research Center in Massachusetts and one of Balch's collaborators. "And if humans are increasing drought frequency, that's going to be a big problem."

Humans, researchers suspect, are increasing the number and intensity of droughts in the Amazon through climate change. It's important to distinguish between "climate" and "weather." **Weather** refers to short-term atmospheric conditions, such as today's temperature, precipitation, wind, humidity, and cloud cover. **Climate** describes the prevailing weather of a specific place over relatively long periods of time (30 years or more). Organisms are more strongly influenced by climate than by any other feature of their environment. On land, for example, features of climate such as temperature and precipitation determine whether a particular region is desert, grassland, or tropical forest.

Climate change, then, is a large-scale and long-term *alteration* in Earth's climate, and it includes such phenomena as global warming, change in rainfall patterns, and increased frequency of violent storms. Although Earth has gone through many changes in its average climate over its 4.6-billion-year history, the speed of the change that has taken place in the past 100 years is without precedent in the climate record. Climate change in recent history has been caused to a large extent by human actions, and its consequences are likely to be negative for people and ecosystems around the world (**Figure 15.5**).

"There are a lot of indications that climate change causes increased drought events, causing increased fire events," says Balch. "It's hard to predict the future, but that's definitely the trend."

Figure 15.4

Scientists take measurements before a planned burn in the Amazon

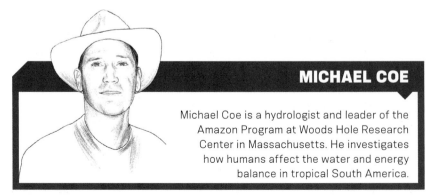

MICHAEL COE

Michael Coe is a hydrologist and leader of the Amazon Program at Woods Hole Research Center in Massachusetts. He investigates how humans affect the water and energy balance in tropical South America.

A Warmer World

"Global warming" and "climate change" are related but not synonymous. **Global warming** is a significant increase in the average surface temperature of Earth over decades or more. Temperature on Earth is generally determined by the angle at

Flooding, melting of glaciers and polar ice, rise in sea level, increase in ocean temperatures, and coral reef bleaching and death.

Trift Glacier, Switzerland
1948   2002   2006

Change in rainfall patterns, drought, ecosystem and habitat destruction, species extinction, agricultural decline, and fishery depletion.

Figure 15.5

Consequences of climate change

> **Q1:** Name two ways in which climate change affects plants, animals, and humans.
>
> **Q2:** Name two ways in which climate change affects the frequency and severity of floods.
>
> **Q3:** How will climate change cause a rise in sea level?

which sunlight strikes the planet. Sunlight strikes Earth most directly at the equator, but at a more slanted angle near the North and South Poles (**Figure 15.6**). For this reason, more solar energy reaches the equator, making it and neighboring tropical regions much warmer than the poles.

Global warming, however, is not dependent on how sunlight strikes the Earth. Instead, it is caused by an increase in **greenhouse gases**. Some gases in Earth's atmosphere, such as carbon dioxide (CO_2), water vapor (H_2O), methane (CH_4), and nitrous oxide (N_2O), absorb heat that radiates away from Earth's surface. These gases are called greenhouse gases because they function much as

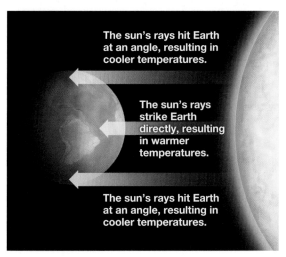

The sun's rays hit Earth at an angle, resulting in cooler temperatures.

The sun's rays strike Earth directly, resulting in warmer temperatures.

The sun's rays hit Earth at an angle, resulting in cooler temperatures.

Figure 15.6

Sunlight strikes Earth most directly at the equator
The angle at which the rays of the sun strike Earth determines how much energy or heat reaches Earth's surface. The more direct the rays are when they strike Earth, the more heat they deliver.

> **Q1:** What is a main determinant of temperature in different areas of Earth?
>
> **Q2:** Why is it colder at the poles than at the equator?
>
> **Q3:** Why is it warmer at the equator than at the poles?

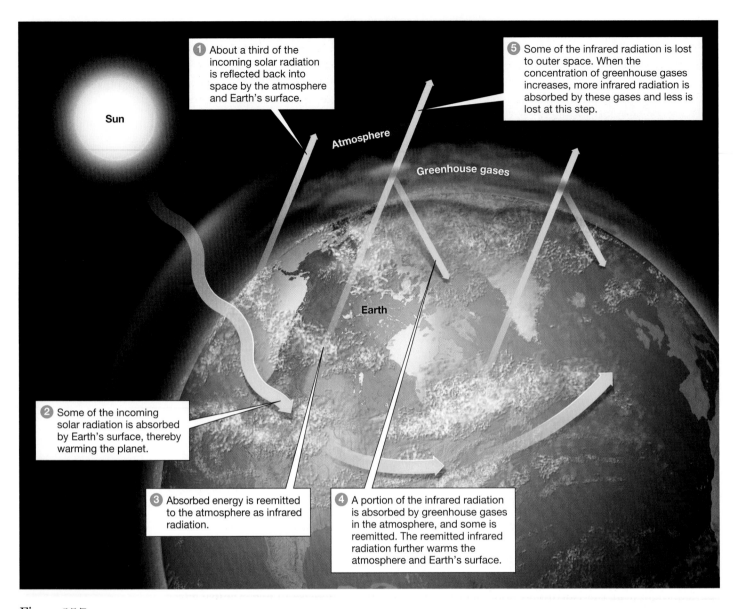

Figure 15.7

How greenhouse gases warm the surface of Earth
Carbon dioxide (CO_2), water vapor (H_2O), methane (CH_4), and nitrous oxide (N_2O) are known as greenhouse gases because they absorb and trap heat that would otherwise radiate away from Earth.

Q1: How much of the incoming solar energy is reflected back to outer space?

Q2: What kind of energy is reemitted to the atmosphere after being absorbed by Earth's surface?

Q3: How are greenhouse gases like a blanket on your bed at night?

the walls of a greenhouse or the windows of a car do: they let in sunlight and trap heat in a process known as the **greenhouse effect** (**Figure 15.7**).

Greenhouse gases are not inherently bad; in fact, they have existed in Earth's atmosphere for more than four billion years, and they play an important part in maintaining temperatures that

are warm enough for life to thrive on Earth. Yet human activities, primarily the burning of fossil fuels, have released an excess of greenhouse gases into the atmosphere, especially the infamous king of greenhouse gases: **carbon dioxide (CO_2).**

Scientists have estimated atmospheric CO_2 levels for both the recent and the relatively distant past, up to hundreds of thousands of years ago, by measuring CO_2 concentrations in air bubbles trapped in ice. This evidence shows a near-perfect historical correlation between CO_2 levels and the surface temperature on Earth. Yet during the last 200 years, levels of atmospheric CO_2 have risen greatly—from roughly 280 to 380 parts per million (**Figure 15.8**). Measurements from ice bubbles show that this rate of increase is greater than even the most sudden increase that occurred naturally during the past 420,000 years. Carbon dioxide levels are now higher than those estimated for any time during that period. In fact, in May 2013, sensors atop a research facility at the Hawaiian volcano Mauna Loa recorded an average daily CO_2 concentration above 400 parts per million—a level that hasn't been reached since well before humans roamed Earth.

The cause of the increasing levels of atmospheric CO_2 has been linked to the burning of fossil fuels like coal and oil, which releases CO_2 into the air. About 75 percent of the current yearly increase in atmospheric CO_2 is due to the burning of fossil fuels. Logging and burning of forests are responsible for most of the remaining 25 percent of the increase, but industrial processes also make a significant contribution. "Humans have a giant impact on the environment," says Coe. "With more than seven billion people on the planet, individual decisions add up to big global changes." (For more on the environmental impact of personal choices, see "How Big Is Your Ecological Footprint?" on the following page.)

All this is to say that the human-caused burning of fossil fuels has led to an increase in carbon dioxide in our atmosphere. That carbon dioxide acts as a greenhouse gas and traps additional heat trying to escape from the planet, causing temperatures on Earth to rise. Since the early twentieth century, Earth's mean surface temperature has increased by about 1.4°F (0.8°C), and it is estimated to rise another 2°F–11.5°F (1.1°C–6.4°C) in the future (**Figure 15.9**).

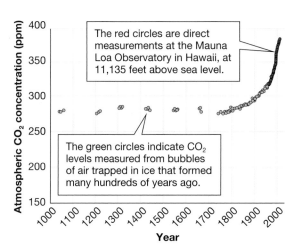

Figure 15.8

Atmospheric CO₂ levels are rising rapidly

Atmospheric CO₂ levels (measured in parts per million, or ppm) have increased greatly in the past 200 years.

Q1: What measurements do the green circles represent?

Q2: What measurements do the red circles represent?

Q3: For approximately how many years has the Mauna Loa Observatory been recording CO₂ levels?

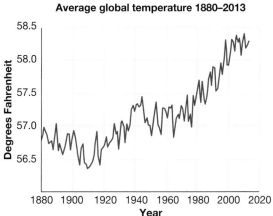

Figure 15.9

Global temperatures are on the rise

Average global temperature has increased greatly over the 140 years since it has been recorded.

Q1: In what years were global temperatures the coolest?

Q2: In what years were global temperatures the warmest?

Q3: What trend is apparent in this graph of actual global temperatures?

How Big Is Your Ecological Footprint?

An action or process is **sustainable** if it can be continued indefinitely without causing serious damage to the environment. The current human impact on the biosphere is *not* sustainable.

Each of us can help build a more sustainable society. We can advocate legislation that fosters less destructive and more efficient use of natural resources, patronize businesses that take measures to lessen their negative impact on the planet, support sustainable agriculture, and modify our own lifestyles. For example, we can increase our use of renewable energy and energy-efficient appliances; reduce all unnecessary use of fossil fuels (for instance, by biking to work or using public transportation); buy seafood from sustainable fisheries; use "green" building materials; and reduce, reuse, and recycle waste. Experts estimate that more than 200 million women around the world wish to limit their family size but have no access to family planning. Those of us who live in developed countries can support aid efforts that provide education, health care, and family-planning services in developing countries.

One measure of sustainability is an **ecological footprint**, which is the area of biologically productive land and water that an individual or a population requires to produce the resources it consumes and to absorb the waste it produces. Scientists compute an ecological footprint using standardized mathematical procedures and express it in *global hectares* (gha). One gha is equivalent to one hectare (2.47 acres) of *biologically productive* space. Approximately one-fourth of Earth's surface is considered biologically productive; this definition excludes areas such as glaciers, deserts, and the open ocean.

According to recent estimates, the ecological footprint of the average person in the world is 2.7 gha, which is about 30 percent higher than the 2.1 gha that would be needed to support each of the world's 7 billion people in a sustainable manner. An ecological footprint can also be expressed in **Earth equivalents**, the number of planet Earths needed to provide the resources we use and absorb the wastes we produce. Currently, the global population uses 1.5 Earth equivalents each year (as shown at the bottom of the figure on the facing page).

Overall, such estimates suggest that, since the late 1970s, people have been using resources faster than they can be replenished—a pattern of resource use that, by definition, is not sustainable. As the world population grows, the amount of biologically productive land available per person continues to decline, increasing the speed at which Earth's resources are consumed.

The per capita consumption of Earth's resources by different countries is most directly related to energy demand, affluence, and a technology-driven lifestyle. As people in populous countries such as China and India become wealthier, their ecological footprints are growing rapidly.

What is *your* ecological footprint? If you are a typical college student, your footprint is probably close to the U.S. average of 8.0 gha. It would take about five planet Earths to support the human population if everyone on Earth enjoyed the same lifestyle that you do (see the top row of the accompanying figure). Your ecological footprint depends on four main types of resource use:

1. *Carbon footprint*, or energy use;

2. *Food footprint*, or the land and energy and water it takes to grow what you eat and drink;

3. *Built-up land footprint*, which includes the building infrastructure (from schools to malls) that supports your lifestyle;

4. *Goods-and-services footprint*, which includes your use of everything from home appliances to paper products.

If you drive a gas guzzler, live in a large suburban house, routinely eat higher up on the food chain (more beef than grains or veggies/fruits), and do not recycle much, your footprint is likely to be higher than that of a person who uses public transport, shares an apartment, eats mostly plant-based foods, and sends relatively little to the local landfill. Most of us can significantly reduce our ecological footprint with little reduction in our quality of life, while bestowing an outsized benefit on our planet.

Our globe is warming. And the effects of that warming on the biosphere are now evident, especially in the Amazon, where trees dominate the landscape. Trees act as a layer connecting the atmosphere above them and the ground beneath them, absorbing water from the ground and releasing oxygen into the air. Any change in tree cover can change the local climate, and increasing wildfires are devastating tree populations.

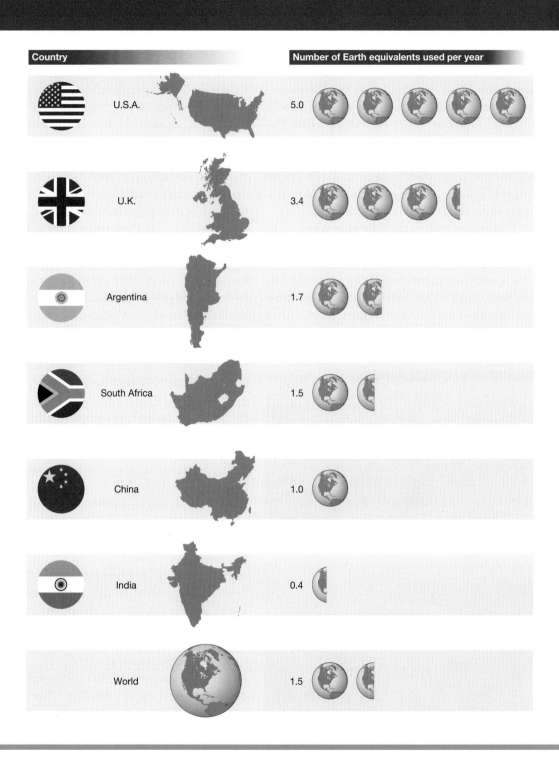

Country			Number of Earth equivalents used per year	
U.S.A.			5.0	
U.K.			3.4	
Argentina			1.7	
South Africa			1.5	
China			1.0	
India			0.4	
World			1.5	

As ecologists, Balch and Coe are interested in how the tropical rainforest is responding to climate change and the associated increase in fire. "The big question is, what's the limit for these tropical forests? How much fire can they withstand before they convert to something else?" asks Balch. That something else would be grasslands, also called savanna. If major parts of the Amazon rainforest were to convert to savanna, the climate would be significantly affected, and there would be no easy way—perhaps no way at all—to revert that land back to rainforest.

Fire and Water

During her first burn in 2004, Balch was surprised to see that the plants in the two experimental plots were fairly resistant to fire. The Amazon's tall, dense tree canopy creates humid air below it, protecting the forest from most fires, which tend to burn slowly under such conditions. For this reason, Balch's experimental plots initially suffered little damage.

That wasn't true in the following years of the experiment. Remember that the Amazon saw a major drought in 2005. Droughts cause tree death, allowing more sunlight to reach the forest floor and dry the leaves and shrubs there, priming the forest for fire (**Figure 15.10**). After the drought, the burn sites experienced major

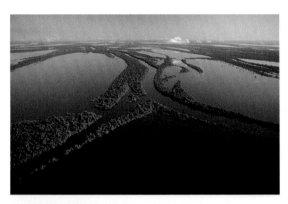

Figure 15.10

Before and after the Amazon drought of 2005
The drought devastated the forest to this extent over much of the forest's range.

damage from the burns. The majority of the trees on the plots died, and when they did, grasses took their place—particularly at the forest edge where it was hottest and driest.

In both the one-year and the three-year burned plots, grasses invaded from adjacent pasturelands. This experimental result suggests that wildfires do indeed push the Amazon toward a drier, grassier savanna-like ecosystem (**Figure 15.11**). "This could be the trigger for changing forest to savanna," says Balch. "The message is that repeated disturbances, such as multiple fires or drought and fire, cause forest to reach its tipping point, to shift into grassland."

While Balch and her team observed changes in plants after fire, ecologist Michael Coe investigated how fire affects its nemesis: water. "The forest is an incredibly important part of the **hydrologic cycle**," the circulation of water from the land to the sky and back again, says Coe (**Figure 15.12**).

Near the equator, direct sunlight causes water to evaporate from Earth's surface. The warm, moist air rises because heat causes it to expand, making it less dense and lighter than air that has not been heated. Then, the warm, moist air cools as it rises. Because cool air cannot hold as much water as warm air can, much of the moisture from a cooling air mass is "wrung out" and falls as rain.

For this reason, most tropical regions, including the Amazon, receive ample rainfall. Earth has four giant **convection cells** in which warm, moist air rises and cools, releasing moisture as rain or snow depending on temperature, and then sinks back to the ground as dry air (**Figure 15.13**). These convection cells, in combination with the angle of sunlight striking the Earth, play a large role in the creation of the major biomes on Earth, such as rainforests and deserts.

In the Amazon, trees are very important players in the water cycle, adds Coe. "The trees pull water out of the soil and evaporate it into the atmosphere in the process of photosynthesizing (through transpiration), so they're the mediators between the rainfall and the streams," he says. "Burning the trees greatly reduces the amount of water getting back into the atmosphere."

Every year of Balch's study, Coe joined Balch at the test site. There, his team had dug 10-meter-deep soil pits: long, dark caverns in which they inserted instruments to measure the moisture

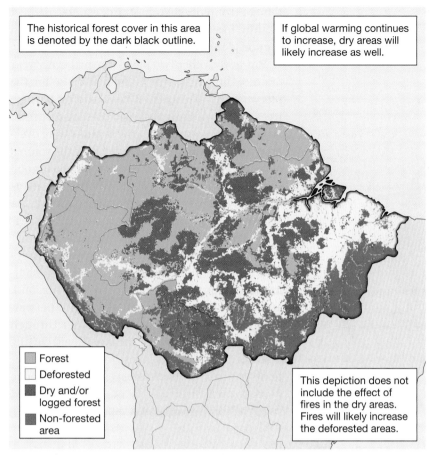

The historical forest cover in this area is denoted by the dark black outline.

If global warming continues to increase, dry areas will likely increase as well.

Forest
Deforested
Dry and/or logged forest
Non-forested area

This depiction does not include the effect of fires in the dry areas. Fires will likely increase the deforested areas.

Figure 15.11

A projection of the Amazon rainforest in 2030

This map of future Amazon rainforest coverage is based on the assumption that global warming will persist but not increase.

Q1: Where will fire most seriously affect the Amazon rainforest?

Q2: Where will fire be the least damaging to the Amazon rainforest?

Q3: This map does not include an increase in pasturelands for grazing animals. Do you think more or less pastureland will be needed in 2030? Explain.

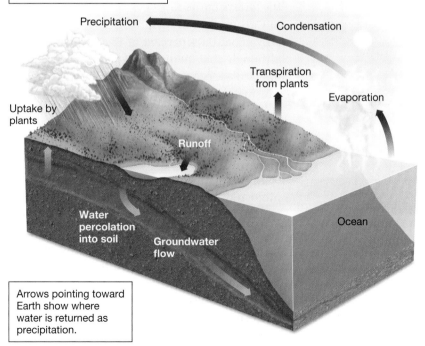

Arrows pointing toward the atmosphere show where evaporation occurs from water sources or is released from plants through transpiration to form humidity or clouds.

Transpiration is the process of plants absorbing water through their roots and releasing this water through their leaves into the atmosphere.

Precipitation
Condensation
Transpiration from plants
Evaporation
Uptake by plants
Runoff
Water percolation into soil
Groundwater flow
Ocean

Arrows pointing toward Earth show where water is returned as precipitation.

Figure 15.12

The hydrologic (water) cycle

Arrows point in the direction of water flow.

Q1: What is transpiration?

Q2: Why is transpiration important to the water cycle?

Q3: If there are fewer plants and therefore less transpiration in a given area, what will happen to the humidity or cloud cover in this area?

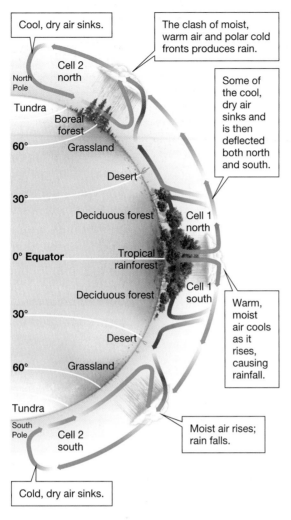

Cool, dry air sinks.

The clash of moist, warm air and polar cold fronts produces rain.

Some of the cool, dry air sinks and is then deflected both north and south.

North Pole

Cell 2 north

Tundra

Boreal forest

60°   Grassland

Desert

30°

Deciduous forest

Cell 1 north

0° Equator

Tropical rainforest

Deciduous forest

Cell 1 south

30°

Warm, moist air cools as it rises, causing rainfall.

Desert

60°   Grassland

Tundra

South Pole

Cell 2 south

Moist air rises; rain falls.

Cold, dry air sinks.

Figure 15.13

Earth has four giant convection cells

Two giant convection cells are located in the Northern Hemisphere and two in the Southern Hemisphere.

Q1: What pattern emerges when you compare rainfall in the Northern and Southern Hemispheres?

Q2: What pattern emerges when you compare the major biomes of the Northern and Southern Hemispheres?

Q3: What happens at the equator to make this region so wet?

content of the soil. In a healthy forest ecosystem, trees absorb a lot of the water in the soil, leaving it nice and dry, with only minimal water runoff into streams. This is what Coe observed in the control plot where nothing was burned. "A healthy forest uses up almost all the water [in the soil]," says Coe.

But in the other two plots—burned every year or every three years—he found that the soil in the pits was wet to the touch. When the forest burned, trees died, so nothing absorbed the moisture from the soil. Consequently, nearby streams were overflowing with water—up to four times the volume of water seen in healthy forests. "That's not a good thing," says Coe. "We're circumventing the natural cycle. Instead of this water going back into the atmosphere, creating more rain and driving vegetation, it's flushing the water out of the system."

And on the burned plots, the invasive grasses that took the place of the trees have very shallow roots, absorb less moisture from the ground, and evaporate less water into the air. In this way, deforestation of large areas—whether through unintentional wildfires or the intentional cutting down of trees—results in less rainfall and hotter temperatures. "If you deforest enough of it, you're going to really decrease the rainfall over a broad swath of this region," says Coe.

The Carbon Games

In addition to measuring the hydrologic cycle and the expansion of grasses into the region, the teams led by Balch and Coe measured how much carbon was released into the atmosphere when the forests burned.

Carbon, in the form of CO_2 gas, makes up only about 0.04 percent of Earth's atmosphere, although that percentage has been creeping upward every year for the last 200 years and causing global warming, as we've seen. Carbon is also found in Earth's crust, where carbon-rich sediments and rocks formed from the remains of ancient marine and terrestrial organisms. Carbon is present in every living thing.

Living cells are built mostly from organic molecules—molecules that contain carbon atoms bonded to hydrogen atoms. After oxygen, carbon is the most abundant element in cells by weight; every one of the main macromolecules in an organism has a backbone of carbon atoms. Living organisms, in both aquatic and terrestrial ecosystems, acquire carbon mostly through photosynthesis. Aquatic producers, such as photosynthetic bacteria and

Forest Devastation

Agriculture is the main driver of tropical deforestation, and Brazil clears more land, by far, than any other country in the world. Since 1990, Brazil has cleared over 42 million hectares of forest, an area the size of California. Unless drastic measures are taken, deforestation will continue as farmers and companies clear land to meet rising global food demands.

Causes of deforestation

- 1%
- 3%
- 2%
- 27%
- 67%

- Agriculture (commercial)
- Agriculture (local/subsistence)
- Infrastructure (e.g., roads)
- Mining
- Urban expansion

Drivers of forest degradation

- 3%
- 8%
- 16%
- 73%

- Timber logging
- Uncontrolled fires
- Wood for fuel and charcoal
- Livestock grazing in forest

Top 5 countries that cleared the most forest between 2000 and 2005

Deforested area in thousands of hectares

Brazil	Indonesia	Sudan	Myanmar	Zambia
15,515	9,357	2,945	2,332	2,224

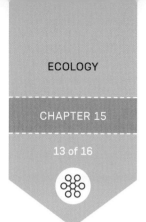

algae, absorb dissolved CO_2 and convert it into organic molecules using sunlight as a source of energy. Plants, the most important producers in terrestrial ecosystems, absorb CO_2 from the atmosphere and transform it into food with the help of sunlight and water.

The transfer of carbon within biotic communities, between living organisms and their physical surroundings, and within the abiotic world is known as the global **carbon cycle** (**Figure 15.14**). One way that carbon is transferred between the biotic and abiotic worlds is through combustion—the burning of carbon-rich materials, living or not.

Some of the organic matter from ancient organisms has been transformed by geologic processes into deposits of fossil fuels such as petroleum, coal, and natural gas. When we extract these fossil fuels and burn them to meet our energy needs, the carbon that was locked in these deposits for hundreds of millions of years is released into the atmosphere as carbon dioxide.

Plants also release carbon back into the atmosphere when they are burned. To assess the

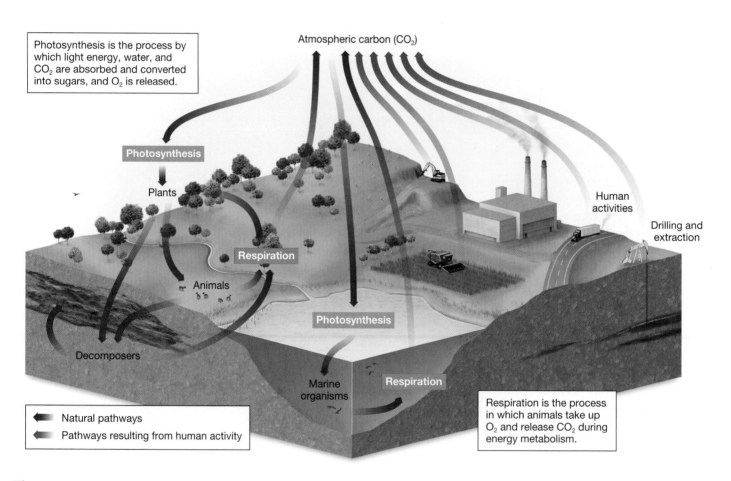

Figure 15.14

The carbon cycle
Arrows point in the direction of carbon flow.

Q1: What are three ways that carbon is released into the atmosphere?

Q2: Are all of the pathways you listed for question 1 affected by human activity?

Q3: What are two biotic reservoirs of carbon?

amount of carbon released by a burning rainforest, Balch and her team measured the amount of leaf litter and branches—the biomass—on the forest floor both before and after a burn, as well as the number of trees affected. Half of that biomass is carbon, so half of the difference between the biomass before and after the burn is the amount of carbon that was released into the atmosphere. And it was a lot.

The very first burn, for example, released "about 20 tons of carbon per hectare," says Balch. That's approximately equal to the carbon emitted by a passenger car driven 864,000 miles (burning about 40,000 gallons of gasoline). And burns have another, indirect, detrimental effect: They destroy trees that normally absorb carbon dioxide from the air. In a normal year, trees in the Amazon absorb approximately 1.5 billion tons of carbon dioxide, helping to slow climate change. In this way, they act as a **carbon sink**, a natural or artificial reservoir that absorbs more carbon than it releases.

But during the 2005 drought, as trees died and rotted, the rainforest stopped absorbing more carbon dioxide than it gave off and actually released carbon *into* the atmosphere. In 2005, five billion tons of carbon were released. Instead of acting as a carbon sink, the Amazon became a **carbon source**, a reservoir that releases more carbon than it absorbs (**Figure 15.15**). And with increasing temperatures, decreasing rainfall, and increasing wildfires, ecologists fear that the Amazon could transition more and more into a carbon source instead of a carbon sink. If that were to happen, our planet would lose one of its greatest buffers against future climate change.

Waiting and Watching

Balch and her crew have now stopped burning the plots and have begun to record what happens to the plots as they recover. In addition to making direct observations of which plants grow back and how their growth affects the water and carbon cycles, the team is also conducting new experiments on the burned plots to see whether it is possible to help prevent the invasion of grasses and encourage a return to rainforest.

In one area, the scientists are planning to plant different types of trees to see whether particular species can successfully reestablish themselves in a mat of grasses and retake that land. Another experiment, also still in the planning phase, will test different amounts of watering to see whether the lack of precipitation is preventing the growth of new trees. If any strategies work well on the small plots, perhaps they can be adapted to save larger areas of rainforest affected by fire.

"We're hoping we find ways to help the forest recover," says Balch. Still, the researchers worry about the future of the region. "Have we crossed a threshold where this part of the forest isn't going to grow back as rainforest?" asks Coe, worry in his voice. If that is true for the small experimental plots, it bodes dangerously for large regions of the Amazon that border agricultural land and are threatened by wildfires.

A carbon sink absorbs more CO_2 than it gives off.

A carbon source gives off more CO_2 than it absorbs.

Figure 15.15

Carbon sinks and sources

Q1: How does a carbon source contribute to global warming?

Q2: How does a carbon sink protect against global warming?

Q3: How can trees act as both a source and a sink?

REVIEWING THE SCIENCE

- **Ecology** is the study of interactions between organisms and their environment.

- All ecological interactions occur in the **biosphere**, which consists of all living organisms on Earth, together with the environments they inhabit.

- **Climate**, the prevailing weather of a specific place over relatively long periods of time, has a major effect on the biosphere. Climate is determined by incoming solar radiation, global movements of air and water, and major features of Earth's surface.

- **Weather** consists of the short-term atmospheric conditions in a given area, such as temperature, precipitation, wind, humidity, and cloud cover.

- The concentration of **carbon dioxide (CO_2)** gas in the atmosphere is increasing at a dramatic rate because of an increase in **carbon sources**, such as the release of CO_2 through the burning of fossil fuels, and a loss of **carbon sinks**, such as the absorption of CO_2 through photosynthesis in large forests.

- Carbon dioxide (CO_2) acts as a **greenhouse gas** and contributes to the **greenhouse effect** by reflecting more heat than usual back to the Earth's surface.

- The greenhouse effect has led to the rise in average global temperature, a phenomenon known as **global warming**.

- **Climate change** is a large-scale and long-term alteration in Earth's climate, and it includes such phenomena as global warming, change in rainfall patterns, and increased frequency of violent storms.

- Both the **carbon cycle** and the **hydrologic (water) cycle** are affected by global warming, disrupting the natural cycling of these molecules in the biosphere.

- The area of the biosphere required to produce the resources and to absorb the waste produced by an individual or population is known as the **ecological footprint**.

- The ecological footprint is often expressed in **Earth equivalents**, the number of planet Earths needed to provide the resources and absorb the waste of an individual or population.

THE QUESTIONS

The Basics

1 The biosphere is

(a) all organisms on Earth, together with their physical environments.

(b) crucial to human survival and well-being.

(c) a source of food and raw materials for human society.

(d) a web of interconnected ecosystems.

(e) all of the above

2 The term "climate" refers to

(a) the average temperature at a given time in a specific area.

(b) the average temperature of an area over time.

(c) the average rainfall in an area at a specific time.

(d) the average rainfall in an area over time.

(e) the average temperature and rainfall of an area over time.

3 Most scientists think that three of the following four statements related to global warming are correct. Which one is *not* correct?

(a) The concentration of greenhouse gases in the atmosphere is not increasing.

(b) Dozens of species have shifted their geographic ranges to the north.

(c) Plant growing seasons are longer now than they were before 1980.

(d) Human actions, such as the burning of fossil fuels, contribute to global warming.

4 Greenhouse gases function by

(a) blocking sunlight but letting out heat from Earth to outer space.

(b) trapping heat that radiates from Earth that would otherwise escape to outer space.

(c) trapping heat that radiates from the sun toward Earth.

(d) releasing heat that radiates from Earth to outer space.

5 A carbon sink consists of

(a) all organisms on Earth that use CO_2 and other factors that remove it from the atmosphere.

(b) only the organisms on Earth that use CO_2.

(c) all organisms on Earth and the environments in which they live.

(d) all organisms on Earth that produce CO_2 and other factors that release it to the atmosphere.

Try Something New

6 You have invested a considerable sum of money in space travel and have come across a planet that has four stable regions ("air cells") in which warm, moist air rises and cool, dry air sinks back to the surface of the planet, creating regional ecosystems on that planet. These are extremely similar to the four air cells found on Earth that are known as

(a) temperate cells.

(b) latitudinal cells.

(c) rain shadow cells.

(d) convection cells.

7 While wandering through the Amazon rainforest, you come across a patch of trees that recently burned. The burning of trees acts as a

(a) carbon sink.

(b) carbon source.

(c) water source.

(d) kitchen sink.

(e) none of these

8 A biology professor is bragging about how small an area of biologically productive land and water she requires to produce the resources she consumes and to absorb the waste she produces. What exactly has she calculated?

(a) ecosystem

(b) sustainability

(c) ecological footprint

(d) habitat

(e) global warming

9 An ecology professor is studying an area of forested land that experiences a forest fire every five years. In the years between forest fires, the area of study absorbs 300 million tons of CO_2. If a forest fire releases 200 million tons of CO_2 in this area, does this forested area act as a carbon sink or a carbon source? Why?

(a) carbon sink because more carbon is absorbed in years between the fires than a fire produces

(b) carbon source because more carbon is absorbed in years between the fires than a fire produces

(c) carbon sink because less carbon is absorbed in years between the fires than a fire produces

(d) carbon source because less carbon is absorbed in years between the fires than a fire produces

10 Several of your friends claim to be ecologists in their spare time by volunteering for environmentally friendly charities. Which of the below activities qualifies for acting as an ecologist under the definition provided in this chapter?

(a) picking up trash along the beach and recycling the materials

(b) planting trees in a recently burned forested area

(c) a community beautification project involving planting wildflowers and cleaning up trash

(d) a project that includes studying an area to determine the best species of native plants to incorporate into the landscape based on the climate and other plants and animals living in the area

Leveling Up

11 **What do *you* think?** The future magnitude and effects of global warming remain uncertain. Do you think we should take action now to address global warming, despite those uncertainties? Or do you think we should wait until we are more certain what the ultimate effects of global warming will be? Support your answer with facts already known about global warming.

12 **Life choices** You can estimate your impact on the planet by using the many "footprint calculators" on the Internet. Take several of the online quizzes available from organizations such as the Global Footprint Network, the Center for Sustainable Economy, and The Nature Conservancy. Each site calculates your ecological footprint a little differently. Compare and contrast the different sites you used. Which one do you think is the most accurate? Why? Which one do you think is the most superficial? Why? Write down a list of ways you can decrease your ecological footprint, and try to adhere to your new lifestyle!

13 **Write Now biology: global warming** Watch the movie *An Inconvenient Truth* (2006), directed by Davis Guggenheim and featuring Al Gore. Write an essay reflecting on what you learned from the film and this chapter about global warming. Use your knowledge of your ecological footprint to reflect on how important it is for all citizens to be aware of this global crisis.

China's One-Child Policy Grows Up

Researchers challenge the need for China's unprecedented program of extreme birth control.

After reading this chapter you should be able to:

- Articulate the difference between population size and population density.
- Calculate a population's doubling time and determine whether a population exhibits logistic or exponential growth.
- Explain the concept of carrying capacity.
- Identify the differences between density-dependent and density-independent changes in population size.
- Discuss China's one-child policy using the facts presented by this chapter.

CHAPTER

16

GROWTH OF
POPULATIONS

In 1980, Joan Kaufman arrived in China. Kaufman, a young American scholar of Chinese language and culture, had been hired by the United Nations Population Fund as part of a small group starting work for the UN in the country. She was there to help conduct a nationwide census, increase contraceptive production, and train aspiring demographers. Demographers are researchers who study the characteristics of human populations, such as size, distribution, and density.

But Kaufman's work would not be part of an ordinary population study, because China was no ordinary country. During the twentieth century, China's population had erupted. The **population size**—the total number of individuals in the population—rose rapidly from 540 million in 1949 to 940 million in 1976. (Comparatively, during this same time period, the United States' population grew from 149 million to 218 million.)

Today, China remains the most populated country in the world. Its **population density** (the number of individuals per unit of area) is 137 people per square kilometer, an area about the size of three city blocks. To calculate population density, the total population size is divided by the corresponding area of interest. In this case, 137 people per square kilometer is the population density for China as a whole, including sparsely populated rural areas. One could instead ask for the population density of just China's eastern coastal region, which includes cities like Beijing and Shanghai. There, the population density is 320 people per square kilometer (**Figure 16.1**). Keep in mind that population density is often difficult to measure, because it depends on an accurate count of the population size. Individuals may be hard to detect, may move between populations, and may inhabit a complex, hard-to-define area.

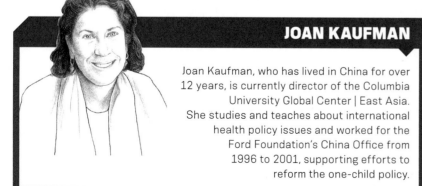

JOAN KAUFMAN

Joan Kaufman, who has lived in China for over 12 years, is currently director of the Columbia University Global Center | East Asia. She studies and teaches about international health policy issues and worked for the Ford Foundation's China Office from 1996 to 2001, supporting efforts to reform the one-child policy.

After its midcentury population explosion, China experienced a different type of government-influenced population change. In 1978, Deng Xiaoping took leadership of the Chinese Communist Party. Deng believed that decreasing the birth rate would improve living standards for citizens and make the country stable and strong.

In general, a **population** is a group of organisms of the same species in a defined area. Populations of living things tend to change in size over time—sometimes increasing, sometimes decreasing. Whether the size of a population increases or decreases depends on the number of births and deaths in the population, as well as on the number of individuals that enter or leave the population. Birth and immigration increase population size; death and emigration reduce it. Environmental factors also influence these characteristics and have a strong impact on population size. In China, for example, public health improvements reduced infant mortality and lengthened life expectancy, leading to an increase in the size of China's population.

Deng decided to try to control population growth by focusing on birth, which increases population size. If births were restricted, he reasoned, the rate at which the population was increasing could be slowed.

In the early 1970s, the Chinese government had already begun encouraging its citizens to take a "later, longer, fewer" approach to having children: later marriage, longer intervals between births, and fewer children. But in 1979 and 1980, when Deng was in power, the government took a more extreme stance on birth control. A new program, called the **one-child policy**, was put in place. It was trumpeted in national proclamations and in the media. The one-child policy restricted most couples to having only one child—and levied expensive fines, up to years' worth of salary, on families that had more than one child (**Figure 16.2**). The goal of the policy, said Deng, was to reach a total population of 1.2 billion and a population growth rate of zero by the year 2000.

Joan Kaufman arrived in China just as the one-child policy was being implemented. At the time, she found the Chinese people to be cautious, nervous. Chinese citizens were already treated as subjects of the state and were restricted in their interactions with foreigners.

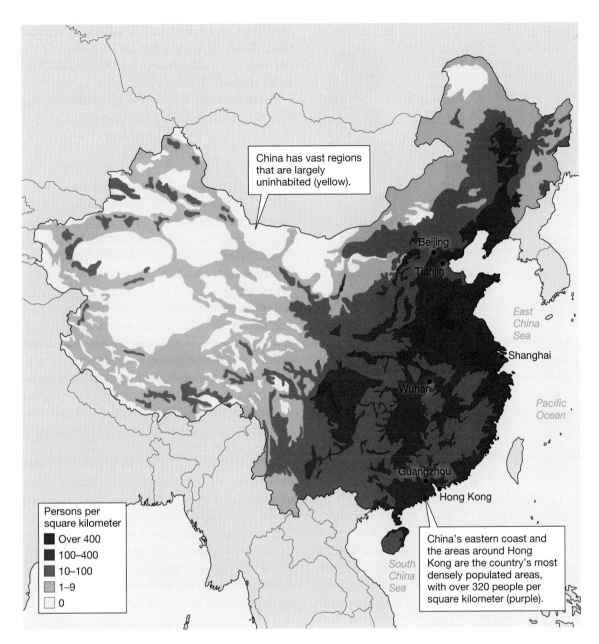

Figure 16.1

Population density of China

Q1: This map does not specify the population densities of individual cities. How do the rural areas around Shanghai compare to the rural areas around your city?

Q2: Look up the population density of your state. How does it compare to the most and least populated areas of China?

Q3: Look up the population density of the United States. How do its most and least populated areas compare to China's?

Now the one-child policy added an unprecedented level of government control over their fertility. The one-child policy can be regarded as the largest human experiment in **population ecology**, the study of the number of organisms in a particular place. During her first four years

Figure 16.2

Only one child is allowed per woman in China
China's one-child policy restricts most couples to one child only. Couples who have a second child face fines and penalties.

in China, Kaufman watched as the policy was instituted. It wasn't popular, she recalls; nor was it well-implemented in rural areas, where many people still had two or more children. Citizens believed, however, that the policy would be short-lived. "It was sold mainly as an emergency policy for an economic purpose," says Yong Cai, a Chinese sociologist at the University of North Carolina, Chapel Hill.

That assumption turned out to be false. Today, more than 30 years later, the one-child policy is still in place in China, and it has had major unintended consequences, including a potentially dangerously skewed sex ratio with far more men than women. This is because, in Chinese culture, having a son provides honor and stability to a family. Traditionally, sons provide for their parents as they age, whereas daughters provide for their husbands' parents. Sex-selective abortions are not uncommon and the average infant mortality rate among females is higher than for males in China. Other unintended consequences of the one-child policy include a growing elderly population with fewer children to support them, and human-rights concerns such as forced

abortions and forced sterilizations. In light of these facts, demographers like Kaufman have become increasingly vocal that the one-child policy should be overturned.

All this prompts questions: How effective was the one-child policy at controlling China's population, and why is it still in place?

Gauging Growth

In the second half of the twentieth century, our planet saw the fastest rate of population growth in its history. Before the one-child policy was implemented, nowhere was this growth more prominent than in China. Demographers use **population doubling time**, the time it takes a population to double in size, as a measure of how fast a population is growing. It is estimated that China's population almost doubled from 1949 to 1976, so its population doubling time was about 30 years during this period. In contrast, the population doubling time for the United States during this time was about 54 years, almost twice as long as China's.

By the 1970s, as the Chinese population crept toward one billion, members of China's government became concerned that China had exceeded its **carrying capacity**, the maximum population size that can be sustained in a given environment. In most species, the growth rate of a population decreases as the population size nears the carrying capacity because resources such as food and water begin to run out. Any limiting resource needed for survival, such as habitat, food, or water, will determine the carrying capacity of a specific population. Because different species have unique needs, the same environment can have different carrying capacities for different resident species. For the same reason, two different environments in which the same species lives will have different carrying capacities. At the carrying capacity, the population growth rate is zero (**Figure 16.3**). Carrying capacity doesn't always appear to restrict human populations, however. In some places, such as China, human populations continue to grow to the detriment of the environment or by using resources from other environments, such as food imported from other countries.

A population, constrained by resources, that approaches its carrying capacity will experience

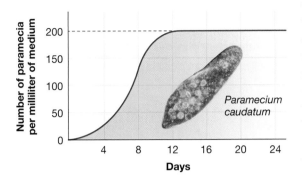

The maximum population size that can be supported indefinitely by the environment is 200 cells per milliliter of growth medium.

Paramecium caudatum

Figure 16.3

Rapidly reproducing single-celled organisms are useful models of population growth

A population of the single-celled protist *Paramecium caudatum* grown in the laboratory increases rapidly at first and then stabilizes at its carrying capacity.

Q1: What features in an environment define its carrying capacity?

Q2: Why is the carrying capacity different for different environments?

Q3: Why is the carrying capacity different for individual species within the same environment?

logistic growth, in which the population grows nearly exponentially at first but then stabilizes at the maximum population size that can be supported indefinitely by the environment. Logistic growth is represented by an **S-shaped growth curve**. However, if a population has no constraint on its resources, it will experience **exponential growth**, which occurs when a population increases by a constant proportion over a constant time interval, such as one year. Exponential growth is represented by a **J-shaped growth curve** (**Figure 16.4**).

Over the last 500 years, Earth's human population has exhibited both logistic and exponential growth. At the end of the last ice age, in approximately 10,000 BCE, there were only 5 million people on Earth. With the advent of agriculture in about 8,000 BCE, the world population began to rise logistically until about 200 years ago. Then, alongside the use of fossil fuels and the industrial revolution, human population growth exploded exponentially. Modern populations have continued to show exponential growth, even in some cases beyond an environment's carrying capacity, to the detriment of the environment. Current estimates of the carrying capacity of Earth range from 2 billion to over 1,000 billion people, with the majority of studies insisting that 8 billion people is the maximum number that Earth can support. At current population growth rates, we will reach this number by the year 2030 (**Figure 16.5**).

The Chinese government believed it needed the restrictive one-child policy to taper exponential growth and achieve an artificial logistic growth curve. But in 2012, two demographic scholars, Yong Cai and Wang Feng, were brave enough to disagree. The one-child policy is "built on the assumption that Chinese cannot control their own fertility," says Cai. "That's just not rational. It's just not true."

Modeling Fertility

Yong Cai was born in China before the one-child policy was put in place. The budding researcher went to Peking University in Beijing, where he studied the history of Chinese society. But when Cai became frustrated by the absence of good population data from the past, he decided to study the present.

Cai came to the United States to get his PhD in sociology at the University of Washington. There he began working with a fellow Chinese scholar, Wang Feng. Both men were interested in China's population and the effect of the one-child policy,

YONG CAI AND WANG FENG

Yong Cai is a sociologist at the University of North Carolina at Chapel Hill. His work examines population growth in China and the effect of the one-child policy. Wang Feng is a scholar at the Brookings-Tsinghua Center for Public Policy, where he studies social inequity and population changes in China.

Figure 16.4

Curves of logistic and exponential growth in natural populations

Q1: What is the carrying capacity in the logistic-growth environment?

Q2: What factor(s) determined the carrying capacity of willow trees in this environment before 1954?

Q3: The exponential-growth graph does not include a carrying-capacity line. Why not?

Logistic growth

At a site in Australia, rabbits heavily grazed young willow trees, preventing willows from growing in the area. The rabbits were removed in 1954.

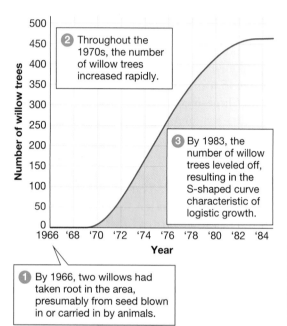

2 Throughout the 1970s, the number of willow trees increased rapidly.

3 By 1983, the number of willow trees leveled off, resulting in the S-shaped curve characteristic of logistic growth.

1 By 1966, two willows had taken root in the area, presumably from seed blown in or carried in by animals.

Exponential growth

In this population of rabbits, each individual produces two offspring, so the population increases by a constant rate—it doubles with each generation.

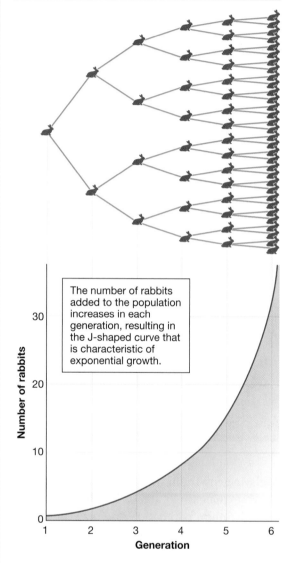

The number of rabbits added to the population increases in each generation, resulting in the J-shaped curve that is characteristic of exponential growth.

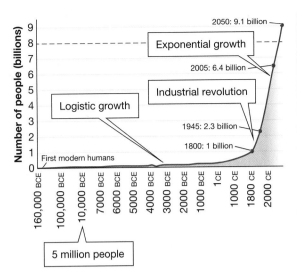

Figure 16.5

Curves of logistic and exponential growth in the world human population
The dashed line indicates the UN's estimated carrying capacity of the Earth.

Q1: According to this graph, approximately when did exponential growth begin?

Q2: What milestone corresponds to the transition from logistic to exponential population growth?

Q3: What is the UN's projected carrying capacity of the Earth, and when will we reach it?

Figure 16.6

Women of rural China
At the weekly bazaar in Liuku, located in the west of China near Tibet, women sell and purchase hens.

but little data existed to quantify the effect of the policy. In 2000, the Chinese government claimed that over 30 years, the policy had prevented 338 million births (more than the current population of the United States). In the decade after that number was published, it was raised to 400 million. But knowing that China's fertility rate—the average number of children born per woman—had already begun to drop with the "later, longer, fewer" effort, and that birth rates were also dropping in other countries around the world without restrictive fertility policies, Cai and Wang were skeptical that the government's claim of 400 million prevented births was true.

In December 2006, Cai and Wang visited China to gather quantitative data about whether the one-child policy was the cause of China's low fertility rate. They visited 50 villages across 6 counties in Jiangsu Province and spoke with more than 3,000 women (**Figure 16.6**). They found that only three percent of women with children had a second child.

Cai and Wang published their findings in 2009, concluding that "the extremely low fertility—total fertility rate of close to 1.0—that currently prevails in this area of China is explained largely by factors other than the government's birth-control policy . . . If China's one-child policy were to be phased out, it is unlikely to lead to an unwanted baby boom in this area of the country."

Three years after their initial visit, the duo went back to Jiangsu to see whether anything had changed. Once again, only about three percent of women with a child had a second child. Cai questioned the women: why did they not have a second child? Overwhelmingly, the women replied that the reason was economic: it's simply too expensive to have more than one child.

In the end, Cai and Wang came to believe that the one-child policy was not as necessary and effective as the Chinese government claimed. Instead, they believed that the social and economic change in China over the previous 30 years had been so dramatic that most people no longer sought to have large families. To support their hypothesis, Cai and Wang developed a set of models to predict a country's future fertility from its fertility history and fertility trends in other countries. They then applied those models to China.

Populations, They Are a-Changin'

China's fertility rate had already begun to drop before the one-child policy was put into place. On average, couples in 1968 were having 5.8 children, but by 1978 that number had declined sharply, to 2.7. Similar trends were happening in developed countries around the globe: with better health care and declining infant mortality rates, parents didn't need to have as many children to ensure that some would survive (**Figure 16.7**).

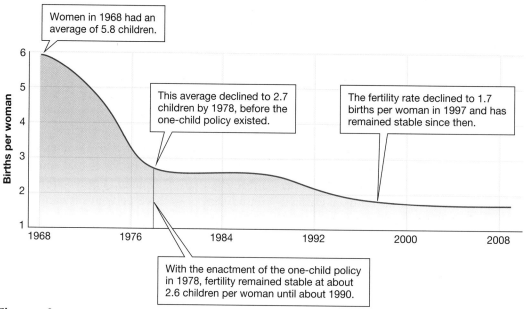

Women in 1968 had an average of 5.8 children.

This average declined to 2.7 children by 1978, before the one-child policy existed.

The fertility rate declined to 1.7 births per woman in 1997 and has remained stable since then.

With the enactment of the one-child policy in 1978, fertility remained stable at about 2.6 children per woman until about 1990.

Figure 16.7

China's fertility rate has declined dramatically

Q1: Within the first 12 years of the one-child policy, did the fertility rates of women in China decrease substantially?

Q2: What is the current fertility rate of women in China?

Q3: From the data plotted on this graph, why do you think it took a while for fertility in China to decline after the one-child policy was enacted, and what, ultimately, do you think caused the decline?

When Cai and Wang applied their model to China, the duo found that fertility would have continued to decline in the country without the one-child policy, due to factors including better access to health care, birth control, and education. "Having kids is not an easy business," says Cai. "When women have more control over their bodies and lives, when education improves and urbanization begins, the population goes down." Cai and Wang estimated that without the one-child policy, women would have actually had fewer children on average than with the policy (**Figure 16.8**). This sounds counterintuitive, but Cai suspects the policy caused anxiety among couples, prompting many to have children at an earlier time. Indeed, in the 1980s, there was a decline in age at first childbearing.

To Cai and Wang, it was clear there were already constraints on family growth in China, even prior to government intervention. Across all types of species, constraints on a population's growth include environmental factors such as food shortage, lack of space, disease, predators, habitat deterioration, weather, and natural disturbances. As the number of individuals in a population increases, fewer resources are available to each individual. As resources diminish near the population's carrying capacity, each individual, on average, produces fewer offspring than when resources are plentiful, causing the birth rate of the population to decrease. In the case of China, as the country's population climbed, resources such as food and living space became more expensive, constraining family growth. Because of that limited access to resources, the carrying capacity seemed to hover at about 1.7 births per woman. Many European countries have also reached that fertility rate, of about 1.7, over the last decade.

Populations, human or otherwise, can change in a density dependent or density-independent manner. **Density-dependent population change** occurs when birth and death rates change as the population density changes. The number of offspring produced and the death rate are often density dependent. Food shortages, lack of space, and habitat deterioration—all these factors influence a population more strongly as it increases in density (**Figure 16.9**). In China, as the population increases, the strain on agriculture and natural resources increases, leading to an increased cost of living, and families have reportedly begun having fewer children to save

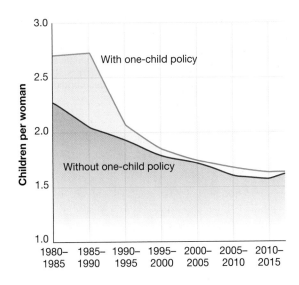

Figure 16.8

What would fertility rates have looked like in China without the one-child policy?
Wang and Cai's estimated fertility rates without the one-child policy (purple line) are actually lower than the United Nations' historical numbers (blue line). This may be because the policy caused anxiety, and couples in the 1980s had children earlier than expected. These estimates suggest that the policy was and is unnecessary.

Q1: How many fewer children would women have had on average without the one-child policy in effect, according to Cai and Wang?

Q2: What factors likely contributed to lower fertility rates of women in China in the last 10 years?

Q3: What do Cai and Wang project will happen if the one-child policy is lifted?

money. In that situation, the population change is density-dependent.

In addition, when a population has many individuals, disease spreads more rapidly (because individuals tend to encounter one another more often), and predators may pose a greater risk (because many predators prefer to hunt abundant sources of food). Disease and predators obviously increase the death rate. These changes are also density dependent.

If a population exceeds the carrying capacity of its environment by depleting its resources, it

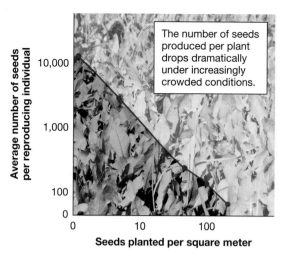

The number of seeds produced per plant drops dramatically under increasingly crowded conditions.

Average number of seeds per reproducing individual (y-axis): 10,000 / 1,000 / 100 / 0

Seeds planted per square meter (x-axis): 0 / 10 / 100

Figure 16.9

Overcrowded conditions result in density-dependent population change
The plantain, shown here, is a small, herbaceous plant.

Q1: What factors may be limiting growth and reproduction in these crowded conditions?

Q2: Why are overcrowded conditions considered density-dependent population changes?

Q3: Relate this example of overcrowded conditions to China's one-child policy. How do you think the situations are similar? How are they different?

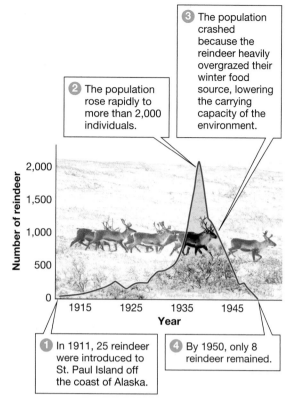

❸ The population crashed because the reindeer heavily overgrazed their winter food source, lowering the carrying capacity of the environment.

❷ The population rose rapidly to more than 2,000 individuals.

Number of reindeer (y-axis): 2,000 / 1,500 / 1,000 / 500 / 0

Year (x-axis): 1915 / 1925 / 1935 / 1945

❶ In 1911, 25 reindeer were introduced to St. Paul Island off the coast of Alaska.

❹ By 1950, only 8 reindeer remained.

Figure 16.10

Habitat destruction results in density-dependent population change

Q1: In what year did the reindeer's numbers begin to rise exponentially?

Q2: In what years was the reindeer's population growth logistic?

Q3: What do you think happened to this population of reindeer? What environmental conditions might support your hypothesis?

may damage that environment so badly that the carrying capacity is lowered for a long time. A drop in the carrying capacity means that the habitat cannot support as many individuals as it once could. Such habitat deterioration may cause the population to decrease rapidly (**Figure 16.10**).

Not all population changes are due to density. **Density-independent population change** occurs when populations are held in check by factors that are not related to the density of the population. Density-independent factors can prevent populations from reaching high densities in the first place. Year-to-year variation in weather, for example, may cause conditions to be suitable for rapid population growth. On the other hand, poor weather conditions may reduce the growth of a population directly (by freezing the eggs of an insect, for example) or indirectly (by decreasing the number of food plants available to that insect).

Other natural disturbances, such as fires and floods, can also limit the growth of populations in a density-independent way. Finally, the effects of environmental pollutants such as DDT are density-independent; such pollutants can threaten natural populations with extinction (**Figure 16.11**). For a comparable

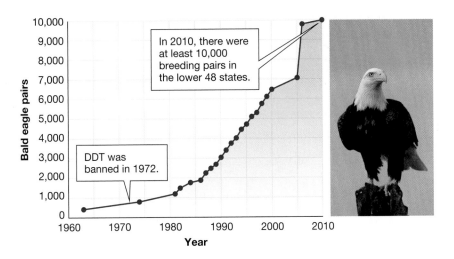

In 2010, there were at least 10,000 breeding pairs in the lower 48 states.

DDT was banned in 1972.

Figure 16.11

Banning the use of the pesticide DDT removed a density-independent population limit

DDT poisoning was directly responsible for declining eagle populations by the middle of the twentieth century. By the early 1960s, population counts indicated that only 417 breeding pairs of bald eagles remained in the lower 48 states—a huge drop from the estimated 100,000 breeding pairs present in 1800. Bald eagle populations increased dramatically after DDT was banned.

Q1: In what year did the bald eagle population rise to more than 2,000 breeding pairs?

Q2: Give some examples of possible density-dependent limits on bald eagle populations.

Q3: Is the population growth of bald eagles more like logistic or exponential growth? Explain why you think so.

threat to human populations, consider the radiation released by nuclear reactor melt-downs and nuclear weapons, which can lead to large, density-independent population changes. Human populations may also change due to severe drought that can destroy agricultural productivity and lead to malnutrition or famine.

Populations of many species rise and fall unpredictably over time. These **irregular fluctuations** in population size are far more common in nature than a smooth rise to a stable population size. In fact, different populations of the same species may experience different patterns of growth.

Populations can also exhibit **cyclical fluctuations**, which are predictable patterns that occur when at least one of two species is strongly influenced by the other. The Canadian lynx, for example, depends on the snowshoe hare for food, so lynx populations increase when hare populations

rise, and they decrease when hare populations drop. Similarly, Native Americans in the Great Plains depended on bison for food and raw materials for tools and shelter. A loss in bison populations resulted in great hardship and decreased population size of the Native Americans that depended on them. In these examples, the population cycles are also density-dependent population changes because each population is affected by the other's numbers. They cycle together due to each other's density (**Figure 16.12**).

One-Child Consequences

While agreeing with Cai and Wang's conclusion that Chinese fertility would have decreased even without China's one-child policy, Joan Kaufman,

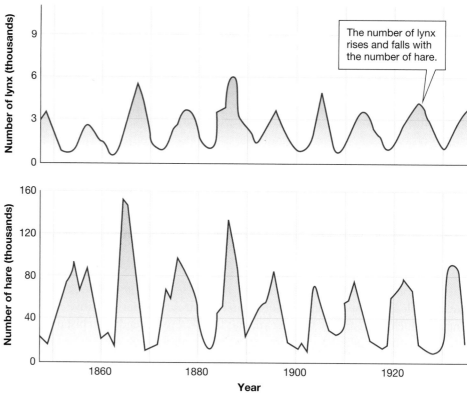

Figure 16.12

Populations of two species may increase and decrease together
The Canadian lynx depends on the snowshoe hare for food, so the number of lynx is strongly influenced by the number of hare.

Q1: During which years did the hare likely have the greatest food supply?

Q2: Besides the number of hare, what other factors might contribute to the number of lynx besides the number of hare?

Q3: Can you draw an average carrying-capacity line on these graphs? Why or why not?

The Cost of a Kid

Middle-class parents of a child born in 2013 can expect to spend a whopping $245,340 to care for that bundle of joy up to the age of 16. The cost estimate includes expenses such as food, childcare, clothing and healthcare. It does not include the cost of a college education. So, how many kids do you want to have?

Expenditures from birth through age 17

1960
Total = $198,560
(in 2013 dollars)

2013
Total = $245,340

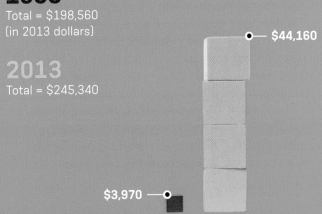

$44,160

$3,970

⬆ **Child care & early education**

$47,650

$39,250

⬇ **Food**

$23,830

$19,630

⬇ **Miscellaneous**

$73,600

$61,550

⬆ **Housing**

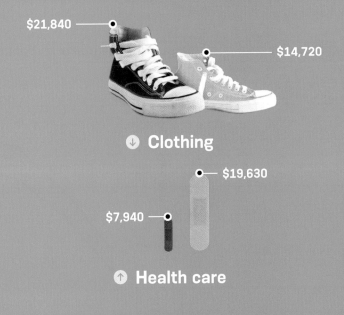

$21,840

$14,720

⬇ **Clothing**

$19,630

$7,940

⬆ **Health care**

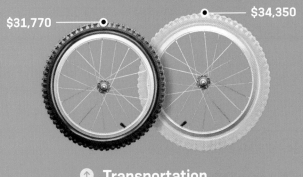

$34,350

$31,770

⬆ **Transportation**

who has studied China's one-child policy and its ramifications since 1980, does believe the policy also had some impact on population growth. In 2011, she compiled a case study about China for the UN Population Fund. In it, she used data from Cai and Wang's work to show that China's total fertility rate has been at about 1.7 children per woman since 2000, and that in 10 of China's 29 provinces and municipalities the fertility rate was 1.1 or lower, rivaling the lowest-fertility countries in the world.

But no matter how the policy has affected the population, Kaufman argues, its most serious consequences are happening now. "China's population policy has achieved its results in reducing births at a huge human cost," she wrote, referring to the distorted sex ratio at birth and instances of forced abortions and other abuses of women's rights. These human-rights abuses have caused many people, in China and abroad, to call for an end to the one-child policy.

Thirty years of the policy has resulted in a skewed sex ratio in the country. China already had a long history of preferring boys over girls, and this preference has been amplified by the one-child policy as many families have aborted daughters through sex-selective abortions (which, though illegal, are widely practiced in China) in order to try again for a boy. The average worldwide sex ratio at birth is 107 male to 100 female births; China's is 120 to 100. Some models have projected that there will be an excess of about 52 million men in China by 2020 (**Figure 16.13**). Of these, 25–30 million men will be of marrying age. By 2040, this could result in 40 million wifeless men in China. Sociologists predict that such a large population of wifeless men might cause social upheaval and gender-based violence, human trafficking, and discrimination against women and girls. Others have speculated that the shortage of women may actually increase the value of females in society.

In addition, China now has an unbalanced age ratio, with a large and growing elderly population supported by a much smaller working population of younger people (**Figure 16.14**). The change is so

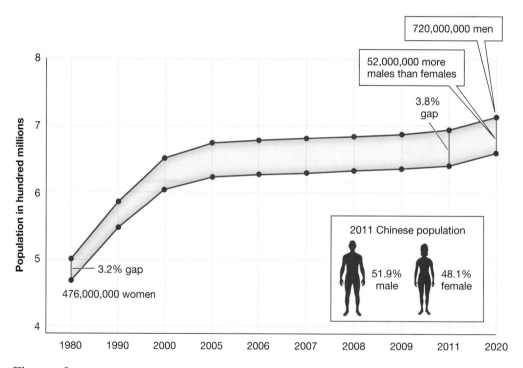

Figure 16.13

Too many boys and not enough girls

More boys than girls are currently being born in China, and the gap is projected to increase by 2040.

pronounced that some of China's primary schools are closing; since 1978 the number of primary-school students has shrunk by two-thirds. Some of those schools have been converted into homes for the elderly.

Arguments for China to relax or abandon the one-child policy continue to heat up. Kaufman, Cai, and Wang have been vocal proponents for abandoning the one-child policy. Yet the Chinese government has been resistant to change. Cai believes one reason is that the National Population and Family Planning Commission, which oversees implementation of the policy, employs some 500,000 people, making it a major bureaucracy with significant power in the country. Another reason, says Kaufman, is that many party officials and people in China believe the country's economic boom is the result of limiting fertility. In the end, "it's easier *not* to change than take a bold step," she says.

But today we may have begun to see inklings of change. In March 2013, China's government announced that the National Population and Family Planning Commission would merge with the Ministry of Health, which many consider a sign of demotion for the organization. And, in November 2013, the one-child policy was officially relaxed, with the announcement that couples in which one parent is an only child may have a second birth.

"I think we're seeing the beginning of the end of this policy," says Kaufman. "I'm hoping the rules will be further relaxed in the near future."

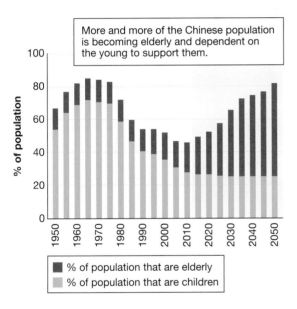

Figure 16.14

An increasingly unbalanced age ratio places a huge burden on the workforce
A decreasing workforce and an increasing elderly population are forecast to strain the Chinese economy in the coming decades.

Q1: What is China's current ratio of children to elderly people?

Q2: What is China's projected ratio of children to elderly people in 2050?

Q3: Why will there be more elderly than children in the future in China?

REVIEWING THE SCIENCE

- A **population** is a group of individuals of a single species located within a particular area.

- **Population size** is the total number of individuals in a population. **Population density** is the population size divided by the area covered by that population.

- Environmental factors such as lack of space, food shortages, predators, disease, and habitat deterioration limit populations. These factors affect the **carrying capacity**, the number of individuals that can live in an environment indefinitely.

- In **logistic growth**, a population grows exponentially at first but then stabilizes after reaching the carrying capacity. It is associated with an **S-shaped growth curve** when graphed.

- In **exponential growth**, a population increases by a constant proportion from one generation to the next generation. It is associated with a **J-shaped growth curve** when graphed.

- **Density-dependent population change** occurs when birth and death rates are affected by population density, which is the case when many individuals occupy the same space and therefore compete for resources.

- **Density-independent population change** occurs when population size is affected by factors that have nothing to do with the density of the population.

- Populations may exhibit **irregular fluctuations** or **cyclical fluctuations** rather than showing a classic logistic growth curve.

THE QUESTIONS

The Basics

1 A group of interacting individuals of a single species located within a particular area is called

(a) a biosphere.

(b) an ecosystem.

(c) a community.

(d) a population.

2 A population that is growing exponentially increases

(a) by the same number of individuals each generation.

(b) by a constant proportion each generation.

(c) in some years and decreases in other years.

(d) none of the above

3 In a population with an S-shaped (logistic) growth curve, after an initial period of rapid increase the number of individuals

(a) continues to increase exponentially.

(b) drops rapidly.

(c) remains near the carrying capacity.

(d) cycles regularly.

4 The growth of populations can be limited by

(a) natural disturbances.

(b) weather.

(c) food shortages.

(d) all of the above

5 Factors that limit the growth of populations more strongly at high densities than at low densities are said to be

(a) density dependent.

(b) density independent.

(c) exponential factors.

(d) sustainable.

6 The maximum number of individuals in a population that can be supported indefinitely by the population's environment is called the

(a) exponential size.

(b) J-shaped curve.

(c) sustainable size.

(d) carrying capacity.

Try Something New

7 A population of plants has a density of 12 plants per square meter and covers an area of 100 square meters. What is the population size?

(a) 120

(b) 1,200

(c) 12

(d) 0.12

8 An artificial pond at a college campus was populated with several mating pairs of pond slider turtles. After many years of slow population growth, campus administrators installed a coin-operated food dispenser at the pond's edge. Students and community members bought and fed the turtles an enormous amount of food each day. Within 5 years the population of turtles in the pond was increasing exponentially. How did the installation of the food dispenser affect the carrying capacity of the turtle population at this pond?

(a) The carrying capacity decreased slightly.

(b) The carrying capacity greatly increased.

(c) The carrying capacity stayed the same.

(d) The carrying capacity increased slightly.

9 About ten years after the coin-operated food dispenser was installed at the pond's edge, campus administrators had it removed. Apparently, complaints of "herds of turtles" crossing streets and parking lots convinced them that feeding the turtles had resulted in too many turtles in the pond. The turtle population has since leveled off and has remained at a relatively stable number. If the pond turtle population growth was graphed and included data points before installation, after installation, and after removal of the coin-operated food dispenser, what would it most closely resemble?

(a)

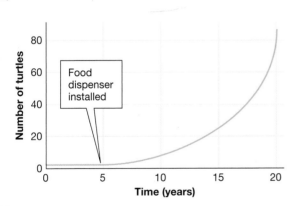

(b)

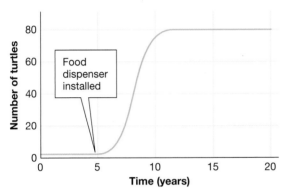

(c)

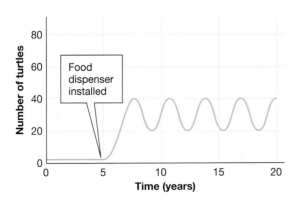

(d)

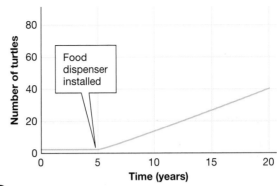

10 Suppose that population ecologists at the college determined that the number of turtles in the pond increased from 6 individuals to 24 individuals in just the first year after installation of the coin-operated food dispenser. What is the population doubling time for the turtles in this population?

(a) 6 years

(b) 6 months

(c) 1 year

(d) 1 month

Leveling Up

11 **What do *you* think?** In the article noted below, Wang Feng discusses what the merger of the National Population and Family Planning Commission with the Ministry of Health might mean for China's one-child policy. Read the article and reflect on it. Do you think this is the end of the one-child policy, or is it only a change in bureaucracy?

"One-Child Policy: Law Still in Effect, but Police, Judges Fired," http://blogs.wsj.com/chinarealtime/2013/03/12/ one-child-policy-law-still-in-effect-but-police-and-judges-fired

12 **Life choices** List five specific actions that you can personally take to limit the growth or impact of the human population on our Earth. How will these actions affect growth or impact?

13 ***Write Now* biology: one-child policy** Forced abortions, including third-trimester abortions, are undoubtedly the biggest controversy surrounding the one-child policy in China. Read both the news article and the opinion piece listed below, and reflect on how these abortions have affected the individuals involved and the population as a whole. What do these articles suggest about human rights in China? What other consequences could have been implemented for those breaking the one-child policy?

- "China Suspends Family Planning Workers after Forced Abortion," http://www.nytimes.com/2012/06/16/world/ asia/china-suspends-family-planning-workers-after-forced-abortion.html

- "China's Brutal One-Child Policy," http://www.nytimes. com/2013/05/22/opinion/chinas-brutal-one-child-policy.html

Of Wolves and Trees

The extermination of wolves in Yellowstone National Park had unforeseen effects on the park's ecosystem. Can the return of this top predator restore order?

After reading this chapter you should be able to:

◆ Understand how species abundance and species richness define diversity in an ecological community.

◆ Describe how the removal of a keystone species disrupts a food web.

◆ Identify the four major species interactions within a community.

◆ Compare and contrast primary and secondary succession.

CHAPTER

17

COMMUNITIES
OF ORGANISMS

Robert Beschta will always remember the day he visited Yellowstone National Park's Lamar Valley in 1996. Beschta, a hydrologist, was there to observe the Lamar River, which winds through the valley's lush lowland of grass and sage (**Figure 17.1**). But on that day as he walked toward the waterway, he noticed something odd: All around the valley, there were not many trees. The few tall, white aspens there looked haggard, their bark eaten away. And there were no young saplings to be seen.

Beschta had studied forestry and rivers for decades, and he knew what a healthy valley was supposed to look like. This was not it. Beschta approached the river and observed that its banks were also devoid of trees. The leafy green cottonwoods and wide willows that had once arched gracefully from the riverbank over the water were absent. And with no tree roots to hold the soil in place, the riverbanks themselves were jagged and eroding. "I was dumbstruck," recalls Beschta. He had never seen anything like it. Something unprecedented was happening in Yellowstone.

Beschta returned to Oregon State University (OSU), where he worked, and gave a seminar on his observations. He showed pictures of aspens with their bark stripped away and empty riverbanks where saplings should have been growing. In the audience, William Ripple sat up a little straighter. Also a scientist at OSU, Ripple studied forest ecology. He was particularly interested in aspen trees, which grow as tall as 70 feet and live up to 150 years. From what Beschta was saying, aspens were no longer growing in Yellowstone. And Ripple wanted to know why.

"It was a scientific mystery as to what was the cause of the decline," recalls Ripple. In that moment, listening to Beschta, Ripple knew exactly what his next research project would be: to document the extent of the aspen decline and determine why it was occurring.

Within a year, Ripple and one of his graduate students, Eric Larsen, traveled to the Lamar Valley. There, they drilled small holes into the centers of aspens and removed from each a small plug of wood, the diameter of a pencil (a process that does not damage the tree). Then they counted the growth rings in each plug—one for every year the tree has been in existence—to determine the age of each tree. They found that most of the aspens had begun to grow *prior* to 1920. After 1920, almost no new trees had started growing.

"We started scratching our heads at that point," says Ripple. He, Beschta, and Larsen began to brainstorm reasons why the trees might have stopped regenerating. They looked for environmental changes in Yellowstone in the 1920s that could have done it: a fire that killed off saplings, or a change in climate that reduced the trees' ability to reproduce. But nothing lined up—until an ordinary moment gave Ripple an extraordinary idea.

Standing in a gift shop in Grand Teton National Park, just south of Yellowstone, Ripple looked up at a poster on the wall. It featured a grove of tall, white aspen trees in the winter. In the middle of the trees, its paws covered in snow, stood a large gray wolf, similar to the one in this chapter's opening photo. "That was an 'aha' moment," says Ripple. "I thought, 'Maybe the wolf protects the aspen.'"

Figure 17.1

The Lamar River flows through Yellowstone National Park

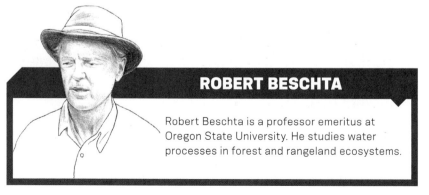

ROBERT BESCHTA

Robert Beschta is a professor emeritus at Oregon State University. He studies water processes in forest and rangeland ecosystems.

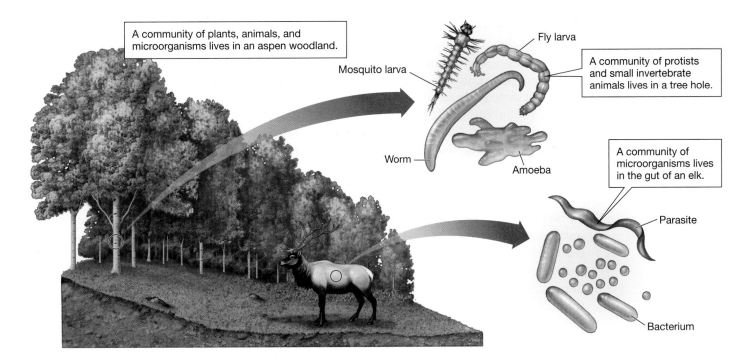

A community of plants, animals, and microorganisms lives in an aspen woodland.

Mosquito larva

Fly larva

A community of protists and small invertebrate animals lives in a tree hole.

Worm

Amoeba

A community of microorganisms lives in the gut of an elk.

Parasite

Bacterium

Figure 17.2

Ecological communities come in all sizes

Smaller communities can be nested within a larger community. This aspen woodland community contains the smaller communities of a temporary pool of water in a tree hole and an elk's gut, among others.

Q1: What is an ecological community?

Q2: Of what community could this aspen woodland be a smaller part?

Q3: What other small communities could be found within this larger community?

The idea of wolves protecting aspens initially seems nonsensical. Why would meat-eating predators protect trees? They wouldn't—at least not directly. But Ripple surmised that wolves in Yellowstone might have had an unintended effect on the community. As an ecologist, Ripple had long studied **ecological communities**, associations of species that live in the same area. Communities vary in size and complexity, from a small group of microorganisms in a temporary pool of water to the whole of Yellowstone Park, home to an estimated 322 species of birds, 67 species of mammals, 1,349 species of plants, and an uncounted number of insects (**Figure 17.2**).

Ripple knew that an ecological community is characterized by the diversity of species that live there, and that diversity is governed by two things: the **relative species abundance**, how common one species is when compared to another; and the

species richness, the total number of different species that live in the community (**Figure 17.3**). Ripple also knew that communities are subject to constant change, and that something must have changed in Yellowstone.

Ecological communities change naturally as a result of interactions between and among species and as a result of interactions between species and

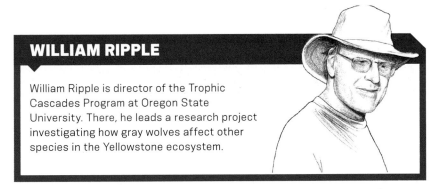

WILLIAM RIPPLE

William Ripple is director of the Trophic Cascades Program at Oregon State University. There, he leads a research project investigating how gray wolves affect other species in the Yellowstone ecosystem.

their physical environment. Ripple knew that both the relative abundance and the richness of species had changed in Yellowstone during the previous century. The relative species abundance had changed because aspen trees and willows were in decline, elk and coyote numbers increased, and bison and beaver populations decreased. Many of the common inhabitants of Yellowstone changed in abundance, therefore the relative species abundance of this community had changed. The species richness had also changed because early in the twentieth century a keystone species, the gray wolf, had gone missing. The total number of species in the community had gone down by one.

A **keystone species** is a species that has a disproportionately large effect on a community, relative to the species' abundance. There are few wolves compared to, say, rabbits, yet the wolves have a stronger effect on the community.

This community has higher relative species abundance for white-barked trees than the community below...

...while this community has higher species richness than the community above.

Figure 17.3

Does greater relative species abundance in a community mean greater species richness?

The characterization of an ecological community is determined by the diversity of species that reside there.

Q1: How does relative species abundance compare between the two communities in this figure?

Q2: How does species richness compare between the two communities?

Q3: How do relative species abundance and species richness define the species diversity of a forest community?

Keystone species are often recognized only when they go missing and their disappearance results in dramatic changes to the rest of the community (**Figure 17.4**). And that is exactly what happened in Yellowstone in the 1920s: the mighty gray wolf disappeared. Or, to put it more accurately, the mighty gray wolf was exterminated.

A Key Loss

In the early 1900s, ranchers and homesteaders killed wolves all across the United States. As a result, wolves were eliminated in much of the Eastern states. Then, in 1915, the U.S. government entered the fray, and a truly systematic slaughter began. In that year, the U.S. government began subsidizing wolf extermination programs all over the country. Wolves, feared and hated by private landowners for killing livestock, were trapped, shot, and skinned. Under the national program, states paid bounties of up to $150 for individual wolf pelts. The extermination happened quickly: the last known wolf den in Yellowstone was destroyed in 1926. At least 136 wolves, maybe more, were killed during the eradication campaign in Yellowstone. Once it was done, the park was wolf-free.

Seven decades later, staring at the poster of a wolf standing among the aspens, Ripple wondered if the loss of that keystone species had led to the decline of aspen trees. He immediately looked up the historical records to see whether the timing matched. Lo and behold, the last wolf had been killed at about the same time the aspens stopped regenerating, in the mid-1920s. Suddenly, it seemed obvious why the aspens had declined: wolves kill elk and elk eat aspens—three species in the same food chain. A **food chain** is a simple list of who eats whom. In scientific terms, it is the direct path by which nutrients are transferred through the community. A **food web**, on the other hand, is a more complex diagram of all the food chains in a single ecosystem and how they interact and overlap (**Figure 17.5**).

In the wolf-elk-aspen food chain, aspen are the **producers**, the organisms at the bottom of the food chain that use energy from the sun to produce their own food through photosynthesis. In Yellowstone and on land all over the Earth, photosynthetic plants like trees, grasses, and shrubs are the major producers (**Figure 17.6**).

In aquatic biomes, photosynthetic plankton are the major producers, as we'll see in Chapter 18.

Further up the food chain are **consumers**: organisms that obtain energy by eating all or parts of other organisms or their remains. Elk and wolves are both consumers: elk eat aspens, and wolves eat elk. In our Yellowstone food chain, elk are **primary consumers**: they eat producers. Wolves are **secondary consumers** because they eat primary consumers. This sequence of organisms eating organisms can continue: a bird that eats a spider that ate a beetle that ate a plant is a **tertiary consumer**; a killer whale that eats a leopard seal that ate a sea bass that ate a krill that ate a phytoplankton is a **quaternary consumer**. We will explore the flow of energy up the food chain in Chapter 18.

The more they discussed it, the more Ripple, Larsen, and Beschta believed that the loss of wolves in Yellowstone allowed the elk population to flourish, eating so many young trees that the aspen population could not regenerate. "We developed a hypothesis that maybe the killing of wolves actually affected the reproduction of aspen trees," says Ripple. Any change in species diversity will have a ripple effect (no pun intended) throughout the community, and the wolves of Yellowstone were no exception. Ripple and Larsen published their hypothesis in 2000, suggesting that the loss of wolves led to increased elk populations and altered elk movements and browsing patterns. In other words, with wolves gone, elk were free to find and eat young aspen shoots whenever they wanted, with no fear of wolves.

In 2001, Beschta returned to the valley and collected data on another species of tree, the cottonwood, which can live more than 200 years. He documented the same trend as before—in the 1920s, cottonwoods suddenly stopped generating new, young trees. In fact, since the 1970s, not a single new cottonwood had been established. "This was dramatic," recalls Beschta. "It's a big deal when you can't have a single cottonwood in this large valley make it to a mature tree."

If elk populations had continued to grow unchecked, they could have reduced the aspen and cottonwood populations to zero, disrupting the ecological community permanently. It is possible for a consumer to eat a species to extinction. But before that could happen, something surprising occurred in Yellowstone: humans brought the wolves back.

How a predator maintains diversity

A *Pisaster* sea star feeding on a mussel. Without this keystone species, the mussels crowd out other species in their community.

Loss of a keystone species reduces diversity

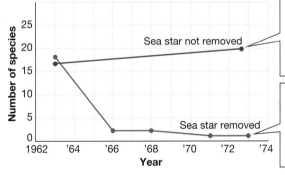

Sea stars completely eliminated mussels in submerged areas of this marine community, enabling other intertidal species to thrive there.

When the sea star *Pisaster* was removed from a community experimentally, the number of species dropped from 18 to 1, a mussel.

Figure 17.4

The star of the community

The sea star *Pisaster ochraceus* is an example of a keystone species. In an experiment conducted along the Pacific coast of Washington state in 1963, sea stars were removed from one site while an adjacent site was left undisturbed.

Q1: How many species were left in 1966 when sea stars were *not* removed from a community?

Q2: How many species were left in 1966 when sea stars were removed from a community?

Q3: How do your answers to questions 1 and 2 demonstrate the importance of a keystone species for the maintenance of diversity in a community?

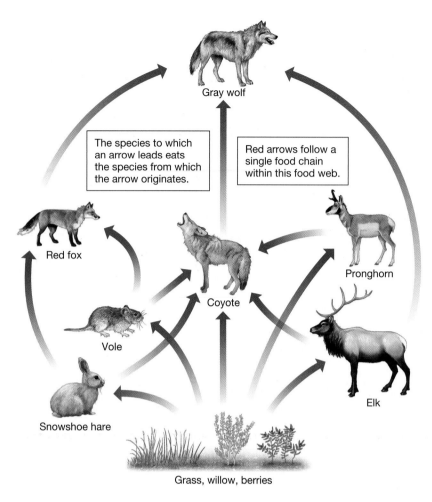

The species to which an arrow leads eats the species from which the arrow originates.

Red arrows follow a single food chain within this food web.

Gray wolf

Red fox

Coyote

Vole

Snowshoe hare

Pronghorn

Elk

Grass, willow, berries

Figure 17.5

Food webs show how energy moves through a community
Food webs are composed of many food chains that show one species eating another.

Q1: What is the difference between a food chain and a food web?

Q2: What species eat the coyote?

Q3: What species does the coyote eat?

A Second Ripple Effect

In 1973, wolves became the first animals to be protected under the Endangered Species Act. It was the dawn of the modern conservation movement, and the idea of returning wolves to Yellowstone grew in popularity. It took some time, but eventually lawmakers agreed to the plan,

and between 1995 and 1997, 41 wild wolves were captured in Canada and released in Yellowstone. It didn't take long for the wolf populations to recover: by 2007, an estimated 170 wolves lived in and around Yellowstone; today that estimate is 400 (**Figure 17.7**).

The loss, and then return, of the wolves had significant impacts on **species interactions** in the park. There are four central ways in which species in a community interact: *mutualism*, *commensalism*, *exploitation*, and *competition*. The classification is based on whether the interaction is beneficial, harmful, or neutral to each species involved. These interactions affect where organisms live and how large their populations grow. Species interactions also drive natural selection and evolution, thereby changing the composition of communities over short and long periods of time.

With wolves back in the park, Ripple suddenly had a way to test his hypothesis about aspen decline: if the loss of wolves was responsible for aspen loss, then the return of the wolves should incite a revival of aspens (and other woody plants, including cottonwoods). But he needed a way to quantify that change. "It's a scientifically difficult task to connect a wolf to a plant. Obviously wolves don't consume plants, so instead we had to connect the dots. There are data that wolves affect elk, and other data that elk affect plants," says Ripple.

Ripple and Beschta went into the field in 2006 and again in 2010 to take measurements. In addition to recording the ages of the trees, they looked for and documented signs of elk browsing on aspens—such as scars where branches or buds had been bitten off—and measured the heights of young aspens. Beschta did the same with the cottonwoods. All were eager to see which species interactions would occur now that wolves were back.

The first type of species interaction is **mutualism**, which occurs when two species interact and both benefit (**Figure 17.8**). For example, Yellowstone is home to 4,600 bison, the largest land mammals in North America, which have a mutualistic relationship with the black-billed magpie. Pests such as ticks burrow into a bison's short, dense hair to suck the beast's blood, but hungry little magpies perch on top of the bison and eat those ticks. Thus, both the bison and the magpie

In tropical rainforests, the abundant producers (plants) store chemical energy harvested from sunlight, which in turn is readily available to consumers.

In deserts, because of the sparse plant life, relatively little chemical energy is available for consumers.

Figure 17.6

Producers are the energy base of a food chain

There are producers in all communities and ecosystems, but they vary in abundance depending on the ecosystem.

Q1: What do producers produce?

Q2: Where do producers acquire the energy they need to perform their function in the food chain?

Q3: Why are producers necessary for life on Earth?

benefit from close interaction with one another. Mutualism is common and important in ecosystems all over Earth: many species receive benefits from, and provide benefits to, other species. These benefits increase the survival and reproduction of both interacting species.

When they aren't perched atop bison, black-billed magpies can be found in large nests atop deciduous or evergreen trees, where they reproduce once a year. These trees, another member of the community, share a commensal relationship with the magpies. **Commensalism** happens when one partner benefits while the other is neither helped nor harmed (**Figure 17.9**). In this case, the magpie benefits from having a safe place to lay eggs, and the interaction has no effect on the tree.

But, as you might have guessed, not all species interactions are as pleasant as those among bison, birds, and trees. In two other types of interactions, at least one of the two species is harmed: *competition*—which we will return to in a moment—and *exploitation*.

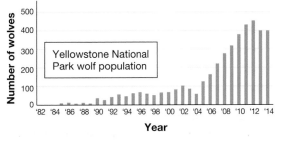

Yellowstone National Park wolf population

Figure 17.7

Wolves today in Yellowstone

The clownfish gains protection from predators by hiding within the sea anemone's stinging tentacles. The clownfish is not harmed because of a thick mucus that covers its body.

The anemone is protected from grazing predators by the clownfish and absorbs the fish's nutrient-rich excrement.

Figure 17.8

Mutualism: friends in need
Both the clownfish and the sea anemone benefit from their relationship.

Q1: How does the clownfish help the anemone?

Q2: How does the anemone help the clownfish?

Q3: Describe what might happen to an anemone without a resident clownfish.

Exploitation occurs when one species benefits and the other is harmed. The species doing the exploiting falls into one of two groups: parasites or predators. **Parasites** live in or on the organisms they harm, called **hosts**. An important group of parasites is pathogens, which cause disease in their hosts. Viruses that cause influenza and Ebola are pathogens, for example, as are the bacteria that cause strep throat, tuberculosis, and pneumonia. Many organisms have evolved mechanisms to avoid being hosts, such as immune systems to help fight off parasitic diseases and infections.

But an immune system can't save prey from a predator. **Predators** are consumers that eat part or all of other animals. There are two types of predators, herbivores and carnivores. Elk are **herbivores,** animals that eat plants. Yellowstone elk feed on the shoots, saplings, and new branches of woody plants like aspen, cottonwood, and willow, especially in winter when other plants are scarce. Wolves are **carnivores,** animals (and, in rare cases, plants) that kill other animals for food. Yellowstone wolves predominantly eat elk, especially in winter, but they also eat deer and any small mammals they can catch, notably beaver.

The animals eaten by predators are called **prey.** All of the animal residents of Yellowstone (and most plants) are prey for other species, except for grizzly bears, mountain lions, eagles, and gray wolves. These animals are all at the top of the food chain. The eating of prey by a predator is considered a form of exploitation because

Figure 17.9

Commensalism: a whale of a ride!
This gray whale's rostrum (snout) is covered in barnacles.

Q1: How is commensalism different from mutualism?

Q2: How do barnacles benefit from living on a whale?

Q3: Do you think a whale could avoid being colonized by barnacles? Why or why not?

The barnacles on a whale's snout enjoy a continuous stream of nutrient-rich water flowing across them from which they can feed. The whale is neither harmed nor helped by these stowaways.

the predator benefits from a tasty meal while the prey is harmed, by being eaten (**Figure 17.10**).

Now that wolves, a top predator in the community, were back in Yellowstone, how would their prey, the elk, react? And how would that reaction affect the elk's food, the trees?

Using the plant measurements taken in the park and comparing those measurements to historical data, Beschta and Ripple found that between 1998 and 2010, as the wolf population in the park grew, elk browsing decreased. In 1998, essentially 100 percent of the young aspen plants were being preyed upon, but by 2010, only 18–24 percent were being eaten. In addition, average aspen heights increased for all areas that the scientists observed. The same revival occurred with cottonwoods. In the 1970s, cottonwoods had entirely stopped adding new young saplings, but by 2012, some 4,660 young cottonwoods were growing over 2 meters high.

Together, the aspen and cottonwood data sets convinced Ripple and Beschta that the wolves were responsible for a cascade of species interactions leading to the restoration of aspen and cottonwood populations. "With wolves back, young, woody plants are doing better and growing taller," says Ripple. "Plant communities are beginning to recover." Different species are growing and spreading at different rates of recovery, he adds, "but there's enough new growth that we suggest it is in support of our basic hypothesis, that the presence or absence of the top predator—the wolf in this case—makes a difference in these plant populations."

Since Ripple and Beschta's discovery, the scientific community has been debating two potential reasons why the plants are flourishing with the return of the wolves. The most straightforward possibility is that the elk population has decreased: wolves kill elk, so there are fewer elk to consume the plants. Yet some of the tree populations seemed to recover faster than the drop in the elk populations would suggest. So a second possibility is that the presence of wolves led to a change in elk behavior called a *fear effect*. Often the presence of a predator in an ecosystem can affect the behavior of its prey. In this case, it is possible, but speculative, that elk stopped grazing in areas where they could easily be seen by wolves, such as along the banks of the Lamar River.

"These two mechanisms, the population density and fear behavior, are difficult to tease apart,

This cheetah is a predator whose many prey species include the African hare.

The parasite *Hippobosca longipennis*, the louse fly, lives off carnivores like the cheetah by sucking their blood, but does not kill its host.

Figure 17.10

Exploitation: predators can also be hosts for parasites
Predation and parasitism harm the prey and the host, respectively.

Q1: What type of harm usually comes to prey species?

Q2: What type of harm usually comes to the hosts of parasites?

Q3: Are the elk that graze on aspen tree saplings predators or parasites? Why?

and we're working on that," says Ripple. "Many today believe it is due to a combination of the two." Beschta agrees: "In my opinion, they've both been going on."

Safety in Numbers, and Colors

Elk may avoid lingering at streams as a way to evade their predators, but other prey have far more elaborate strategies to avoid being consumed. The poison dart frog, for example, is among the most toxic animals on Earth, and it evolved bright colors as **warning coloration** to alert potential predators to the dangerous chemicals in its tissues (**Figure 17.11**, bottom). Such warning coloration can be highly effective. Young blue jays, for example, quickly learn not to eat brightly colored monarch butterflies, which contain chemicals that cause nausea and, at high doses, death.

Mimicry

The viceroy butterfly (left) mimics the color and pattern of the monarch butterfly (right), which contains toxic compounds.

Camouflage

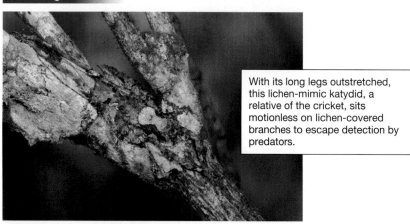

With its long legs outstretched, this lichen-mimic katydid, a relative of the cricket, sits motionless on lichen-covered branches to escape detection by predators.

Warning coloration

The bright colors of the poison dart frog warn potential predators of the deadly chemicals contained in its tissues.

Figure 17.11

Adaptive coloration responses to predation

Prey species have adapted many elaborate strategies to avoid being eaten by predators, including warning coloration, mimicry, and camouflage.

Q1: How do predators know that brightly colored prey are usually toxic?

Q2: Do you think mimicry works if the toxic species is in low abundance? Why or why not?

Q3: Why is camouflage considered an adaptive response to predation?

Then there are species that, though not poisonous, evolved coloration to make them look as if they are. Through **mimicry**, the viceroy butterfly, which is not poisonous, imitates the color and pattern of the monarch butterfly (**Figure 17.11**, top). That "borrowed" coloration scares away blue jays and other birds that may have felt sick the last time they ate a monarch.

Another mechanism to avoid being eaten is **camouflage**, any type of coloration or appearance that makes an organism hard to find or hard to catch (**Figure 17.11**, middle). The pygmy sea horse, leaf-tailed gecko, and stone flounder are all masters of camouflage.

Finally, many prey, from massive elephants to tiny ants, have evolved a different strategy to avoid becoming dinner: living together. By group living, these animals are able to act together to warn each other when a predator is about to attack and even repel attacks as a united front (**Figure 17.12**).

Predator-prey interactions are a type of exploitation in which the predator benefits. But there is one final type of species interaction where no one benefits: **competition**, in which both interacting species are negatively affected.

Competition most often occurs when two species share an important but limited resource, such as food or space. In Yellowstone, beavers and elk both eat woody plants. Woody plants are part of the **ecological niche** for both species—the set of conditions and resources that a population needs in order to survive and reproduce in its habitat. Because the niches of the beavers and the elk overlap, these species compete. When two (or three or more) species compete, each has a negative effect on the other because one is using resources that the other then cannot access. (If resources are abundant, however, there may be no competition between species, even if their niches overlap.)

There are two main categories of competition, *exploitative* and *interference*. In **exploitative competition**, species compete indirectly for shared resources, such as food. In this case, each species reduces the amount of the resource that is available for the other species, but they do not directly interact or come in contact with each other (**Figure 17.13**). When wolves returned and elk populations declined, beavers had less competition for food, especially willow trees, and the number of beaver colonies in Yellowstone rose from 1 to 12.

The success of goshawk attacks on wood pigeons decreases greatly when there are many pigeons in a flock.

Although a single musk ox may be vulnerable to predators such as wolves, a group forming a circle makes a difficult target.

Figure 17.12

Safety in numbers

Animals that live in groups are better able to warn each other and sometimes fend off attacking predators.

Q1: What percentage of pigeons are caught when they are alone and not in a flock?

Q2: For wood pigeons, what is the minimum number of individuals that provides protection from goshawks?

Q3: Why do you think a group of musk oxen versus a lone musk ox would be safer from a pack of wolves?

Figure 17.13

Exploitative competition: a new species moves in

Two different species of wasps—*Aphytis lingnanensis* and *Aphytis chrysomphali*—share an ecological niche but do not physically come in contact with one another.

Q1: Which species is the superior competitor?

Q2: In what ecological niche do these wasp species compete?

Q3: Why is this example considered exploitative competition? How do you know?

The wasp *Aphytis lingnanensis* was introduced to southern California in 1948.

1948

Both species of wasps prey on the same types of insects that damage citrus crops. With two wasp species eating the same prey, there was less food for both species.

1959

Cause and Effect

Wolves were reintroduced to Yellowstone National Park in 1995, and their return had a major impact on other species in the park, especially elk and aspen. Elk began avoiding areas where wolves could easily prey upon them, such as near riverbanks and especially riverbanks with downed logs, where escape would be difficult. With decreasing elk presence, aspen flourished in these areas. The effects depicted here are a powerful example of the influence of a keystone predator on an ecological community.

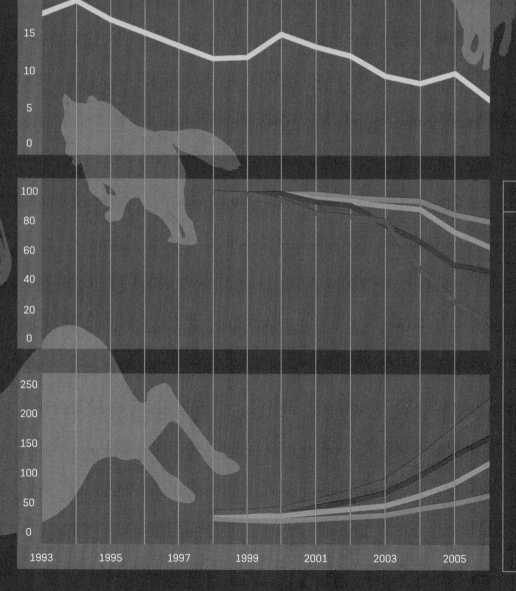

Wolf population

Elk population
(in thousands)

Percentage of aspen trees grazed upon

Height of aspen trees
(in cm)

Legend

Trees in uplands with downed logs

Trees in uplands without logs

Trees on riverbanks with downed logs

Trees on riverbanks without logs

1993 1995 1997 1999 2001 2003 2005

Elk and bears also interact through exploitative competition. Elk eat the leaves and branches on shrubs, resulting in a decrease in the number of berries the shrubs produce. That's not good for grizzly bears, which love to eat berries. Knowing of this relationship, Ripple hypothesized that a decrease in elk would result in an increase in berry-producing shrubs, and that bears would be eating more berries. To see whether this was true, Ripple, Beschta, and others spent two years analyzing grizzly scat that had been collected in the park. They compared the percentage of fruit in current scat to that of scat data that had been collected and saved before 1995, prior to the wolf return. Over a 19-year period, they found that the percentage of fruit in the grizzly diet went up as elk populations went down.

Organisms also compete through **interference competition**, in which one organism directly excludes another from the use of a resource (**Figure 17.14**). In Yellowstone, for example, bears and wolves often fight over the carcass of an animal. In interference competition, one individual physically gets in the way of another individual that is trying to obtain a resource.

A Community Restored

Today, the return of the wolf is having a clear and significant impact on the ecological community of Yellowstone, from the rebirth of aspens and cottonwoods to growing beaver populations. These are signs of **succession**, the process by which the species in a community change over time. "We're on a very important upward trend," says Beschta.

All ecological communities change over time, sometimes because of human intervention, as in Yellowstone, but also because of natural changes in species composition—the number of individuals in a population often changes as the seasons change, for example—and natural disturbances such as fires, floods, and windstorms. In addition, communities can broadly change by the slow loss or gain of populations of species over long periods of time through natural selection.

There are two major types of succession that ecologists have observed in nature. **Primary**

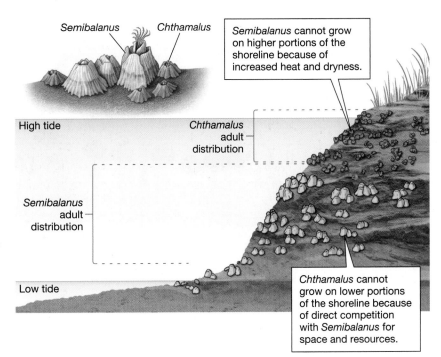

Semibalanus cannot grow on higher portions of the shoreline because of increased heat and dryness.

Chthamalus cannot grow on lower portions of the shoreline because of direct competition with *Semibalanus* for space and resources.

Figure 17.14

Interference competition keeps species apart

On the rocky coast of Scotland, the larvae of *Semibalanus* and *Chthamalus* barnacles settle on rocks on both high and low portions of the shoreline.

Q1: How does interference competition work?

Q2: Does interference competition happen in the higher portions of the shoreline or the lower?

Q3: Which species is the better competitor?

succession occurs in newly created habitats, such as on an island just emerged from the sea or on the soil left behind a retreating glacier (**Figure 17.15**). A new habitat begins with no species. Often the first species to colonize an area alters the habitat in ways that enable later-arriving species to thrive. A specific type of flowering plant may grow on a new island, for example, and then a species of bee that feeds on that flower subsequently joins the habitat.

Secondary succession occurs after a disturbance within a community—such as the loss of a keystone species like the wolves in Yellowstone or the devastation caused by a forest fire—has reduced the number of species in a community. During secondary succession, communities

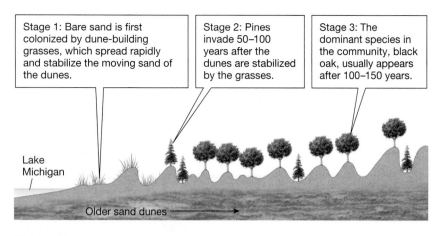

Stage 1: Bare sand is first colonized by dune-building grasses, which spread rapidly and stabilize the moving sand of the dunes.

Stage 2: Pines invade 50–100 years after the dunes are stabilized by the grasses.

Stage 3: The dominant species in the community, black oak, usually appears after 100–150 years.

Lake Michigan

Older sand dunes ⟶

Stage 4: Climax communities of black oak have lasted up to 12,000 years.

Figure 17.15

Primary succession: from nothing to climax community
Sand becomes woodland near Lake Michigan.

Q1: What species represents the first colonizers of the sand dunes?

Q2: What species is the intermediate species, and how does it become the dominant species?

Q3: What species is the mature, climax community species and how does it become the dominant species?

usually regain the successional state that existed before the disturbance. This type of succession does not take as long as primary succession, because some species still exist in the community.

Luckily, communities can and do bounce back from disturbances, but the time required to regain a previous state varies from years to decades to centuries. Yellowstone is currently experiencing a secondary succession as wolf, aspen, and beaver populations slowly return. It will likely still be a while before the park returns to its previous state, says Ripple. "We were 70 years without wolves, and now we're less than 20 years since wolves have returned, so this is going to take time." But he and others are hopeful that Yellowstone will return to its status as a **climax community**, a mature community whose species

composition remains stable over long periods of time (**Figure 17.16**).

Yet even as Yellowstone's recovery is underway, ecological communities around the planet are being threatened by the loss of other keystone species, especially large carnivores. In 2013, Ripple and colleagues analyzed 31 carnivore species around the globe, including leopards, lions, cougars, and sea otters. They found that more than 75 percent of those 31 species are declining, and that 17 of them now occupy less than half of their former ranges. Because of all the species interactions discussed here, it is clear that those changes will have major effects on ecological communities. "Humans are affecting predators around the globe in a major way," says Ripple. "It's a worldwide issue."

In 1988, a large fire destroyed a portion of the mature lodgepole-pine forest in Yellowstone National Park.

By 1992, the lodgepole-pine forest regrowth was gaining momentum.

An example of a mature, climax community lodgepole-pine forest.

Figure 17.16

Secondary succession: from disturbance to climax community
There has been a slow but steady regrowth of lodgepole-pine forest in Yellowstone National Park.

Q1: What other types of disturbances could you imagine destroying a forest?

Q2: How is secondary succession different from primary succession?

Q3: What is a climax community?

REVIEWING THE SCIENCE

- An **ecological community** can be characterized by its species composition, or diversity. This diversity has two components: **relative species abundance** (the number of individuals of each species that exist in the community) and **species richness** (the total number of different species that live in the community).

- **Keystone species** have a disproportionately large effect, relative to their own abundance, on the richness and abundances of the other species in a community. The removal or disappearance of these keystone species results in dramatic changes to the rest of the community.

- A **food chain** is a single direct line of who eats whom among species in a community. A **food web** depicts how overlapping food chains of the community are connected.

- **Producers**, organisms found at the bottom of a food chain that use light energy to produce their own food, are eaten by **consumers**. Consumers are classified as **carnivores** or **herbivores** depending on whether they eat animals or plants, respectively.

- **Species interactions** in a community can be beneficial, harmful, or without benefit or harm to each of the interacting species.

  - In **mutualism**, both species benefit.

  - In **commensalism**, one species benefits at no cost to the other.

  - In **exploitation**, one species benefits and the other is harmed. **Predation** and **parasitism** are types of exploitation.

  - In **competition**, both species are harmed. Competition occurs when **ecological niches** overlap and includes **exploitative competition** and **interference competition**.

- **Succession** establishes new communities (**primary succession**) and replaces disturbed communities (**secondary succession**). In stable environments without disturbances, called mature communities or **climax communities**, species composition remains stable over long periods of time.

THE QUESTIONS

The Basics

1 A single sequence of feeding relationships describing who eats whom in a community is a

(a) life history.

(b) keystone relationship.

(c) food web.

(d) food chain.

2 The process of species replacement over time in a community is called

(a) global climate change.

(b) succession.

(c) competition.

(d) community change.

3 Organisms that can produce their own food from an external source of energy without having to eat other organisms are called

(a) suppliers.

(b) consumers.

(c) producers.

(d) keystone species.

4 A low-abundance species that has a large effect on the composition of an ecological community especially when removed is called a

(a) predator.

(b) herbivore.

(c) keystone species.

(d) dominant species.

5 A cheetah eats an antelope that ate some grass. The cheetah is a _____, while the antelope is a _____.

(a) tertiary consumer; secondary consumer

(b) secondary consumer; primary consumer

(c) primary consumer; producer

(d) secondary consumer; producer

Try Something New

6 Cattle egrets trail livestock, sometimes perching on their backs for a better view. These birds eat insects stirred up by the grazing animals. This relationship is an example of

(a) parasitism.

(b) predation.

(c) commensalism.

(d) mutualism.

7 A grandmother has sent her twin grandchildren a single bag of candy to share. The kids agree to share the candy equally, but at night, one of them sneaks into the kitchen and swipes some extra pieces of candy. The other kid swipes candy from the bag during her sibling's piano lesson. This scenario is an example of

(a) exploitative competition.

(b) interference competition.

(c) mutualism.

(d) commensalism.

(e) parasitism.

8 When a female cat comes into heat and is ready to mate, she urinates more frequently and in a large number of places. Male cats from the neighborhood congregate near urine deposits and fight with each other for the female's attention and breeding rights. In what interaction are the male cats engaging?

(a) commensalism

(b) parasitism

(c) interference competition

(d) exploitative competition

(e) mutualism

9 Rabbits can eat many plants, but they prefer some plants over others. Assume that the rabbits in a grassland community containing many plant species prefer to eat a species of grass that happens to be a superior competitor. If the rabbits were removed from the region, which of the following do you think would most likely happen?

(a) The grassland community would have higher relative species abundance and lower species richness.

(b) The grassland community would have lower relative species abundance and higher species richness.

(c) The grassland community would remain largely unchanged in relative species abundance and species richness.

(d) The grassland community would have lower relative species abundance and lower species richness.

10 A field formerly planted with corn was abandoned by the landowners and allowed to grow wild. Weeds and grasses grew in the field first and were followed by small trees. What type of succession is happening in this abandoned field?

(a) primary

(b) secondary

(c) tertiary

(d) It is impossible to tell from the information provided.

11 Chose the statement that best describes a climax community in Rocky Mountain National Park.

(a) A mountain slope that has been cleared of evergreen trees and is now sprouting aspen trees.

(b) A mountainside with mature evergreen trees that have dominated the landscape for generations.

(c) An area of soil, sand and rocks left behind after a dam burst and flooded the area.

(d) Lichens and mosses growing on bare rock at upper elevations due to an increase in average yearly temperatures.

12 Analyze the food web below and answer the following questions:

(a) Which species does NOT have a predator shown on the food web?

(b) Which species shown on the food web performs photosynthesis (captures light energy to make their own food)?

(c) Which species has only one predator and only one prey shown on the web?

(d) Which species has the most predators shown on the web?

Leveling Up

13 **Doing science** Citizen science is an amazing way for anyone and everyone to get involved in scientific research. Search the Internet for "citizen science" projects relevant to this chapter's topics, using keywords like "parasite," "predator," "prey," and "group living," to name a few. Examples include MonarchHealth, a project in which volunteers sample wild monarch butterflies for a protozoan parasite to track its prevalence across North America. Participate in a project and, in writing, reflect on what you learned.

14 **Doing science** How can you affect the preservation and reintroduction efforts for North American wolf species in Yellowstone and in other areas? How can you help other efforts to reintroduce species? Write a letter explaining how wolves (or other species) are beneficial to our wilderness ecosystems using concepts learned in this chapter. Search the Internet for appropriate contacts where you can send your letter(s).

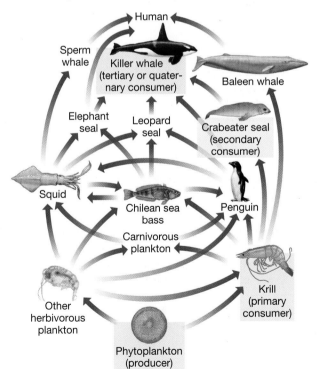

Here and Gone

Researchers discover an alarming decline of plankton in the ocean.

After reading this chapter you should be able to:

- Define an ecosystem and give examples.
- Describe why nutrients are recycled in an ecosystem but energy is not.
- Use an energy pyramid to assign trophic levels to a food chain.
- Calculate an ecosystem's net primary productivity.

As a teenager, Daniel Boyce made extra cash working as a deckhand on fishing boats. As he got older, Boyce decided to turn his interest in oceans and science into a career, and in 2007 he joined the lab of marine biologist Boris Worm at Dalhousie University in Nova Scotia, Canada.

In 2003, Worm had published a study showing that the industrialized fishing boom that began in the 1950s had decimated predatory fish communities. Boyce decided to follow up on his mentor's work by studying how that loss of predatory fish reverberated down the food chain. Specifically, he wanted to see how the loss of ocean predators affected plankton, a diverse group of free-floating organisms that drift around the ocean and are the primary producers for the ocean ecosystem, supporting virtually all marine animals.

When Boyce proposed this study, little did he know that it would uncover a profound shift in ocean ecosystems around the world—a finding so shocking that just suggesting it would plunge him and his collaborators into a public controversy.

Going Green

Boyce's initial goal, similar to that of William Ripple's work studying wolves in Yellowstone (see Chapter 17), was to show how the loss of a top predator affects an ecosystem. In that story, a variety of species living in Yellowstone National Park interacted to form what we call an *ecological community*. A group of communities interacting with one another and with the physical environment they share is an **ecosystem** (**Figure 18.1**). To say it another way, an ecosystem is characterized by the interactions of organisms in the **biotic** world ("biotic" means "pertaining to life") with the **abiotic** (nonliving) world. The abiotic world includes the atmosphere, water, and Earth's crust.

An ecosystem may be small or large; a puddle teeming with protists is an ecosystem, as is the Atlantic Ocean. And smaller ecosystems can be nested inside larger, more complex ecosystems. This variety means that ecosystems do not always have sharply defined physical boundaries. Instead, ecologists often define an ecosystem by the distinctive ways in which it functions, especially the means by which *energy* and *nutrients* are acquired and distributed by the biotic community (**Figure 18.2**).

The activity of primary producers, in particular, profoundly influences the characteristics of an ecosystem. Ecologists often demarcate an ecosystem according to the types of producers it contains and the community of consumers that the producers support. A duckweed-covered pond, a tallgrass prairie, and a beech-maple woodland are all examples of ecosystems that can be defined by the specific types of producers that capture and supply energy to consumers.

To see how overfishing was affecting food chains in the ocean, Boyce first looked to the

Figure 18.1

Ecosystems in Nova Scotia, Canada

Overfishing has decimated fish populations in Nova Scotia, affecting all members of the food chains in which these fish reside—all the way down to phytoplankton.

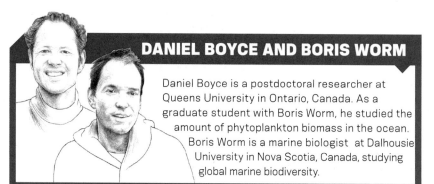

DANIEL BOYCE AND BORIS WORM

Daniel Boyce is a postdoctoral researcher at Queens University in Ontario, Canada. As a graduate student with Boris Worm, he studied the amount of phytoplankton biomass in the ocean. Boris Worm is a marine biologist at Dalhousie University in Nova Scotia, Canada, studying global marine biodiversity.

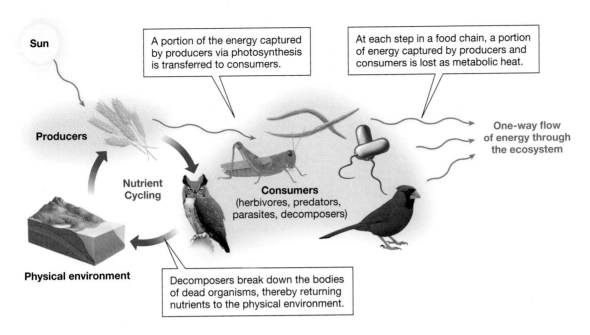

Sun

A portion of the energy captured by producers via photosynthesis is transferred to consumers.

At each step in a food chain, a portion of energy captured by producers and consumers is lost as metabolic heat.

Producers

One-way flow of energy through the ecosystem

Nutrient Cycling

Consumers
(herbivores, predators, parasites, decomposers)

Physical environment

Decomposers break down the bodies of dead organisms, thereby returning nutrients to the physical environment.

Figure 18.2

Overview of energy and nutrient flow in an ecosystem
Energy flows through an ecosystem in a single direction, as indicated by the red arrows. Nutrients are cycled between organisms and the physical environment, as depicted by the blue arrows.

Q1: What process do producers use to capture energy from the sun?

Q2: What class of organisms breaks down the dead bodies of other organisms?

Q3: What happens to most of the energy in an ecosystem?

producers of the ocean ecosystem: phytoplankton. These small, floating microalgae come in a fantastic array of shapes and sizes, from smooth orbs to segmented spirals to pointy crescents (**Figure 18.3**). Phytoplankton are primarily microscopic, but in large groups they form the green color often seen in water. The more phytoplankton in the water, the greener the water is; the less phytoplankton, the bluer it is.

Phytoplankton are green because they are photosynthetic: They convert light energy from the sun into chemical energy using chlorophyll, the green pigment critical to the process of photosynthesis. Because phytoplankton are photosynthetic, these water-living organisms inhabit the top layer of water in the ocean (and almost every body of freshwater as well), a location that gives them easy access to sunlight. Thanks to their ability to photosynthesize sunlight, phytoplankton are primary producers and the central

means through which energy enters the ocean ecosystem.

Bottom of the Pyramid

Two main components flow through ecosystems: energy and nutrients. First, consider the path of energy. Producers like phytoplankton capture energy from the sun and transform it into fuel energy, and that energy is passed up the food chain as one organism eats another. An **energy pyramid** represents the amount of energy available to organisms in an ecosystem. Each level of the pyramid corresponds to a step in a food chain and is called a **trophic level**. For instance, in the ocean, phytoplankton are on the first trophic level. Zooplankton, which are larger,

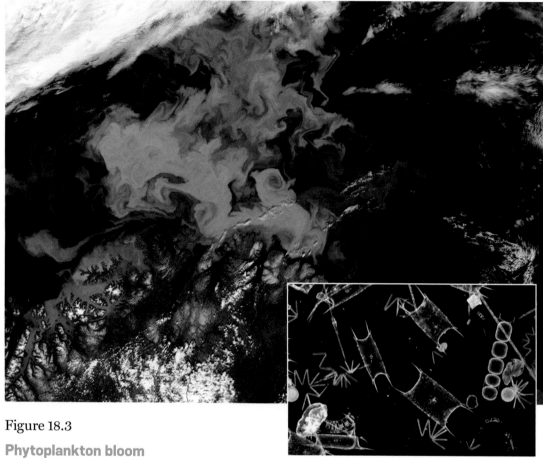

Figure 18.3

Phytoplankton bloom

The turquoise area in this aerial photo is a phytoplankton bloom occurring off the coast of Norway. When a population of phytoplankton (inset) increases rapidly, it discolors the water in which it resides.

multicellular plankton that feed on phytoplankton, are the second trophic level. Small fish such as herring are the third level, and large fish such as tuna are the fourth (**Figure 18.4**).

At each trophic level, a portion of the energy captured by producers is lost as **metabolic heat**, the heat released as a by-product of chemical reactions within a cell, especially during cellular respiration. Organisms lose a lot of energy as metabolic heat, as revealed by the fact that a small room crowded with people rapidly becomes hot; that warmth is the result of metabolic heat leaving our bodies. On average, roughly 10 percent of the energy at one trophic level is transferred to the next trophic level. The remaining 90 percent of the energy that is not transferred is either not consumed (for example, when we eat an apple, we eat only a small part of the apple tree), is not taken up by the consumer's body (for example, we cannot digest the cellulose contained in the apple), or is lost as metabolic heat.

Because of this steady loss of heat, energy flows in *only one direction* through ecosystems. It enters Earth's ecosystems from the sun (in most cases) and leaves them as metabolic heat. Every unit of energy captured by producers is eventually lost from the biotic world as heat. Therefore, energy cannot be recycled within an ecosystem. It travels up an energy pyramid, never down.

In contrast, **nutrients**—chemical elements required by living organisms—*are* recycled and reused within and across ecosystems. While Earth receives a constant stream of light energy from the sun, our planet does not acquire more nutrients on a daily basis: a constant and finite pool of nutrients cycles through the land, water, and air. If nutrients were not cycled between organisms and the physical environment, life on Earth would not exist.

Nutrients pass through the abiotic world, from rocks and mineral deposits into soil and water, and then on to the biotic world via absorption by

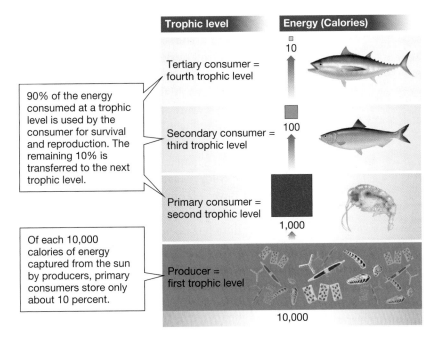

Figure 18.4

Energy pyramid

The levels of the energy pyramid correspond to steps in a food chain.

Q1: What percentage of the original 10,000 calories is available to a shark that might eat the tuna in this figure?

Q2: At what trophic level would we categorize the tuna?

Q3: Why is energy "lost" at each trophic level?

producers and cycled among consumers for varying lengths of time. Phytoplankton, for example, require the nutrients nitrogen, phosphorus, iron, and silicon for growth. When zooplankton eat phytoplankton, they take up those nutrients, and so on up the food chain (**Figure 18.5**).

Nutrients are eventually returned to the abiotic world when **decomposers** break down the dead bodies of other organisms, both consumers and producers. In some ecosystems, decomposers break down 80 percent of the *biomass*, or biological material, made by producers. Without decomposers, nutrients could not be repeatedly reused, and life would cease because all essential nutrients would remain locked up in the bodies of dead organisms. In this way, decomposers are the "cleaners" of an ecosystem. Bacteria and fungi are important decomposers in the ocean, as are hagfish, worms, and others.

Ecologists and earth scientists use the term "nutrient cycle" to describe the passage of a chemical element through the abiotic and biotic worlds (**Figure 18.6**). The nutrient cycle and the flow of energy are two of four processes that link the biotic and abiotic worlds in an ecosystem. These **ecosystem processes** also include the water cycle and *succession*, the process by which the species in a community change over time (as discussed in Chapter 17).

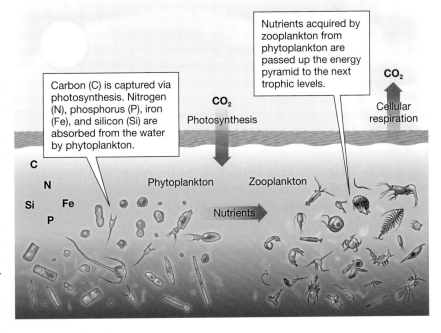

Figure 18.5

Carbon and nutrient passage

Nutrients pass from the abiotic world to the biotic world, where they are absorbed by producers and cycled among consumers.

Q1: Which organisms are the producers in this ecosystem?

Q2: Which nutrient is captured through photosynthesis?

Q3: Name the biotic factors and the abiotic factors shown in this figure.

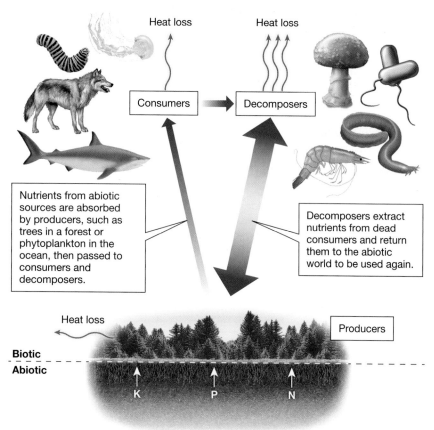

Figure 18.6

Nutrient cycling

Nutrients, including potassium (K), phosphorus (P), and nitrogen (N), are constantly cycled between the abiotic and biotic worlds.

Nutrients from abiotic sources are absorbed by producers, such as trees in a forest or phytoplankton in the ocean, then passed to consumers and decomposers.

Decomposers extract nutrients from dead consumers and return them to the abiotic world to be used again.

Q1: What is the main difference between a consumer and a decomposer?

Q2: Which of the things shown in the figure are abiotic?

Q3: Describe all the points at which heat is lost in this figure.

A Multitude of Measurements

For over 100 years, researchers around the globe have studied ecosystems containing phytoplankton. Boyce tapped into that wealth of research to document past and present levels of phytoplankton in the oceans.

The amount of phytoplankton biomass in a given area can be estimated by the concentration of chlorophyll found there (**Figure 18.7**). For decades, nearly all ocean studies have used chlorophyll concentration as a reliable metric of phytoplankton biomass. Chlorophyll concentration is measured by detecting the color of water. Water takes on deeper shades of green as the amount of chlorophyll increases. When there is no chlorophyll, water appears clear.

Ideally, Boyce would have used satellite data to detect chlorophyll and thus phytoplankton concentrations, since satellites today take high-resolution color measurements of the ocean surface. Yet Boyce planned to review phytoplankton levels over the past 100 years, and high-quality satellite data has been available for only the last decade. He needed another source of data.

In a first-of-its-kind analysis, Boyce, together with his adviser Boris Worm and the oceanographer Marlon Lewis, combined two types of chlorophyll measurements. The first type, dating all the way back to 1899, were recorded with nothing more than a rope and a disk.

In 1865, the Pope asked priest and astronomer Pietro Angelo Secchi to measure the clarity of water in the Mediterranean Sea for the purposes of the papal navy. Secchi designed one of the simplest measurement devices ever used: a dinner plate–sized disk painted with black and white stripes attached to a rope. The disk is lowered into water until the white stripes disappear (as they become obscured by chlorophyll from phytoplankton), and the depth at that point is

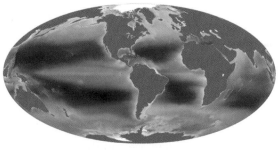

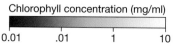

Chlorophyll concentration (mg/ml)

0.01   .01   1   10

Figure 18.7

Average chlorophyll concentration in the oceans

Phytoplankton are most abundant in high latitudes, along coastlines and continental shelves, and along the equator in the Pacific and Atlantic Oceans (yellow)—but are scarce in remote oceans (dark blue).

Figure 18.8

Secchi disks indirectly measure chlorophyll concentrations
The Secchi disk is lowered into water until its white stripes become obscured by the chlorophyll in phytoplankton.

recorded (**Figure 18.8**; see also the chapter-opening photo). Chlorophyll concentrations derived from Secchi disk measurements have recently been corroborated by satellite data, so scientists know they are reliable.

In addition to gathering Secchi disk data, scientists at sea regularly use lab tools to directly measure the quantity of chlorophyll in the water (as opposed to observing its color and relating that to chlorophyll concentration). Boyce found hundreds of thousands of these direct chlorophyll measurements online in open-source databases. "There's been a huge increase in the amount of publicly available oceanographic data out there," says Boyce.

But to use the data, Boyce first had to separate the wheat from the chaff. "With any big database, there are bound to be measurements that are entered incorrectly, for whatever reason," says Boyce. He, Worm, and Lewis ruled out measurements that were inappropriate for their study, such as those taken where the ocean floor was less than 25 meters deep, because changes of water transparency in those cases could be caused by sediment or runoff from landmasses nearby rather than by phytoplankton.

The team analyzed each data set separately—Secchi disk measurements and direct chlorophyll measurements—and then together. To combine the two, they converted all the Secchi measurements into the same units as those used for direct chlorophyll concentrations. In total, the blended data set included 445,237 chlorophyll measurements collected between 1899 and 2008.

Using two different methods of analysis, Boyce found a significant decline in phytoplankton levels in 60–80 percent of Earth's oceans where data were available during the last century. Overall, the team found that phytoplankton had declined by about one percent of the global average each year. One percent sounds like a small number, but one percent every year just since 1950 translates into a staggering total phytoplankton decrease of 40 percent in the ocean.

The Precious One Percent

A 40 percent loss in the main producer in any ecosystem is a worrisome number, but especially with respect to phytoplankton. Phytoplankton support fisheries, produce half the oxygen we breathe, and take in carbon dioxide from the atmosphere, which helps offset the greenhouse effect and global warming.

An ocean with less phytoplankton will function differently because ecosystems depend on **energy capture**, the trapping and storing of solar energy by the producers at the base of the ecosystem's energy pyramid. Herbivores, predators, and decomposers all depend indirectly on energy capture. If an ecosystem has an abundance of producers, it can often support more consumers at higher trophic levels. In a tropical forest, for example, an abundance of plants capture energy from the sun, and the forest teems with life. On the flip side, relatively little energy is captured in an environment with few producers. In tundra or desert regions, for example, less food is available, and fewer animals can live there. These significant differences have prompted ecologists to categorize large areas of Earth's surface into distinct regions, called **biomes**, that are defined by their unique climatic and ecological features (**Figure 18.9**).

The **boreal forest** is the largest terrestrial biome. It is dominated by coniferous trees that grow in northern or high-altitude regions with cold, dry winters and mild summers. The soil is thin and nutrient-poor, so while plants generally receive adequate moisture during the growing season, plant diversity is relatively low. Large herbivores include elk and moose. Small carnivores, such as weasels, wolverines, and martens, are common. Larger carnivores include lynx and wolves.

The **tundra** biome is found at the poles and on mountaintops. Trees are absent or scarce because of the short growing season. The vegetation is dominated by low-growing flowering plants, and the boggy landscape is covered in mosses and lichens, important food sources of herbivores. Rodents provide food for carnivores like foxes and wolves. Bears and musk oxen are among the few large mammals.

The **chaparral** is a shrubland biome dominated by dense growths of drought-resistant plants in regions with cool, rainy winters and hot, dry summers. These conditions make the chaparral exceptionally susceptible to wildfires. Common vegetation in the California chaparral includes scrub oak, pines, mountain mahogany, manzanita, and the chemise bush. Rodents such as jackrabbits and gophers are common, and there are many species of lizards and snakes.

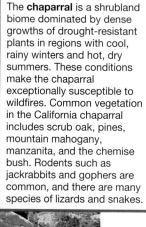

■	Tundra
■	Boreal forest
■	Temperate deciduous forest
■	Grassland
■	Chaparral
■	Desert
■	Tropical forest

Ecosystems within the **freshwater** biome are heavily influenced by the terrestrial biomes that they border or through which their water flows. Lakes are landlocked bodies of standing freshwater. Rivers are bodies of freshwater that move continuously in a single direction. Wetlands are characterized by standing water that is shallow enough for rooted plants to emerge above the water surface. A bog is a freshwater wetland with stagnant, oxygen-poor water, and low productivity and species diversity. In contrast, grassy marshes and tree-filled swamps are highly productive wetlands with a high diversity of organisms.

The **tropical forest** biome is characterized by warm temperatures, about 12 hours of daylight each day, and either seasonally heavy or year-round rains, which tend to leach nutrients from the soil. Soils in this biome further tend to be nutrient-poor because a large percentage of nutrients are locked up in the living tissues (biomass) of organisms. Tropical rainforests, which may receive in excess of 200 cm (80 inches) of rain annually, are some of the most productive ecosystems on Earth, with a rich diversity of organisms.

The **grassland** biome cannot sustain vigorous tree growth, but its moisture levels are not as low as in deserts. Grasslands are found in both temperate and tropical latitudes and are dominated by grasses and herbaceous plants such as coneflower and shooting star, although scattered trees are found in some, such as the tropical grasslands known as the savanna. Burrowing rodents like voles and prairie dogs may aerate the soil, thereby improving growing conditions. Many grasslands have been converted to agriculture.

The **temperate forest** biome is dominated by trees and shrubs adapted to relatively rich soil, snowy winters, and moist, warm summers. These forests display greater species diversity than do the tundra and boreal forest biomes: oak, maple, hickory, beech, and elm are common. Herbivores include squirrels, rabbits, deer, raccoons, and beavers, while bobcats, mountain lions, and bears make up the carnivores. Amphibians and reptiles are common.

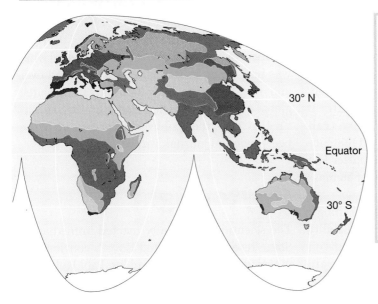

30° N

Equator

30° S

The **desert** biome makes up one-third of Earth's land surface. Because desert air lacks moisture, it does not retain heat well. As a result, temperatures can be above 45°C (113°F) in the daytime and then plunge to near freezing at night. Desert plants have small leaves to minimize water loss. Succulents, such as cacti, store water in their fleshy stems or leaves. Most animals in the desert are nocturnal, hiding in burrows during the heat of the day and emerging at night to feed.

Estuaries are the shallowest but most productive of the aquatic biomes. They are tidal ecosystems where rivers flow into the ocean. They have a constant ebb and flow of fresh and salt water, and organisms must be able to tolerate daily changes in salt water concentrations. The plentiful light, the abundant supply of nutrients delivered by the river system, and the regular stirring of nutrient-rich sediments by water flow create a rich and diverse community of photosynthesizers. Grasses and sedges are the dominant vegetation in most estuaries.

The **marine** biome, characterized by salt water, is the largest biome on our planet. The *coastal region* stretches from the shoreline to the edge of the continental shelf and is highly productive because of the availability of nutrients and oxygen. A majority of Earth's marine species live in the coastal region. The intertidal zone, closest to the shore, is a challenging environment where organisms such as seaweeds, worms, crabs, sea stars, sea anemones, and mussels are submerged and exposed to dry air on a twice-daily basis. The relatively nutrient-poor *open ocean* begins about 40 miles offshore and is much less productive than coastal waters.

Figure 18.9

Amazing biomes

Biomes do not begin and end abruptly, but rather transition into one another. Terrestrial biomes are categorized by temperature, precipitation, and altitude; aquatic biomes are determined by proximity to shorelines.

Assessing the overall amount of energy captured by producers is important in determining how an ecosystem works, because energy capture influences the amount of food available to other organisms. The **net primary productivity** (**NPP**) of an ecosystem is the energy, usually measured in units of carbon, acquired through photosynthesis over a particular time period that is available for the growth and reproduction of producers. NPP is the amount of energy captured by photosynthetic organisms minus the amount they expend on cellular respiration and other maintenance processes. NPP is typically estimated by measuring the amount of new biomass produced by the photosynthetic organisms in a given area during a specified period of time.

According to estimates, the NPP of all producers on Earth exceeds 100 billion tons of carbon biomass per year. Roughly half of this productivity comes from phytoplankton in the ocean. Therefore, phytoplankton produce roughly 50 billion tons of carbon per year. So if Boyce's calculations are right, a loss of 1 percent of that biomass is 500 million tons of organic matter lost from the oceans *each year*. That's a lot of biomass to lose.

Net primary productivity relies on four things: sunlight, water, temperature, and the availability of nutrients. The most productive ecosystems on land are tropical forests; the least productive are deserts and tundra (including some mountaintop communities). The most productive ecosystems in water are estuaries—regions where rivers empty into the sea—because nutrients drained off the land stimulate the growth and reproduction of phytoplankton and other producers, which in turn nourish large populations of consumers. The least productive aquatic biome is the deep ocean, where sunlight does not penetrate.

Despite similarities between the NPP requirements on land and in the ocean, the global pattern of NPP differs between the two. On land, the NPP is highest at the equator and decreases toward the poles. But in the ocean, the general pattern relates not to latitude but to distance from shore: the productivity of marine ecosystems is often high in ocean regions close to land and relatively low in the open ocean (**Figure 18.10**). This is because nutrients needed by aquatic photosynthetic organisms are in better supply near land, thanks to delivery from streams and rivers. Wetlands such as swamps and marshes, which

trap soil sediments rich in nutrients and organic matter, can be so productive that they match the productivity levels of tropical forests.

A loss of 500 million tons of phytoplankton each year could potentially affect the ocean's NPP and ocean life. "Almost all biological life in the ocean depends on phytoplankton. A reduction in the biomass of phytoplankton will result in less secondary production in the oceans," says Boyce. That means fewer sharks, whales, fish, eels—you name it.

The team's discovery was shocking, to say the least. No one else had documented a global decline in phytoplankton before. So, to be confident in their results, Boyce and Worm did several more rounds of data analysis, checking again and again to make sure they were using the right numbers in ways that correctly represented what was happening in the natural world. And over and over, they came back with the same results: global phytoplankton declined over the last century. In 2010, they published that finding in the peer-reviewed journal *Nature*.

Phyto-Fight

The scientific community reacted immediately. Some researchers doubted Boyce's conclusions; others were outright incredulous. Paul Falkowski at Rutgers University told a *New York Times* reporter that he had not found the same trend in a long-term analysis of the North Pacific (though Boyce contends that their trends were very similar), and another team had actually seen an increase starting around 1978 in the central North Pacific. Falkowski called Boyce's paper "provocative" but said he'd "wait another several years" to see whether satellite data would back up the finding.

Then, in 2011, three separate research teams published formal critiques of the work. One suggested that the declining trend was an error resulting from the use of two different types of measurements: Secchi disk readings and direct chlorophyll measurements. A second team echoed that idea, reanalyzed the data in a way that showed an increase in phytoplankton, and then bluntly concluded, "Our results indicate that much, if not all, of the century-long decline reported by [Boyce] is attributable to this [sampling bias] and not to a global decrease in phytoplankton biomass."

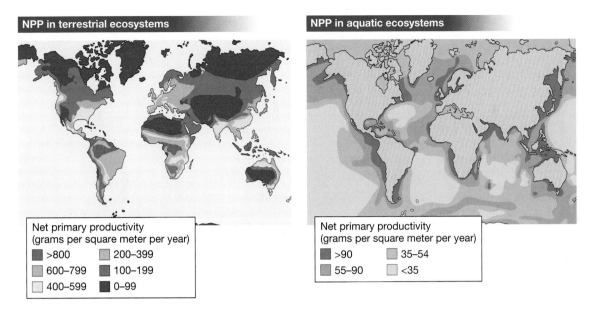

| NPP in terrestrial ecosystems | NPP in aquatic ecosystems |

Net primary productivity (grams per square meter per year)
- >800
- 600–799
- 400–599
- 200–399
- 100–199
- 0–99

Net primary productivity (grams per square meter per year)
- >90
- 55–90
- 35–54
- <35

Figure 18.10

Global variation in net primary productivity
NPP can be measured as the number of grams of new biomass made by producers each year in a square meter of each biome's area. NPP varies greatly across both terrestrial and aquatic ecosystems.

Q1: Which terrestrial biome is represented by the red color?

Q2: Which biome is represented by the blue color?

Q3: Which biomes in each panel have the lowest NPP?

The third team noted that Boyce's finding conflicted with eight decades of data on phytoplankton biomass collected by a large project called the Continuous Plankton Recorder (CPR) survey, which monitors the Northeast Atlantic Ocean. The CPR survey, started in 1931, employs a unique instrument pulled through the ocean by commercial fishing vessels to collect millions of samples of plankton. The CPR survey found that over the last 20–50 years, phytoplankton biomass increased in the Northeast Atlantic, says Abigail McQuatters-Gollop, a researcher at the Sir Alister Hardy Foundation for Ocean Science, which operates the survey.

After reading the critiques, Boyce, Worm, and Lewis went back to the data. First they applied a correction factor suggested by the critics, in the hopes of removing any bias between the two types of data. "We did that, and the trends remained similar," says Boyce. Next they again estimated changes over time individually for the two data sources. "That didn't change the trends either," says Boyce. Next, they incorporated additional suggestions by their peers and created a new, expanded database of chlorophyll measurements from which to work. Finally, they re-estimated changes in chlorophyll using this new database and their revised analysis methods, but the phytoplankton still seemed to be declining, independently of the type of data or how the data were analyzed. The three researchers published their reanalysis in a series of three papers in 2011,

ABIGAIL MCQUATTERS-GOLLOP

Abigail McQuatters-Gollop is a science and policy researcher at the Sir Alister Hardy Foundation for Ocean Science, home of the Continuous Plankton Recorder survey, a study that has monitored plankton in the North Atlantic and North Sea since 1931.

Productive Plants

Data on net primary productivity (NPP)—the total energy available in an ecosystem for the growth and reproduction of primary producers—is sparse for whole biomes, but researchers have been able to estimate the NPP of those listed here using information about different vegetation types and carbon sources in each area.

Net primary productivity for selected biomes

(in grams per square meter per year)

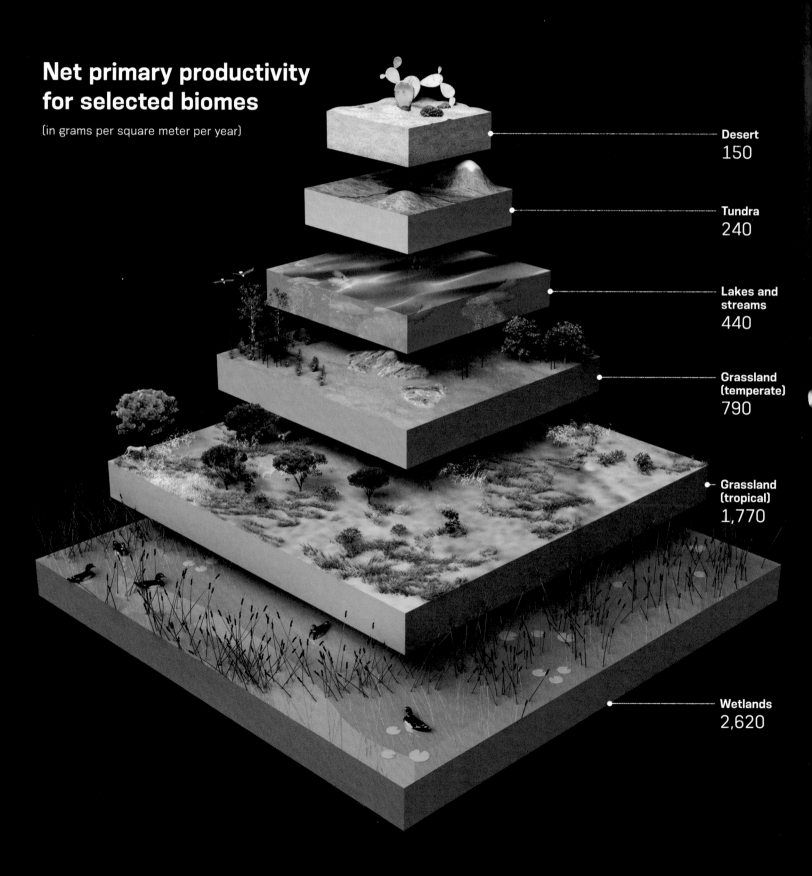

Desert
150

Tundra
240

Lakes and streams
440

Grassland (temperate)
790

Grassland (tropical)
1,770

Wetlands
2,620

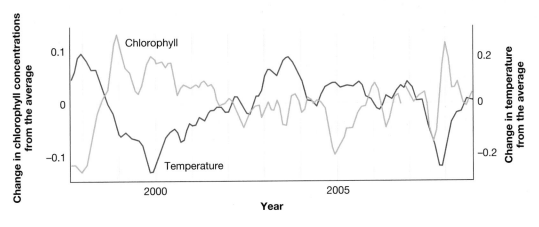

Figure 18.11

As ocean temperature increases, chlorophyll decreases

Between late 1997 and mid-2008, satellites observed that warmer-than-average temperatures (red line) were correlated with below-average chlorophyll concentrations (green line).

Q1: In what years were chlorophyll levels the highest?

Q2: In what years were the temperature changes from the average the greatest?

Q3: Are there any points where a temperature increase did not cause a chlorophyll decrease?

2012, and 2014, demonstrating the same decline again and again.

But the additional work has not silenced the critics. "It's still pretty hotly debated," admits Boyce. "The story is not over by any means." In 2011, in fact, Worm traveled to an international plankton conference where he, McQuatters-Gollop, and others debated the topic in front of a live audience. "It was an amicable meeting that generated loads of discussion," says McQuatters-Gollop. At the meeting, the researchers agreed that the best thing to do was to combine as many data sets as possible. Still, they could not agree on whether phytoplankton populations have increased or decreased in the ocean.

Heating Up

If Boyce, Worm, and Lewis are right—that a phytoplankton decline is occurring—the finding raises the important question of why. As part of his research, Boyce investigated possible causes of the downward trend. In one study, Boyce compared changes in sea surface temperature to changes in chlorophyll levels, and he noted a strong correlation: over the last 100

years, chlorophyll concentrations declined *and* ocean temperatures increased. This correlation has been closely followed in recent years (**Figure 18.11**).

But correlation does not prove causation, and much more work needs to be done to support the idea that global warming caused the decline. It is a logical hypothesis, however, because as the planet warms, water in the oceans mixes less, limiting the nutrients delivered to the surface from decomposers in the deep sea. As a result, phytoplankton do not receive the nutrients they need for growth and reproduction. In 2014, Boyce published experimental data supporting the hypothesis. Working with marine scientists in Germany, he found that warming the water in a controlled, experimental ocean water enclosure led to reduced phytoplankton biomass.

Another pressing question is how a phytoplankton decline affects the planet. Consequences could include altering the carbon cycle between the ocean and the atmosphere, changing heat distribution in the ocean, and causing a decrease in the supply of food in the ocean. Whatever the case, if phytoplankton populations are decreasing, says Boyce—and he is sure they are—there will be profound effects.

REVIEWING THE SCIENCE

- An **ecosystem** consists of communities of organisms and the physical environment in which those communities live. It is the sum of all **biotic** factors interacting with **abiotic** factors. Energy, materials, and organisms can move from one ecosystem to another.

- An **energy pyramid** represents the amount of energy available to each **trophic level** of a food chain in an ecosystem.

- Energy enters an ecosystem when producers capture it from an external source, such as the sun. A portion of the energy captured by producers is lost as **metabolic heat** at each trophic level. As a result, energy flows in only one direction through ecosystems.

- **Nutrients** are the chemical elements required by living organisms. Unlike energy, nutrients are recycled and reused within and across ecosystems. Earth has a fixed amount of nutrients.

- **Decomposers** break down the dead bodies of other organisms, both consumers and producers. Without decomposers, nutrients would remain locked up in the bodies of dead organisms.

- Four **ecosystem processes** link the biotic and abiotic worlds in an ecosystem: nutrient cycling, energy flow, water cycling, and succession.

- Ecosystems depend on **energy capture**, the trapping of solar energy by producers via photosynthesis and its storage as chemical compounds, such as carbohydrates, in their bodies.

- Earth is categorized into ten **biomes**, regions defined by their climatic and ecological features.

- The **net primary productivity (NPP)** is the energy acquired through photosynthesis that is available for growth and reproduction to producers in an ecosystem. NPP is determined by the amount of biomass produced in a given area during a specified period of time.

THE QUESTIONS

The Basics

1 The movement of nutrients between organisms and the physical environment is called

(a) nutrient cycling.

(b) ecosystem services.

(c) net primary productivity.

(d) decomposition.

2 How much energy is transferred up the energy pyramid from one trophic level to the next?

(a) 90 percent

(b) 50 percent

(c) 10 percent

(d) levels vary from 10 to 50 percent

3 How much of the energy consumed by organisms is used for survival and reproduction?

(a) 90%

(b) 10%

(c) 50%

(d) 80%

4 Which organisms are considered the "recyclers" of our planet?

(a) consumers

(b) producers

(c) phytoplankton

(d) decomposers

5 The terrestrial biome that receives the most consistent year-round rainfall is

(a) wetland.

(b) boreal forest.

(c) tropical forest.

(d) chaparral.

Try Something New

6 I am the biome that is characterized by shrubs and nonwoody plants that grow in regions with cool, rainy winters and hot, dry summers. Who am I?

(a) wetland

(b) boreal forest

(c) tropical forest

(d) chaparral

7 The pika is a small mammal that is adapted to cold, high altitude habitats above the tree-line. If its mountainous habitats were destroyed, where would the next best biome be for this densely-furred critter?

(a) desert

(b) rain forest

(c) tundra

(d) boreal forest

8 In an energy pyramid, an owl, a cardinal, and a grasshopper are the fourth, third, and second trophic levels, respectively. If the grasshopper passes 100 Calories to the cardinal when eaten, how many grasshoppers would the cardinal have to eat to obtain 10,000 Calories?

(a) 10

(b) 100

(c) 1,000

(d) 10,000

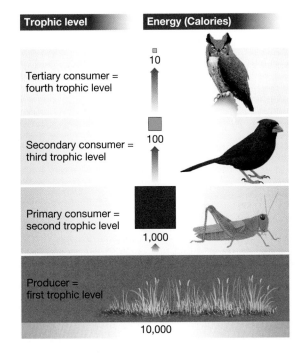

Trophic level	Energy (Calories)
Tertiary consumer = fourth trophic level	10
Secondary consumer = third trophic level	100
Primary consumer = second trophic level	1,000
Producer = first trophic level	10,000

9 In the energy pyramid described in question 8, how many cardinals would the owl have to eat to obtain 10,000 Calories?

(a) 10

(b) 100

(c) 1,000

(d) 10,000

10 While sloshing around in the swampy wetlands of a nearby forest preserve, a small child notices a rotting tree branch covered in fungi. What are these organisms doing on this tree branch?

(a) These are producers acquiring energy through photosynthesis.

(b) These are consumers acquiring energy from the wood and releasing nutrients back to the earth.

(c) These are decomposers acquiring energy from the wood and releasing nutrients back to the earth.

(d) These are decomposers releasing nutrients back to the earth, but acquiring no energy for themselves in the process.

Leveling Up

11 *Write Now* **biology: human-caused biome shifts** The location of Earth's different biomes depends on climate and altitude, for the most part. However, human activities play a role in the conversion of one biome to another, as has been seen many times in history. Research one major change in a biome category based on human activity, and describe how and why this change happened. Speculate on how this change could have been specifically avoided. (*Note:* Do not analyze a change via deforestation to agricultural land, since agricultural land is not a natural biome. *Hint:* Take a look at Easter Island as one example.)

12 **Doing science** Join forces with millions of others classifying phytoplankton on your computer. Do an Internet search on "citizen science phytoplankton" and sign in as a citizen scientist, complete the tutorial, and start helping researchers quantify the phytoplankton in our oceans.

13 **What do *you* think?** Some people think the current U.S. Endangered Species Act should be replaced with a law designed to protect ecosystems, not species. The intent of such a law would be to focus conservation efforts on what its advocates think really matters in nature: whole ecosystems. Given how ecosystems are defined, do you think it would be easy or hard to determine the boundaries of what should and should not be protected if such a law were enacted? Give reasons for your answer.

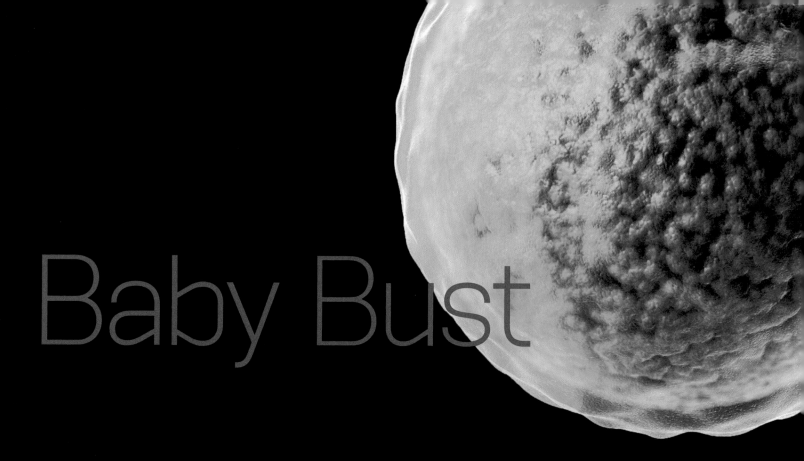

Baby Bust

*Facing dwindling births, Denmark searches
to resolve problems of infertility.*

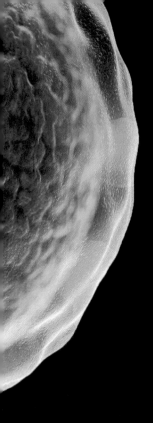

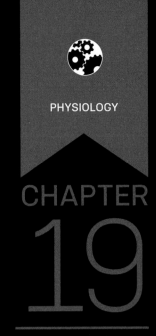

CHAPTER

19

HOMEOSTASIS,
REPRODUCTION,
AND
DEVELOPMENT

A woman slides a bra strap off her naked shoulder. In Danish, a seductive man's voice asks, "Can sex save Denmark's future?"

The commercial, which aired on Danish television, then switches to a view of an empty playground. The commercial, produced by a travel agency, encourages responsible Danes to book a romantic holiday with the company's "ovulation discount" and try to conceive while on vacation. The ad concludes with a large banner proclaiming, "Do it for Denmark" (**Figure 19.1**).

Birth rates reached a 27-year low in the year 2014 in Denmark, a Nordic country in northern Europe. When asked, most Danish couples said they would like to have two or three children, yet the present fertility rate is only 1.73 children per family—not high enough to maintain Denmark's current population. Infertility is now considered an epidemic in the country. In fact, one in 10 children in Denmark is conceived using reproductive technologies. "We see more and more couples needing to get assisted fertility treatment," Bjarne Christensen, secretary general of Sex and Society, Denmark's leading family-planning association, told *Bloomberg News* in October of 2014. "We see a lot of people who don't succeed in having children."

Commercials like the one described here make a patriotic, if whimsical, appeal to Danish citizens to have more children, but experts debate whether falling fertility rates are a cause for concern or celebration. Some applaud the decrease as a way to slow global population growth and reduce human consumption of limited natural resources. (See Chapter 16 for more on population growth.) But others in countries facing declining fertility rates—including Denmark, France, the United Kingdom, and recently the United States—worry about the burdens that will be placed on a generation smaller than the one before it (**Figure 19.2**). Government officials fear that this trend will produce a smaller workforce and fewer young people to care for retirees, reducing their countries' economic strength.

To counteract falling birth rates, some governments have begun taking pro-fertility stances, including providing free postnatal care and subsidized day care. In Denmark, Sex and Society, a nongovernmental organization, provides sex education materials for most of Denmark's schools, and it recently unveiled a new series of lesson plans entitled, "This is how you have children!" Instead of focusing on contraception and how to avoid becoming pregnant, these new classes educate students about what fertility is, how aging affects fertility, and when the best times to have children may be.

"We have for many years addressed the very important issues of how to avoid becoming pregnant, how to avoid sexual diseases, how they have a right to their own bodies, but we totally forgot

Figure 19.1

A print advertisement in the "Do It for Denmark" campaign

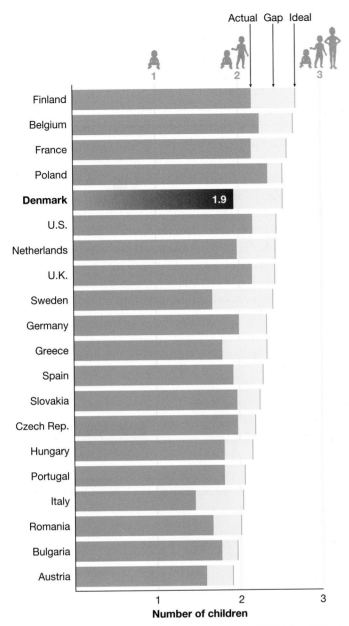

Actual Gap Ideal

1   2   3

Finland
Belgium
France
Poland
Denmark — 1.9
U.S.
Netherlands
U.K.
Sweden
Germany
Greece
Spain
Slovakia
Czech Rep.
Hungary
Portugal
Italy
Romania
Bulgaria
Austria

1   2   3
Number of children

Sources: European data from 2011 Eurobarometer, U.S. data from 2006 and 2008 General Social Survey

Figure 19.2

Women in Europe and the United States want more children than they have

U.S. and European women aged 40–54 were asked, "What do you think is the ideal number of children for a family to have?" Their answers were consistently higher than actual birth rates.

Q1: In which of the countries shown in the graph did women want the most children, on average? The least?

Q2: Where is the largest gap between the number of children women would like to have and the number they actually have?

Q3: Which country had the highest average number of children? The lowest?

to tell the kids that we cannot have children forever," Søren Ziebe, head of Copenhagen University Hospital's fertility clinic, told reporters. "There is a biological limit."

Seeking Stability

That biological limit may be the cause of Denmark's falling birth rate. Population studies show that, on average, couples are waiting longer to have children, and conception can be more difficult after age 30.

But other causes are also possible: Laboratories are investigating whether chemicals or other environmental factors might be affecting the human **reproductive system**, the parts of the body responsible for reproduction.

Our bodies are highly efficient and well-coordinated communities of 200 different specialized cell types. **Anatomy** is the study of the structures that make up a complex multicellular body, and **physiology** is the science that focuses on the functions of anatomical structures. Through physiological research, scientists hope to determine why fertility rates are falling in Denmark, and whether anything can be done about it.

Their first step has been to examine the reproductive system, which is made up of cells, tissues, and organs. **Tissues** are made up of cells that act in an integrated manner to perform a common set of functions (**Figure 19.3**). A tissue may be composed of just one cell type, or it may contain multiple cell types; in either case, the cells that compose a tissue cooperate to perform the distinctive functions of that tissue. An **organ** has more than one tissue type and forms a functional unit with a distinctive shape and location in the body. An **organ system** is composed of two or more organs that work in a closely coordinated manner to perform a distinct set of functions in the body.

The human body has 11 major organ systems, including the reproductive system (**Figure 19.4**). Each system will be covered in more detail in the following chapters, along with plant organ systems in Chapter 23. Each organ system and

its organs are unique in function and form—from the beating, blood-filled heart of the circulatory system to the electrical, thread-like nerves of the nervous system—but they share one important commonality: For proper function, an organ system requires a stable internal environment.

Most biological processes take place only within a certain temperature range, with the right amount of water, at the appropriate pH, and at a particular concentration of chemicals. In the reproductive system, for example, the male testes need to maintain a temperature about 1–2 degrees lower than the usual body temperature of 98.6°F (37°C) for the successful production and storage of sperm. And the female vagina requires a pH between about 3.8 and 4.5 to keep out bacteria while still maintaining a healthy environment for fertility.

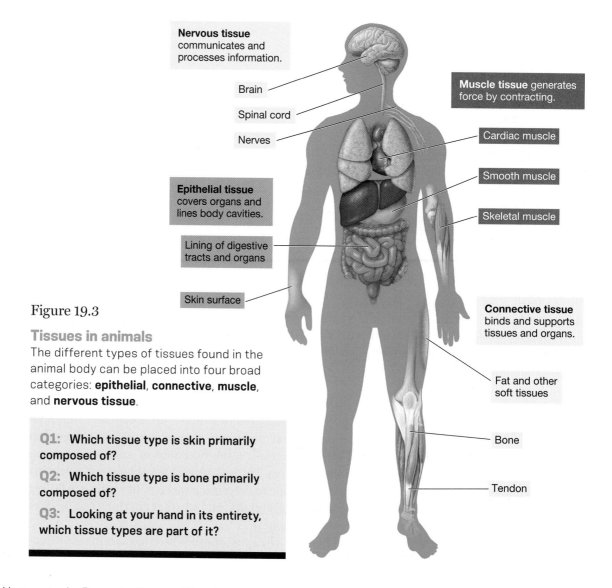

Figure 19.3

Tissues in animals

The different types of tissues found in the animal body can be placed into four broad categories: **epithelial**, **connective**, **muscle**, and **nervous tissue**.

Q1: Which tissue type is skin primarily composed of?

Q2: Which tissue type is bone primarily composed of?

Q3: Looking at your hand in its entirety, which tissue types are part of it?

The **urinary system** removes excess fluid from the body, along with waste products and toxins.

In the **digestive system**, large molecules of food are broken down in the mouth, stomach, and small intestine, and nutrients are absorbed in the small and large intestines.

The **circulatory system** whisks oxygen from the lungs to the heart, which then pumps oxygen-rich blood to the rest of the body through a closed network of vessels.

The **respiratory system** brings in oxygen and expels carbon dioxide through the lungs.

The **endocrine system** works closely with the nervous system to regulate all other organ systems. It consists of a number of glands and secretory tissue.

The **integumentary system** is the largest organ system in the human body, covering and protecting the surface of the body.

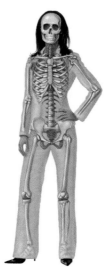

The **nervous system** is a key player in sensing the external world and the body's internal state, and it communicates with all of the other organ systems.

The **skeletal system** provides an internal framework to support the body of vertebrates. It consists of bone, cartilage, and ligaments.

The **muscular system** produces the force that moves structures within the body. It works closely with the skeletal system.

The **immune system** defends the body from invaders such as viruses, bacteria, and fungi.

The **reproductive system** generates gametes and, depending on the animal group, may also support fertilization and prenatal development.

Figure 19.4

Organ systems

The 11 major organ systems of the human body work in an integrated manner.

Q1: Which organ system defends the body from infectious diseases such as the common cold or flu?

Q2: Which organ systems transport oxygen to cells?

Q3: Which organ systems regulate the activities of the other organ systems?

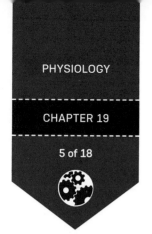

These environments are regulated through **homeostasis**, the process of maintaining a relatively constant internal state despite changes in the external environment. Homeostasis enables an organism to continually sense its internal state and rapidly adjust. In this way, despite large fluctuations in the outside world, homeostasis maintains the internal conditions best suited for life processes (**Figure 19.5**).

Homeostasis occurs via **homeostatic pathways**, sequences of steps that reestablish homeostasis if there is any departure from the genetically determined normal state (also called the **set point**). Homeostatic pathways continually monitor the physical and chemical characteristics of the internal environment and trigger regulatory processes within the body if this monitoring system detects any deviation from the set point.

Homeostatic pathways depend on **feedback loops**. A **negative feedback** loop turns off or reduces the output of a process. For example, if a person drinks a large milk shake, the level of glucose in the blood rises. In response, cells in the pancreas produce insulin, which allows glucose

A structure in the brain launches homeostatic pathways if core body temperature deviates from the set point (about 37°C).

Temperature receptors in the skin function as sensors in the human body. If body temperature deviates from its set point, regulatory processes are activated.

Skin receptors say body temperature is too high.

Skin receptors say body temperature is too low.

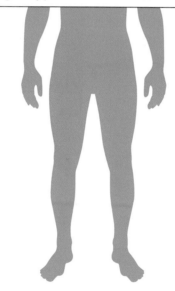

If body temperature is too high, regulatory processes are activated that bring it down to its set point. Increased sweating and blood flow to the skin release heat to the environment.

If body temperature is too low, regulatory processes are activated that bring it up to its set point. Decreased blood flow to the skin reduces heat loss to the environment, and shivering generates extra metabolic heat.

Figure 19.5

Homeostasis maintains stable internal conditions

Because of homeostatic pathways, large fluctuations in external conditions produce little or no change in the overall state within the animal body. For example, even when the outside temperature is very hot or cold, the human body's internal temperature stays within the narrow range required for survival.

Q1: Why is it important to maintain a stable body temperature?

Q2: Which organ system is involved as a temperature sensor? (See Figure 19.4 for an overview of organ systems.)

Q3: Give another example of a homeostatic pathway in humans.

to enter cells. Thus, blood glucose concentration declines.

A **positive feedback** loop, on the other hand, increases the output of a process. Blood clotting triggered by a broken blood vessel, for example, is a positive feedback loop: The process of clotting releases chemicals that lead to accelerated clotting, continuing until the clot plugs the break in the blood vessel wall. In this feedback loop, the chemicals increase the amount of clotting.

Overall, negative feedback loops are more common than positive feedback loops in homeostatic pathways. Examples of homeostatic pathways include *thermoregulation* (the control of heat gain and loss) and *osmoregulation* (the control of internal water content and solute concentration by an organism). Homeostasis is critical to the maintenance of our organ systems, including the reproductive system—which brings us back to doing it in Denmark.

All in the Timing

Danish researchers have long suspected that the primary reason for declining fertility rates is that women are waiting longer to have children. In the 1970s, on average, a Danish woman gave birth to her first child at 24 years old. But in 2014, that average age was 29, with more and more women waiting until they were over 35 to have a first child. And the older a woman gets, the more difficult it is for her to conceive.

Humans reproduce through **sexual reproduction**, in which haploid gametes from a male (the **sperm**) and a female (the **egg**, or ovum) combine to form a diploid **zygote**, which develops into a multicellular individual that is genetically unique and different from either parent. Here, we will primarily review human sexual reproduction, but keep in mind that sex in other animals is more variable than our human perspective might lead us to expect (**Figure 19.6**). In particular, recall that other animals can also reproduce via **asexual reproduction**, in which cells from only one individual produce the offspring, so all of the offspring's genes come from that parent. (For more on the benefits and pitfalls of sexual and asexual reproduction, see "Why Sex?" on page 205 of Chapter 12.)

Human eggs develop through a series of cell divisions called **oogenesis**, the production of mature eggs capable of being fertilized (**Figure 19.7**, right). Oogenesis begins before birth,

Sea stars can reproduce by breaking off an arm, which then regenerates into a new individual. Although some animals rely exclusively on asexual reproduction, most asexually reproducing species switch between sexual and asexual reproduction, depending on environmental conditions.

In frogs and other species, females release their eggs, which males then cover with sperm. External fertilization is common among aquatic animals; internal fertilization is more common among land animals.

Clownfish begin life as males, but the largest fish in a group will change to female. Other species begin as females and may then switch to males, while still other species are both male and female at the same time. Individuals that produce both functional testes and functional ovaries—and are therefore both male and female—are called **hermaphrodites**.

Figure 19.6

Animals display a rich variety of reproductive systems
Some animals reproduce by cloning themselves through asexual reproduction (left). Others fertilize gametes externally (middle), while others still are both male and female either simultaneously or sequentially (right).

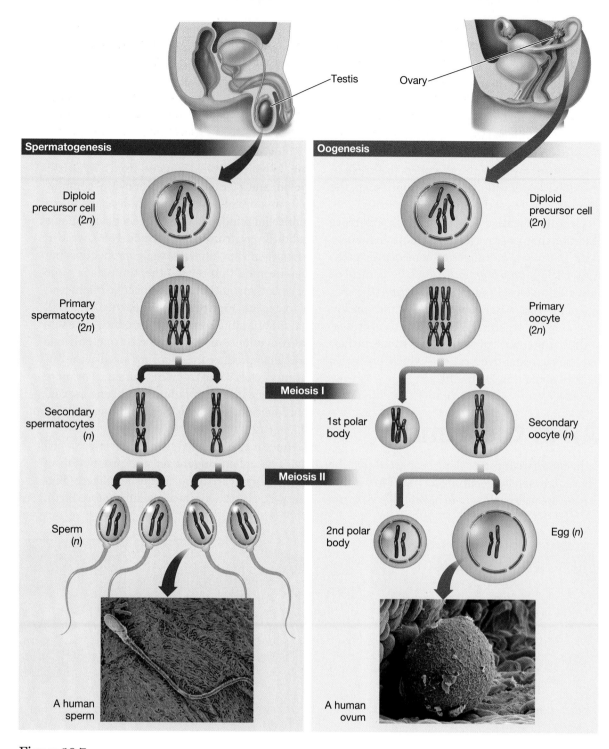

Testis

Ovary

Spermatogenesis

Diploid
precursor cell
(2n)

Primary
spermatocyte
(2n)

Secondary
spermatocytes
(n)

Sperm
(n)

A human
sperm

Oogenesis

Diploid
precursor cell
(2n)

Primary
oocyte
(2n)

Meiosis I

1st polar
body

Secondary
oocyte (n)

Meiosis II

2nd polar
body

Egg (n)

A human
ovum

Figure 19.7

Sexual reproduction requires the production of haploid gametes
Spermatogenesis produces haploid sperm, and oogenesis produces haploid eggs (ova).

Q1: Identify one way in which spermatogenesis and oogenesis are the same.

Q2: How many sperm are produced from each precursor cell? How many eggs are produced from each precursor cell?

Q3: How much time elapses between the appearance of a precursor cell and the formation of a sperm? How does this process differ for an egg? (You will need to read ahead to answer this question.)

when germ line cells multiply and develop into immature, diploid egg cells called **primary oocytes**. At birth, the ovaries of a female already contain her entire lifetime supply of primary oocytes, about 1–2 million cells. These cells remain in a state of suspended development until the production of hormones at puberty stimulates one, or occasionally two, of them to mature each month in preparation for ovulation. By the time she reaches puberty, at about 10–12 years of age, approximately 400,000 viable primary oocytes remain—still more than she will use in her lifetime.

Human females do not produce mature eggs continuously. Instead, individual eggs mature and are released in a hormone-driven sequence of events known as the **menstrual cycle** (**Figure 19.8**). The menstrual cycle averages about 28 days, but cycle lengths from 21 to 35 days are considered normal.

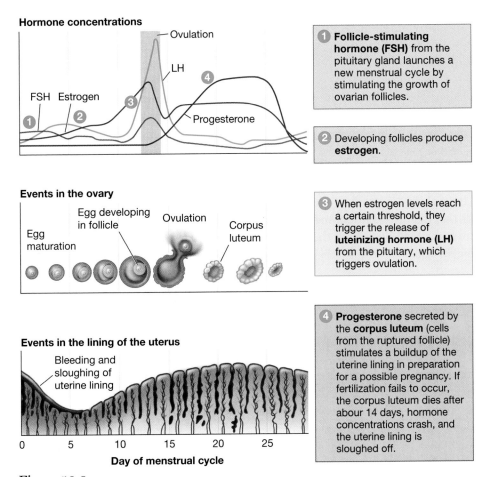

Hormone concentrations

Ovulation
LH
4
FSH Estrogen
3
1 2
Progesterone

① **Follicle-stimulating hormone (FSH)** from the pituitary gland launches a new menstrual cycle by stimulating the growth of ovarian follicles.

② Developing follicles produce **estrogen**.

Events in the ovary

Egg developing in follicle   Ovulation
Egg maturation   Corpus luteum

③ When estrogen levels reach a certain threshold, they trigger the release of **luteinizing hormone (LH)** from the pituitary, which triggers ovulation.

Events in the lining of the uterus

Bleeding and sloughing of uterine lining

0   5   10   15   20   25
Day of menstrual cycle

④ **Progesterone** secreted by the **corpus luteum** (cells from the ruptured follicle) stimulates a buildup of the uterine lining in preparation for a possible pregnancy. If fertilization fails to occur, the corpus luteum dies after abour 14 days, hormone concentrations crash, and the uterine lining is sloughed off.

Figure 19.8

The human menstrual cycle

A menstrual cycle begins with the first day of bleeding, which marks the end of the previous cycle. Over the next few weeks, a succession of hormones stimulates the release of an egg and signals the uterine lining to grow and thicken in preparation for a potential pregnancy. If pregnancy does not occur, hormone levels plummet and the lining is sloughed off as menstrual flow, ending that menstrual cycle.

Q1: Which hormones important for the menstrual cycle are produced in the pituitary gland?

Q2: Which hormone is involved in producing the uterine lining?

Q3: How is the egg follicle involved in producing hormones?

A woman has more primary oocytes than she will use in her lifetime, but evidence suggests that those eggs decline in quality as a woman ages—a conclusion supported by the increased risk of birth defects in children born to older mothers. In addition, if eggs from younger women are implanted into women over 40, the pregnancy rate equals the rate associated with the younger women who donated the eggs: In other words, young eggs result in young pregnancy rates.

When a woman passes 40 years of age, she shows a clear drop in her ability to produce normal eggs and bear children. Human females reach menopause—the end of their reproductive lives—around the age of 50.

A decline in egg quality is one suggested cause of Denmark's fertility problem: Women today are waiting longer to have children, so their eggs are older and their fertility has decreased. Age also affects the fertility of men, although males do not undergo the clearly identifiable menopause that is characteristic of females. Instead, their sex drive and their ability to produce sperm slowly decrease as they age. In addition, men produce fewer sperm as they age, decreasing their chances of fertilizing an egg.

Spotlight on Sperm

While parents' increasing age at conception likely contributes to Denmark's fertility problem, researchers suspect that other factors are also at play—particularly, the quality of male sperm. The production of male gametes and the production of female gametes differ in several important ways. The supply of a female's primary oocytes is limited, and once a primary oocyte develops into a mature ovum, it is lost from the supply. In contrast, sperm precursor cells in males constantly

replenish the pool of sperm. It is also noteworthy that in a normal menstrual cycle, a human female produces only one mature egg cell, while a male produces hundreds of millions of sperm cells every day.

Another difference is that ova are typically much larger than sperm. The human egg is visible (just barely) to the naked eye, but individual sperm can only be seen under a microscope. Sperm contain little substance beyond chromosomes and the cellular machinery needed to move up the female reproductive tract, attach to an egg, and propel the sperm's chromosomes into the egg's cytoplasm. Sperm are simple packages with valuable information inside. An ova, on the other hand, is a plump, complex cell full of organelles.

The first reports of falling male sperm counts came in 1992 from Niels Skakkebæk at the University of Copenhagen. Looking at 61 different papers describing the semen quality of almost 15,000 men—both the density and volume of sperm in semen—Skakkebæk and his team found that sperm counts around the world had dropped by 50 percent from about 1940 to 1990. In a peer-reviewed paper, they concluded, "As male fertility is to some extent correlated with sperm count, the results may reflect an overall reduction in male fertility." In other words, a decrease in the quality of semen could be lowering male fertility.

In human males, meiosis occurs inside structures called *seminiferous tubules*: twisty, spaghetti-like tubes that permeate the **testes**. In response to male hormones that surge at the onset of puberty, diploid germ line cells in the tubules start dividing to form sperm in a sequence of steps known as **spermatogenesis** (Figure 19.7, left). The average man produces about 300 million sperm each day. Surplus sperm that accumulate over time are degraded and reabsorbed by the cells that line the tubules.

Skakkebæk's 1992 study of declining semen quality sparked a lot of interest. Hundreds of studies followed, as researchers sought the cause for the decline. Many suspected an environmental cause, such as a toxin.

"We all wanted to study it in more detail," says Tina Kold Jensen, a researcher at the University of Southern Denmark. But instead of looking at past studies, Jensen and her collaborators decided it was necessary to track these trends in real time. "We decided if we wanted to get closer to the truth, we had to collect our own data."

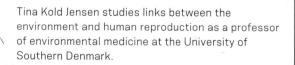

TINA KOLD JENSEN

Tina Kold Jensen studies links between the environment and human reproduction as a professor of environmental medicine at the University of Southern Denmark.

Preventing Pregnancy

The Food and Drug Administration approved the first birth control pill for sale in the United States in 1960. It became immediately popular as a way for women to control their fertility, and has been said to have contributed to a sharp increase in college attendance and graduation rates for women. Here is an overview of some of today's most popular forms of birth control, with data on just how effective each form is.

Pregnancies per 100 women in a year

Intrauterine Device
T-shaped plastic device must be positioned inside the uterus by a health care professional.

○ *Hormonal*

○ *Non-hormonal*

.2
.8

.5
.15

○ **Female Sterilization**
The oviducts are sealed with clamps or by other surgical means.

○ **Male Sterilization**
The tubes that carry sperm are sealed surgically by a health care professional.

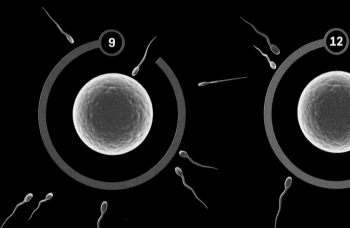

○ **The Pill**
Hormones suppress ovulation, preventing the release of an egg, by mimicking the hormonal levels of pregnancy.

9

12

○ **Diaphragm**
Dome-shaped latex cup filled with spermicide, inserted before intercourse, covers the cervix and keeps sperm out of the uterus.

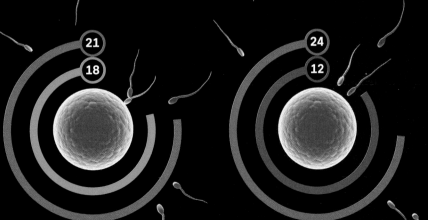

○ **Female Condom**
Plastic pouch, inserted before intercourse, lines the vagina to prevent sperm from entering.

○ **Male Condom**
Plastic or latex pouch covers the penis to keep sperm from entering the vagina.

21
18

24
12

Sponge
A sponge containing spermicide is inserted deep in the vagina prior to intercourse.

○ *Women who have given birth before*

○ *Women who have never given birth*

Jensen partnered with Skakkebæk and Niels Jørgensen at Copenhagen University Hospital. Starting around 1995, the scientists began recruiting young male volunteers at government-required physical exams, asking each to provide a sperm sample. By 2010 the team had amassed sperm from over 5,000 volunteers. When analyzed, the data seemed to contradict Skakkebæk's initial finding: Over the 15 years of the study, semen quality did not decline—in fact, it improved slightly. Sperm concentrations rose from 43 million per milliliter in 1996–2000 to 48 million per milliliter in 2006–10.

Still, the rise was not enough to suggest that fertility improved during that period. "Although we see a slight rise, only 23 percent of the young men had optimal semen quality," said Jørgensen in a statement when the study was released. "In fact, the semen quality of 27 percent of the men was so poor, it will probably take these men longer to make their partner pregnant," he added. "Furthermore, for 15 percent of the men, semen quality was so poor, they are likely to need fertility treatment in order to conceive."

Driven by Hormones

Jørgensen and Jensen have gone on to study the factors that might have caused such low levels of semen quality. In May of 2014, their team found that 98 percent of 308 young men had detectable urinary levels of *bisphenol A* (BPA), a chemical found in plastics that disrupts the body's endocrine system (see Chapter 5 for more on BPA). Men with higher BPA levels, they discovered, also had higher levels of testosterone and other hormones. That correlation suggests that BPA could be affecting hormone feedback loops, but

additional research is needed to identify possible mechanisms for such an effect.

Hormones regulate nearly all aspects of reproduction in animals, from mating behaviors to the development and birth of offspring. The emergence of sex-specific characteristics in the fetus and the maturation of reproductive organs during puberty are examples of long-term aspects of reproduction that are controlled by hormones. The regular stimulation of sperm production in males and the monthly cycle of menstruation in females are also regulated by hormones (see Figure 19.8 for the role of hormones in the menstrual cycle, and Chapter 22 for more on hormones).

Testes and ovaries produce three major types of hormones: estrogens, progestogens, and androgens. Both males and females produce all three, but in different ratios; for example, males have more androgens than estrogens, and females have more estrogens than androgens. **Estrogens** play a role in determining female characteristics such as wide hips, a voice that is pitched higher than that of males, and the development of breast tissues. The primary estrogen is **estradiol**. **Progestogens** have a number of functions in the female body, including thickening the lining of the uterus and increasing the blood supply to it to create a suitable environment for a developing fetus. **Progesterone** is the most important of the progestogens. **Androgens** stimulate cells to develop characteristics of maleness, such as beard growth and the production of sperm. The primary androgen is **testosterone**.

Testosterone, together with another closely related androgen, directs the development of internal reproductive structures such as the sperm ducts and prostate gland. A third androgen directs the development of external structures such as the penis.

In addition to uncovering the link between BPA and hormone levels, the Danish research team recently found evidence that regular alcohol consumption may affect semen quality, possibly by changing testosterone levels. In particular, large amounts of alcohol significantly lowered semen quality: Men who consumed 40 or more drinks in a week had a 33 percent reduction in sperm as compared with men who drank just 1–5 drinks per week. In their conclusion, the authors went so far as to warn young men to "avoid habitual alcohol intake."

NIELS JØRGENSEN

Niels Jørgensen is a member of the Department of Growth and Reproduction at Copenhagen University Hospital in Denmark. He studies male infertility.

"Do It for Denmark"

Jørgensen, Jensen, and others continue to seek the causes for declining semen quality, while demographers track social reasons for decreasing fertility rates, such as the increasing age of mothers. "Of course there are a lot of factors involved, including social factors," says Jensen. "But it's not all due to social factors. When we talk about fertility rates, we need to think about biology as well."

In the meantime, as we saw earlier, Danish officials are working to encourage young people to reproduce. After years of teaching how pregnancy can be prevented through abstinence—refraining from sexual intercourse—and **contraceptives** such as the Pill, teachers will now teach how fertilization occurs (**Figure 19.9**).

A woman releases an egg from her ovaries, or *ovulates*, about once every 28 days. The egg moves down the **oviduct** or *fallopian tube*, a four-inch-long tube that connects the **ovary** to the **uterus**. If the woman has sexual intercourse, the man's **penis** ejaculates almost 300 million sperm into her **vagina**. The sperm swim from the vagina into the uterus through an opening called the **cervix**, and then up the oviduct in response to a chemical signal released by the ovary.

Only a few hundred sperm manage to reach the egg in the oviduct, and only one of those lucky sperm fertilizes the egg to create a zygote. Although both parents contribute equally to the genetic material of the zygote, its organelles and other cellular machinery come almost entirely from the female.

Human development in the uterus averages about 38 weeks and is divided into three stages known as **trimesters**, each about three months long (**Figure 19.10**). During the first trimester, the zygote develops from a single cell into an **embryo** possessing all the main tissue types. All organ systems are established by the third month, and the developing individual is now known as a **fetus**. During the next three months, the second trimester, the organs develop and the fetus increases in size. By the start of the third trimester, fetal development has progressed to the point that, with the help of modern technology,

the fetus has reasonably good odds of surviving outside the mother's body. It gains a good deal of weight during the third trimester, and its circulatory and respiratory systems prepare for living in a gaseous atmosphere rather than the watery world of the amniotic fluid.

By the end of the third trimester, the fetus is ready for its sudden transition from the uterus to the outside world—childbirth (**Figure 19.11**). The last few weeks of pregnancy are marked by hormonal changes. Specifically, higher estrogen levels in the mother's blood make the muscles of her uterus more sensitive to **oxytocin**, a hormone secreted by the fetus and, later in the birth process, by the mother's pituitary gland. Oxytocin stimulates the uterine muscles and causes the placenta to secrete prostaglandins, which reinforce the contractions. Labor begins when the muscles of the uterus begin to contract in response to these hormones.

In a positive feedback loop, more contractions cause the production of more oxytocin, and the strength of the contractions increases as more oxytocin is produced. The cervix begins to open, and the increasingly strong contractions eventually push the fetus out of the mother's body. At this point the positive feedback ends, and the contractions subside as oxytocin levels decrease. The placenta, often referred to as the "afterbirth," is expelled during the last stage of childbirth.

At birth, a baby becomes physically independent of its mother. It no longer obtains its oxygen and nutrients directly from her blood, and it must eat, breathe, and maintain homeostasis on its own. Development does not end when an animal is born. Humans spend about a quarter of their lives reaching full adult size. Most of this growth occurs during childhood, before sexual maturity.

Today, all signs point to birth rates continuing to fall in developed nations, including the United States. In May of 2014, the U.S. government reported that American fertility had fallen to a record low of just 1.86 births per woman, putting us on the same page as Denmark, with a birth rate below the amount needed to keep the country's population from falling. As of yet, however, the United States has not adopted policies designed to encourage women to have more children—though the idea is not far-fetched. Is "Do It for America" in our future?

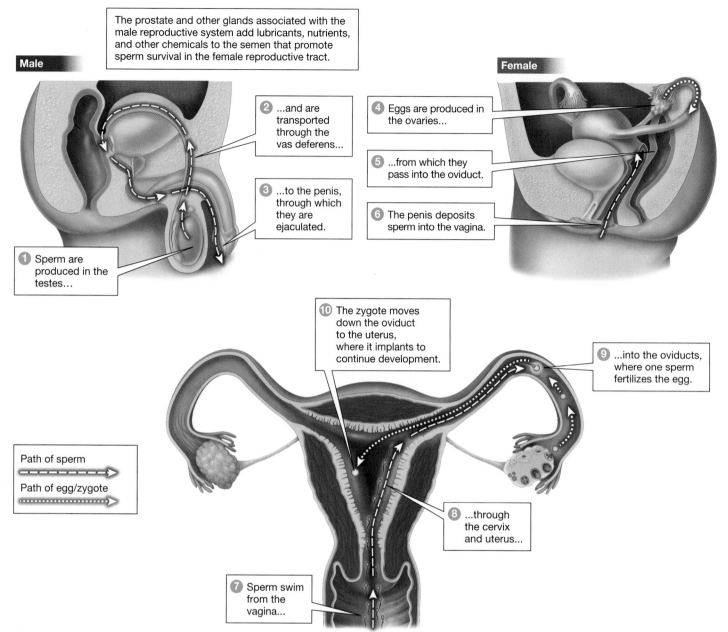

The prostate and other glands associated with the male reproductive system add lubricants, nutrients, and other chemicals to the semen that promote sperm survival in the female reproductive tract.

Male

2 ...and are transported through the vas deferens...

3 ...to the penis, through which they are ejaculated.

1 Sperm are produced in the testes...

Female

4 Eggs are produced in the ovaries...

5 ...from which they pass into the oviduct.

6 The penis deposits sperm into the vagina.

10 The zygote moves down the oviduct to the uterus, where it implants to continue development.

9 ...into the oviducts, where one sperm fertilizes the egg.

Path of sperm

Path of egg/zygote

8 ...through the cervix and uterus...

7 Sperm swim from the vagina...

Figure 19.9

Fertilization takes place in the oviduct
Fertilization results in a zygote that can develop in the sheltered environment of the uterus.

Q1: If an egg is released but no sperm enter the oviduct, what is likely to occur?

Q2: If sperm enter the oviduct but no egg is present, what is likely to occur?

Q3: Sperm can live for up to five days inside a woman's body. Furthermore, eggs may be released at any point in the cycle, although mid-cycle ovulation is the norm. If you are trying *not* to become pregnant, when is it safe to have unprotected intercourse?

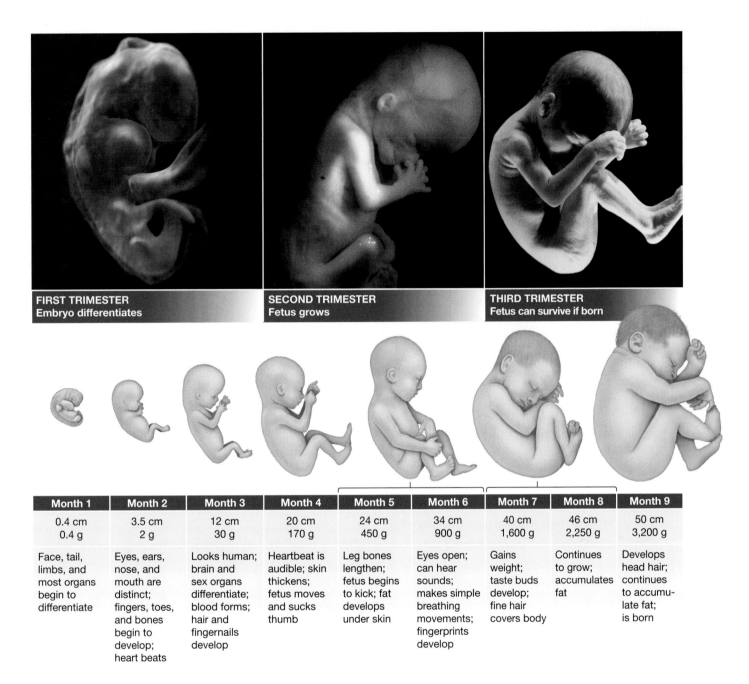

Month 1	Month 2	Month 3	Month 4	Month 5	Month 6	Month 7	Month 8	Month 9
0.4 cm	3.5 cm	12 cm	20 cm	24 cm	34 cm	40 cm	46 cm	50 cm
0.4 g	2 g	30 g	170 g	450 g	900 g	1,600 g	2,250 g	3,200 g
Face, tail, limbs, and most organs begin to differentiate	Eyes, ears, nose, and mouth are distinct; fingers, toes, and bones begin to develop; heart beats	Looks human; brain and sex organs differentiate; blood forms; hair and fingernails develop	Heartbeat is audible; skin thickens; fetus moves and sucks thumb	Leg bones lengthen; fetus begins to kick; fat develops under skin	Eyes open; can hear sounds; makes simple breathing movements; fingerprints develop	Gains weight; taste buds develop; fine hair covers body	Continues to grow; accumulates fat	Develops head hair; continues to accumulate fat; is born

FIRST TRIMESTER
Embryo differentiates

SECOND TRIMESTER
Fetus grows

THIRD TRIMESTER
Fetus can survive if born

Figure 19.10

Nine months in the womb

Although it is a critical step in human reproduction, fertilization marks only the beginning of a nine-month-long period of development within the mother's uterus.

Q1: Place these terms in the correct order of development: embryo, fetus, infant, zygote.

Q2: In what trimester is the fetus most likely to survive outside its mother's body?

Q3: The first trimester is the most sensitive time for exposure to mutagens. Why might that be?

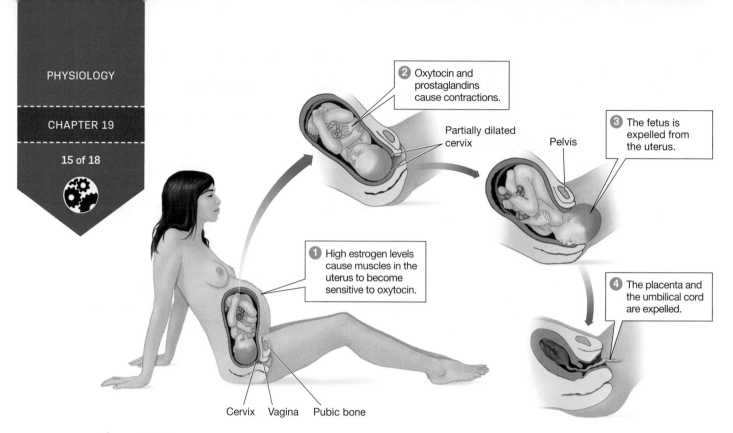

2 Oxytocin and prostaglandins cause contractions.

Partially dilated cervix

Pelvis

3 The fetus is expelled from the uterus.

1 High estrogen levels cause muscles in the uterus to become sensitive to oxytocin.

4 The placenta and the umbilical cord are expelled.

Cervix   Vagina   Pubic bone

Figure 19.11

Childbirth Is orchestrated by hormones

Childbirth occurs in stages, driven by the hormone oxytocin. Oxytocin signals uterine muscles to contract. The contractions become stronger as the amount of oxytocin increases. The mother's cervix opens, and the fetus is eventually expelled from the uterus, followed by the placenta soon after.

Q1: What is the role of estrogen in childbirth?

Q2: Explain how the involvement of hormones in childbirth is an example of a positive feedback loop.

Q3: If a woman has been pregnant for more than 40 weeks, her doctor might give her an injection of oxytocin to precipitate labor. How would that bring about labor?

REVIEWING THE SCIENCE

- Cells that work in an integrated manner to perform a common set of functions constitute a **tissue**. Four main types of tissues are found in vertebrates: **epithelial**, **connective**, **muscle**, and **nervous**. An **organ** is made up of more than one tissue type and forms a functional unit with a distinctive shape and location in the body. An **organ system** is composed of two or more organs that work in a closely coordinated manner to perform a distinct set of functions.

- **Homeostasis** is the process of monitoring the internal environment of an organism. **Homeostatic pathways** have two basic features: sensors that monitor the internal environment, and regulatory processes that attempt to restore the normal internal state when deviations from optimal conditions are detected.

- Many homeostatic pathways are controlled by **feedback loops**. In **negative feedback** loops, the results of a process cause that process to slow down or stop. In **positive feedback** loops, the results of a process cause it to speed up.

- **Sexual reproduction** involves the organs of the **reproductive system** and the union of male and female gametes. Most animals produce offspring in this way, although some can individually produce genetically identical offspring via **asexual reproduction**.

- **Oogenesis** and **spermatogenesis** are the production of **eggs** and **sperm**. Fertilization fuses the haploid sperm and haploid egg to produce a diploid **zygote**. Males produce sperm in **testes**, and females produce eggs in **ovaries**.

- Approximately monthly, one egg released from a woman's ovary moves into the **oviduct**, where it can be fertilized. During sexual intercourse, the man's **penis** releases into the woman's **vagina** nearly 300 million sperm, only one of which can fertilize the egg.

- During the first **trimester** of human development in the **uterus**, embryonic cells rapidly differentiate into the various organs and structures present at birth. From the ninth week of development on, the developing human is called a **fetus**. During the second and third trimesters, the fetus grows rapidly.

- Childbirth occurs in stages. The hormone **oxytocin** signals uterine muscles to contract. The contractions become stronger as positive feedback increases the amount of oxytocin produced.

- Pregnancy can be prevented by **contraceptives**. Some contraceptives can also help prevent the transfer of sexually transmitted diseases, but most do not.

- Development for most animals continues after birth. Unlike most other animals, modern humans live part of their lives beyond their reproductive years.

THE QUESTIONS

The Basics

1 Denmark's birth rate is dropping because

(a) women are now older when they begin a family.

(b) men are now older when they begin a family.

(c) exposure to chemicals is decreasing the sperm count in men.

(d) all of the above

2 Tissues

(a) are composed of cells that work in an integrated manner.

(b) have a distinctive shape and location in the body.

(c) are composed of multiple organs.

(d) are composed of only one cell type.

3 Homeostasis does *not* maintain

(a) cellular pH.

(b) body temperature.

(c) environmental temperature.

(d) blood oxygen levels.

4 _____ stimulate cells to develop the characteristics of maleness.

(a) Estrogens

(b) Spermatogens

(c) Progestogens

(d) Androgens

5 What is the typical order of events to produce an embryo?

(a) gamete development, ovum release, intercourse, fertilization, implantation

(b) gamete development, ovum release, fertilization, implantation, intercourse

(c) gamete development, ovum release, intercourse, implantation, fertilization

(d) ovum release, gamete development, intercourse, fertilization, implantation

6 Link each of the following terms with its definition.

SPERMATOGENESIS	1. Cells from only one parent produce offspring.
OOGENESIS	2. The process of producing sperm.
SEXUAL REPRODUCTION	3. The process of producing eggs (ova).
ASEXUAL REPRODUCTION	4. Gametes from two parents combine to produce offspring.

7 Circle the correct terms in the following sentence:

Homeostasis maintains a constant internal state through [**homeopathic, homeostatic**] pathways that trigger regulatory processes when there is movement away from the body's [**set point, initiator**]. For example, body temperature homeostasis is maintained through a [**negative, positive**] feedback loop—a process known as [**thermoregulation, osmoregulation**].

8 Beginning with the final day of bleeding, identify the correct order of events in the human menstrual cycle by numbering them from 1 to 5.

___ a. The corpus luteum produces progesterone for about 14 days.

___ b. FSH stimulates the growth of egg follicles.

___ c. The egg is released from the follicle (now corpus luteum).

___ d. The uterine lining sloughs off (menstruation occurs) with decreased progesterone.

___ e. Estrogen levels increase and trigger LH.

Try Something New

9 The actual effectiveness of birth control is often significantly lower than the theoretical effectiveness, or "effectiveness if used as directed." For example, there is a 9% annual pregnancy rate of women on the Pill, although its effectiveness is given as over 99%. Similarly, although the condom is considered 98% effective, the associated annual pregnancy rate is between 10% and 18%. Would you expect a difference between theoretical and actual effectiveness for sterilization (female tubal ligation and male vasectomy)? If you do expect a difference, what would be the cause of that difference?

10 In maintaining homeostasis, small animals face more challenges than do large animals. Large animals tend to exchange water, solutes, and heat with their environment more slowly than small animals do, because a large animal has a larger volume relative to its surface area than a smaller animal has. The ratio between these two quantities—surface area and volume—determines how quickly or slowly an animal can gain or lose water, solutes, or heat. When the ratio of surface area to volume is relatively high, gains and losses are rapid; when the ratio is relatively low, gains and losses happen more slowly. Explain how this ratio indicates the need to be extra careful to protect newborns from temperature extremes. What else does the ratio of surface area to volume suggest we need to consider when caring for a newborn?

Leveling Up

11 **Looking at data** To answer the following questions, refer to the accompanying graph below, which shows data collected for the National Survey of Family Growth by the U.S. CDC (Centers for Disease Control and Prevention).

a. What does the y-axis show?

b. How many children, on average, does a woman with a bachelor's degree (or higher) have?

c. How many children, on average, does a man with a high school diploma or GED have?

d. Describe in your own words what the graph shows about the relationship between education and reproduction in the United States.

e. State a scientific hypothesis that might explain the relationship you described in part (d).

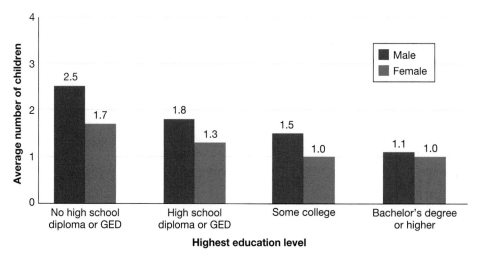

NOTE: GED is General Educational Development high school equivalency diploma.

12 *Write Now* **biology** You are the aide to a U.S. senator, who has just sent you an e-mail including an attached article from a constituent in her (and your) home state. The constituent is concerned that people are having fewer children than in the past and thinks that the Senate should pass a law making it illegal to sell birth control to married couples. The senator asks you to write a two-page "white paper" (500 words) that she will use to respond to the e-mail and possibly also to propose legislation. The senator is extremely busy, and she neglected to take a biology class in college, so she is relying on you to give her a clear, concise, and accurate summary of the issue raised by the e-mail, as well as possible actions to take.

Write a position paper addressing the following points (using about half a page, or 125 words, on each point). Begin by reading the article that was sent to the senator: http://national. deseretnews.com/article/1522/the-potential-impact-of-falling-fertility-rates-on-the-economy-and-culture.html. You may use other resources, including your textbook.

a. Summarize the main points of the article, defining terms (for example, "fertility rate," "replacement rate") as needed.

b. Find the fertility rate for your state, and compare it to the U.S. fertility rate. How do the two rates differ, and what is your best hypothesis for why they differ? Do the points made in the article hold for your state?

c. What are some potential challenges to the legislation proposed in the e-mail? Is there alternative legislation—or another action—that might have the same effect with fewer challenges?

d. In your final paragraph, advise the senator on how she should respond to the letter and whether she should propose the legislation recommended in the e-mail. Also provide an alternative recommendation, which does not have to involve increasing reproduction. Justify your opinion with data and logic.

The Sunshine Vitamin

Is vitamin D the new supernutrient?

After reading this chapter you should be able to:

◆ Explain the roles of the major classes of nutrients in maintaining a healthy body.

◆ For a given vitamin or mineral, describe its main functions and dietary sources, and the possible effects of a deficiency.

◆ Relate the structure to the function of one component of the digestive system.

◆ Demonstrate how flexible and rigid elements of a joint work together to produce controlled motion.

◆ Compare and contrast skeletal, smooth, and cardiac muscle.

◆ Compare and contrast cartilage and bone.

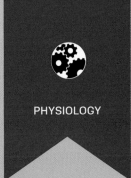

CHAPTER

20

DIGESTIVE, MUSCULAR, AND SKELETAL SYSTEMS

The babies' bones were not growing well. As a PhD student at McMaster University in Ontario, Canada, Hope Weiler was observing infants in a local hospital's neonatal intensive care unit, or NICU. They were being given dexamethasone, an anti-inflammatory steroid.

Prematurely born infants, or "preemies," often have chronic lung disease and cannot breathe on their own (**Figure 20.1**). To help these babies breathe, doctors would treat them with many doses of dexamethasone over the course of a month. But Weiler observed that the medication, while saving the infants' lives, also appeared to interfere with bone formation. Infants treated with dexamethasone had smaller heads, thinner bones, and shorter stature than those who didn't receive the medication.

"We wanted to know more about why that was happening," says Weiler, now the Canada Research Chair in Nutrition, Development, and Aging and a professor at McGill University in Quebec.

Determined to find ways to help preemies grow, Weiler started a study tracking the bone mass of infants. During the study, she discovered that many newborn infants had low vitamin D levels in their cord blood.

Vitamins are small, organic nutrients needed by our bodies, but only in tiny amounts (**Figure 20.2**). They participate in a great variety of essential metabolic processes, such as helping blood cells form and maintaining brain function. Some vitamins bind to enzymes to speed up chemical reactions within a cell. Some act as a delivery service, supplying chemical groups needed in important metabolic reactions. Others act as signaling molecules. And some are even believed to work as antioxidants, substances that protect body tissues from destructive chemicals known as free radicals.

Most vitamins have multiple functions in the animal body, and vitamin D is no exception. Vitamin D, a fat-soluble vitamin, is required to absorb calcium from food and therefore is critical for bone growth and bone remodeling. Most vitamin D is naturally made by our bodies when ultraviolet (UV) rays are absorbed through the skin—an organ system in its own right, the **integumentary system** (see "The Skin We're In" on page 352).

Vitamins are a type of *nutrient*. **Nutrients** are components of foods that an organism needs to survive and grow. These can be micronutrients (vitamins and minerals) or macronutrients. Macronutrients are large organic molecules classified into three main categories: Carbohydrates, lipids, and proteins. Macromolecules serve as sources of energy and furnish the body with chemical building blocks such as sugars, fatty acids, and amino acids. Although an adult human can synthesize some of the 20 amino acids needed to make proteins, we must get 8 of them, called **essential amino acids**, from food. Another type of macronutrient is dietary fiber, which does not contribute amino acids or energy to the body but is critical for survival because it affects how other nutrients are absorbed in the gut.

Minerals are inorganic chemicals that have critical biological functions. Carbon, hydrogen, oxygen, and nitrogen make up about 93 percent of the animal body, so, by convention, these four elements are excluded from the category of dietary

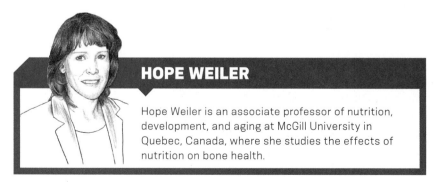

Figure 20.1

A premature baby receiving oxygen via a nose tube

HOPE WEILER

Hope Weiler is an associate professor of nutrition, development, and aging at McGill University in Quebec, Canada, where she studies the effects of nutrition on bone health.

Vitamin C is abundant in fruits and vegetables, and it assists in the maintenance of teeth, bones, and other tissues. Deficiency of this vitamin leads to scurvy, in which teeth and bones degenerate.

Fish is the richest source of **vitamin D**; fortified foods (such as milk, soy milk, and breakfast cereals) are important sources for most people. A deficiency in vitamin D leads to poor formation of bones and teeth.

Folic acid, a **B vitamin**, is abundant in green vegetables, legumes, and whole grains. B_{12}, another B vitamin, is scarce in plant foods but abundant in milk, meat, fish, and poultry. It is important for the maintenance of teeth, bones, and other tissues.

Vitamin E is abundant in nuts, vegetable oils, whole grains, and egg yolk. It protects lipids in cell membranes and other cell components.

Leafy green vegetables and some fruits (e.g., avocado and kiwi) are rich in **vitamin K**, which is also manufactured by intestinal bacteria. A deficiency can cause prolonged bleeding and slow wound healing.

Carotene is responsible for the color of yellow and orange fruits and vegetables. It is converted into **vitamin A** within our bodies. Vitamin A aids in production of the visual pigment needed for good eyesight and is used in making bone.

Figure 20.2

Vitamins needed in the human diet

Humans need nine **water-soluble vitamins** (C and eight different B vitamins) and four **fat-soluble vitamins** (A, D, E, and K) from their diet. Because water-soluble vitamins are easily excreted in urine, they tend not to accumulate in body tissues, which means we must obtain these vitamins from food on a regular basis. Fat-soluble vitamins are not excreted as readily and tend to accumulate in body fat, so excessive consumption can lead to overdosing.

Q1: Which vitamins described in the figure are important for healthy bones?

Q2: Which vitamins are you more likely to overaccumulate?

Q3: In your own diet, are there any vitamins that you may not be eating enough of?

minerals. But more than 20 other elements, such as fluoride, sodium, and iodine, are essential for the normal function of most animals, and these are classified as dietary minerals. Calcium is the most abundant mineral in the body; it makes up a large proportion of your bones.

Low levels of vitamin D have been identified not only in premature infants, but also in patients with a variety of ailments, including osteoporosis (a progressive bone disease), schizophrenia, erectile dysfunction, and more. If lack of vitamin D does contribute to these disorders, then

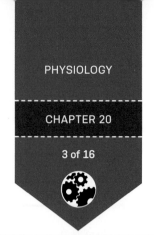
supplementation with the vitamin should be protective. Many supplement manufacturers, in fact, now tout D as a "supervitamin" and "star supplement" that will protect individuals from a wide range of diseases.

But has science shown that claim to be true? Just because low vitamin D and disease are correlated does not mean that one causes the other.

So, scientists have long wondered: Does low vitamin D cause disease, or is it involved in some other way?

You Are What You Eat

In her experiment studying premature infants, Weiler hypothesized that low vitamin D levels were an indirect result of treatment with dexamethasone.

Dexamethasone is a powerful steroid that helps a premature baby's lungs work better, but it has a side effect: It interferes with calcium absorption in the gut. When the drug is present, the intestines do not absorb calcium correctly. To counteract this effect, the babies' bodies began trying to find ways to compensate.

That is where vitamin D became critical: Vitamin D promotes calcium absorption in the intestines and in bones. So the babies were metabolizing (using up) all their vitamin D in an effort to absorb more calcium in the gut, and thus didn't have enough left circulating in their blood to build bones.

Absorbing vitamins is one important function of the **digestive system**, which also processes food and eliminates unusable waste. The digestive system of most animals consists of a long, hollow passageway, known as the digestive tract, and a number of accessory organs, such as the pancreas and liver (**Figure 20.3**).

Eating, or **ingestion**, is the first stage in the processing of food by the digestive system. **Digestion**, the chemical breakdown of food, begins almost immediately after ingestion in many species.

During ingestion, a bite of food—say, a spoonful of cornflakes—is deposited in the **oral cavity**, the mouth. There, an array of different types of teeth, which are shaped to cut, crush, or grind food into smaller pieces, begin to break apart the cornflakes. Many small pieces of food provide a greater surface area for digestive enzymes to work on than do fewer larger pieces.

The muscular tongue mixes the crushed cereal particles with saliva. **Saliva** contains enzymes that start to break down starches—or any carbohydrate—into sugars. If you chew a piece of bread for long enough, for example, its starches

The Skin We're In

The **integumentary system** is the largest organ system in the human body, accounting for almost 15 percent of our weight. It covers the body and protects it from environmental hazards such as extreme temperatures and dangerous pathogens. It also prevents water loss and protects the body from physical damage. Sensory receptors are embedded in the skin (see Chapter 21 for more on the nervous system), and skin is the site of vitamin D synthesis.

The integumentary system consists of the skin and structures embedded in the skin, such as hair and nails in humans, or feathers, hooves, and scales in other vertebrates. The skin is made up of three layers. Moving inward from the outermost layer, the layers are the epidermis, dermis, and hypodermis.

As the figure below shows, the skin contains multiple tissue types. The epidermis is an example of epithelial tissue. The nerve endings are examples of nervous tissue. The arrector pili muscle is composed of smooth muscle tissue. Much of the dermis is made up of connective tissue. And adipose tissue dominates in the hypodermis, a thick insulating sheet under the other layers of the skin.

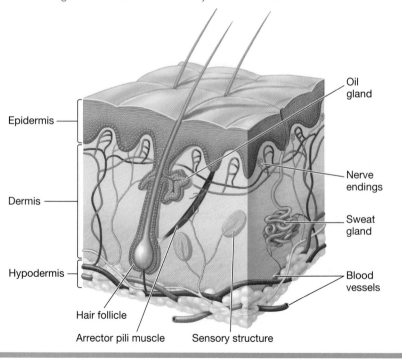

Epidermis

Dermis

Hypodermis

Hair follicle

Arrector pili muscle

Sensory structure

Oil gland

Nerve endings

Sweat gland

Blood vessels

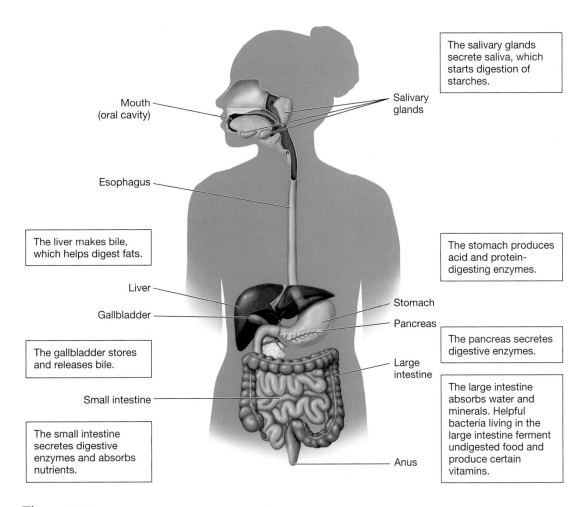

Mouth (oral cavity)

Salivary glands

The salivary glands secrete saliva, which starts digestion of starches.

Esophagus

The liver makes bile, which helps digest fats.

Liver

The stomach produces acid and protein-digesting enzymes.

Gallbladder

Stomach

Pancreas

The gallbladder stores and releases bile.

The pancreas secretes digestive enzymes.

Large intestine

Small intestine

The large intestine absorbs water and minerals. Helpful bacteria living in the large intestine ferment undigested food and produce certain vitamins.

The small intestine secretes digestive enzymes and absorbs nutrients.

Anus

Figure 20.3

The digestive system converts food into absorbable nutrients

As food moves through the digestive system, it is broken down into small molecules that can be absorbed by the lining of the intestine.

Q1: List, in order, the structures of the digestive system that a piece of swallowed food would pass through, beginning with the mouth.

Q2: What part of the digestive system hosts bacteria that produce vitamins?

Q3: What is the shared function of the liver and gallbladder?

are digested to sugar and it will begin to taste sweet. Saliva is also important for turning the crunchy cereal into a moist mass that can slip easily down the throat.

The tongue assists in pushing the now-moist cereal into the throat, or **pharynx**, where the back of the mouth and the nasal cavity come together. The pharynx is the common entryway for both the air tube (the trachea) and the food tube (the **esophagus**). That is why, on occasion, you may cough up food or liquid that "went down the wrong tube"; it accidentally went down the trachea instead of the esophagus.

Normally, the pharynx is very good at separating air and food. When the mushy bite of cereal makes contact with the wall of the pharynx, it stimulates nerves that launch the **swallowing reflex**, in which a flap of tissue, called the epiglottis, seals off the entry into the trachea. The cereal is then pushed into the esophagus.

Waves of muscular contractions carry the cereal down the esophagus and into the **stomach**.

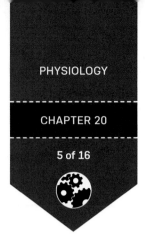
Protein digestion begins in the stomach, which secretes acid and enzymes that break down complex protein molecules. Muscles in the wall of the stomach alternately contract and relax to mix the food particles with the acid and enzymes. The resulting watery mixture is stored in the stomach until it can move into the small intestine.

The **small intestine** is a highly coiled thin tube about 3–4 centimeters in diameter. If straightened, the small intestine would extend about 20 feet (6 meters). The upper and lower regions of the small intestine serve different functions. The upper region, which lies nearest the stomach, uses enzymes secreted by the **pancreas** and by the intestine itself to break down large molecules into simpler forms that the body can absorb. Here, the digestion of proteins, carbohydrates, and lipids, including fats, is completed.

The digestion of fats poses a particular problem because fats are not soluble in water, yet they need to be broken down and made to mix with the watery contents of the digestive tract.

Bile is a fluid that helps digest fats by creating a coating that enables the fat globules to interact with water molecules to partially dissolve them. The large globules break down into tiny droplets, which offer a larger work surface for lipid-degrading enzymes. Bile is produced by the **liver**, an organ that serves a multitude of functions. Some of the bile made by the liver is stored in the **gallbladder**, which dispenses the bile into the small intestine as needed.

The lower region of the small intestine is specialized for **absorption**, the uptake of mineral ions—including calcium—and small molecules by cells lining the cavity of the digestive tract. These small molecules include broken-down sugars, fatty acids, and amino acids. The innermost lining of the small intestine presents a large surface area for that process (**Figure 20.4**).

Vitamins, including D, are also absorbed in the lower region of the small intestine. Only a few foods naturally contain vitamin D, including fatty fish such as salmon and tuna, and egg yolks.

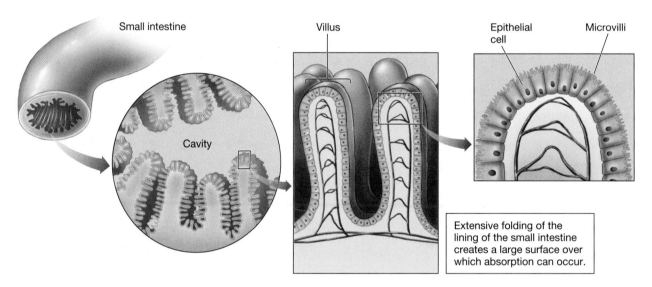

Extensive folding of the lining of the small intestine creates a large surface over which absorption can occur.

Figure 20.4

The small intestine is specialized for absorption

Nutrients are absorbed in the small intestine by large numbers of fingerlike projections called **villi** (singular "villus"). Each villus is about one millimeter long, with a surface consisting of cells specialized for nutrient absorption. The plasma membrane of each of these cells also has many tiny projections, called microvilli. This complex folding of the intestinal lining produces almost 300 square meters of surface area for absorption.

Q1: Why is a larger surface area important for absorption?

Q2: In what way are the epithelial cells lining the villi modified to increase absorption?

Q3: Explain the role of the capillaries within each villus.

Most of our ingested vitamin D comes from fortified foods—foods to which a vitamin or mineral is added. Milk, for example, is often fortified with vitamin D, voluntarily in the United States and by law in Canada. Both countries mandate that all baby formula be fortified with vitamin D. Breakfast cereals often contain added vitamin D as well.

Most of the nutrients absorbed by the digestive tract are sent to the bloodstream, which eventually delivers them to every cell in the body. Whether from the skin or the gut, vitamin D is sent to the liver and then the kidney (**Figure 20.5**). In each organ, the nutrient is chemically modified to become an active form of the vitamin that our cells can use.

Our original bite of cereal contains very few nutrients by the time it arrives in the final segment of the digestive tract. This residual matter is prepared for **elimination**—the removal from the body of solid waste, consisting mostly of indigestible material and bacteria that inhabit the digestive tract—during passage through the

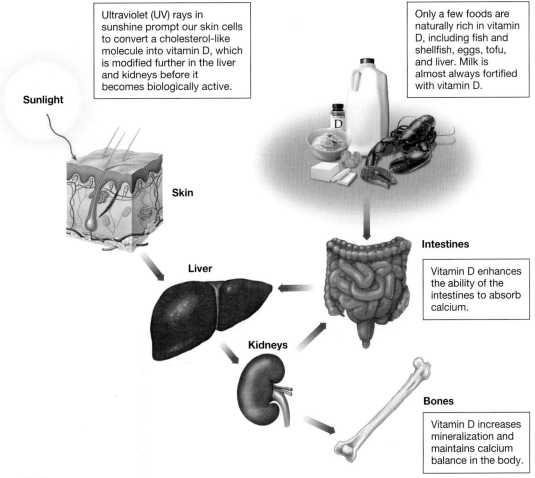

Ultraviolet (UV) rays in sunshine prompt our skin cells to convert a cholesterol-like molecule into vitamin D, which is modified further in the liver and kidneys before it becomes biologically active.

Only a few foods are naturally rich in vitamin D, including fish and shellfish, eggs, tofu, and liver. Milk is almost always fortified with vitamin D.

Sunlight

Skin

Liver

Intestines

Vitamin D enhances the ability of the intestines to absorb calcium.

Kidneys

Bones

Vitamin D increases mineralization and maintains calcium balance in the body.

Figure 20.5

Sources of vitamin D

Vitamin D is the only vitamin that we can manufacture entirely within our tissues, yet many Americans get inadequate amounts of vitamin D.

Q1: What organs of the digestive system are involved in processing vitamin D?

Q2: Do you get the majority of the vitamin D you need from the sun, from natural foods, or from dietary supplements?

Q3: It is true that you can increase your vitamin D levels by visiting a tanning booth. Why is it considered a bad idea to increase vitamin D in this way?

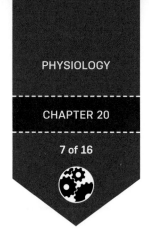
large intestine, or **colon**. The colon absorbs almost all remaining minerals and water from the waste. Then, large numbers of bacteria living in the colon break down the remaining waste to squeeze out the very last nutrients that the body can absorb. These bacteria also produce certain vitamins that are absorbed into the body from the colon. The waste, or **feces**, leaves the body through the **anus**, a muscle-lined opening.

Building Bones

Today, dexamethasone is no longer used to treat premature babies, having been replaced with a safer drug. Weiler's study is one of many that, over the years, have confirmed the cause-and-effect role of vitamin D on bone health in the skeletal system: With low levels of vitamin D, the babies did not grow a healthy skeleton.

Like most vertebrates, humans have a bony internal skeleton that supports the body, gives it shape, and protects soft tissues and organs (**Figure 20.6**). The **axial skeleton**, made up of 80 bones, supports and protects the long axis of the body. It includes the skull, the ribs, and a long, bony spinal column. Although the axial skeleton plays a role in movement, its primary purpose is to protect vital organs. The 126 bones of the arms, legs, and pelvis make up the **appendicular skeleton** ("appendicular" means "relating to an appendage or limb"). These bones have more to do with motion than with protection.

Another important, though often passed over, part of our skeleton is **cartilage**, a dense tissue that combines strength with flexibility. In the human skeleton, cartilage gives form to the nose, the ears, and part of the rib cage. In addition, cartilage is found at nearly every point in the body where two bones would otherwise come into direct contact; it creates a smooth surface that prevents the two bony surfaces from grinding against each other.

Cartilage does contain cells, but it consists primarily of nonliving, extracellular

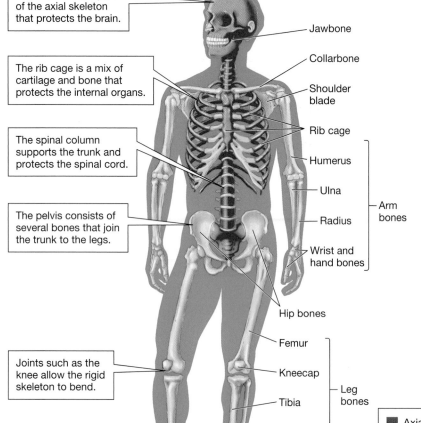

The skull forms a part of the axial skeleton that protects the brain.

The rib cage is a mix of cartilage and bone that protects the internal organs.

The spinal column supports the trunk and protects the spinal cord.

The pelvis consists of several bones that join the trunk to the legs.

Joints such as the knee allow the rigid skeleton to bend.

Jawbone

Collarbone

Shoulder blade

Rib cage

Humerus

Ulna

Radius — Arm bones

Wrist and hand bones

Hip bones

Femur

Kneecap

Tibia — Leg bones

Fibula

Foot bones

■ Axial skeleton
■ Appendicular skeleton
■ Cartilage

Figure 20.6

The human skeleton
The axial human skeleton protects vital organs; the appendicular skeleton facilitates movement.

Q1: The collarbone is part of which skeleton: axial or appendicular?

Q2: Which parts of the skeleton are made of cartilage?

Q3: Which part of the skeleton protects the central nervous system (the brain and spinal cord)?

material—bundles of **collagen**, a tough but pliable protein found in a great variety of tissues, including skin, blood vessels, bones, teeth, and the lens of the eye.

Sharks belong to a small class of vertebrates that do not have a skeleton made of bones; their skeleton is made entirely of cartilage and connective tissue. Keep in mind, too, that not all animals have their skeleton on the inside. While humans and other vertebrates have an internal skeleton, or **endoskeleton**, many other animals, such as lobsters and insects, have an **exoskeleton**, an external skeleton that surrounds and encloses the soft tissues it supports (**Figure 20.7**).

Like cartilage, much of bone is made up of nonliving material. Still, it is a living tissue that has a blood and nerve supply. Specialized bone cells, called **osteocytes**, surround themselves with a hard, nonliving mineral matrix composed largely of calcium and phosphate. Vitamin D maintains calcium and phosphate concentrations in the body that are necessary to form that matrix. Although they are just single cells, osteocytes can live as long as the organism whose skeleton they belong to.

Bones are made of two major types of bone tissue (**Figure 20.8**). **Compact bone** forms the hard, white outer region. **Spongy bone**, honeycombed with numerous tiny cavities, lies inside the compact bone and is most abundant at the knobby ends of our long bones. Long bones and some others, such as the ribs and breastbone, have a hollow interior, which makes them light but strong. The cavities inside hollow bones contain **marrow**, a tissue that, depending on the type of bone, stores fat or produces blood cells.

Vitamin D, as previously noted, promotes calcium absorption in the gut and maintains calcium and phosphate concentrations. It is also needed for calcium absorption into bones, and thus is essential to bone growth and remodeling. Without it, bones can become thin, brittle, or misshapen—symptoms of a condition called rickets.

Rickets causes bones to soften, especially near **joints**, the junctions in the skeletal system that let the skeleton move in specific ways (**Figure 20.9**). Walking, for example, requires movement at the hips and knees, as well as at other joints. The lower jaw connects to the skull at a joint so that it can move relative to the rest of the skull, enabling us to chew and talk.

Figure 20.7

A newly molted cicada emerges from its exoskeleton
Exoskeletons provide a protective armor for many animals and also protect terrestrial invertebrates from excessive moisture loss. The rigidity of an exoskeleton means that immature animals that outgrow their exoskeletons must shed them periodically, a process known as molting.

Joints are held together by collagen-rich ligaments and tendons. **Ligaments** are specialized, flexible bands of tissue that join bone to bone, while **tendons** connect muscle to bone.

Wherever two moving parts rub against each other, as in a joint, wear can erode bone and friction can waste energy. So the joint is lined with a sheet of tissue, called the synovial membrane, that forms a cavity called the **synovial sac**. The

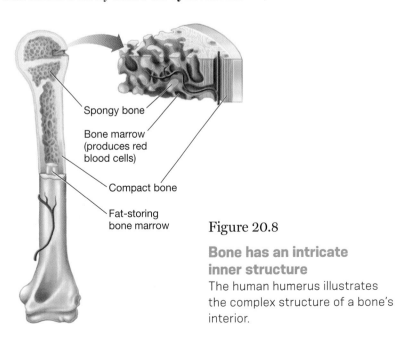

Spongy bone

Bone marrow (produces red blood cells)

Compact bone

Fat-storing bone marrow

Figure 20.8

Bone has an intricate inner structure
The human humerus illustrates the complex structure of a bone's interior.

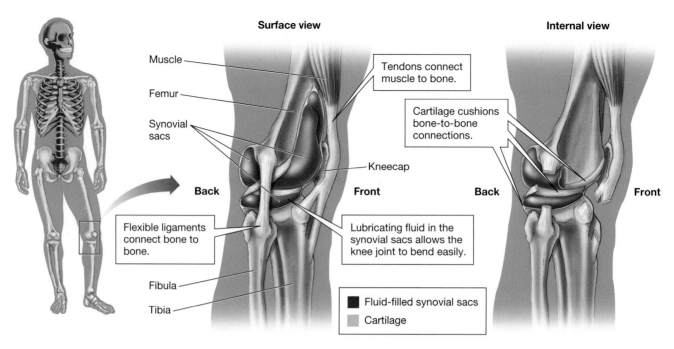

Surface view

Internal view

Muscle

Femur

Synovial sacs

Tendons connect muscle to bone.

Cartilage cushions bone-to-bone connections.

Back

Front

Back

Front

Kneecap

Flexible ligaments connect bone to bone.

Lubricating fluid in the synovial sacs allows the knee joint to bend easily.

Fibula

Tibia

■ Fluid-filled synovial sacs
■ Cartilage

Figure 20.9

The human knee illustrates how rigidity and flexibility together enable movement
Although our knees differ in detail from other joints in the body, they make a good general model for the way flexible and rigid materials are combined in a joint to allow controlled motion.

Q1: What is the function of the synovial sac?

Q2: Compare and contrast ligaments and tendons.

Q3: Knee injuries are some of the most common sports injuries. Why do you think that is?

space inside the synovial sac is filled with a lubricating fluid that reduces friction between the two bony surfaces.

Altogether, these five components—bone, cartilage, ligaments, tendons, and synovial sacs—work to move a joint safely and with precise control.

Show of Strength

There is now no doubt that vitamin D is vital for bone health, but over decades, scientists have also been amassing data on the role of vitamin D in nonskeletal tissues. One of these is muscle tissue, which forms the body's **muscular system**.

Muscle tissue is unique to animals. The skeleton and its joints are the framework for motion, while muscles provide the power necessary for movement. Muscle tissue possesses a crucial property: It can contract and relax. When we walk, run, or jump, we are using *voluntary* contractions. But

even when we are still, we're using *involuntary* muscles: The heart is pumping blood, the lungs are pumping air, and food is being moved along the digestive tract by muscular contractions. Involuntary muscles do their work without our having to think about them (**Figure 20.10**).

In November 2014, researchers in Belgium compiled all the available data on the role of vitamin D in skeletal muscle, which consists of many bundles of long **muscle fibers**. A muscle fiber is a long, narrow cell that can span the length of an entire muscle because it is made up of several muscle cells that fused together during development (**Figure 20.11**). Each muscle fiber is packed with cylindrical structures containing proteins that contract by bracing against each other. Each such cylinder is known as a **myofibril**. Myofibrils are organized into series of contractile units called **sarcomeres**.

To assess the impact of vitamin D supplementation on skeletal muscle health, the Belgian authors identified 30 randomized, controlled trials involving a total of 5,615 individuals. In each trial, an experimental group of people was given vitamin D supplements while a control group was not.

Looking at the data from all of those trials, the researchers found a "small but significant positive effect of vitamin D supplementation on global muscle strength" and saw that supplementation seemed to be more effective on people aged 65-plus than on younger people.

In addition to research connecting vitamin D and bone and muscle health, there are studies correlating low vitamin D levels with almost every other tissue and ailment under the sun, from heart disease to prostate cancer to dementia. But a recent study has cast a large shadow of doubt over whether vitamin D is actually the *cause* of these nonskeletal disorders.

Beyond Bone

To one skin cancer researcher in France, the flurry of studies linking vitamin D to nonskeletal diseases was surprising. Epidemiologist Philippe Autier, vice president of population research at the International Prevention Research Institute (iPRI) in Lyon, France, had been following the vitamin D story from early on. He had grown skeptical of vitamin D health claims—that vitamin D could prevent a wide range of diseases—when the tanning industry began touting vitamin D as a benefit of tanning booths while conveniently glossing over the risks of skin cancer from the beds' UV rays (**Figure 20.12**).

Shortly after the tanning industry started promoting vitamin D's health benefits, numerous studies began to emerge about vitamin D's role in a variety of nonskeletal diseases. "We were surprised by the fact that there were so many relationships between low vitamin D and diseases of all sorts, from brain disease to lung disease to infectious disease to cancer," says Autier. As he began reading the primary literature, he found that for virtually every disease, some correlation between vitamin D and the disease existed, and usually it followed a particular pattern.

"It was almost always that low vitamin D was associated with a greater risk of disease, and high vitamin D was protective," says Autier. But such conclusions were made from observational studies. Much rarer were randomized clinical trials,

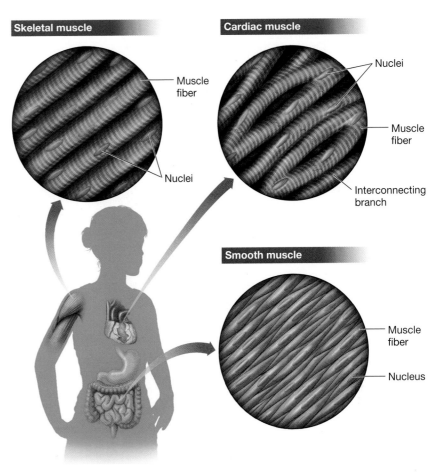

Figure 20.10

Specialized types of muscles for different types of movement

Skeletal muscle has a distinctive banded appearance, brought about by the sarcomeres that make it up (see Fig. 20.11). **Cardiac muscle** is also banded, and its muscle fibers are branched, which helps produce the coordinated contractions known as heartbeats. **Smooth muscle** contracts involuntarily, and through a different mechanism than other muscle types. Without sarcomeres, no bands are seen in smooth muscle.

Q1: Which types of muscles can you voluntarily contract?

Q2: Do the muscles in the heart contract voluntarily or involuntarily?

Q3: You do not have to think about breathing (otherwise, sleeping would be dangerous!), but you *can* increase or decrease your rate of breathing. Are the muscles involved in breathing, then, voluntary or involuntary?

in which one group of people was given vitamin D supplements, another was not, and their likelihood of developing a particular ailment was compared. In those studies, Autier noticed, higher levels of vitamin D did not seem to protect the individuals taking supplements.

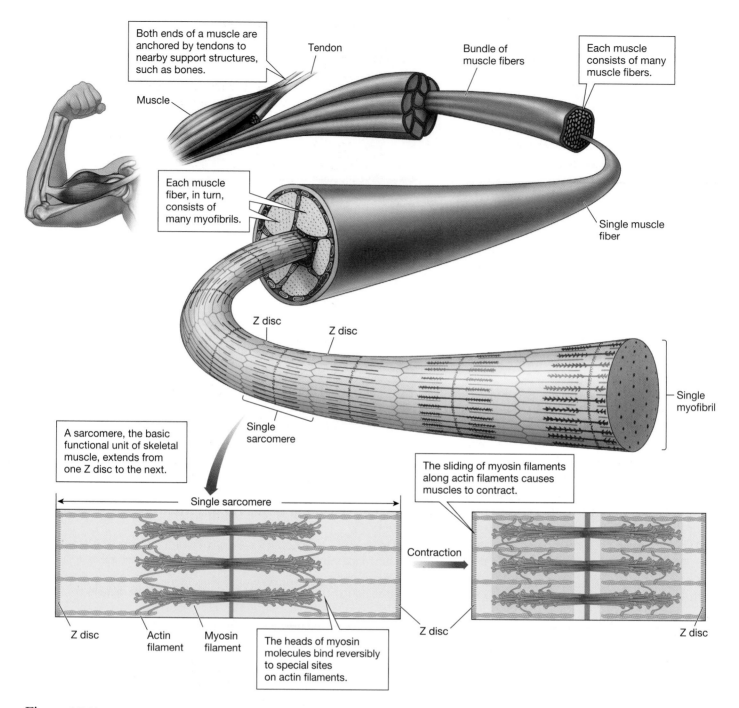

Both ends of a muscle are anchored by tendons to nearby support structures, such as bones.

Tendon

Bundle of muscle fibers

Each muscle consists of many muscle fibers.

Muscle

Each muscle fiber, in turn, consists of many myofibrils.

Single muscle fiber

Z disc

Z disc

Single myofibril

A sarcomere, the basic functional unit of skeletal muscle, extends from one Z disc to the next.

Single sarcomere

The sliding of myosin filaments along actin filaments causes muscles to contract.

Single sarcomere

Contraction

Z disc   Actin   Myosin
               filament   filament

The heads of myosin molecules bind reversibly to special sites on actin filaments.

Z disc

Z disc

Figure 20.11

The microscopic structure of muscle

Muscles contain bundles of muscle fibers, which contain myofibrils made up of sarcomeres. Muscle contraction depends on the movement of actin and myosin filaments within each sarcomere. The end of each sarcomere is composed of two Z discs, each of which contain a large protein that provides anchor points for the actin filaments.

Q1: List muscle structures from smallest to largest, beginning with sarcomeres.

Q2: What are the components of the sarcomere?

Q3: Across animal species, the microscopic structure of muscles is the same. Why, then, are there differences in strength among animals?

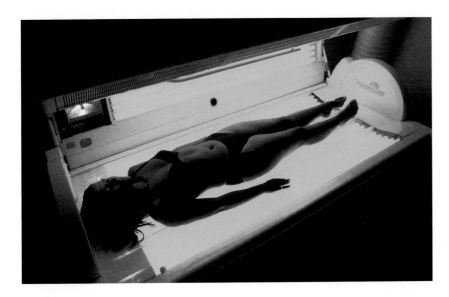

Figure 20.12

Tanning beds are a risky source of vitamin D

Exposure to ultraviolet rays, either outdoors from the sun or indoors with tanning beds or lamps, greatly increases the risk of skin cancer. It can also cause eye cancer, cataracts, and premature wrinkling. About 400,000 Americans a year develop skin cancer due to indoor tanning.

Many people began asking Autier whether they should take a vitamin D supplement to protect themselves. "The question was legitimate," says Autier, "so we had to address it."

With colleagues at iPRI, Autier analyzed data from 290 observational studies and 172 randomized trials examining the effects of vitamin D on nonskeletal disorders. They spent six years gathering then analyzing the data from those studies. "We had a huge database with many servers," says Autier. "You should have seen my office." From the data deluge, they found that high doses of vitamin D did not prevent any of the disorders they looked at. "There was nothing," says Autier.

His team concluded, therefore, that low vitamin D levels are not a cause of ill health. So why, then, do so many sick patients have low levels of vitamin D? One hypothesis is that diseases (especially cancer) are often associated with inflammation, and inflammation can reduce vitamin D concentrations in the body.

Three months after Autier's paper came out, a second, independent research team from the University of Auckland in New Zealand published another meta-analysis that came to a similar conclusion: Vitamin D supplementation does not reduce the risk of nonskeletal diseases.

All in all, these findings suggest that vitamin D supplements may not protect against nonskeletal disorders, though Autier emphasizes that their findings do not apply to pregnant women and young children, for whom vitamin D is critical during development.

Autier would be happy to see more large, randomized trials to confirm his findings. Weiler agrees. "Vitamin D doesn't have the same long history of people focusing on it, like heart disease," she says. "We know some relationships are there, but we haven't much looked at cause and effect."

In the meantime, consumers should not be nervous about taking vitamin D. The current recommended daily allowance from the Institute of Medicine, an impartial, nonprofit health and science policy institute, is 600 international units (IUs) of vitamin D per day to maintain health. (Seniors and breast-fed infants may need more.) Taking vitamin D for long periods of time in doses higher than 4,000 units per day is potentially unsafe, according to the National Institutes of Health, as it may cause excessively high levels of calcium in the blood. When blood levels of calcium are too high, the risk of heart attack and stroke rise dramatically.

Autier encourages other researchers to keep studying vitamin D. The story is not over. "It's quite good that we continue," he says, "because vitamin D levels could tell us a lot about the mechanisms of many diseases."

PHILIPPE AUTIER

Philippe Autier is an epidemiologist at the International Prevention Research Institute in Lyon, France. He researches cancer, specifically the role of UV light in skin cancer.

Nutritional Needs

Vitamins and minerals are essential nutrients that our bodies require in small amounts. Supplement manufacturers sell a range of different multivitamins, but the broad consensus from nutrition experts is that nothing substitutes for a healthy diet. Check out the data below to see how much of each nutrient you should be ingesting in a day, and some of the best foods in which to find them.

- ☐ Men's multivitamin
- ■ Women's multivitamin
- ■ RDA male
- ☐ RDA female

The recommended daily allowance (RDA) is the average daily dietary nutrient intake level that is sufficient to meet the nutrient requirements of nearly all (97–98 percent) healthy individuals in a particular life stage and gender group.

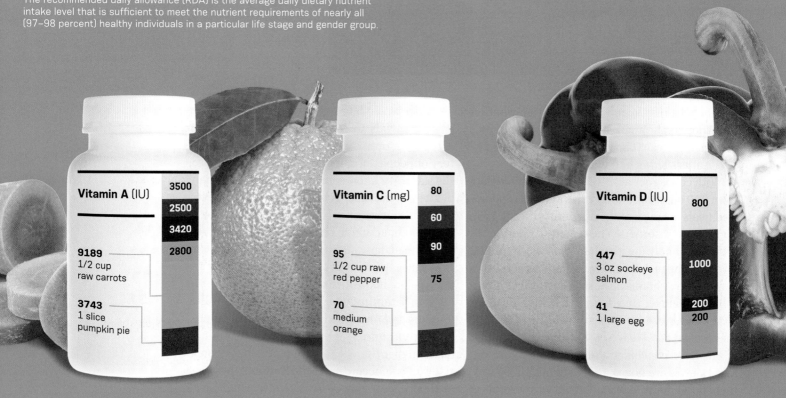

Vitamin A (IU)
3500
2500
3420
2800
9189 1/2 cup raw carrots
3743 1 slice pumpkin pie

Vitamin C (mg)
80
60
90
75
95 1/2 cup raw red pepper
70 medium orange

Vitamin D (IU)
800
1000
200
200
447 3 oz sockeye salmon
41 1 large egg

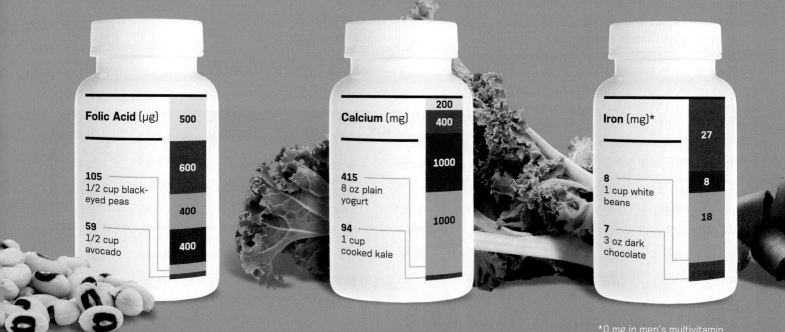

Folic Acid (µg)
500
600
400
400
105 1/2 cup black-eyed peas
59 1/2 cup avocado

Calcium (mg)
200
400
1000
1000
415 8 oz plain yogurt
94 1 cup cooked kale

Iron (mg)*
27
8
18
8 1 cup white beans
7 3 oz dark chocolate

*0 mg in men's multivitamin

REVIEWING THE SCIENCE

- Animals rely on **nutrients** for chemical building blocks and energy.
- **Vitamins** are organic compounds obtained from food that regulate metabolic processes in the animal body. **Minerals** are small, inorganic molecules needed by the body in small amounts.
- The **digestive system** is a tubular passageway that, in conjunction with accessory organs, processes ingested food.
  - After **ingestion**, food is broken down into smaller pieces in the **oral cavity** through the grinding action of teeth.
  - **Saliva** moistens the food and begins the chemical breakdown of starch.
  - Food passes from the oral cavity to the **pharynx**, down the **esophagus**, and into the acidic environment of the **stomach**, where protein **digestion** begins.
  - Partially digested food moves from the stomach into the upper region of the **small intestine**, where enzymes secreted by the **pancreas** and the intestine complete digestion of the food. The **liver** produces **bile** (stored and delivered by the **gallbladder**), which helps digest fats.
  - In the lower region of the small intestine, digested nutrients are absorbed into the body. The lining of the small intestine is highly folded and bears fingerlike projections (**villi**) that present a large surface area for absorbing nutrients.
- From the small intestine, any unabsorbed material moves into the **colon**, where remaining water and minerals are absorbed. Here, bacteria break down the waste and release nutrients that the body can absorb.
- The **axial skeleton** supports and protects vital organs along the long axis of the body. The **appendicular skeleton** is composed of the bones in the arms, legs, and pelvis, and is primarily involved with movement.
- Most bones are made of **spongy bone** surrounded by harder **compact bone**. Hollow bones contain **marrow**.
- **Ligaments** connect bone to bone. **Tendons** connect muscles to bones. Tendons and ligaments are made of **collagen**.
- Muscles provide the power necessary for movement. Muscles consist of **muscle fibers**, each of which is packed with myofibrils. **Myofibrils** contain repeating units called **sarcomeres** that contract.
- Movements are created by specialized muscle types. **Cardiac muscle** is found in the heart, where its contractions pump blood. **Smooth muscle**, which is found in the digestive tract and blood vessels, contracts in waves. **Skeletal muscle** is under conscious control and is banded in appearance.

THE QUESTIONS

The Basics

1 Which nutrient is most important for energy *storage*?

(a) proteins

(b) carbohydrates

(c) lipids

(d) all of the above

2 Which of the following is true of vitamin D?

(a) A deficiency can interfere with the formation of bones and teeth.

(b) Sunlight is the only source.

(c) Leafy green vegetables are an excellent source.

(d) It is a water-soluble vitamin.

3 Vitamin K is

(a) a necessary dietary mineral.

(b) manufactured by intestinal bacteria.

(c) important for good eyesight.

(d) a water-soluble vitamin.

4 Which of the following is *not* true of the small intestine?

(a) It is specialized for absorption.

(b) It absorbs nutrients through villi and microvilli.

(c) It secretes digestive enzymes.

(d) It is specialized for elimination of waste.

5 _____ and _____ are important for controlled movement of the knee.

(a) The axial skeleton; appendicular skeleton

(b) The ulna; radius

(c) Rigidity; flexibility

(d) The femur; tibia

6 Link each of the following terms with the best definition.

SKELETAL MUSCLE	1. Muscle that is consciously controlled.
SMOOTH MUSCLE	2. Type of muscle found in the heart.
CARDIAC MUSCLE	3. Type of muscle used for walking and running.
VOLUNTARY MUSCLE	4. Type of muscle found in the digestive system.
INVOLUNTARY MUSCLE	5. Muscle that works without conscious control.

7 Circle the correct terms in the following sentence:

Within a [**muscle, tendon**], sarcomeres shorten when their [**actin, myosin**] filaments slide over [**actin, myosin**] filaments joined to [**Z, X**] discs on either side of the [**muscle fiber, sarcomere**].

8 Beginning with the mouth, identify the order in which food moves through the digestive system by numbering the events below from 1 to 5.

___ a. Nutrients are absorbed through cells of the villi.

___ b. Food is broken into smaller pieces through chewing.

___ c. Digestive enzymes are released by accessory organs.

___ d. Bacteria help digest food and produce some vitamins before sending wastes out of the body.

___ e. Acids break down proteins for further digestion.

Try Something New

9 Our bones are constantly changing in response to how we live. Physical activity builds stronger bones. The bones in a pitcher's throwing arm, for example, are stronger and have larger ridges to which muscles can attach than do the same bones in the other arm. Inactivity leads to weaker bones because special osteocytes step up the rate at which they remove tissue from the bone when the skeleton is not under physical stress; such weakening is seen, for example, in a person confined to bed by an injury or illness. Which of the following activities would be the least effective in increasing bone strength? Explain why.

(a) walking

(b) dancing

(c) weight lifting

(d) swimming

(e) running/jogging

10 A January 2013 Tumblr post that went viral on the Internet read as follows:

> eating is so bad*** i mean you put something in a cavity where you smash and destroy it with 32 protruding bones and then a meat tentacle pushes it into a vat of acid and after a few hours later you absorb its essence and transform it in(to) energy just wow.

Please identify each element of the digestive system discussed in the post. Do you think this is an accurate description of digestion?

Leveling Up

11 **Looking at data** There are two kinds of dietary minerals: macrominerals, which your body needs in larger amounts; and trace minerals, which your body needs in smaller amounts. Macrominerals include calcium and phosphorus. Trace minerals include iron, copper, iodine, zinc, and fluoride.

(a) According to the graph below, what is the dividing line between macrominerals and trace minerals?

(b) What does the y-axis show?

(c) Describe in your own words what the graph shows.

(d) How much potassium is needed by an average person?

(e) Is the amount of potassium you stated in part (d) the same amount that we need to eat daily?

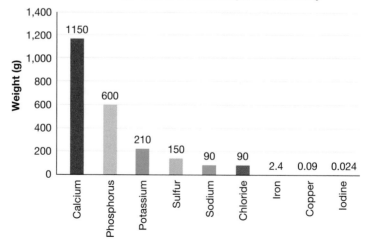

Average Mineral Content Required in the Body

12 Life choices Nutritionists agree that we can obtain all our necessary nutrients in optimal amounts from natural sources, provided we eat a well-balanced diet that includes a variety of foods. Recognizing that relatively few people meet this ideal in reality, some physicians advise their patients to take a multivitamin and mineral supplement as "added insurance." The use of this kind of dietary supplement raises the specter of toxicity from excessive intake, particularly of the fat-soluble vitamins. Most, but by no means all, manufacturers attempt to avoid high levels of the fat-soluble vitamins in their vitamin preparations, but the burden is largely on us to use supplements wisely because this industry is not highly regulated by the government. Review the information provided on dietary supplements by the National Institutes of Health Office of Dietary Supplements (http://ods.od.nih.gov/HealthInformation/DS_WhatYouNeedToKnow.aspx) and the U.S. Food and Drug Administration, or FDA (http://www.fda.gov/Food/DietarySupplements/UsingDietarySupplements/ucm110567.htm). Take notes so that you can answer the following questions.

(a) What aspects of the sale of dietary supplements does the government regulate?

(b) Are you confident that this regulation is sufficient to make supplements safe to use as labeled? Why or why not?

(c) List three recommendations that you found informative or helpful. Will they change your use of dietary supplements? Explain your reasoning.

(d) Do you think that a mobile app for dietary supplements (MyDS at the NIH site) would be helpful to have? Why or why not?

(e) Identify one dietary supplement that you feel would be helpful for you to take. Will you take it? Why or why not?

Body (Re)Building

Could engineered human tissues, brought to life in the lab, replace failing organs in people?

After reading this chapter you should be able to:

- Create a flowchart depicting the movement of blood through a cardiovascular system.
- Describe the different components of blood and the function of each.
- Compare and contrast the three types of blood vessels.
- Diagram the elements of the respiratory system and show how air moves through it.
- Explain how gases are exchanged in the lungs.
- Describe how the structure of a nephron relates to its function in the urinary system.
- Distinguish between the central and peripheral nervous systems.
- Identify the sensory systems active in humans.
- Label a neuron, showing the direction of signal transmission.

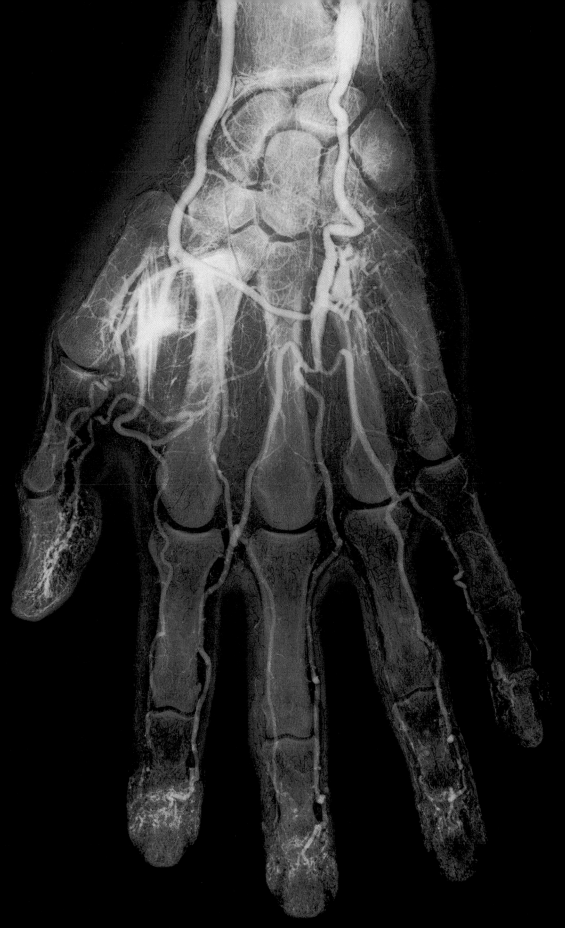

PHYSIOLOGY

CHAPTER

21

CIRCULATORY,
RESPIRATORY,
URINARY, AND
NERVOUS
SYSTEMS

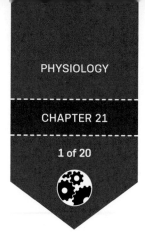

Standing over the operating table, covered head to toe in teal scrubs, physician Jeffrey Lawson lifts a long, white tube out of a bath of clear liquid. Carefully, slowly, he threads the tube through the unconscious patient's upper arm and stitches each side of the tube to an exposed blood vessel.

Bioengineer Laura Niklason stands to the side of the operating table, watching closely (**Figure 21.1**). She and Lawson created the tube being implanted in the patient. Once the operation is complete, Lawson steps back, and Niklason gives him a hug. "Congratulations to you," says Niklason happily. "We're saving the world!"

On June 5, 2013, Lawson, Niklason, and their team transplanted a laboratory-grown blood vessel into a human. It was the first such procedure in the United States and a major feat for the field of *tissue engineering*, the effort to grow or regenerate tissues or organs using engineering materials and principles. Other engineered tissues implanted into humans have included nerve grafts, bladders, and windpipes.

Lawson and Niklason imagine a future in which any type of organ can be constructed in the lab. In an ideal world, "we'll be able to grow all sorts of tissues for patients, so a surgeon can literally reach up on a shelf, pull down a tissue graft, and implant it in a patient," Niklason said in 2013, shortly after the first blood vessel transplant. "That's really going to be a revolution—being able to grow replacement parts for patients, that they don't have to wait for."

Organ shortage is a major concern in health care (**Figure 21.2**). According to the U.S. government, the number of people waiting for an organ would fill a football stadium—twice. And every day, an average of 18 people die waiting for transplants of kidneys, hearts, livers, lungs, and more.

One possible solution to this organ crisis is the Frankenstein-like approach of tissue engineering: building organs in the lab. It sounds farfetched, but research teams have already begun to engineer complicated organs such as kidneys, lungs, even whole hearts. These are just a few of the organs that make up the major organ systems in our bodies, including the circulatory, respiratory, urinary, and nervous systems—each one critical to the healthy functioning of our bodies (see Figure 19.4 for an overview).

That healthy functioning, including growth and homeostasis, depends on the internal transfer of signaling molecules, transport proteins, waste products, and other substances in the body. Distances inside the bodies of most multicellular organisms are far too great for diffusion to be an effective means of distributing these materials, so elaborate organ systems exist to transport them around the body. The **circulatory system** moves oxygen from the lungs to the heart, which then pumps oxygen-rich blood to the rest of the body. The **respiratory system** brings in oxygen and expels carbon dioxide to support cellular respiration. The **urinary system** removes excess fluid from the body,

Figure 21.1

A bioengineered blood vessel

In 2013, a team of doctors at Duke University was the first in the United States to implant a bioengineered blood vessel into a patient.

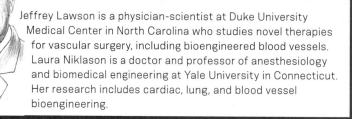

JEFFREY LAWSON AND LAURA NIKLASON

Jeffrey Lawson is a physician-scientist at Duke University Medical Center in North Carolina who studies novel therapies for vascular surgery, including bioengineered blood vessels. Laura Niklason is a doctor and professor of anesthesiology and biomedical engineering at Yale University in Connecticut. Her research includes cardiac, lung, and blood vessel bioengineering.

along with waste products, toxins, and other water-soluble substances that are not needed.

These three systems are interconnected in the body: Carbon dioxide collected by the circulatory system is delivered to the respiratory system for exchange with the outside environment. The circulatory system brings substances dissolved in blood to the urinary system for discharge into the environment. And a unique high-speed communication system called the **nervous system** coordinates the many muscles involved in the functioning of organs, including the heart, lungs, and bladder. The nervous system directs the rapid contractions of muscles and processes information received by the senses, such as touch, sound, and sight. In this way, the nervous system enables an animal to detect food, find a mate, avoid predators, and respond to extremes of heat and cold.

These organ systems, as well as the other organ systems introduced in Chapter 19, govern how your body operates. If one of your organs fails, a whole system can shut down. That's why researchers are exploring how to build new organs, writing a recipe for creating flesh and blood.

Emergency Meeting

Niklason and Lawson met at Duke University Medical Center in 1999 while performing surgery together. Waiting in the operating room to move their patient, the two struck up a conversation. Within moments, they discovered they shared strikingly similar interests: Lawson was a surgeon who spent his days suturing blood vessels and running a laboratory where he grew different types of blood cells, and Niklason was an anesthesiologist and bioengineer who had recently opened her own lab to build blood vessel tubes. Both were dreaming of the same goal—the creation of a blood vessel from scratch—and they had complementary skills. "It made us a perfect partnership," says Lawson.

That first conversation bloomed into a friendship and working partnership. Together, Lawson and Niklason started building prototypes of blood vessels, using his knowledge of blood cells and her knowledge of the engineering forces at

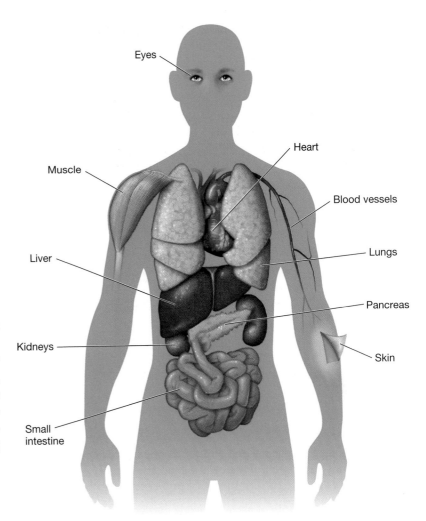

Figure 21.2

Organs needed for donation

work in blood vessels. **Blood vessels** are a critical part of the **cardiovascular system**, a closed circulatory system consisting of a muscular *heart*, a complex network of *blood vessels* that collectively form a closed loop, and *blood* that circulates through the heart and blood vessels. Almost all of our cells lie within 0.03 millimeters of a blood vessel with which they exchange materials by diffusion. Carrying blood so close to all the trillions of cells in our bodies requires an extensive network of vessels.

Lawson and Niklason built and then implanted blood vessels created of various materials and cell types in rats to see whether blood would successfully flow through the tubes. **Blood** is composed of cells and cell fragments that float in a fluid known as **plasma** (**Figure 21.3**). Blood plasma itself has a low capacity for

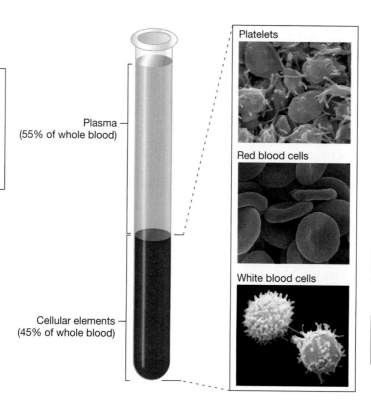

Blood plasma is 92% water and contains dissolved gases, ions, and molecules that are critical for homeostasis, as nutrients, or as signaling molecules. Much of the carbon dioxide carried in the blood is dissolved in plasma.

Plasma (55% of whole blood)

Cellular elements (45% of whole blood)

Platelets

Platelets are small cell fragments. They can clump together to help stop the loss of blood if a blood vessel is damaged. Platelets release substances that stimulate plasma proteins to create a meshwork of protein strands, platelets, and blood cells to collectively form a blood clot.

Red blood cells

A mature red blood cell has no nucleus, and its cytoplasm is packed with oxygen-binding proteins called hemoglobin. Each hemoglobin molecule can carry up to four oxygen molecules. Because each human blood cell contains about 280 million hemoglobin molecules, a single one of these cells can bind over a billion molecules of oxygen.

White blood cells

Several different kinds of white blood cells help defend the body from invading organisms.

Figure 21.3

Human blood consists of fluid and several different cell types

Whole blood consists of plasma and different kinds of cells and cell fragments, three of which are shown here. Red blood cells account for about 95 percent of the cells in blood.

Q1: Where is the majority of carbon dioxide carried in the blood?

Q2: Where is the majority of oxygen carried in the blood?

Q3: What would happen if your red blood cells carried a mutation that made the hemoglobin less effective at binding to oxygen (as in sickle-cell disease)?

transporting dissolved oxygen, but **red blood cells** in the plasma carry significant amounts of oxygen, greatly increasing the oxygen-carrying capacity of blood.

In 2005, the duo hit upon a technique that seemed to work. They collected smooth muscle cells from the blood vessels of organ donors and grew those muscle cells on a biodegradable frame shaped like a blood vessel. The cells worked like little machines, churning out proteins that formed a three-dimensional scaffold of connective tissue called the *extracellular matrix*. Then the original, biodegradable frame dissolved, leaving behind a sturdy tube of cells and extracellular matrix.

There was still a catch. The immune system of a human body rejects cells from another person

(see Chapter 22 for more on the immune system), so Lawson and Niklason had to wash away the original donor muscle cells to leave behind just the tubular extracellular matrix. "It's like we kept the mortar but the bricks all washed away," says Lawson. They tested their creation in rats. It worked. They had made a functioning blood vessel.

The vessels were first tested in humans in Poland in December 2012. The researchers found that once an engineered tube is implanted into a patient, the patient's own blood and muscle cells take up residence in the tube, filling in the cracks. "What started off as our structure now becomes your tissue," says Lawson. "And it all seals together so it doesn't leak."

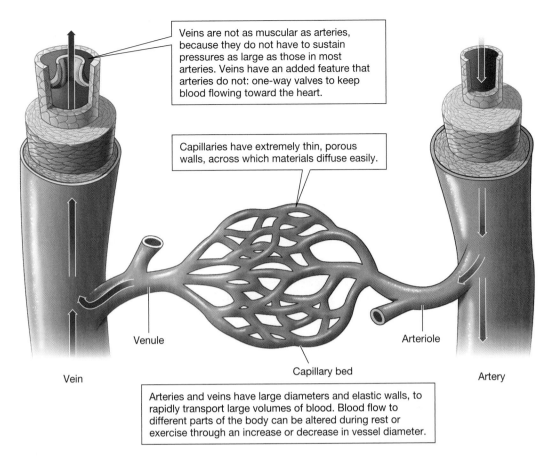

Veins are not as muscular as arteries, because they do not have to sustain pressures as large as those in most arteries. Veins have an added feature that arteries do not: one-way valves to keep blood flowing toward the heart.

Capillaries have extremely thin, porous walls, across which materials diffuse easily.

Venule

Arteriole

Capillary bed

Vein

Artery

Arteries and veins have large diameters and elastic walls, to rapidly transport large volumes of blood. Blood flow to different parts of the body can be altered during rest or exercise through an increase or decrease in vessel diameter.

Figure 21.4

Arteries, veins, and capillaries

Arteries carry blood away from the heart. Veins carry blood toward the heart. Arterioles and venules are smaller arteries and veins, respectively. The narrowest arteries and veins connect with each other in a fine network known as a capillary bed.

Q1: Why are arteries more muscular than veins?

Q2: What structural feature(s) of capillaries enable easier diffusion into and out of surrounding tissues?

Q3: Why do you think that capillaries are not typically transplanted?

Importantly, the engineered vessel has the structural strength of a normal blood vessel in the body, so it can withstand the force of the blood pulsing through it. The human body has three major kinds of blood vessels through which blood flows: *arteries*, *veins*, and *capillaries* (**Figure 21.4**). **Arteries** are large vessels (.1–10 mm) that transport blood away from the heart. **Veins** (.1–2 mm) are large vessels that carry blood back to the heart. **Capillaries**, the smallest vessels at .005 to .01 mm, exchange materials by diffusion with nearby cells.

The large vessels—arteries and veins—are built for mass transport of blood. Currently, Niklason and Lawson have built blood vessels with diameters of six millimeters (about the width of a pencil) and three millimeters. But they also expect to be able to make and transplant larger vessels, such as the aorta, the main artery of the body. Capillaries, however, are very small and are not typically transplanted. They are built for slower movement of blood, and their large surface area facilitates the exchange of materials with surrounding cells.

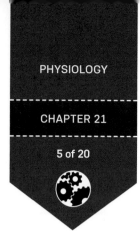

Niklason and Lawson's blood vessels are not the only tissues in the cardiovascular system that researchers are trying to engineer. Scientists are attempting to grow the **heart**, a muscular organ the size of a fist in humans that works as the body's circulatory pump (**Figure 21.5**). Like the hearts of all other mammals, the human heart is divided into four chambers that form two distinct pumping units, which are independent but coordinated.

The left atrium receives oxygenated blood from the lungs and pumps it to the left ventricle, which pumps it through the **systemic circuit** to cells performing cellular respiration. The right atrium receives blood low in oxygen and laden with CO_2 returning from the systemic circuit, and pumps it to the right ventricle, which pumps it through the **pulmonary circuit** for gas exchange in the lungs.

Together, these chambers pump some 7,000 liters of blood per day; about 1,850 gallons. **Heart rate** is the number of times a heart beats per minute; an average resting (not exercising) human heart beats about 60–100 times per minute. The force of blood pushing through blood vessels is called **blood pressure**. As the heart contracts to push blood out, blood pressure is at its highest, and when the heart relaxes after each contraction, blood pressure is at its lowest. The first is referred to as **systole**, and is the top number in a blood pressure reading. The second is **diastole**, and is the bottom number in a blood pressure reading. The human circulatory system adjusts the heart rate and patterns of blood distribution according to the body's needs.

The heart might seem like too complex an organ to replicate using tissue engineering techniques, but researchers have demonstrated some success using a tactic similar to the one Niklason and Lawson took with blood vessels, starting with an entire organ as a scaffold (such as a pig or rat heart, of which there is a ready supply, or a heart from an organ donor). They use detergents to strip away all the original cells that would cause an immune reaction, then repopulate the heart with cells that are a better match for the patient. Researchers have created pumping rat hearts in a dish using this technique, though these bioengineered hearts are still too primitive to work when transplanted into an animal.

Since the initial implant of Niklason and Lawson's blood vessel, 60 more implants of bioengineered blood vessels have been performed in the United States. And in November of 2013, a company testing the blood vessels released preliminary safety results: Of the first 20 patients in Poland who received engineered blood vessels, none of the transplants had become infected, and none showed evidence of rejection.

"We've got a tube that works to put into people, so now we have the groundwork to make things more complicated than a blood vessel," says Lawson. "There is still an unlimited amount of science to do."

Breathe In, Breathe Out

In 2010, after she had moved from Duke to Yale University, Niklason expanded her research program from creating simple blood vessels to attempting to build a lung. **Lungs** are the main organs of the *respiratory system*, which carries air from the nose (or mouth) to the lungs through a series of tubular passageways. These airways allow air to move between the external environment and the inside of the body—specifically, the gas exchange surfaces in the lungs.

The process of taking air into the lungs (inhaling) and expelling air from them (exhaling) is called **breathing** (**Figure 21.6**). The air we inhale is about 21% oxygen and contains little carbon dioxide and water vapor. The air we exhale has less oxygen (15%) and contains about 4% each of carbon dioxide and water vapor. That's because these gases are exchanged at the surface of the cells that line our lungs: Oxygen is removed from the inhaled air and sent to the bloodstream, while carbon dioxide and water vapor are removed from the bloodstream and added to the air that is exhaled.

At Yale, Niklason's team bioengineered a rat lung and implanted it in a rat. There, it briefly supported gas exchange for the animal before filling up with fluid, suggesting that more work needs to be done. Any engineered lung needs to be highly reliable, as lungs are vital to transporting oxygen throughout the body. Although humans can live for more than a week without

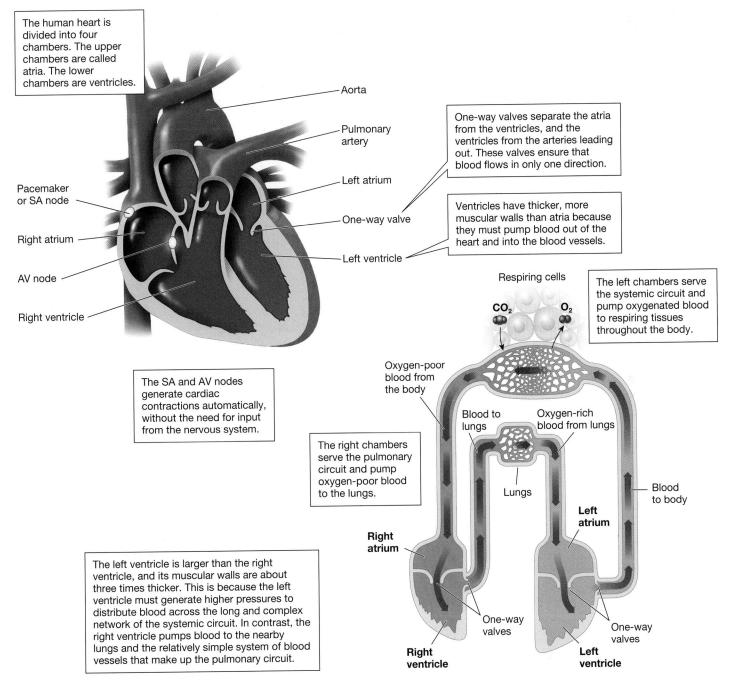

The human heart is divided into four chambers. The upper chambers are called atria. The lower chambers are ventricles.

Aorta

Pulmonary artery

Left atrium

One-way valve

Left ventricle

One-way valves separate the atria from the ventricles, and the ventricles from the arteries leading out. These valves ensure that blood flows in only one direction.

Ventricles have thicker, more muscular walls than atria because they must pump blood out of the heart and into the blood vessels.

Pacemaker or SA node

Right atrium

AV node

Right ventricle

The SA and AV nodes generate cardiac contractions automatically, without the need for input from the nervous system.

The left chambers serve the systemic circuit and pump oxygenated blood to respiring tissues throughout the body.

Respiring cells

CO_2 O_2

Oxygen-poor blood from the body

Blood to lungs

Oxygen-rich blood from lungs

Lungs

Blood to body

The right chambers serve the pulmonary circuit and pump oxygen-poor blood to the lungs.

Right atrium

Left atrium

The left ventricle is larger than the right ventricle, and its muscular walls are about three times thicker. This is because the left ventricle must generate higher pressures to distribute blood across the long and complex network of the systemic circuit. In contrast, the right ventricle pumps blood to the nearby lungs and the relatively simple system of blood vessels that make up the pulmonary circuit.

One-way valves

One-way valves

Right ventricle

Left ventricle

Figure 21.5

The human heart

The right and left sides of the heart function as two separate pumps, although the two upper chambers, or **atria** (singular "atrium"), contract in unison, as do the two lower chambers, the **ventricles**. This unified contraction begins with a signal from the **pacemaker**, or **sinoatrial (SA) node**. The signal causes both atria to contract; it also causes the **atrioventricular (AV) node** to pass the signal on to the ventricles about a tenth of a second later. The short delay allows the atria to empty completely.

Q1: Beginning with the left atrium, list the locations of a drop of blood as it moves through the circulatory system.

Q2: Why is the left ventricle larger than the right ventricle, and why are its walls thicker and more muscular?

Q3: Some people have an artificial pacemaker implanted in their heart. What is its function?

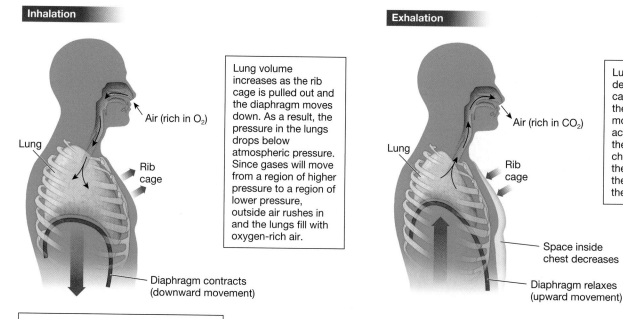

Lung

Air (rich in O₂)

Rib cage

Lung volume increases as the rib cage is pulled out and the diaphragm moves down. As a result, the pressure in the lungs drops below atmospheric pressure. Since gases will move from a region of higher pressure to a region of lower pressure, outside air rushes in and the lungs fill with oxygen-rich air.

Diaphragm contracts (downward movement)

The diaphragm is a thick sheet of muscle that forms the floor of the chest cavity.

Lung

Air (rich in CO₂)

Rib cage

Lung volume decreases as the rib cage is pulled in and the diaphragm moves up. This action compresses the space in the chest cavity, raising the pressure inside the lungs and forcing the air out of them.

Space inside chest decreases

Diaphragm relaxes (upward movement)

Figure 21.6

Breathing

Breathing involves two main steps: inhalation, when air is pulled into the lungs, and exhalation, when air is pushed out of the lungs. Most of the time, breathing is controlled automatically by sensory systems located in the heart and brain. If we choose, we can also control our breathing with a system of muscles, the most important of which are the rib muscles and the diaphragm.

Q1: When is air richer in oxygen—as it enters the body or as it exits the body?

Q2: The figure shows air entering through the nose. Where else can air enter the respiratory system?

Q3: Explain how the body creates a change in air pressure during breathing.

food and for a few days without water, a mere four minutes without oxygen results in irreversible brain damage. And lung tissue is notoriously bad at repairing itself, so lung transplants are often the only option in the case of lung damage from disease or trauma.

The respiratory system can be divided into two parts (**Figure 21.7**). The **upper respiratory system** includes airways in the nose, mouth, and throat. When we inhale, air enters through the nostrils and moves into each nasal cavity. Next the air enters the throat, or **pharynx**, an area where the back of the mouth and the two nasal cavities join together into a single passageway. From the pharynx, air moves into the **larynx**, or voice box, which forms the entryway to the windpipe, or **trachea**.

The trachea is the start of the **lower respiratory system**. Within the chest, the trachea branches into two smaller tubes called **bronchi** (singular "bronchus"). Each bronchus leads to one of the paired lungs, the organs where gases are exchanged. Together, the trachea, bronchi, and lungs make up the lower respiratory system.

Inside the lungs is where the respiratory system gets intricate: The bronchi divide into **bronchioles**, a series of branching, ever-smaller tubes. The tiniest bronchioles open into the **alveoli** (singular "alveolus"), small clusters of sacs that resemble a bunch of grapes. Gases are exchanged across the moist surface of the thin layer of cells that line each alveolar sac (**Figure 21.8**). Niklason continues to try to achieve

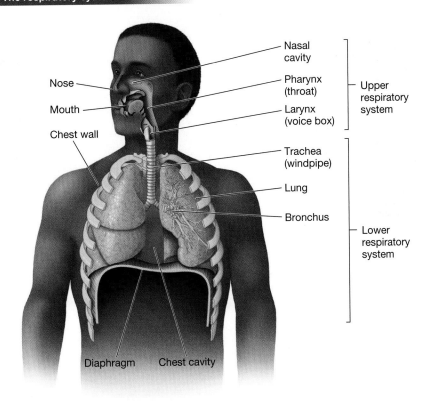

The respiratory system

Nasal cavity

Nose

Pharynx (throat)

Mouth

Larynx (voice box)

Upper respiratory system

Chest wall

Trachea (windpipe)

Lung

Bronchus

Lower respiratory system

Diaphragm   Chest cavity

Figure 21.7

The human respiratory system

The respiratory system consists of the upper respiratory system (nose, mouth, throat, and larynx) and the lower respiratory system (the trachea, bronchi, and the lungs).

Q1: Beginning with the nose, list the locations of a molecule of oxygen as it moves through the respiratory system.

Q2: From the lung, where would the oxygen molecule move?

Q3: How might an upper respiratory infection (URI) like the common cold affect your respiratory system and thus your breathing?

this gas exchange in an engineered lung. So far, her rat lung transplants have been able to do so for about two hours before failing.

These organs, however, have yet to be successfully transplanted.

Waste Not

At Massachusetts General Hospital in Boston, Harald Ott, a pioneer in the field of organ bioengineering, has engineered whole rat, pig, and human lungs by cleaning donated organs and repopulating them with new cells (**Figure 21.9**).

HARALD OTT

Harald Ott is a thoracic surgeon at Massachusetts General Hospital and leads a laboratory focused on organ engineering and regeneration at Harvard Medical School.

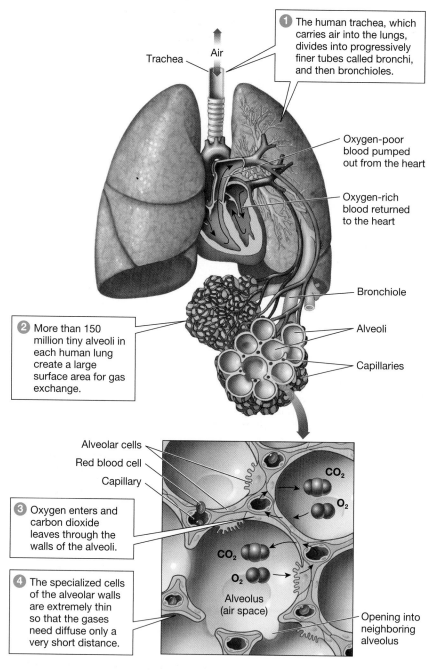

① The human trachea, which carries air into the lungs, divides into progressively finer tubes called bronchi, and then bronchioles.

Trachea

Air

Oxygen-poor blood pumped out from the heart

Oxygen-rich blood returned to the heart

Bronchiole

Alveoli

② More than 150 million tiny alveoli in each human lung create a large surface area for gas exchange.

Capillaries

Alveolar cells

Red blood cell

Capillary

CO₂

O₂

③ Oxygen enters and carbon dioxide leaves through the walls of the alveoli.

CO₂

O₂

④ The specialized cells of the alveolar walls are extremely thin so that the gases need diffuse only a very short distance.

Alveolus (air space)

Opening into neighboring alveolus

Figure 21.8

Gas exchange

The structure of the lungs speeds the diffusion of oxygen and carbon dioxide into and out of the body by providing a large surface area for gas exchange.

Q1: Why is a large surface area important for gas exchange?

Q2: Does carbon dioxide move into or out of the alveoli? Into or out of capillaries at the surface of the alveoli?

Q3: When a person has pneumonia, the alveoli may fill with fluids. Why would this be a problem?

Figure 21.9

Bioengineered rat lungs

Researchers at Massachusetts General Hospital have bioengineered the lungs of rats, pigs, and humans. None have yet been capable of functioning within a living individual.

In 2013, Ott's lab used the scaffolding technique to engineer rat kidneys, which were then transplanted into rats. In vertebrates, **kidneys** are a set of paired organs that maintain water and solute homeostasis, serving as key components of the *urinary system*.

All animals must regulate the concentrations of solutes, such as sodium and calcium, in their body fluids, but terrestrial animals face the additional challenges of conserving water and retaining vital solutes. The solute composition of an animal is affected by metabolic activities within the body: As cells metabolize macromolecules, they use up chemicals dissolved in body fluids and produce new ones. This process results in waste products that must somehow be removed from the body.

Kidneys filter and regulate the composition of the blood as it moves through them. When blood leaves the kidneys and returns to the circulatory system, it is cleansed of metabolic wastes and carries water and solutes in normal amounts. The volume of blood leaving the kidneys is slightly smaller than the entering volume because some water is lost to make **urine**, the waste-carrying solution that is expelled from the body.

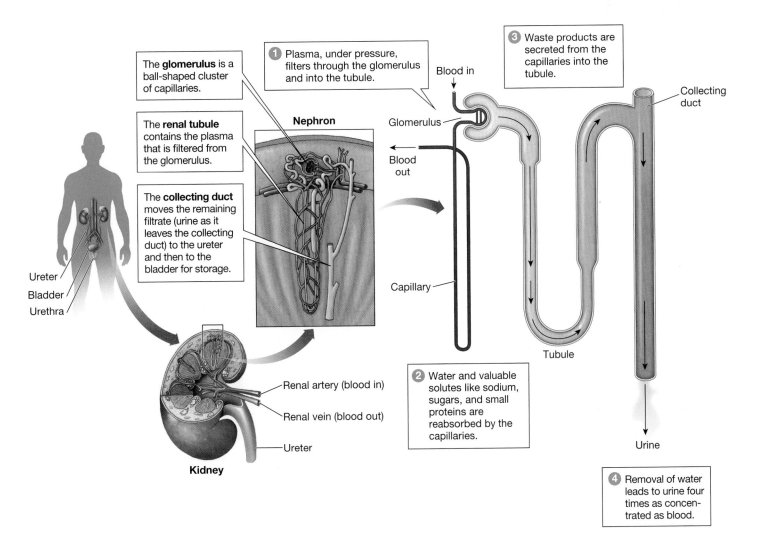

Figure 21.10

The human urinary system

The kidney regulates internal water content, balances solute concentrations, and removes toxic wastes. All of this work occurs within the nephron.

Q1: What is released from the kidney, and where does it go after release?

Q2: What is the difference between reabsorption and secretion?

Q3: Alcohol suppresses the kidney's ability to reabsorb water. What common consequence of drinking alcohol is related to this fact, and how might you alleviate this problem?

The blood-cleansing work of the kidneys—**filtration**—is performed by the kidney's basic functional unit, the **nephron** (**Figure 21.10**). Each human kidney has about a million of these tiny filtration units. When the kidneys fail, an individual can no longer filter waste. Approximately 100,000 individuals in the United States currently await kidney transplantation. As they wait, these patients receive dialysis three times per week, during which a machine filters their blood for them. This lifesaving technology requires a significant time commitment, and dialysis patients have a high risk of infection.

An engineered kidney would provide a permanent solution for these patients. But an engineered kidney would need to do more than just

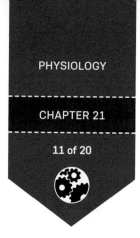

Figure 21.11

A bioengineered kidney: before and after

The Ott lab at Massachusetts General Hospital bioengineered rat kidneys that function when transplanted. Before implanting, the kidney is flushed of its donor cells and repopulated (seeded) with recipient cells so that it will not be rejected by the recipient's immune system when transplanted.

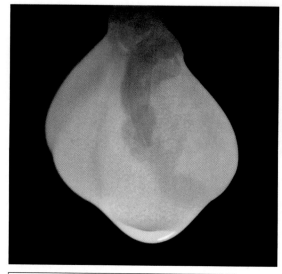

A rat kidney with all its cells removed, leaving only the extracellular matrix made of collagen.

The collagen "scaffold" has now been reseeded with cells.

remove waste from the blood. Filtration is only one of three parts of the kidney's job. A second important function is the **reabsorption** of water and valuable solutes such as sodium, chloride, and sugars before they leave the kidneys. A third function of the kidney is **secretion**: The kidney actively transports excess quantities of substances such as potassium and hydrogen ions, and some medications and toxins, from the blood into the liquid passing through the kidney.

The concentrated fluid that results from the combination of filtration, reabsorption, and secretion is urine. Urine from the many collecting ducts in a kidney drains into a long tube, the *ureter*, which delivers the fluid to the urinary bladder for storage. In urination, the bladder empties through a tube called the *urethra*. In Ott's lab in Boston, the bioengineered rat kidneys successfully produced urine when transplanted into rats (**Figure 21.11**).

Coming to Your Senses

To date, the kidney is the most complex organ recreated in the lab and successfully transplanted into an animal. Meanwhile, bioengineering research in a different organ system—the nervous system—has already yielded astonishing improvements in patients' lives, including the life of navy corpsman Edward Bonfiglio Jr.

At 5:00 a.m. on August 27, 2009, Edward Bonfiglio awoke to the sound of the phone ringing. It was his son, Edward Jr., a member of the U.S. Navy serving on active duty in Afghanistan. "Dad?" Edward said, "I got shot. Don't tell Mom."

During a routine foot patrol, Edward's unit had been ambushed, and he'd been shot in his left leg. The bullet hit his sciatic nerve, a long, thick bundle of nerve fibers that runs from the lower back down the leg. Edward immediately

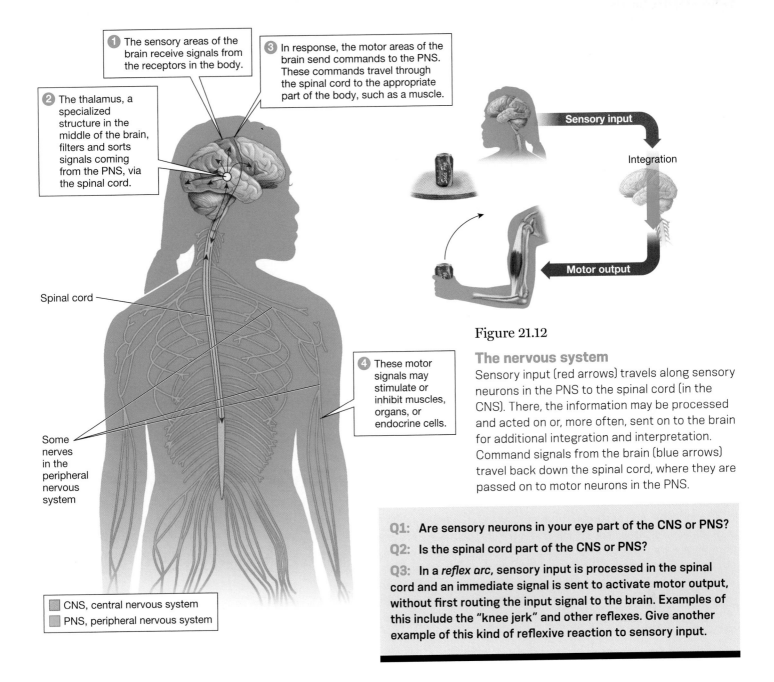

1 The sensory areas of the brain receive signals from the receptors in the body.

2 The thalamus, a specialized structure in the middle of the brain, filters and sorts signals coming from the PNS, via the spinal cord.

3 In response, the motor areas of the brain send commands to the PNS. These commands travel through the spinal cord to the appropriate part of the body, such as a muscle.

Spinal cord

Some nerves in the peripheral nervous system

4 These motor signals may stimulate or inhibit muscles, organs, or endocrine cells.

Sensory input

Integration

Motor output

◼ CNS, central nervous system
◼ PNS, peripheral nervous system

Figure 21.12

The nervous system

Sensory input (red arrows) travels along sensory neurons in the PNS to the spinal cord (in the CNS). There, the information may be processed and acted on or, more often, sent on to the brain for additional integration and interpretation. Command signals from the brain (blue arrows) travel back down the spinal cord, where they are passed on to motor neurons in the PNS.

Q1: Are sensory neurons in your eye part of the CNS or PNS?

Q2: Is the spinal cord part of the CNS or PNS?

Q3: In a *reflex arc*, sensory input is processed in the spinal cord and an immediate signal is sent to activate motor output, without first routing the input signal to the brain. Examples of this include the "knee jerk" and other reflexes. Give another example of this kind of reflexive reaction to sensory input.

lost all feeling and function below his left knee. Doctors found that the bullet had sliced a five-centimeter gap in the sciatic nerve. Without that nerve intact, the leg would not function.

After he arrived home in the United States, doctors gave Edward two options: either amputate the leg or repair the damaged nerve with a new kind of nerve graft, a bioengineered tube that could potentially reconnect the two ends of his severed nerve and bring feeling back to his leg. Edward chose the latter—to try to repair the nerve in his leg, part of his peripheral nervous system.

The *nervous system* of vertebrates is a communication system that transmits signals among various parts of the body. It can be divided into two main units: the **central nervous system**, or **CNS**, and the **peripheral nervous system**, or **PNS** (**Figure 21.12**). The CNS consists of the brain and spinal cord. The **brain** has a large capacity for processing diverse types of sensory information, and it controls and coordinates nerve signals throughout the body (see "What's in Your Head?" on page 380). The **spinal cord** is a thick central nerve cord that is continuous with the brain, acting as a filter between the brain and sensory neurons.

The PNS consists of **sensory organs**—such as the eyes and ears—plus the nerves (except for the retinal, optic, and olfactory nerves, which are

What's in Your Head?

The human brain is mind-bogglingly complex. With an estimated 100 billion nerve cells, it is the epicenter of the nervous system. The brain is the organ that most distinctly sets humans apart from other species, giving us our capacity to reason, feel, and remember. The brain is made up of *gray matter* (the cell bodies of neurons) and *white matter* (the branching network of neurons' projecting dendrites and axons, winding tendrils that connect one neuron to many other neurons).

Weighing in at three pounds, the human brain has three sections: *forebrain*, *midbrain*, and *hindbrain*. The forebrain contains the *thalamus*, a central switchboard that processes and directs incoming sensory information, and the *cerebrum*, whose outer layer (the *cerebral cortex*) handles most of the brain's actual information processing. The forebrain also contains the *hypothalamus*, *hippocampus*, and *amygdala*, structures that make up the *limbic system*, our emotional control center. The limbic system regulates emotions and motivations, and participates in memory formation.

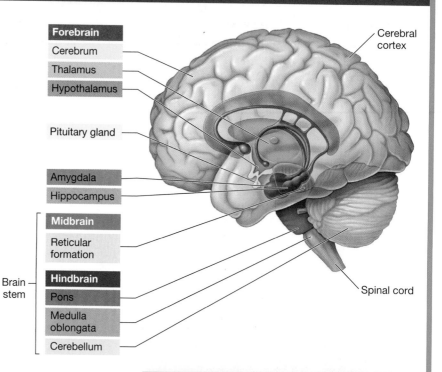

The midbrain and hindbrain together make up the *brain stem*. The midbrain coordinates sensory information from the thalamus with the peripheral nervous system for simple physical movement. The main structure in the midbrain is the *reticular formation*, which regulates consciousness. Finally, the hindbrain, evolutionarily the oldest part of the brain, controls the most basic functions including balance, heart rate, and breathing. The hindbrain consists of the *medulla oblongata*, *pons*, and *cerebellum*.

Neuroscience is an extremely active area of research. Scanning technologies such as diffusion MRI, which tracks the diffusion of water through white matter, and functional MRI, which detects brain activity by measuring changes in blood flow, are helping us map the brain at an unprecedented level of detail. On the right is a scan from a diffusion MRI (artificially colored) depicting the elegant structure of white matter fibers twisting through the brain.

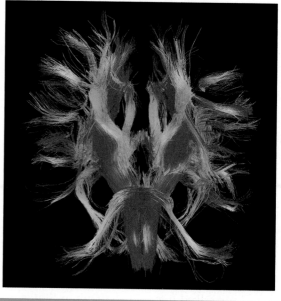

CHRISTINE SCHMIDT

Christine Schmidt heads the biomedical engineering department at the University of Florida. She develops biomaterials to guide nerves to regenerate.

part of the CNS), including the nerve in Edward Bonfiglio's leg. The PNS converts stimuli from the sensory organs into **sensory input**—signals that are received, transmitted, and processed by the CNS. Animals are constantly bombarded with sensory input from their external and internal environments. Five main classes of sensory receptors in humans receive this input; some additional categories, not known to be active in humans, are found in other animals (**Table 21.1**).

Through sensory input, the PNS gathers information from the external and internal environments and sends it to the CNS. For

Table 21.1

Different Ways to Sense the World

Receptor Type	Stimulus	Sense(s)
Chemoreceptors	Chemicals	Taste, smell
Photoreceptors	Light	Vision
Mechanoreceptors	Physical changes	Touch, hearing, proprioception (body position), balance
Thermoreceptors	Moderate heat and cold	Thermoreception (gradations of heat and cold)
Pain receptors	Injury, noxious chemicals, chemical and physical irritants	Pain, itch
Electroreceptors*	Electrical fields (especially those generated by muscle contractions of other animals)	Electrical sense
Magnetoreceptors*	Magnetic fields	Magnetic sense

*The sensory receptors listed here are found in many animals, including most vertebrates. Electroreceptors and magnetoreceptors, however, are not known to be active in humans.

example, imagine placing your hand near a hot stove; the heat input to your skin is transmitted as a signal from the PNS to the CNS. The CNS integrates and processes the information and generates a signal in response, dictating a particular action, or **motor output** such as moving your hand away from the stove. The PNS then relays that signal to the body part that will complete the action: The hand moves away from the stove.

All this action in the nervous system is conducted by specialized cells called **neurons** that transmit signals from one part of the body to another in a fraction of a second (**Figure 21.13**). The structure of a neuron reflects its unique function, and different types of neurons respond to different types of stimuli.

Neurons in the PNS are fragile and can be damaged easily. For instance, if you've ever hit your "funny bone," you've actually caused trauma to your ulnar nerve at the elbow, resulting in numbness and tingling. A more severe blow—such as trauma from a broken bone, or a bullet wound as in Edward's case—can cause loss of motor or sensory function.

Unlike lungs, kidneys, or the heart, nerves do have the ability to repair themselves, but they need help. "If you have an injury to a nerve, we can get those nerves to regenerate, but they need a pathway, like a sidewalk, to migrate on," says Christine Schmidt, a biomedical engineer at the University of Florida. As a postdoctoral fellow, Schmidt started looking for biomaterials to act as sidewalks to guide severed neurons to migrate.

The traditional approach to nerve repair is to remove a piece of nerve from somewhere else in a patient's body, such as a nerve in the leg that receives sensory input from the top of the foot, and stitch it into the damaged nerve to restore function. However, this process requires surgery to take the nerve from the leg, and one loses feeling on the top of the foot. And in some cases, using one's own nerves is not an option. Edward's gap was five centimeters wide—far too big to use a nerve from another part of his body.

Another option is a synthetic graft, a hollow plastic tube inserted as a way to get nerve cells to grow from one location to another. Yet these tubes can be used only in very small places and, like artificial blood vessels, can be rejected by the immune system.

"There was really a need for grafts that could provide better guidance for neurons, without using one's own nerves," says Schmidt. Schmidt's

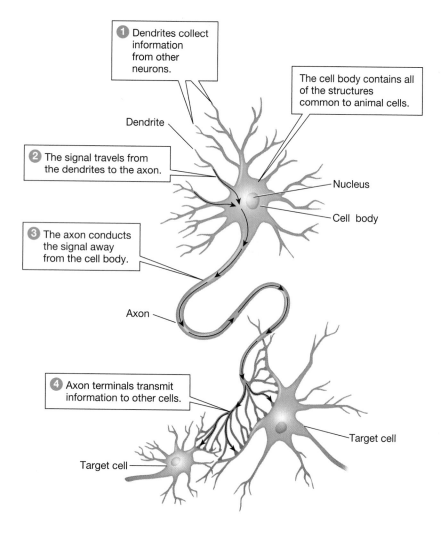

Figure 21.13

Neurons transmit signals from one cell to another
A neuron receives information from other cells, including other neurons, through one or more **dendrites**. The neuron pictured here has a single long **axon** with branched endings. Axons carry signals away from the cell body and to another cell. Two target cells are shown here.

The diagram labels:

1. Dendrites collect information from other neurons.
2. The signal travels from the dendrites to the axon.
3. The axon conducts the signal away from the cell body.
4. Axon terminals transmit information to other cells.

Dendrite

The cell body contains all of the structures common to animal cells.

Nucleus

Cell body

Axon

Target cell

Target cell

Q1: How do neurons look different from other cells you have learned about in this book?

Q2: What is the function of these differences?

Q3: Some people are born without the capacity to feel pain. Although it might initially sound nice not to feel pain, it is in fact quite dangerous. Describe a situation in which it would be dangerous not to feel pain.

lab obtained nerves from animal cadavers and spent three years testing different methods of washing the nerves to remove cells that could cause immune rejection. "It was like a cooking experiment," says Schmidt, to find the right combination of chemicals and physical forces, such as rinsing and swirling the nerves, to retain the nerve's architecture but remove all the original cells. Her team perfected the process in 2004 and published the results.

Right away, a company involved in nerve repair called and asked Schmidt about using her washing process on human tissue. Together, she and the company tested the process on human cadaver tissues. After a successful demonstration in those tissues, the company soon began selling off-the-shelf nerve grafts for use in hospitals. In late 2009, Edward Bonfiglio had one of these grafts implanted into his leg (**Figure 21.14**). Within months of his surgery, he wiggled his toes. It was "one of the greatest moments I had in my entire life," he later said.

"It's rewarding," says Schmidt. "It's pretty neat to take something all the way from the bench to impacting patients." Still, she adds, there's plenty more bioengineering to be done in the nervous system. For instance, no nerve grafts can currently repair injuries to the spinal cord.

As for Edward, after months of intense physical therapy, he was able to walk with a cane, then walk on his own, then jog. Today he is a student at Penn State University, studying to be a physician's assistant and actively training for the Paralympics.

Figure 21.14

Using bioengineering to repair a damaged nerve
Edward Bonfiglio chose experimental surgery over amputation, and today can walk and run without assistance.

Have a Heart

Over 100,000 people in the United States are right now waiting for an organ transplant. For many of them, the phone call that a suitable donor organ is available may never come. On average, 21 people die each day in the U.S. while awaiting a transplant. Consider registering to be an organ donor: By donating your organs after you die, you could save or improve as many as 50 lives.

Organs needed vs. organs transplanted

Waiting list registrations and completed transplants for 2013

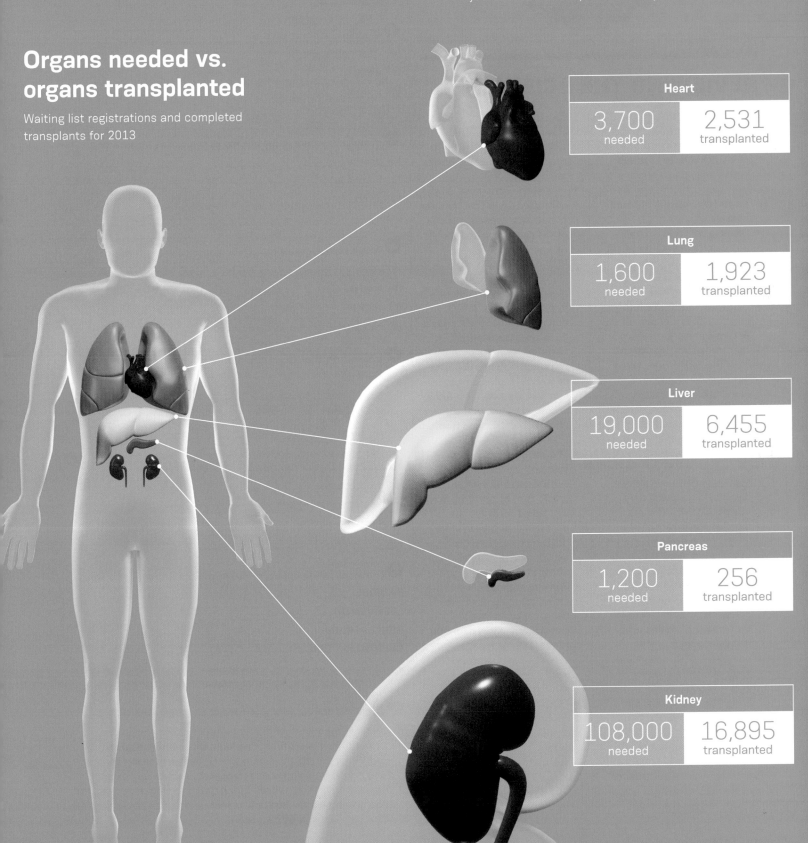

Heart

| 3,700 needed | 2,531 transplanted |

Lung

| 1,600 needed | 1,923 transplanted |

Liver

| 19,000 needed | 6,455 transplanted |

Pancreas

| 1,200 needed | 256 transplanted |

Kidney

| 108,000 needed | 16,895 transplanted |

REVIEWING THE SCIENCE

- Humans and other vertebrates have a **cardiovascular system**, a closed **circulatory system** with a chambered heart that pumps blood through a complex network of **blood vessels**.
- **Blood** is composed of cells and cell fragments that float in a fluid known as **plasma**. **Red blood cells** are the main oxygen carriers in the blood.
- Large vessels are built for mass transport of blood. **Arteries** are large vessels that transport blood away from the heart. **Veins** are large vessels that carry blood back to the heart. **Capillaries**, the smallest vessels, facilitate the exchange of materials with surrounding cells.
- The human cardiovascular system has two main circuits. In the **pulmonary circuit**, oxygen-deficient blood is pumped to the lungs. In the **systemic circuit**, oxygenated blood returning from the lungs is pumped out to body tissues.
- The mammalian **heart** is composed of four chambers that make up two separate muscular pumps, each composed of an **atrium** and a **ventricle**. The left atrium and ventricle pump blood to the body; the right atrium and ventricle pump blood to the lungs.
- The human **respiratory system** carries air from the nose (or mouth) to the **lungs**, eventually reaching clusters of tiny sacs in the lungs called **alveoli**. Gas exchange takes place in the alveoli, where oxygen diffuses into the blood and carbon dioxide diffuses out of it.
- Inhalation and exhalation are controlled by the contraction of muscles, especially those of the **diaphragm** and the rib cage.
- In the **urinary system** of many animals, including humans, **kidneys** regulate body water and solute concentrations. The kidney's basic unit, the **nephron**, performs three functions: **filtration, reabsorption**, and **secretion**. The resulting concentrated solution, **urine**, is carried by ducts to the bladder and excreted from the body.
- The vertebrate **nervous system** is divided into the **central nervous system** (**CNS**), consisting of the **brain** and **spinal cord**, and the **peripheral nervous system** (**PNS**), consisting of the sensory organs and all the remaining nerves.
- **Sensory organs** convert environmental stimuli into nerve impulses that are carried by sensory **neurons** to the CNS. All human senses rely on **sensory input** from just five types of sensory receptors.
- The CNS integrates sensory information and seeds an output signal, often a **motor output**, through the PNS to the appropriate bodypart.

THE QUESTIONS

The Basics

1 Blood plasma transports

(a) waste products.

(b) water.

(c) solutes.

(d) all of the above

2 In humans, where is the gas exchange surface located?

(a) pharynx

(b) bronchus

(c) alveolus

(d) bronchiole

3 Which blood vessels carry blood back toward the heart?

(a) veins

(b) arteries

(c) ventricles

(d) capillaries

4 In a neuron, the _____ conducts signals to other cells.

(a) dendrite

(b) axon

(c) nucleus

(d) cell body

5 What sensory system is *not* active in humans?

(a) pain receptor

(b) electroreceptor

(c) mechanoreceptor

(d) chemoreceptor

6 Link each of the following terms with the correct definition.

SA NODE

LEFT VENTRICLE

RIGHT ATRIUM

PULMONARY CIRCUIT

1. Sends blood to the right ventricle.
2. Pumps blood to the systemic circuit.
3. The heart's pacemaker.
4. Pumps oxygen-poor blood to the lungs.

7 Circle the correct terms in the following sentence:

Kidneys conduct their primary work of filtering the blood through their basic functional unit, the [**neuron, nephron**]. Besides filtration, the kidney must [**reabsorb, secrete**] water and important solutes, and [**reabsorb, secrete**] toxins and overabundant substances. Urine drains from the kidneys into the bladder via the [**urethra, ureter**] and empties from the bladder via the [**urethra, ureter**].

8 Beginning with sensory input from a mechanoreceptor in your toe, place the response events of your nervous system in the correct order by numbering them from 1 to 5.

____ a. Command signals travel from the brain to the spinal cord.

____ b. Input is sent to the brain.

____ c. A sensory signal is sent to the spinal cord.

____ d. The muscle responds to the signal.

____ e. Processing occurs in the brain.

Try Something New

9 If an organism has a greater concentration of CO_2 in its lungs than in its blood, will there be net transport of CO_2 from the lung air space to the alveolar capillaries, or from the alveolar capillaries to the lung air space? Explain your reasoning.

10 The number of times our hearts beat per minute is referred to as heart rate. Each heartbeat lasts a little less than 1 second and consists of a series of events called the *cardiac cycle*. The blood pressure measured in a doctor's office reflects the pressure in the arteries leading to the body from the left ventricle. A blood pressure reading of 120/80 (systole/diastole), for example, means that contraction of the left ventricle generates 120 millimeters of mercury (mm Hg) of pressure in the arteries, followed by a drop to 80 mm Hg when the ventricles relax and refill.

(a) How would you predict that heart rate changes with exercise in the short term, and in the long term?

(b) "White coat hypertension" refers to the situation in which patients display higher blood pressure at the doctor's office than in normal daily life. How would you explain this phenomenon?

Leveling Up

11 **What do *you* think**? In the highly competitive world of endurance sports, any means of improving performance offers a significant advantage. Of the 24 Tour de France winners since 1961, 13 have either tested positive for performance-enhancing drugs or admitted to using them. Several cyclists have said they injected the hormone erythropoietin (EPO). Made naturally in the kidneys, EPO increases the blood's oxygen-carrying capacity, making it valuable in endurance sports such as marathon races, cross-country skiing, and bicycle racing. EPO increases the oxygen-carrying capacity of blood by stimulating red blood cell production. In addition to increasing the production of red blood cells, EPO stimulates the growth of capillaries that carry oxygen to tissues.

A synthetic form of EPO developed by drug companies is used to treat patients with anemia, kidney damage, and malaria. Athletes inject synthetic EPO, a process called "blood doping" that has many health risks. Too many red blood cells can make the blood so thick that it clots or fails to flow easily through the heart. Nearly two dozen endurance athletes are thought to have died of heart attacks caused by doping with EPO. Because EPO is a naturally occurring hormone, identifying synthetic EPO in blood or urine samples of athletes has been difficult.

(a) What is EPO, and why does it offer a performance advantage in sports, especially endurance events such as cycling and rowing? How could taking EPO kill a person?

(b) Some cyclists increase their red blood cell counts by training at high altitudes. The low oxygen content of mountain air triggers the natural release of EPO. Other athletes have accomplished the same thing by spending time in special low-oxygen tents. Do you think either of these approaches is more acceptable than injecting EPO or cells engineered to express EPO? Where would you draw the line, and why?

12 **Life choices** As of 2014, almost half of Americans (more than 117 million people) have identified themselves as organ donors in the event of their death. Unfortunately, many more people are in need of an organ than can be helped, and this number continues to rise. Every day, about 80 people receive organ transplants but another 18 people die, waiting for a transplant that never came. Refer to the maps on the following pages to answer the first three questions below.

(a) Which state has the highest number of organ donors? The lowest number?

(b) Which state has the highest percentage of organ donors? The lowest?

(c) Where does your state fall, in terms of the percentage of organ donors? Does this surprise you? Why or why not?

(d) Are you an organ donor? Why or why not?

(e) Suggest one way that a state (yours or another) could increase the number of its residents who identify themselves as organ donors.

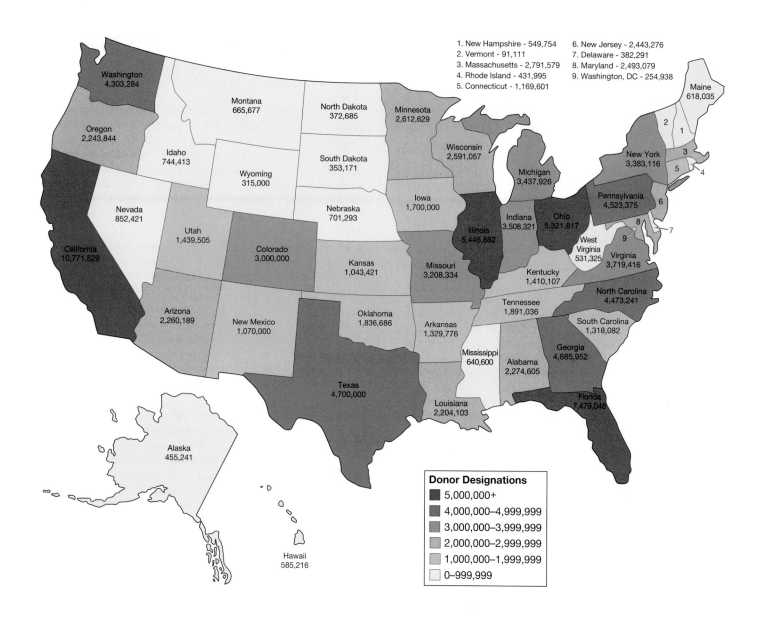

1. New Hampshire - 549,754
2. Vermont - 91,111
3. Massachusetts - 2,791,579
4. Rhode Island - 431,995
5. Connecticut - 1,169,601
6. New Jersey - 2,443,276
7. Delaware - 382,291
8. Maryland - 2,493,079
9. Washington, DC - 254,938

Washington
4,303,284

Oregon
2,243,844

Montana
665,677

North Dakota
372,685

Minnesota
2,612,629

Maine
618,035

Idaho
744,413

Wyoming
315,000

South Dakota
353,171

Wisconsin
2,591,057

Michigan
3,437,926

New York
3,383,116

Nevada
852,421

Utah
1,439,505

Colorado
3,000,000

Nebraska
701,293

Iowa
1,700,000

Illinois
5,446,882

Indiana
3,508,321

Ohio
5,321,617

Pennsylvania
4,523,375

California
10,771,529

West
Virginia
531,325

Virginia
3,719,416

Arizona
2,260,189

New Mexico
1,070,000

Kansas
1,043,421

Missouri
3,208,334

Kentucky
1,410,107

Oklahoma
1,836,686

Tennessee
1,891,036

North Carolina
4,473,241

Arkansas
1,329,776

South Carolina
1,318,082

Mississippi
640,600

Alabama
2,274,605

Georgia
4,685,952

Texas
4,700,000

Louisiana
2,204,103

Florida
7,479,048

Alaska
455,241

Hawaii
585,216

Donor Designations

- 5,000,000+
- 4,000,000–4,999,999
- 3,000,000–3,999,999
- 2,000,000–2,999,999
- 1,000,000–1,999,999
- 0–999,999

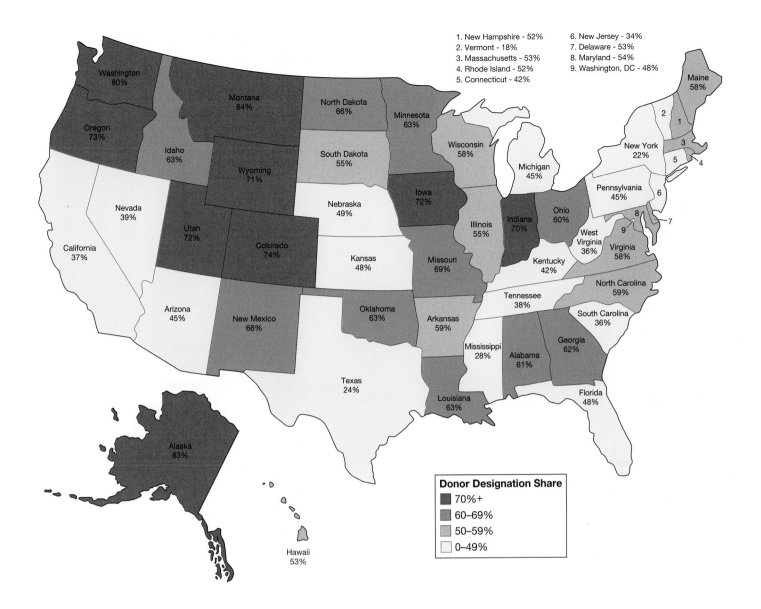

1. New Hampshire - 52%
2. Vermont - 18%
3. Massachusetts - 53%
4. Rhode Island - 52%
5. Connecticut - 42%

6. New Jersey - 34%
7. Delaware - 53%
8. Maryland - 54%
9. Washington, DC - 48%

Washington 80%
Montana 84%
North Dakota 66%
Minnesota 63%
Maine 58%
Oregon 73%
Idaho 63%
Wisconsin 58%
New York 22%
Nevada 39%
Wyoming 71%
South Dakota 55%
Michigan 45%
Pennsylvania 45%
California 37%
Utah 72%
Iowa 72%
Indiana 70%
Ohio 60%
Nebraska 49%
Illinois 55%
West Virginia 36%
Virginia 58%
Colorado 74%
Kansas 48%
Missouri 69%
Kentucky 42%
North Carolina 59%
Arizona 45%
New Mexico 68%
Oklahoma 63%
Arkansas 59%
Tennessee 38%
South Carolina 36%
Mississippi 28%
Alabama 61%
Georgia 62%
Texas 24%
Louisiana 63%
Florida 48%
Alaska 83%
Hawaii 53%

Donor Designation Share
70%+
60–69%
50–59%
0–49%

Testing
the Iceman

A Dutch daredevil claims he can fend off disease with his mind. Two skeptical scientists take the case.

After reading this chapter you should be able to:

- Explain how cells communicate with each other via the endocrine system.
- Describe two different ways a hormone can act on a target cell.
- Identify the immune system's first, second, and third lines of defense.
- Compare and contrast the role of white blood cells in the innate and adaptive immune systems.
- Diagram the processes of inflammation and blood clotting.
- Distinguish between a primary and a secondary adaptive immune response.
- Create a flowchart depicting the sequence of events as a vertebrate immune system responds to a pathogen.

CHAPTER

22

ENDOCRINE AND IMMUNE SYSTEMS

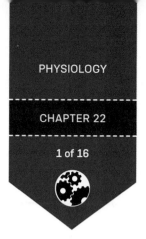

The scantily clad young men lie on the ground, looking up toward the sky. With sunglasses on and hands propped behind their heads, they look as if they're tanning at the beach—but there are no piña coladas or warm sand here. Instead, these 18 men, wearing only swim trunks, are lying on cold, white snow in the mountains of Poland. And lying with them is the Iceman.

Wim Hof, a Dutch daredevil known as the "Iceman" who holds numerous world records for cold exposure, breathes deeply, leading the youths in an exercise. Over four days, he will train them to tolerate extreme cold. During his rigorous program, they will swim in near-freezing water every day and climb a snow-covered mountains in just shorts (**Figure 22.1**). Hof claims that exposure to the cold, combined with meditation and breathing exercises, will enable the men to fend off illness and disease.

Matthijs Kox stands to the side of the Iceman's trainees, taking notes. Kox, a researcher in intensive-care medicine at Radboud University Medical Center in the Netherlands, first met Hof in 2010, when the Iceman was visiting another laboratory at the university. A team in the physiology department was measuring Hof's ability to regulate his core temperature while standing in an ice bath (**Figure 22.2**). The scientists were surprised to find that rather than decreasing as expected, Hof's core temperature actually increased, and his metabolism climbed. While standing in the ice bath talking to his examiners, Hof mentioned that he could also consciously modulate his autonomic nervous system and immune system.

It was an unbelievable claim. The autonomic nervous system operates body functions that humans cannot voluntarily control, such as heartbeat and blood pressure. The **immune system**—a remarkable defense system that protects us against most infectious agents—has also long been known to be involuntary.

Figure 22.1

"Iceman" Wim Hof trains volunteers under extreme conditions

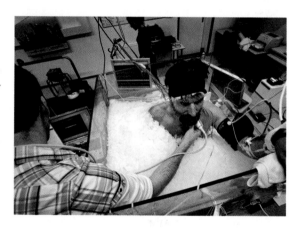

Figure 22.2

Hof's vital signs are monitored while he is immersed in ice

But Hof had a history of doing the unbelievable. He had claimed the Guinness World Record for longest ice bath by staying immersed in ice for 1 hour, 52 minutes, and 42 seconds. He had climbed part of Mount Everest wearing nothing but shorts. He had run a marathon through the snow at −20°C (−4°F), again wearing only shorts.

Hof's testers in the physiology unit told him that a Radboud University professor named Peter Pickkers had a way to measure a person's immune response. So Hof hoofed it to Pickkers's office, shook his hand, and said, "I can modulate my immune system. I heard you can measure it. Will you measure mine?"

WIM HOF

Wim Hof, better known as the "Iceman," is a Dutch celebrity who holds numerous world records for withstanding extreme cold.

Hormonal Changes

Pickkers was skeptical of Hof's claim, which had the whiff of pseudoscience. But Hof was an interesting character, so Pickkers went online and watched videos of his feats. "There were remarkable things I did not know of—things that, if you had asked me beforehand, I would have said, 'That's not possible. It's not possible to run half a marathon barefoot in the snow,'" says Pickkers. "But he did that."

Pickkers raised the idea of testing Hof to Kox, who was one of Pickkers's PhD students at the time, studying how the brain and immune system interact. Pickkers and Kox discussed the possibility at length, and they decided to give Hof a chance to prove his claim. But they were going to do it while adhering strictly to the principles of the scientific process. "You can imagine some people wondered what we were doing with this guy," says Kox. "So we really focused on doing this in a very sound, precise manner, with no doubt about the scientific integrity of the project."

Hof claimed that the regimen for consciously controlling his immune system required three components: cold exposure, meditation, and breathing exercises. So, the team tested Hof's blood before and after an 80-minute full-body ice bath while Hof performed breathing and meditation exercises. Each time the scientists took blood, they went back to the lab and exposed the blood cells to molecules of endotoxin, a substance found in the cell walls of bacteria that activates an immune response in the human body. After the regimen of ice, breathing, and meditation, Hof's cells had a far more subdued immune system response, showing very low levels of proteins associated with activation of the immune system, compared to similar cells before the regimen. The cause of that subdued immune response was unclear, but the researchers suspected stress hormones played a role.

Hormones are signaling molecules that tell other cells what to do under specific situations or at certain times in the life cycle of the individual. Hormones are produced by specialized secretory cells of the **endocrine system** (**Figure 22.3**).

These secretory cells are often organized into discrete organs called **endocrine glands**. Major endocrine glands are located throughout the human body. Unlike other glands, such as

Thyroid

Pancreas

Adrenal gland

Testes

Female:

Ovary

The hypothalamus is the main coordinator of the endocrine system. It also integrates the endocrine system with the nervous system.

Some organs, such as the pancreas, function as endocrine glands and also as ducted (*exocrine*) glands.

Endocrine cells are also scattered throughout the lining of the stomach and intestine.

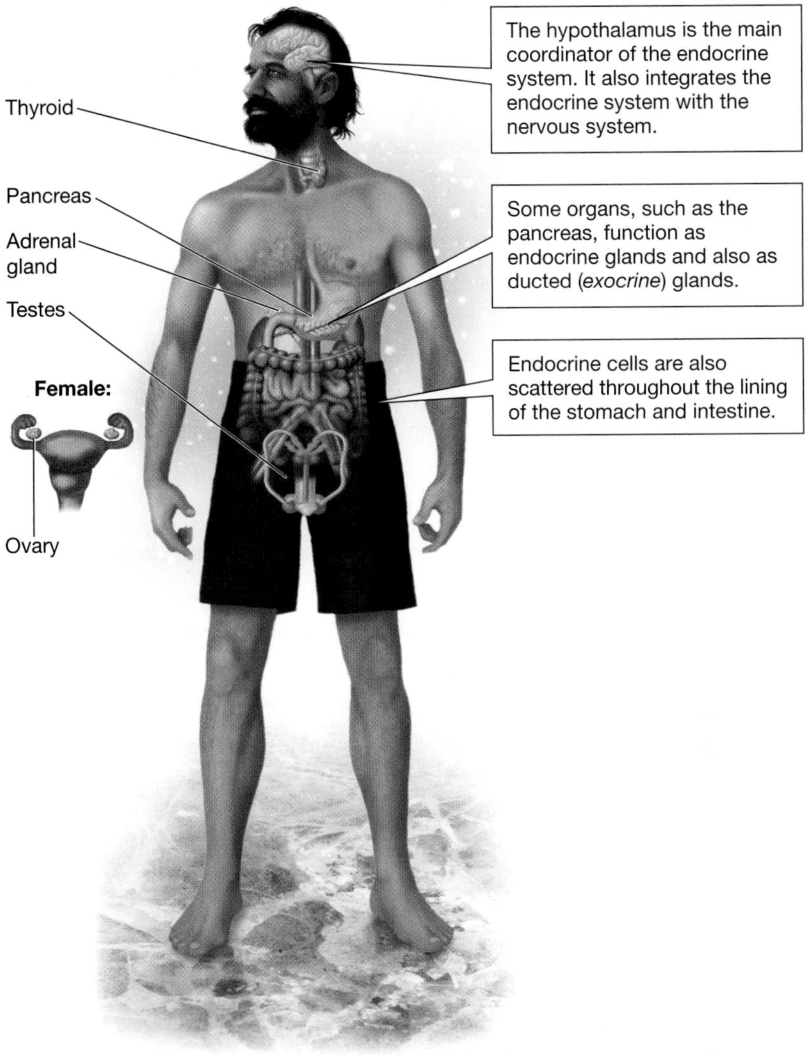

Figure 22.3

The endocrine system is composed of hormone-secreting cells

The endocrine system consists of cells organized into ductless glands, plus scattered endocrine cells embedded in other tissues or organs. These cells all release hormones directly into the circulatory system.

Q1: What organ coordinates the endocrine system?

Q2: How does an endocrine gland differ from an exocrine gland?

Q3: How do male and female endocrine systems differ?

tear ducts, endocrine glands do not have ducts or tubes that deliver secretions from the gland directly to the site of action. Instead, endocrine glands release hormones into body fluids such as blood, which carries these chemical messengers throughout the body (**Figure 22.4**). In a

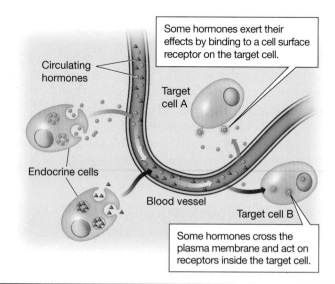

Figure 22.4

Hormones enable cells to communicate with one another
Hormones released by endocrine cells travel through the circulatory system to produce a response in target cells often located at a distance in the body.

Q1: How do hormones travel to target cells?

Q2: Distinguish between an endocrine cell and a target cell.

Q3: Why is a hormone called a signaling molecule?

Some hormones exert their effects by binding to a cell surface receptor on the target cell.

Circulating hormones

Target cell A

Endocrine cells

Blood vessel

Target cell B

Some hormones cross the plasma membrane and act on receptors inside the target cell.

subsequent experiment, Kox measured the levels of a stress hormone called cortisol in Hof's blood. Hof's blood after the ice, breathing, and meditation regimen contained far higher levels of cortisol than before.

In the human body, most hormones can travel only as fast as the blood moves, which means they take several seconds or more to arrive at their target cells. Hormones coordinate functions that take place over timescales ranging from seconds (such as immediately increasing one's heartbeat in reaction to fear) to months (such as preparing a uterus to contract during the birthing process; see Figure 19.11).

Typically, hormones become greatly diluted after they are released into the circulatory system. They must therefore be able to exercise their effects at very low concentrations, as vitamins do. Hormones are effective in small amounts because they bind to their targets with great specificity. Cortisol, for example, binds to a very specific receptor present on the surface of almost every cell in the body.

Cortisol, adrenaline (also called epinephrine), and noradrenaline (norepinephrine) are three hormones produced by the **adrenal glands**, a pair of endocrine glands that sit atop the kidneys. The release of these hormones launches a number of rapid physiological responses, including boosting blood glucose levels.

When a single hormone molecule binds to its receptor, it sets in motion a chain of events that may ultimately activate thousands of protein molecules in the target cell (**Figure 22.5**). When cortisol binds its receptor, it initiates a pathway that results in the regulation of genes involved in development, metabolism, and immune response. This signal amplification—from a single hormone molecule to the activation of many proteins and genes—means that just a few hormone molecules can have a substantial impact on a target cell. Through its effects on many cells, a hormone can exert a profound influence on the body as a whole.

Some hormone-secreting cells are not organized into distinct glands like the adrenal glands, but are instead embedded as single cells or clusters of cells within other specialized tissues and organs. For example, the main role of the kidneys is to filter blood, yet some cells in the kidneys produce hormones that stimulate red blood cell production. Altogether, the endocrine glands and the endocrine cells embedded in other organs, such as the kidneys, make up the endocrine system.

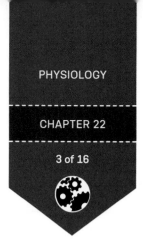

PETER PICKKERS AND MATTHIJS KOX

Peter Pickkers (left) is a professor of experimental intensive-care medicine who studies the innate immune system. Matthijs Kox (right) is a researcher in intensive-care medicine. Both work at Radboud University Medical Center in the Netherlands.

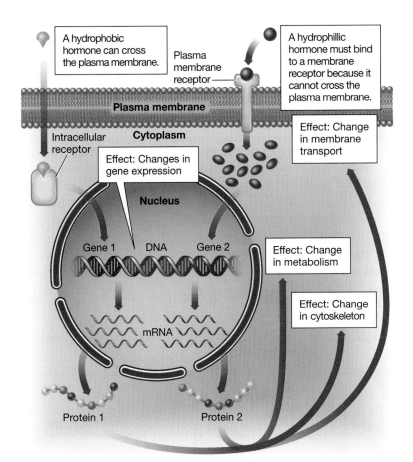

A hydrophobic hormone can cross the plasma membrane.

Plasma membrane receptor

A hydrophillic hormone must bind to a membrane receptor because it cannot cross the plasma membrane.

Plasma membrane

Intracellular receptor

Cytoplasm

Effect: Change in membrane transport

Effect: Changes in gene expression

Nucleus

Gene 1   DNA   Gene 2

mRNA

Effect: Change in metabolism

Effect: Change in cytoskeleton

Protein 1   Protein 2

Figure 22.5

Hormonal signals are amplified within the cell
Hormones are effective at low concentrations because of their specificity and because tiny amounts of a hormone can generate a large internal signal within a target cell.

Q1: Describe the two ways that a hormone outside a cell can exert its effect on a cell.

Q2: Within the cell, how does a hormone bring about a change in cell activity?

Q3: It takes very little of the hormone cortisol to have large effects throughout the body. Explain why.

Brain-Body Connection

Testing Hof's cells alone wasn't enough for Kox and Pickkers. They wanted to measure his entire body's immune response, and wanted to know if the breathing and meditation techniques made any difference. After the ice bath, they asked Hof to perform his meditation and breathing techniques while they injected him directly with the endotoxin. In previous experiments, healthy volunteers injected with endotoxin experienced fever, headaches, and shivering, accompanied by high levels of signaling proteins, called **cytokines**, that immune system cells use to communicate when an invader is present.

At various times before and after the injection, Kox measured Hof's blood levels for hormones and cytokines. Kox then compared Hof's results to those of a control group of 112 healthy volunteers who had previously taken the same test. To the scientists' surprise, as soon as Hof began practicing his breathing techniques, his adrenaline levels skyrocketed. And unlike the other volunteers, Hof reported almost no flu-like symptoms. Topping it off, the number of cytokines in his blood—indicative of an immune response—was less than half that of the control group. Hof appeared to have suppressed his immune system—voluntarily.

How was that possible? Kox considered the possibilities. First, a tiny region at the base of the vertebrate brain, the **hypothalamus**, coordinates the endocrine system and integrates it with the nervous system (refer back to Figure 22.3). The hypothalamus contains neurons, which interact with the brain, and endocrine cells, which produce hormones. It is a literal brain-body connection.

One well-known part of that connection involves the adrenal glands. In response to stress messages from the brain, the adrenal glands release adrenaline into the blood. If a man sees a rattlesnake in front of him, for example, he is likely to jump back or at least freeze in place, his heart racing. This quick response is due to the connection between the nervous system and the

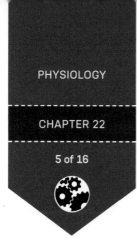

adrenal glands (**Figure 22.6**): The nervous system processes visual information (*Snake!*) and transmits an alarm signal to the adrenal glands within a fraction of a second. The adrenal glands kick in right away, pouring adrenaline and noradrenaline into the blood. Adrenaline stimulates glycogen breakdown in liver and skeletal muscle cells, causing glucose to be released into the bloodstream. It also speeds up the heartbeat and the force with which the heart contracts, so that glucose is delivered throughout the body more rapidly. In this way, glucose becomes available to fuel a rapid response to a stressful situation.

Within just a few seconds, then, these hormones increase the pumping of blood and trigger the release of glucose, all of which support the next move: fight or flight. In the case of an encounter with a snake, that may mean either arming oneself with a stout stick or running away. It turns out, however, that adrenaline plays another role aside from triggering glucose delivery: Research has shown that it also subdues the activity of immune system cells.

Hof appeared to be able to consciously activate his nervous system (in the absence of real stress) to prompt the release of adrenaline, thereby suppressing his immune response to the endotoxin voluntarily. It was a "remarkable" finding, says Pickkers, but he wasn't ready to jump to conclusions. The case study of a single individual is weak scientific evidence for any phenomenon. Perhaps Hof was simply an outlier: "Everyone can play a

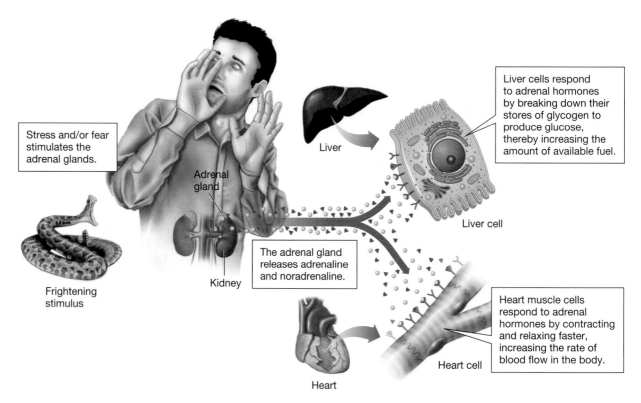

Stress and/or fear stimulates the adrenal glands.

Frightening stimulus

Adrenal gland

Kidney

The adrenal gland releases adrenaline and noradrenaline.

Liver

Liver cells respond to adrenal hormones by breaking down their stores of glycogen to produce glucose, thereby increasing the amount of available fuel.

Liver cell

Heart

Heart cell

Heart muscle cells respond to adrenal hormones by contracting and relaxing faster, increasing the rate of blood flow in the body.

Figure 22.6

Adrenal hormones produce a rapid response to stress or fear
The adrenal glands produce adrenaline (epinephrine) and noradrenaline (norepinephrine), which trigger the rapid release and delivery of stored energy.

Q1: Describe an event (other than the one illustrated in the figure) that might cause the release of adrenaline.

Q2: What organs does adrenaline affect?

Q3: What do you think would happen if your adrenal glands were constantly releasing adrenaline?

little baseball, but there is only one Derek Jeter," says Pickkers. It could have been that Hof had a unique genetic mutation or another factor that enabled him to control his autonomic nervous and immune systems.

But Hof claimed that he was not an outlier, that he could teach his technique to anyone. "I'm sure everybody is able do this," Hof told Pickkers. Pickkers challenged him to prove it. For scientific validation, Hof needed to teach his method to a group of healthy volunteers so that Pickkers could then compare that group's immune responses to those of an untrained control group of volunteers. In this controlled way, Hof might produce stronger scientific evidence for his claim.

Innate Defenders

If Hof was right—if it was possible to voluntarily control the immune system—the discovery would do more than change our understanding of the immune system; it would offer hope to people with autoimmune diseases, individuals in whom the immune system is overactive.

When healthy, the immune system protects animals from most infectious agents, called **pathogens**. Human pathogens include viruses, bacteria, and protists, as well as some fungi and multicellular animals such as parasitic worms. A well-known example is human immunodeficiency virus or HIV, the virus responsible for AIDS (see "What Makes HIV so Deadly?").

Pathogens infect animals only if they can find a way into the body. An animal's first line of defense against pathogens is **external defenses**, which reduce the likelihood that a harmful organism or virus will gain access to internal tissues. Linings that separate the "outside" from the "inside" of the body—the skin and the linings of the lungs, for example—act as a physical barrier to keep out most pathogens. Other external defenses include chemical agents (such as enzymes) and chemical environments (such as acidic conditions) that keep the invaders from attaching to or growing on body surfaces (**Figure 22.7**).

Although external defenses do a good job of keeping out most pathogens, the body is still vulnerable. Wounds, in the form of cuts, abrasions, and punctures, are common, and many pathogens will take advantage of breaks in the skin to gain entry to their hosts.

What Makes HIV So Deadly?

In the early 1980s, doctors in the United States began to notice that gay men were dying of a variety of rare diseases, including a skin cancer called Kaposi's sarcoma, an unusual kind of pneumonia, and other infections that most people ordinarily shake off. By the mid-1980s it was clear that patients with the syndrome—named acquired immunodeficiency syndrome, or AIDS—had broken immune systems, the result of infection by a virus called the human immunodeficiency virus, or HIV.

In North America and Europe, the number of new cases rapidly increased, claiming the lives of tens of thousands of people each year. Initially, most new cases were limited to gay men, intravenous drug users, and people who had received blood transfusions. The common denominator was contact with the blood or body fluids of others: Couples during sex, drug users when sharing used needles, and surgical and hemophilia patients who received blood transfusions contaminated with HIV.

In time, safe-sex education and clean-needle programs reduced the rates of infections among gay men and blood transfusion patients, but the virus spread to other populations. Globally, 39 million people have died of AIDS, leaving over 17 million orphaned children. In just one year, 2013, 1.5 million people died of AIDS and 2.3 million more became infected.

Inside the bloodstream, individual HIV enters immune system cells and reproduces inside them, eventually killing so many immune cells that the body's defenses are crippled. In the short term, remaining immune system cells track down HIV-infected cells and destroy them. Because they do such a good job of killing HIV-infected cells in the blood, most people with HIV have about a decade of normal health before they become ill, even without any treatment.

Over time, however, the HIV viruses in the body evolve. As HIV evolves, the immune system cells no longer recognize and kill the virus. The population of HIV viruses increases and begins destroying immune system cells faster than they can multiply. The body no longer has the immune system cells it needs to fight off infections by bacteria, yeasts, and other viruses. Once the immune system collapses, a person is vulnerable to almost any infection.

So far, there is no effective vaccine or cure for HIV. But a variety of new drugs enable people with AIDS to live years longer with fewer symptoms. "HIV cocktails," as the standard mixture of therapeutic drugs are called, prevent the viral genetic material from replicating or prevent the virus from merging with plasma membranes and entering cells. But the drugs can cost hundreds or thousands of dollars a month. Because treatment is so costly, only one in five AIDS patients in Africa and Asia receives effective treatment. For now, the best way to slow the spread of the disease remains safe-sex education, the free availability of condoms, and clean-needle programs.

Once inside, pathogens confront a second line of defense—the cells and defensive proteins of the **innate immune system**. To mount an internal defense that kills, disables, or isolates invading pathogens, the body first must recognize that

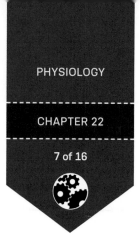

Eyes: Tears contain antibacterial enzymes

Nose: Hairs and mucus trap microbes

Ears: Earwax inhibits bacterial growth

Mouth: Saliva contains antibacterial enzymes; mucus traps microbes

Skin: Physical barrier to pathogens

Digestive system: Low pH in stomach kills pathogens; "good" bacteria in intestines outcompete "bad" bacteria, fungi, and viruses

Figure 22.7

The immune system's first line of defense is preventing the entry of pathogens
Our skin and the linings of our respiratory and digestive systems form physical and chemical barriers against pathogens.

Q1: What is the main physical barrier that animals use to keep out pathogens?

Q2: Give an example of a chemical defense within the digestive system.

Q3: Explain why rubbing your eyes and nose during flu and cold season is not recommended.

self from nonself, they mistakenly attack the body's own cells, leading to autoimmune diseases such as rheumatoid arthritis (in which immune cells attack the lining of the membranes that surround joints) or type 1 diabetes (in which immune cells attack the pancreas, which makes insulin).

If there were a way, as Hof claims, to subdue the immune system at will, people with these diseases would have another avenue, aside from expensive drug therapies, by which to control their rebellious immune systems.

Team Effort

Despite Hof's personal achievement, Pickkers and Kox didn't really think he would be able to teach others to voluntarily control their innate immune systems. "We thought it would be a negative result," says Pickkers. If nothing else, Hof had been performing his technique for 30 years, so even if he could teach it, Pickkers doubted he could teach a novice enough to influence his or her own immune system in just a few days. Hof disagreed, arguing that a short training regimen would be sufficient to impart the ability. Kox didn't think that most of the volunteers would even make it through the training.

It was no easy study for the participants. Over four days, 18 healthy, young male volunteers were taken into the mountains of Poland and exposed to the cold in various ways: standing in the snow barefoot for 30 minutes, lying in the snow bare-chested for 20 minutes, swimming in ice-cold water each day for several minutes, hiking up a snowy mountain in nothing but shorts and shoes. Hof also taught them his meditation and breathing techniques, including deep inhalations and exhalations.

Contrary to Kox's expectation, all 18 participants completed the training, and 12 of them were then randomly selected to come back to the lab for the final part of the experiment: Exposure to the endotoxin, to test whether they could consciously regulate their immune response. Their results would then be compared to those of 12 healthy controls who had also been exposed to the endotoxin but had not received Hof's training. Now the scientists would finally find out whether Hof could teach someone to control the activity of the innate immune system.

an invader is present. Although a person is not consciously aware of it, a healthy body can distinguish foreign invaders (nonself) from its own cells (self). If the internal defenders fail to tell

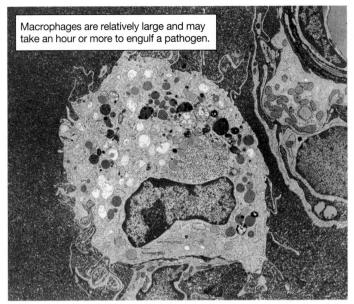

Macrophages are relatively large and may take an hour or more to engulf a pathogen.

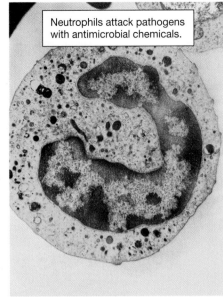

Neutrophils attack pathogens with antimicrobial chemicals.

Figure 22.8

Phagocytes destroy pathogens by engulfing them

Phagocytes are a kind of white blood cell, a family of defense cells found in body fluids including blood, where they intercept invading pathogens. Two different kinds of phagocytes—a macrophage and a neutrophil—are seen in these colorized transmission electron micrographs (TEMs).

Q1: Place these terms in order from most to least inclusive: neutrophil, white blood cell, phagocyte, innate immune system.

Q2: Compare and contrast macrophages and neutrophils.

Q3: Why would it be a problem if your innate immune system identified the insulin-producing cells in your pancreas as "nonself?"

The innate immune system reacts to cells or molecules that do not belong in the body by activating defense cells and proteins to eliminate the unwelcome guests. A suite of pathogen-recognizing cells called **phagocytes**, a type of white blood cell, mark and destroy foreign invaders by engulfing and digesting them (**Figure 22.8**).

This immune response is said to be innate (inherent) because the necessary components are constantly at the ready for deployment against an invading pathogen. The innate response can be local, occurring at the point of entry, or global, involving the whole body. Like the external defense system, innate immunity is indiscriminate as to which foreign invaders it repels, so it is considered a **nonspecific response**. The innate immune system is an ancient defense mechanism found in both invertebrates and vertebrates.

In addition to defending against invaders, the innate immune system plays two other critical roles. First, it responds to tissue damage from a pathogen invasion or wound by mounting an immediate and coordinated sequence of events known as **inflammation** (**Figure 22.9**). Cytokines are a clear marker of inflammation, and thus a good way to measure the action of the immune system. The second role of the innate immune system is clotting blood to close a wound. Sealing an open wound reduces blood loss and restores the integrity of external defense barriers (**Figure 22.10**).

To test the innate immune response of the 24 study participants (the 12 trained volunteers and the 12 untrained controls), the scientists injected

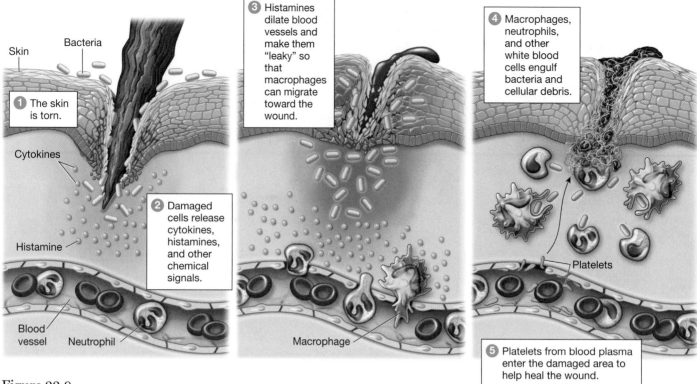

Figure 22.9

① The skin is torn.

Skin
Bacteria
Cytokines
Histamine
Blood vessel
Neutrophil

② Damaged cells release cytokines, histamines, and other chemical signals.

③ Histamines dilate blood vessels and make them "leaky" so that macrophages can migrate toward the wound.

Macrophage

④ Macrophages, neutrophils, and other white blood cells engulf bacteria and cellular debris.

Platelets

⑤ Platelets from blood plasma enter the damaged area to help heal the wound.

The inflammatory response acts against invading pathogens

Inflammation occurs when the innate immune system swings into action after cellular damage is detected, cleaning up damaged tissues and preventing the spread of pathogens. Inflammation can occur anywhere inside the body. Here we see an inflammatory response following a puncture wound to the skin.

Q1: What is the role of white blood cells in inflammation?

Q2: What would happen if histamines were not produced during inflammation?

Q3: Why is inflammation called a "nonspecific" immune response?

Figure 22.10

Blood clots help prevent pathogens that may be present in a wound from spreading

Sticky cell fragments, or **platelets** (shown here in light blue), and clotting proteins (yellow) form a gel-like mesh that traps blood cells, creating a blood clot that seals broken skin. Clotting can begin as quickly as 15 seconds after tissue damage occurs. Growth of new tissue eventually repairs the wound more permanently.

Q1: Why is blood clotting an important immune response?

Q2: How are the inflammatory response and blood clotting similar?

Q3: Some people have a genetic disorder in which their blood cannot clot. Why would this be a problem?

each participant with endotoxin and monitored them for six hours. Hof visited his trainees during the experiment, coaching them through his breathing techniques.

The results were clear: After being injected, and while performing the breathing techniques, the trainees showed higher adrenaline levels than the controls—higher even than the adrenaline produced by a person's first bungee jump. "They produced more adrenaline just lying in bed than somebody standing in front of an abyss going to jump in fear for the first time," says Hof. "That means direct control of your hormone system, and your hormones have a direct relationship to the immune system."

In addition, the trainees had fewer flu-like symptoms and lower fevers, and their cytokines—the signaling proteins of the immune system and markers of inflammation—were at less than half the level of the control group. "We were very surprised," says Kox. "It was impressive that these guys could do all this cold exposure training, but I still thought the chances were slim they'd be able to modulate their immune systems. But the results were so convincing."

Adapting to the Enemy

Kox and Pickkers's study did not address the human immune system's third line of defense. In contrast to the nonspecific responses of external defenses and the innate immune system, the more complex **adaptive immune system** is tailored against specific invaders.

Adaptive immunity goes beyond simply recognizing something as nonself. Instead, specialized defense cells are trained to recognize only one strain of pathogen and to activate a **specific response**. The adaptive immune system is based in the lymphatic system and relies on two main weapon systems: *Antibody-mediated immunity* and *cell-mediated immunity* (**Figure 22.11**).

Antibody-mediated immunity uses powerful Y-shaped proteins called **antibodies** to recognize and attack invaders. Antibodies recognize nonself markers—**antigens**—on the pathogen and mark the pathogen for destruction. **B cells**, specialized lymphocytes created and matured in the bone marrow, produce thousands of antibodies per second aimed specifically at the pathogen that has been recognized.

Cell-mediated immunity recognizes cells that have been infected by a pathogen such as a virus, as well as cancer cells. **T cells**, lymphocytes created in the bone marrow and matured in the thymus, recognize markers on the surface of a cell that has been infected. The T cell then kills the infected cell so that it cannot spread the disease to other cells.

Compared to innate immunity, adaptive immunity is slow to mobilize. However, it is the most sophisticated and effective of animal defense systems because of the amazing selectivity with which the adaptive immune system attacks a particular invader.

Adaptive immunity occurs in two stages. The very first time a person is exposed to a particular pathogen, the **primary immune response** is activated. This response takes time—more than two weeks, sometimes—to reach full steam. Because of that slow start, and because pathogens multiply so rapidly, people infected with an aggressive pathogen for the first time can sometimes lose the race, becoming ill and dying. Therefore, any pathogen that is new to humans is particularly dangerous, and the nonspecific response of innate immunity may be more beneficial. However, the combined action of innate immunity and the primary response of adaptive immunity prevails against most pathogens.

A distinctive feature of the adaptive immune system is **immune memory**, the capacity of this defense system to remember a first encounter with a specific pathogen and to mobilize a speedy and targeted response to future infection by the same strain. This "memory" is what enables us to become immune to attacks by the same strain after we suffer the disease a first time. Once you've had measles, for example, you never get sick from the measles virus again, because the adaptive immune system recognizes the virus and quickly eradicates it the next time. Keep in mind that this means you are not born with immune memory. Each individual must build one over time by being exposed to various pathogens.

Because of immune memory, the adaptive immune system produces a faster, more dramatic response to pathogens when it encounters them a second time. The second encounter, when the adaptive immune system is poised and ready to respond, is the **secondary immune response**.

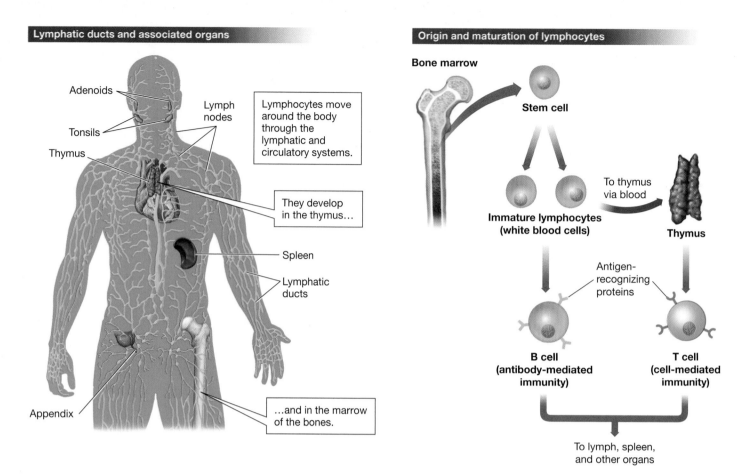

Lymphatic ducts and associated organs

Adenoids

Tonsils

Thymus

Lymph nodes

Lymphocytes move around the body through the lymphatic and circulatory systems.

They develop in the thymus…

Spleen

Lymphatic ducts

Appendix

…and in the marrow of the bones.

Origin and maturation of lymphocytes

Bone marrow

Stem cell

Immature lymphocytes (white blood cells)

To thymus via blood

Thymus

Antigen-recognizing proteins

B cell (antibody-mediated immunity)

T cell (cell-mediated immunity)

To lymph, spleen, and other organs

Figure 22.11

Adaptive immunity resides in the lymphatic system

(Left) The **lymphatic system** consists of lymphatic ducts, lymph nodes, and associated organs. (Right) **Lymphocytes** originate from stem cells in bone marrow. B cells mature in the bone marrow; T cells mature in the thymus. Lymphocytes circulate in the lymphatic and circulatory systems and accumulate in lymph nodes and other organs, such as the spleen, appendix, and tonsils.

Q1: Why are B and T cells so named?

Q2: In what way is this immune system "adaptive"?

Q3: Why is the adaptive immune response considered the third layer of the immune system?

We acquire immunity in two ways: either actively or passively. We acquire **active immunity** to a particular pathogen when our own bodies produce antibodies against that pathogen; they are not received from an outside source. This happens naturally when we're exposed to certain pathogens, such as the measles virus. We can also acquire active immunity to certain diseases through vaccination.

We acquire **passive immunity** by receiving antibodies that were not made by our own bodies. A human fetus acquires antibodies from exchanges between its blood and its mother's blood. This antibody sharing continues after birth: Mother's milk is rich in antibodies because the mother's immune system has encountered many antigens and made many antibodies in her lifetime. Thanks to that antibody-rich milk, a nursing baby receives passive immunity to a broad range of pathogens. Passive immunity produces no memory cells, so it wears off as the received antibodies degrade, usually within a few weeks or months. There is no evidence, yet, that Wim Hof is able to control his adaptive immune response.

Driven by Hormones

Hormones regulate not only the classical "fight or flight" response, but also the sleep and wake cycles of all people. These signaling molecules in the blood also become elevated or depressed during periods of stress (cramming for a test, anyone?) and exercise. What are your levels of melatonin, epinephrine, and cortisol right now?

Hormones in the human body

Melatonin

Regulates sleep timing, blood pressure, and more. Levels fluctuate over a 24-hour cycle, peaking at night.

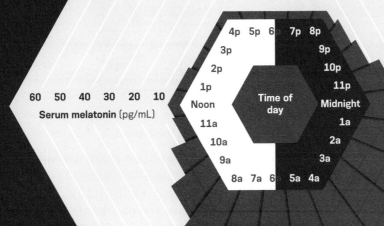

60 50 40 30 20 10
Serum melatonin (pg/mL)

Time of day

4p 5p 6p 7p 8p 9p 10p 11p Midnight 1a 2a 3a 4a 5a 6a 7a 8a 9a 10a 11a Noon 1p 2p 3p

Stress hormone that quickens the heartbeat, among other effects. Levels increase as exercise intensity goes up.

Stress hormone that regulates homeostasis in the body. Levels drop during prolonged high-intensity exercise.

Epinephrine

(Adrenaline)

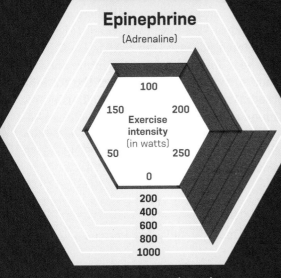

100
150   200
Exercise intensity
(in watts)
50   250
0

200
400
600
800
1000

Plasma epinephrine (pg/mL)

Cortisol

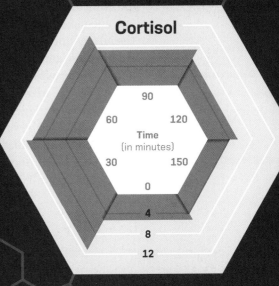

90
60   120
Time
(in minutes)
30   150
0

4

8

12

Cortisol (µg/dL)

Figure 22.12

Hof continues to train recruits to use his novel methods

The Iceman Cometh

In 2014, Pickkers and Kox published the results of their experiment in the *Proceedings of the National Academy of Sciences*, one of the world's most highly respected and cited peer-reviewed journals. They are conducting follow-up studies to determine whether one or more of the three parts of Hof's technique—cold exposure, breathing, and meditation—is primarily responsible for the adrenaline release and subsequent immune suppression, and exactly how these effects come about. Kox suspects that the breathing techniques are the main factor, since Hof's breathing appears to trigger the release of hormones, but he cannot yet be sure.

"It needs to be studied a whole lot more," agrees Hof, who is eager to continue putting his method under the magnifying glass of the scientific process (**Figure 22.12**). "By meticulous experiments and measurements—not speculation—we want to show this works. I'm very thankful to the professors at Radboud who dared to go into this."

Whether Hof's technique can help individuals with autoimmune disorders is still up for debate. Although the training worked for young, physically fit men, the team does not know whether it will work for older people with autoimmune diseases who already have compromised organ systems. "We would not advise people to do this [right now]," says Pickkers, until there are additional studies. "We have to be careful there are no unwanted side effects or risks."

But he is optimistic about the future and still sounds surprised about how well the training worked. "We confirmed that, indeed, using the techniques of Wim Hof, humans are able to modulate their autonomic nervous system and influence their immune response," says Pickkers. "It is remarkable."

REVIEWING THE SCIENCE

- A **hormone** is a signaling molecule distributed through the body by the circulatory system. Because hormones move only as quickly as the blood moves, they tend to coordinate functions that are slower and longer-lasting than those under the influence of the nervous system.

- A single hormone may affect many different kinds of target cells, potentially triggering a different response in each. Hormones act on target cells either by moving through the plasma membrane to the cell's interior or by acting on receptors embedded in the plasma membrane.

- The **endocrine system** is made up of the glands and specialized cells that produce hormones. The **hypothalamus** coordinates the endocrine system and integrates it with the nervous system. The **adrenal glands** produce hormones responsible for the fight-or-flight response.

- The vertebrate **immune system** possesses three layers of defenses against **pathogens**.

  - The first layer consists of **external defenses**: physical and chemical barriers, including the skin and the linings of the respiratory and digestive systems.

  - The second line of defense is the **innate immune system**. Several types of blood cells and molecules produce the **nonspecific responses** of the innate immune system, including **phagocytes** such as macrophages and neutrophils, which engulf and destroy pathogens. Tissue damage stimulates **inflammation** and blood clotting.

  - The third line of defense is the **adaptive immune system**, providing long-term defenses in the form of **specific responses** to pathogens and parasites. These responses are mediated by powerful proteins called **antibodies**, or by cells.

- The **lymphatic system** provides the primary sites for adaptive immunity. White blood cells called **lymphocytes** confer specific immunity. Immature lymphocytes differentiate into **B cells** in the bone marrow and **T cells** in the thymus. Each lymphocyte has special membrane proteins that bind only to a specific antigen of a specific pathogen.

- The **primary immune response** from the adaptive immune system is relatively slow and mild. The **secondary immune response** is a faster, stronger response to a pathogen that has been encountered one or more times.

- **Active immunity** can be acquired through natural exposure to a pathogen or through a vaccine. **Passive immunity** comes from receiving antibodies that were not made by our own bodies, such as when a fetus acquires antibodies from its mother.

THE QUESTIONS

The Basics

1 Hormones are

(a) secretory cells.

(b) endocrine glands.

(c) signaling molecules.

(d) target cells.

2 Which of the following is *not* true of hormones?

(a) They are distributed through body fluids.

(b) They must be present in large amounts to be effective.

(c) They are produced by specialized cells.

(d) They act on target cells.

3 Adrenaline

(a) is produced in the adrenal glands.

(b) increases the amount of glucose in the bloodstream.

(c) suppresses immune system activity.

(d) all of the above

4 The _____ is the immune's system second line of defense.

(a) innate immune system

(b) adaptive immune system

(c) combination of physical and chemical barriers to pathogen entry

(d) all of the above

5 Which of the following is/are *not* a part of the innate immune system?

(a) phagocytes

(b) antibodies

(c) inflammation

(d) clotting

6 Link each of the following terms with the correct definition.

ADAPTIVE IMMUNE RESPONSE

1. The glands and specialized cells that produce hormones.

ENDOCRINE SYSTEM

2. The blood cells and molecules that provide a nonspecific response to pathogens.

INNATE IMMUNE RESPONSE

3. The organ that coordinates the endocrine system and integrates it with the nervous system.

HYPOTHALAMUS

4. Long-term defense against pathogens centered in the lymphatic system.

7 Circle the correct terms in the following sentence:

The first time you are exposed to a pathogen, the [**primary, secondary**] immune response is activated. The [**primary, secondary**] immune response to a pathogen is stronger and more rapid. You acquire [**active, passive**] immunity to a pathogen when your own body creates the antibodies against that pathogen. [**Active, Passive**] immunity comes from the antibodies produced by another person, such as your mother when you were in utero or nursing. Vaccines are an example of [**active, passive**] immunity.

8 Beginning with a perceived threat (for example, a spider), identify the correct order of events by numbering them from 1 to 5.

____ a. Target cells amplify the hormonal signal to produce a response.

____ b. The liver breaks down glycogen to glucose, and the heart increases its rate and the force of its contractions.

____ c. Adrenaline reaches target cells in the liver and heart.

____ d. The hypothalamus signals the adrenal glands that a threat is present.

____ e. The adrenal glands release adrenaline into the bloodstream.

9 Identify which of the below are characteristics of antibody-mediated (A) or cell-mediated (C) immunity.

____ a. relies on Y-shaped proteins to identify pathogens.

____ b. B-cells produce proteins specific to a pathogen

____ c. lymphocytes matured in the thymus identify infected cells.

____ d. Antigens on the pathogen allow it to be identified as non-self.

____ e. Infected cells are destroyed so that an infection cannot spread to other cells

Try Something New

10 Wim Hof and his trainees had increased levels of the stress hormone cortisol and decreased immune function during the experiments described in this chapter. How might these changes negatively affect their endocrine and immune systems over the long term?

11 Describe in your own words what a B cell is.

12 Increased body temperature (fever) is part of the body's innate immune response. Fever is uncomfortable and can be dangerous if very high. It is often treated with over-the-counter medicines like acetaminophen, ibuprofen, naproxen, or aspirin. What are possible negative effects of this treatment?

Leveling Up

13 **Life choices** While clotting is an important component of the innate immune response, it can also be dangerous. For example, a blood clot may block arteries to the heart or brain, leading to a heart attack or stroke. Aspirin reduces blood clotting by interfering with the body's production of a lipid called thromboxane A2. This lipid normally helps platelets clump together (see Figure 22.10), so aspirin, by inhibiting its production, reduces clotting and "thins the blood." Some doctors may prescribe a daily dose of aspirin for patients at risk of heart attack or stroke. Review the U.S. Preventive Services Task Force recommendations on daily aspirin therapy (http://www.uspreventiveservicestaskforce.org/uspstf/uspsasmi.htm), and then answer the questions below.

(a) Do you fall into one of the categories for which daily aspirin therapy is recommended? If yes, which one? If no, is there an aspirin therapy category that you think you'll be in eventually?

(b) How strong is the evidence supporting aspirin therapy in the category you identified in question (a), if any? (See the "Grade" column in the task force recommendations.)

(c) With this information in hand, do you plan to take aspirin daily?

(d) Will you speak with your doctor before taking aspirin daily? Why or why not?

(e) Do you know anyone who has had a heart attack or stroke? Do they take aspirin daily?

14 Doing science Pickkers and Kox say that their next experiment will attempt to determine whether one or more of Hof's techniques—cold exposure, meditation, and breathing exercises—is primarily responsible for the increased cortisol release and decreased immune response. Kox predicts that the breathing exercises will prove to be most important. (You can see a video of the trainees undergoing cold exposure and a trainee during an experiment at http://www.pnas.org/content/111/20/7379.full.) Imagine that it is your responsibility to design the next experiment for Pickkers and Kox. Please include answers to the following questions in the description of your experimental design.

(a) What are your experimental hypotheses?

(b) Give at least one prediction for each of your hypotheses.

(c) Identify your control group and treatment group(s). How many subjects will be in each group? Justify your sample size.

(d) Give a detailed description of the treatment for each group.

Amber Waves
of Grain

We've been growing the same domesticated crops for thousands of years. To survive the future, we're going to need new ones.

After reading this chapter you should be able to:

- Identify the structure and explain the function of plant tissues, organs, and organ systems.
- Give an example of how plants use hormones or other chemical means to survive, grow, and reproduce.
- Compare and contrast how plants and animals grow.
- Identify a nutrient needed for plant growth and reproduction.
- Diagram the alternation of generations in a plant life cycle.
- Explain the variety of ways in which plants are pollinated and disperse their seeds, and how this relates to the structure of flowers and fruit.

CHAPTER

23

PLANT
PHYSIOLOGY

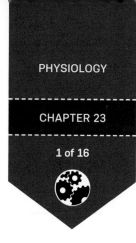

Lee DeHaan wanders through a field of grain, golden knee-high stalks brushing against his faded jeans. Tall, with an angular face and goatee, DeHaan looks the part of a farmer, with a floppy hat and tanned arms to boot. But though he grew up on a corn farm in Minnesota, DeHaan is now an *agronomist*, a researcher studying the science of producing and using plants for food, energy, land preservation, and more. In these fields around his office at the Land Institute in Salina, Kansas, DeHaan is domesticating a new crop—a grain bred from humble prairie grass, a grain with the potential to transform agriculture (**Figure 23.1**).

The way we plant and harvest crops is unsustainable. Nearly 70 percent of the freshwater used by humans is put toward irrigation; farmland is losing productivity because of deforestation, overgrazing, and poor agricultural practices that lead to erosion and pollution; and staple crop production worldwide has leveled off. All this does not bode well when our population is projected to grow by another three billion people over the next half century.

Figure 23.1

Harvest at the Land Institute

Researchers gather ripe stalks of a prairie grass that holds hope for feeding our growing population.

LEE DEHAAN

Lee DeHaan is an agronomist at the Land Institute in Salina, Kansas, where he leads a new crop domestication program.

We need new crops. The common crops grown in the United States today—corn, wheat, and soybeans—were domesticated by our ancestors thousands of years ago to produce high yields and be easily harvested and replanted. Yet these crops are failing to meet our sustainability needs. They require large amounts of water, fertilizer, and pesticides, and they are vulnerable to weather changes, pests, and diseases. They are inefficient and delicate; we need food crops that are hardy and resilient.

Unfortunately, humans stopped domesticating new crop plants long ago. So, despite Earth's rich diversity of plants—over 300,000 species, including more than 50,000 species of edible plants—we rely on fewer than 20 of them to provide 90 percent of our food.

More than 250,000 of Earth's plant species are *flowering plants* (also called *angiosperms*). Worldwide, people get over 80 percent of their calories from flowering plants such as grasses (wheat, rice, corn), legumes (peas, beans, peanuts), potatoes, and sweet potatoes. But as climate change affects agriculture—increasing temperatures and severe weather, and causing more disease—and as the global population grows, we're going to need tough, plentiful crops. That's no easy requirement to meet.

Perfecting Plants

In 2001, as a young and ambitious plant breeder, DeHaan joined a research team at the Land Institute, a 600-acre research center devoted to developing sustainable alternatives in agriculture. At the time, he was the young scientist on the team, so his bosses handed him an ambitious long-term project: to create a new type of grain by domesticating a wild grass that grows year after year.

Plants can be grouped into three categories on the basis of their life cycle: *annuals*, *biennials*, and *perennials*. **Annuals** complete their entire life cycle in one year. In flowering plants, an annual has one year to grow from a seed into a mature plant, produce flowers, and make the seeds that will start the next generation. Annual crops must be replanted every year.

A **biennial** plant, in contrast, grows and matures for a year but does not initially reproduce.

Reproduction takes place in the second year of growth. After the second year, biennial crops must be replanted.

Finally, many flowering plants are **perennials**, which live three years or more, and sometimes for hundreds or even thousands of years. Perennials, which once made up much of the natural grasslands that dominated Earth, are alive year-round and are efficient at nutrient cycling and water management.

Wheat, corn, rice, soybeans—these are all annuals. Our ancestors saw the advantages of breeding annuals: Compared with perennials, annuals produce more seed, and replanting them every season speeds the process of domestication.

But perennials have advantages too, and it was those advantages that DeHaan wanted to tap into by breeding a perennial as a food crop. First, perennials do not waste energy by regrowing **roots** each year, as annuals do. Instead, they grow long, deep roots that anchor in the soil and remain there year after year. These roots enable perennials to absorb water and nutrients more efficiently than annuals do. The roots also outcompete weeds, so less weed killer is needed

to grow perennials. Deep roots also hold carbon in the soil, acting as a *carbon sink*. Grain-producing plants such as wheat, rice, and corn are **monocots**, with fibrous, branching roots. Other plants, including soybeans and oak trees, are **dicots**, which grow straight, thick taproots (**Figure 23.2**).

Roots are part of the plant body, which is relatively simple in its organization compared to the bodies of vertebrate animals (**Figure 23.3**). Plant bodies are made of three basic tissue types: dermal, ground, and vascular tissues. **Dermal tissues**, which form the outermost layer of the plant, protect the plant from the outside environment and control the flow of materials into and out of the plant. **Ground tissues**, which form the intermediate layer, make up the bulk of the plant body and perform a wide range of functions, including support, wound repair, and photosynthesis. Finally, in or near the center of the plant body are the **vascular tissues**, *phloem* and *xylem*, which contain stacks of long cells forming continuous tubes that run throughout the plant body, linking all organs of the root and shoot systems. **Phloem** ships sugars from the

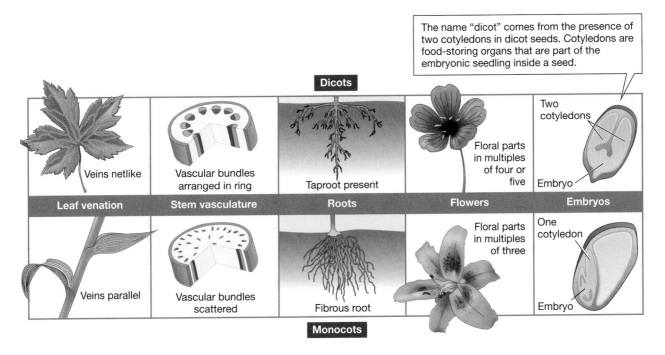

The name "dicot" comes from the presence of two cotyledons in dicot seeds. Cotyledons are food-storing organs that are part of the embryonic seedling inside a seed.

Dicots

Veins netlike | Vascular bundles arranged in ring | Taproot present | Floral parts in multiples of four or five | Two cotyledons / Embryo

Leaf venation | **Stem vasculature** | **Roots** | **Flowers** | **Embryos**

Veins parallel | Vascular bundles scattered | Fibrous root | Floral parts in multiples of three | One cotyledon / Embryo

Monocots

Figure 23.2

Monocots and dicots

Flowering plants have traditionally been classified into two main groups—*dicots* and *monocots*—on the basis of their external form and internal structure. The dicots are the larger of these two informal categories and make up about 175,000 species, including beans, squash, oak trees, and roses. Monocots include all the grasses, members of the lily family, palm trees, and banana plants.

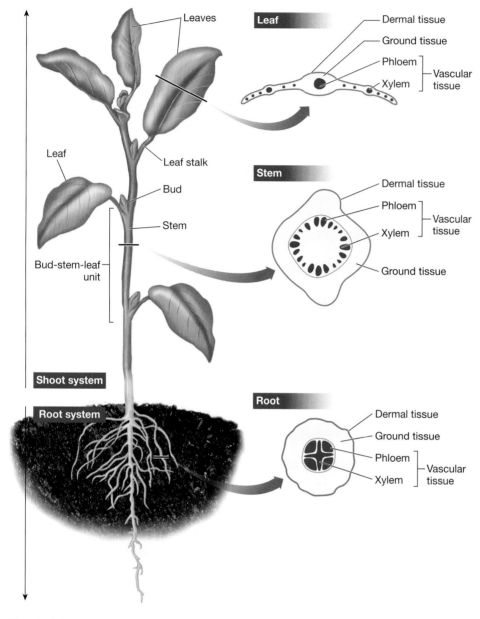

Figure 23.3

How plants are built

Plants have three basic tissue types—dermal, ground, and vascular—which make up three types of organs: roots, stems, and leaves. Belowground, plants grow by extending old roots and producing new lateral roots. Aboveground, plants grow by adding new bud-stem-leaf units.

Q1: What is the function of the vascular tissue?

Q2: A plant organ is green if the cells within it contain chloroplasts. In the figure, which plant organ does not contain chloroplasts, and why do you think that is?

Q3: Which tissue type has chloroplast-containing cells? Why?

leaves, where they are produced, to living cells in every part of the plant. **Xylem** transports water and minerals, absorbed from the soil, upward from the roots and outward to the leaves.

DeHaan's attempt to domesticate intermediate wheatgrass (*Thinopyrum intermedium*)—a wild, flowering perennial plant—started as a side project. Widely used for hay and pasture, intermediate wheatgrass grows wild across the western United States and Canada. It has tall, thin shoots that grow green in the fall and turn golden brown in the spring and summer, and long, deep roots stretching belowground.

The body of a flowering plant, whether intermediate wheatgrass, corn, or roses, can be divided into those two basic organ systems: the belowground *root system* and the aboveground *shoot system*. These two organ systems are specialized for life in two very different environments: roots in soil, shoots in air.

The **root system** anchors the plant, absorbs water and nutrients from the soil, transports food and water, and may store food (**Figure 23.4**). *Root hairs* greatly increase the surface area through which plants can absorb water and mineral nutrients. Annuals typically grow short roots. Perennials grow longer roots, which last year after year, depending on the species.

Roots are one of plants' three basic organs. The other two are stems and leaves, which form the **shoot system** (**Figure 23.5**). **Stems** provide the plant with structural support, transport food and water, and hold leaves up to intercept light. Although cells in the stems of many plants can perform photosynthesis, most sugars are produced by photosynthesis in the **leaves**. Wheat and intermediate wheatgrass have long, pointy leaves that grow from the stems. At the tip of each shoot, and the base of many leaves, is

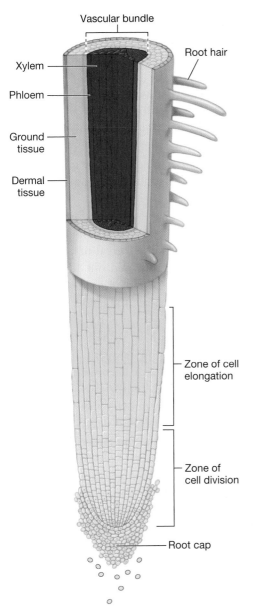

Figure 23.4

The root system

Roots produce numerous outgrowths from dermal cells called *root hairs*, which aid in water and nutrient absorption. Plant roots have a region of active cell division, protected by the root cap, and a region of cell elongation, in which cells increase in size and complete their development.

Q1: How do root hairs increase the amount of water and nutrients that a plant can absorb?

Q2: How do roots make a plant more stable?

Q3: Given question 2, which do you predict would be more stable in harsh weather conditions—annuals or perennials?

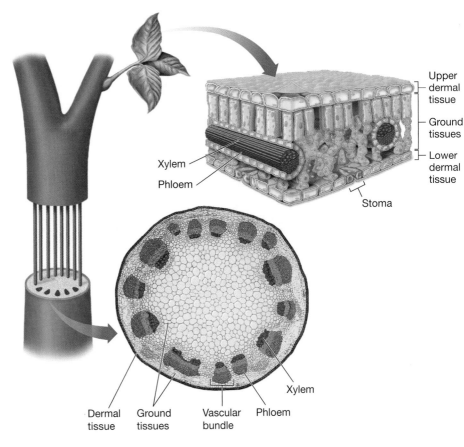

Figure 23.5

The shoot system

Leaves produce the majority of the plant's food through photosynthesis. Stems provide structural support and may perform a limited amount of photosynthesis.

Q1: Which part of the shoot system—the bud, stem, or leaf unit—produces flowers?

Q2: Find the stoma in the figure. How is its location on the leaf important for its function?

Q3: What nutrients does phloem move from the leaves to other parts of the plant?

a bud. Under the right conditions, buds produce new shoots or *flowers*.

Flowers house the structures that produce male and female gametes and, in many species, also facilitate the delivery of the sperm-bearing pollen to the female reproductive organs (**Figure 23.6**). Portions of the flower develop into fruit, which helps disperse seeds in a highly effective way.

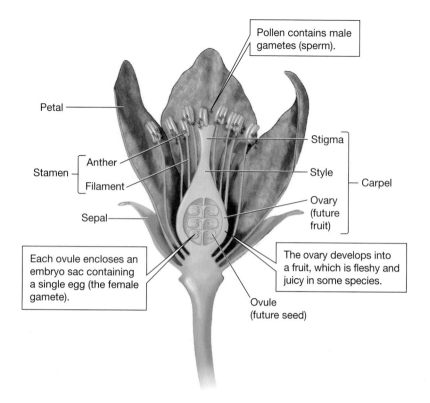

Pollen contains male gametes (sperm).

Petal

Stamen — Anther
Filament

Sepal

Stigma

Style

Carpel

Ovary (future fruit)

Ovule (future seed)

Each ovule encloses an embryo sac containing a single egg (the female gamete).

The ovary develops into a fruit, which is fleshy and juicy in some species.

Figure 23.6

Four whorls make a flower

The various parts of a flower are arranged in concentric rings, or whorls. From the outermost whorl inward, a typical flower consists of four whorls: Sepals, petals, stamens, and carpels. All of the petals in a flower are collectively known as the *corolla*. The collective term for all of the sepals is the *calyx*.

The outer layer of leaves is made up of dermal tissues, which include the regulated pores known as **stomata** (singular "stoma") that control gas exchange. Plants open stomata to let in carbon dioxide needed for photosynthesis. When open, stomata release oxygen and lose water through evaporation. Most plants open their stomata in the daytime and close them at night, conserving water when photosynthesis is not an option. A plant experiencing water stress due to an inadequate water supply will close its stomata to conserve water, no matter the time of day. Intermediate wheatgrass and other deep-rooted perennials are less likely to experience water stress because of their extensive underground network of roots to extract water from the ground. Yet they do often become dormant and close their stomata during the hot summer.

Breeding Begins

Intermediate wheatgrass first came to the attention of agronomists in 1983, when the Rodale Institute, a Pennsylvania research institute that studies organic agriculture, evaluated close to 100 species of perennial grasses for traits including seed size, flavor, and harvestability. Intermediate wheatgrass emerged the winner: It is hardy and currently not susceptible to any major pest or disease.

Disease resistance is important because crops are at the mercy of pathogens. About 20 years ago, for example, a fungal disease called Fusarium head blight wiped out Minnesota's barley crop. Because plants are under attack by pathogens and predators, they have developed a rich variety of mechanisms to deter attacks (**Figure 23.7**). In addition to physical defenses such as thorns, many plants contain chemical substances that are toxic to herbivores. Nicotine, the addictive chemical in cigarette smoke, protects tobacco plants from insect predators. Caffeine protects the leaves and seeds of coffee and tea plants from potential predators. Plants also circulate and secrete antimicrobial chemicals to kill pathogens that could infect them.

In addition to using chemicals to protect themselves, plants produce chemicals called **hormones** to coordinate internal activities necessary for growth and reproduction. Hormones are active at very low concentrations.

After intermediate wheatgrass proved to be resistant to pests and produced better seeds than other perennial grasses, the Rodale Institute began collecting strains of the plant from all around the world. Over 12 years, researchers bred the strains, selected the best 20, then identified the top 14 individual plants.

At the Land Institute, DeHaan received seed from the 14 Rodale plants and began breeding them, tinkering with the possibilities of intermediate wheatgrass. In his first year, DeHaan planted some 3,000 plants in an effort to create a large group of diverse individuals. Then, he watched the grass grow.

Unlike most animals, which grow early in their lives and then level off, plants grow throughout their lives—a pattern called **indeterminate growth**. Plants can increase in length by **primary growth**, the repeated, modular addition

1 A tough outer surface protects the plant from water loss and heat and cold stress, and serves to block pathogens from entry. Thorns and other defenses keep predators away.

2 Plants deploy nonspecific chemical defenses, which involve a range of broadly targeted anti-pathogen and anti-herbivore chemicals, including hormones that signal other parts of the plant to manufacture chemical weapons and have them "battle ready."

3 Plants also have specific chemical defenses. For example, they have a number of genes that respond to complementary genes in a pathogen. If a specific pathogen is detected, protective defenses including toxic chemicals are unleashed.

1 **Auxins** are necessary for cell division and the formation of organs such as roots. They are involved in *phototropism*, the growth of shoots toward the light, and in *gravitropism*, the growth of roots toward the ground and the growth of shoots away from the ground.

2 **Cytokinins** are necessary for cell division, and they promote shoot formation. The levels of both cytokinins and auxins decline sharply just before plants drop their leaves in the fall season.

3 **Abscisic acid (ABA)** mediates adaptive responses to drought, cold, heat, and other stresses.

4 **Ethylene** stimulates the ripening of some fruits, including apples, bananas, avocadoes, and tomatoes. Ethyiene activates enzymes that convert starches into sugars, resulting in a sweeter fruit.

Figure 23.7

Plants use chemicals to survive, grow, and reproduce

Plants use chemicals to protect themselves from both living and nonliving threats in their environment. And like animals, plants use hormones to coordinate internal activities necessary for growth and reproduction.

Q1: Give an example of how plants use chemicals to defend themselves.

Q2: Of the plant hormones introduced in this figure, which promote growth?

Q3: What is the first line of plant defense? Is this a chemical defense?

of a bud-stem-leaf unit aboveground or new lateral roots belowground. They can also grow in thickness, called **secondary growth**. Secondary growth includes the thickening of both stems and roots.

Indeterminate and modular growth habits give plants great flexibility to respond to changing environmental conditions, such as high or low levels of sunlight, water, or nutrients. Plants tend to add many new parts when conditions are favorable and few new parts when conditions are not. This flexibility also enables plants to replace damaged tissues and organs. In fact, plants are so flexible in their development that most living cells in the adult plant body can generate whole new plants.

When it was time to harvest the intermediate wheatgrass, DeHaan and his team went around their field and recorded traits of the plants, such as their heights and how early they flowered. Then the team plucked the heads off the plants and placed them in barcoded bags. Back in the lab, technicians analyzed the seeds from each plant, recording size and weight—the larger the seed, the more grain can be produced for food—and noting how easily the seeds shed their outer husks, because the less sticky a husk is, the easier it is to process a grain.

DeHaan's first experimental crop grew plants of many sizes and traits. "It's like scratching off lottery tickets," DeHaan told National Public Radio in 2009. "Maybe there's something amazing in there. We'll see." As he recorded traits, DeHaan found that most of his best performers—those with larger seeds and deeper roots—came from a few families, so DeHaan went back into the field and uprooted those specific plants. These were brought into the warm greenhouse for the winter and bred with each other. Their offspring were then planted the following fall, and the process repeated.

But the more DeHaan bred these particular plants, the more he restricted their gene pool, creating a genetic bottleneck by excluding genes that had been in the original population of 3,000 plants. He began to worry that he might have lost genes needed to improve certain traits. One particular trait he struggled to identify and retain in his crop was the ability to mature early in the season.

Many plants, especially species native to temperate regions, perceive the seasons by sensing the length of the day. This is possible because day length varies with the season: Days are shorter in winter and longer in summer. This sensing of the duration of light and dark in a 24-hour cycle is known as **photoperiodism**. Plants use day length to sense when conditions are favorable for flowering and seed germination. The dormancy of buds through fall and winter, as well as their regrowth in spring, is also influenced by photoperiodism. DeHaan wanted to identify wheatgrass that flowered early in the spring, so he obtained hundreds of wild collections of wheatgrass that mature early from the U.S. Department of Agriculture (USDA) and he crossed those that flowered early with his own plants.

By 2010, DeHaan had something good: He had doubled both the seed yield and the weight of seeds of his intermediate wheatgrass. "It was succeeding a lot faster than we thought," says DeHaan. "It was really easy to grow."

Unfortunately, "intermediate wheatgrass" doesn't sound like a tasty grain one might like in cereal or bread. So DeHaan and his team renamed the newly domesticated grain "Kernza," a combination of the world "kernel" and the name of a native tribe of the region, Kanza (which also inspired the state name "Kansas").

Kernza is different from wild intermediate wheatgrass in several ways. First, the seed size is larger. "When I started working with it, the typical seed weighed 3.5 milligrams," DeHaan said in 2010. "Now, our best seeds are 10 milligrams." That's progress, but there's more to be done: An average wheat seed weighs 35 milligrams.

Then there are the roots. Kernza's roots grow down to 10 feet (**Figure 23.8**). The deep roots make the plant very efficient at sucking water out of the soil. That makes it more resistant to climate change than annual crops, and the plant has already been shown to fare better in drought

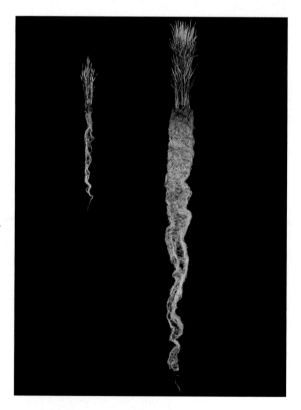

Figure 23.8

Kernza roots grow deeper than wheat roots
Kernza's longer roots (right) make the plant better able to collect water from surrounding soil, and to store carbon and nitrogen that would otherwise be lost to the air or waterways.

GM Crops Take Root

Since 1996, the total land area used to grow genetically modified (GM) crops has expanded dramatically. Most of these crops, including the varieties depicted here, are enhanced with the insertion of a gene that imparts insect resistance or herbicide tolerance. Broad scientific consensus has concluded that food derived from GM crops poses no greater risk to health than food from non-modified crops.

Global area of genetically modified crops

Area shown in millions of hectares

Soybean   Maize   Cotton   Canola

14
3
3
8
1998

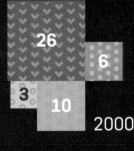

26
6
3
10
2000

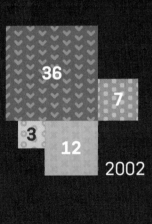

36
7
3
12
2002

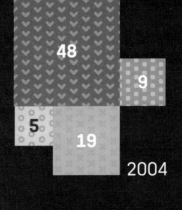

48
9
5
19
2004

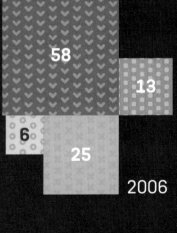

58
13
6
25
2006

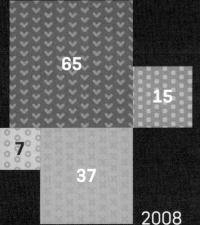

65
15
7
37
2008

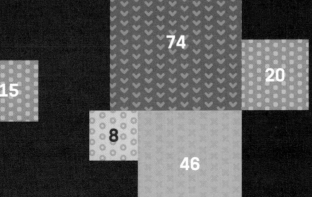

74
20
8
46
2010

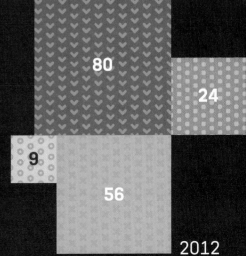

80
24
9
56
2012

conditions than traditional wheat does. Furthermore, the long roots hold soil together, preventing erosion, which carries away fertile soil, fertilizers, and pesticides with it. Major crop-producing countries all around the world lose tens of billions of tons of topsoil every year to erosion.

Kernza's long roots have additional benefits. A recent study found that within two years of planting, Kernza beat annual wheat at accessing groundwater, accumulating carbon in the soil, and absorbing nitrogen fertilizer. To grow, plants need CO_2, water, and mineral nutrients, especially nitrogen, phosphorus, and potassium. Most of the dry weight of a plant comes from CO_2 absorbed from the air and converted into carbohydrates by photosynthesis. Plants require **macronutrients** (nitrogen, phosphorus, potassium, calcium, sulfur, and magnesium) in relatively large amounts and **micronutrients** (including iron, zinc, and copper) in small amounts. Carbon, oxygen, and hydrogen are obtained from air or water; the rest of the nutrients plants need must be obtained from soil. In agriculture, farmers add most of the nutrients to the soil as fertilizer, but fertilizer can run off into water, polluting it. Perennial plants make better use of fertilizer than annuals by efficiently retaining and recycling it. In the same study mentioned earlier, Kernza reduced the amount of nitrogen leached into the nearby ecosystem by up to 99 percent compared with wheat.

Crop Collaboration

Large seeds and deep roots are good traits, but they aren't enough to make Kernza a successful crop. Other specialty grain crops yield about 1,000 pounds per acre. In Kansas, Kernza was yielding less than 300 pounds per acre. So DeHaan sought help. Around 2007, DeHaan took his Kernza plants north to the University of Minnesota, where he had once been a student.

DONALD WYSE

Donald Wyse develops and promotes winter annual and perennial crops at the University of Minnesota, where he is a professor of agronomy and plant genetics.

There, he began working with agronomist Donald Wyse and other wheat breeders at the university. Planting in Minnesota turned out to be a boon: Kernza had larger yields there because intermediate wheatgrass is a cool-season grass and performs better in a cooler place.

Wyse was an ideal collaborator. He had spent years breeding perennial and winter annual crops for agriculture in Minnesota, including perennial flaxseed and perennial sunflower. One of the great values of a perennial, in addition to its having deep roots that prevent erosion and requiring less fertilizer and pesticide, says Wyse, is that its roots and ground tissues are active during a longer period of the year, absorbing more solar energy and thus increasing an area's net primary productivity (see Chapter 18 for a discussion of NPP). This, he says, is "high-efficiency agriculture."

In Minnesota, for example, the roots of corn and soybeans are active for just two and a half months a year, usually after the heaviest rains. That means all the sunlight from the other nine and a half months of the year is wasted. But perennials can perform photosynthesis from the moment the snow melts in the spring until the first snowfall of the winter, says Wyse. With perennials, he says, "we're harvesting more energy."

In 2011, DeHaan, Wyse, and the Minnesota team planted a field of more than 2,000 intermediate-wheatgrass plants from 69 families and then measured traits like biomass, grain yield, and seed shape. They've continued this process every year, sometimes pairing complementary strengths and weaknesses. In the summer of 2014, they planted 14,000 seedlings.

The domestication continues, and it is not easy. DeHaan estimates it will be another 10 years before Kernza is ready to economically compete with wheat. One of the challenges of breeding intermediate wheatgrass has to do with how it reproduces. Many plants, such as dandelions and poplar trees, can reproduce asexually, when a parent plant forms a genetically identical clone. But staple crops, including grains, reproduce sexually.

The overall principle of sexual reproduction in plants is similar to that in animals: A haploid male gamete (*sperm*) fuses with a haploid female gamete (*egg*) to give rise to a diploid cell, the *zygote*, which undergoes mitosis to create a multicellular diploid *embryo*. In time, the embryo develops into an individual offspring, which represents the next generation (**Figure 23.9**).

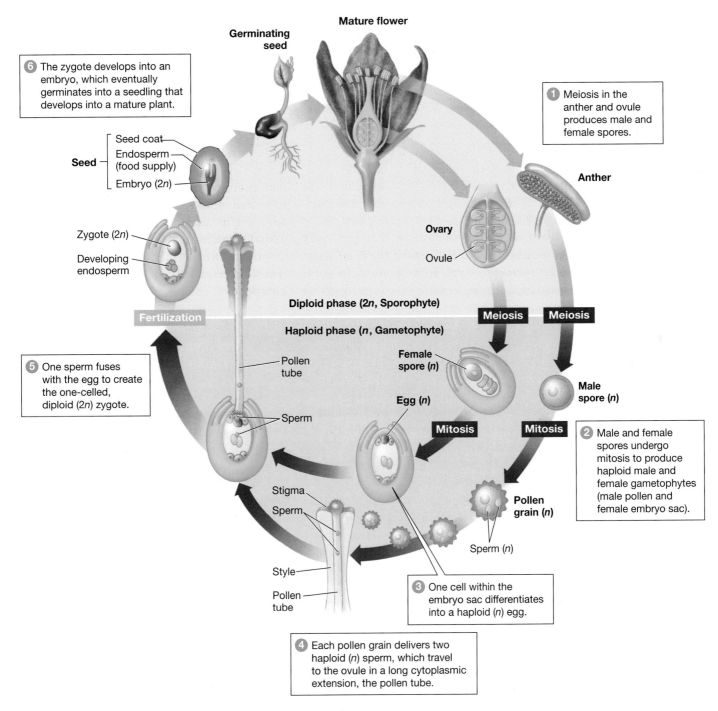

Mature flower

Germinating seed

⑥ The zygote develops into an embryo, which eventually germinates into a seedling that develops into a mature plant.

① Meiosis in the anther and ovule produces male and female spores.

Anther

Seed {
 Seed coat
 Endosperm (food supply)
 Embryo (2n)
}

Ovary

Ovule

Zygote (2n)

Developing endosperm

Diploid phase (2n, Sporophyte)

Fertilization

Haploid phase (n, Gametophyte)

Meiosis   **Meiosis**

⑤ One sperm fuses with the egg to create the one-celled, diploid (2n) zygote.

Pollen tube

Female spore (n)

Egg (n)

Male spore (n)

Sperm

Mitosis   **Mitosis**

② Male and female spores undergo mitosis to produce haploid male and female gametophytes (male pollen and female embryo sac).

Stigma

Sperm

Pollen grain (n)

Sperm (n)

Style

Pollen tube

③ One cell within the embryo sac differentiates into a haploid (n) egg.

④ Each pollen grain delivers two haploid (n) sperm, which travel to the ovule in a long cytoplasmic extension, the pollen tube.

Figure 23.9

From generation to generation

The life cycle of flowering plants is marked by **alternation of generations**. Haploid (n) stages of the life cycle are shown in purple; diploid (2n) stages, in orange.

Q1: Are eggs and sperm haploid or diploid? Are they part of the sporophyte or gametophyte generation?

Q2: Why is the plant life cycle, but not animal life cycle, referred to as an "alternation of generations"?

Q3: How does asexual reproduction differ from the plant life cycle diagrammed here?

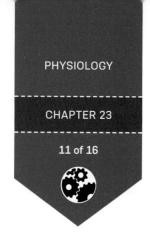

Plant and animal life cycles differ in one key respect: In animals, meiosis creates gametes and nothing but gametes. In plants, meiosis generates haploid cells called **spores**. Each spore undergoes mitotic divisions to create a haploid, multicellular **gametophyte**. Specialized cells in the gametophyte differentiate to produce sperm or egg cells.

In an animal, gametes are the only haploid cells, and animals have no such thing as a gametophyte. However, gametophytes are part of the life cycle of every plant. In flowering plants like Kernza, the gametes are contained in flowers.

Once fertilized, the zygote develops into an embryo within the ovule, which is contained within the ovary, at the base of the flower. The outer layers of the ovule harden into a protective seed coat. Each seed contains the ingredients for growing a young plant of the next generation: a mature embryo, a food source, and the seed coat.

The ovary surrounding the seed forms a *fruit*, yet another plant organ. Once a fruit is formed, the embryo enters dormancy, and the seed is dispersed from its parent (**Figure 23.10**). Seeds are often dispersed by wind, or by attaching to or being eaten (and excreted) by animals. Dormancy ends when the embryo is stimulated by favorable conditions to start growing again. Then the seed germinates, and a seedling grows into a plant that will mature and produce flowers and fruits, the beginning of another generation. This adult plant is a **sporophyte**, analogous to an individual animal in that both are multicellular, diploid organisms.

Annual wheat plants *self-pollinate*: They reproduce by fertilizing themselves with their own pollen to create a zygote before their flowers

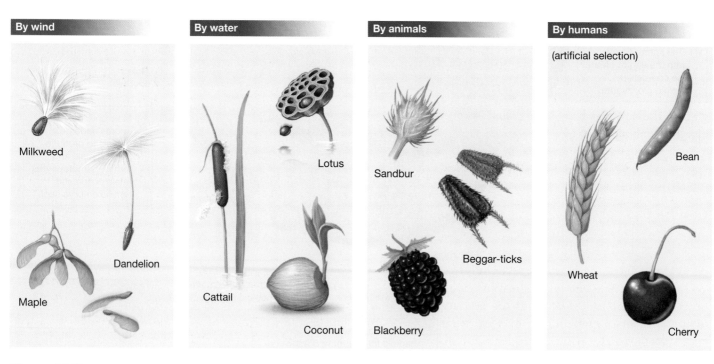

Figure 23.10

Plants spread their seeds in a variety of ways
Many plants spread their seeds via the wind or water, but others use modified seeds and fruit to attract animals or attach to their fur. Humans have artificially selected the fruit and seeds of many species to make them more edible.

Q1: What kinds of seeds would you expect to be contained within a sugary fruit: seeds spread by water or by animals? Why?

Q2: What kinds of seeds would you expect to weigh less: those spread by air or those artificially selected by humans? Why?

Q3: How does the height of a tree affect its ability to spread its seeds widely?

even open. But intermediate wheatgrass, and many other plants, do not self-pollinate. They require a mate. Given that plants cannot travel to find mates, how does sperm-containing pollen from one plant reach the eggs of another plant?

In some species, like grasses and pine trees, pollen is transported by wind. Wind can carry pollen long distances, but most pollen blown by wind lands in inhospitable places (such as parking lots or lakes), not on the flower of another plant of the same species. Many flowering plants sidestep this problem by having their pollen transported by animals such as insects, birds, and even mammals (**Figure 23.11**). Plants attract these **pollinators** with brightly-colored, sweet-smelling flowers filled with sugary nectar. It is a mutualistic relationship, in which both species benefit.

DeHaan and his team need to mate specific plants together in order to breed for specific traits. So, instead of relying on wind or pollinators, they put a bag over the flowering stalks of two plants they want to breed so that pollen from one plant passes directly to the second plant in the bag, and no others (**Figure 23.12**). "We call them 'plant condoms,'" says DeHaan with a laugh. "The bags keep crosses we don't want to happen from happening."

And because an intermediate-wheatgrass plant can breed with only another plant—rather than with itself, as annual wheat can do—the offspring are genetically diverse; no child is *exactly* like its parent. In fact, a child can be very different from its parents, because there is a great deal of variation in the genome of intermediate wheatgrass. Each gene may have six or more versions, or alleles, in a single plant, says DeHaan. So, when breeding new plants, the team can never match the exact genotype and phenotype of a parent but rather must constantly juggle traits to try to capture the right set of traits in a single line of plants. For example, DeHaan is currently working to grow shorter plants, because taller plants tend to tip over, which makes them difficult to harvest.

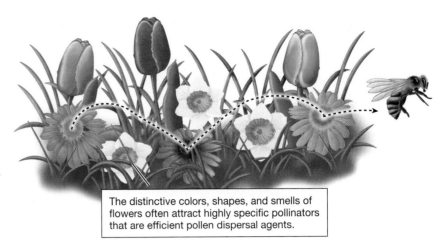

The distinctive colors, shapes, and smells of flowers often attract highly specific pollinators that are efficient pollen dispersal agents.

Honeybees carry the pollen that dusts their bodies from flower to flower as they search for food.

Birds have good color vision and favor red flowers with long floral tubes.

Figure 23.11

Bribing animals to do the work
Pollinators provide stationary plants with a way of transporting sperm to eggs. The spectacular colors, shapes, and odors of flowers, in combination with food rewards such as nectar, lure pollinating animals into visiting several flowers of the same species, incidentally transferring pollen in the process.

Q1: Many crops, including apples, peppers, and tomatoes, depend on insect pollinators. What would happen if no pollinator visited a plant?

Q2: Although they look very different, flower petals are actually modified leaves. Do flowers still perform the main function of leaves? Explain your reasoning.

Q3: Some flowers look like an insect. How might this resemblance attract pollinators?

Perennial Pancakes

Early on, DeHaan estimated that it would take some 50 years for a new plant to be domesticated to match the yield of wheat, but now USDA and University of Minnesota researchers are supplementing his hands-on breeding work with gene sequencing efforts. These scientists have been associating genetic markers with particular traits, such as seed size. Once they have enough traits mapped on the genome, they will no longer have to wait for a plant to mature to observe its

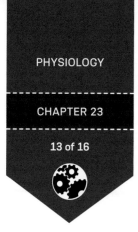

Figure 23.12

Breeding the next generation of Kernza plants
The flowering stalks of Kernza plants are enclosed in bags to prevent the pollen from spreading to any other plants. Scientists identify plants with the traits they want to pass on to the next generation and selectively breed them in this way.

Q1: What traits are scientists selecting for in each new generation of Kernza?

Q2: What would happen if the Kernza plants were allowed to freely pollinate, rather than being selectively pollinated by hand?

Q3: After the flowers have been hand-pollinated, the scientist covers the flower stalk with the bag seen in the photo. Why?

phenotype; instead they will simply sample the DNA of young seedlings and predict a phenotype on the basis of genetic markers.

"Within a very short period of time, we'll be able to identify key genes in the domestication process, and at a very low cost," says Wyse. With these tools, Wyse thinks the breeders won't take too long to match the yield of wheat. "It may only take half a decade to see dramatic improvements in these new crops," says Wyse. "We think crops of this type are going to have great value."

But there's more to a food crop than how well it grows and can be harvested. It also has to taste good. DeHaan has used Kernza flour to make cookies, cakes, scones, bread, and more (**Figure 23.13**). He and other Land Institute staff have enjoyed them, though at least one food scientist says the flour will need some flavor improvement to go mainstream. A team at the University of Minnesota is currently experimenting with Kernza's food traits, including taste, texture, and ability to rise.

Both beer brewers and sustainable food companies have expressed an interest in using the new grain. DeHaan and Wyse also imagine that the crop by-product could be used as hay or to produce biofuel. "We'd love this to be a dual-use crop," says DeHaan. Farmers, for example, could harvest the grain for a food company and then use the "residue" plant material in the field, typically the leftover bottom of the stalk, to make biofuel. "Those two things together could result in an economically viable crop," he says.

Figure 23.13

Kernza flour
Scientists are finding uses for Kernza in traditional baking, and using it to make alcohol, animal feedstock, and, hopefully in the future, even biofuel.

REVIEWING THE SCIENCE

- Plant bodies are made of three basic tissue types: dermal, ground, and vascular tissues. **Dermal tissues** control the flow of materials into and out of the plant and protect the plant from attack. **Ground tissues** make up the bulk of the plant body and participate in support, wound repair, and photosynthesis. **Phloem** and **xylem** are the two types of **vascular tissue**, transporting sugars and water, respectively, through the entire plant.

- The plant body contains two basic *organ systems*: the belowground **root system** and the aboveground **shoot system**. Plant organs include *roots*, *stems*, *leaves*, *flowers*, and *fruits*. **Roots** absorb water and mineral nutrients, anchor the plant, and store food. **Leaves** produce the majority of the plant's food through photosynthesis. **Stems** support the plant and may perform a limited amount of photosynthesis.

- Plants defend themselves against the external environment with a tough outer covering, physical weapons such as thorns, and an arsenal of toxic chemicals.

- Like animals, plants use **hormones**, chemicals produced at low concentrations, to coordinate growth and reproduction.

- Plants have **indeterminate growth**, meaning that they can grow throughout their lives. Aboveground, plants increase in length by **primary growth**, the addition of basic bud-stem-leaf units. Plants increase in thickness by **secondary growth**, the thickening of stems and roots.

- To grow, plants need CO_2, water, and mineral nutrients. Plants need **macronutrients** (especially nitrogen, phosphorus, and potassium) in large quantities.

- Many plants can reproduce asexually, but most can also reproduce sexually. Meiosis in plants produces a single-celled **spore**, which divides through mitosis to create a haploid, multicellular **gametophyte**. Cells in the gametophyte differentiate into egg or sperm. The fertilization of the egg by sperm generates a diploid zygote, which gives rise to the diploid, multicellular **sporophyte**. This **alternation of generations** (gametophyte and sporophyte) is a hallmark of plant life cycles.

- In flowering plants, male and female reproductive parts are contained in flowers. Flowers attract animal **pollinators**, which provide immobile plants with a way of transporting sperm-containing pollen to eggs.

- Fertilization produces a zygote, which divides and develops into an embryo. The embryo is located within the ovule, which hardens into a protective seed coat. Each seed contains the ingredients for growing a young plant of the next generation: a mature embryo, a food source, and the seed coat.

THE QUESTIONS

The Basics

1 Water is transported through the plant body by

(a) xylem.

(b) phloem.

(c) stomata.

(d) root hairs.

(e) flowers.

2 Which of the following is *not* a plant organ?

(a) stem

(b) fruit

(c) vascular bundles

(d) root

(e) flower

3 Which *micro*nutrient do plants need to absorb from the environment to survive and grow?

(a) carbon

(b) sugar

(c) potassium

(d) nitrogen

(e) zinc

4 Which of the following is a chemical defense strategy used by plants?

(a) maintaining a tough dermal layer

(b) storing poisons in leaves

(c) signaling plant shoots to grow toward the light

(d) stimulating the ripening of fruit

(e) none of the above

5 Which plant tissue controls the flow of materials into and out of the plant and protects the plant from attack?

(a) vascular tissue

(b) ground tissue

(c) dermal tissue

(d) phloem

(e) xylem

6 What kind of seed is best for being dispersed by wind?

(a) dandelion

(b) apple

(c) coconut

(d) Kernza

(e) seeds held within a bur

7 Link each of the following terms with its definition.

MODULAR GROWTH

INDETERMINATE GROWTH

PRIMARY GROWTH

SECONDARY GROWTH

1. Increases in length by the division of cells located at the tip of each stem and each root.
2. Grows throughout its life.
3. Grows in size by repeatedly adding the same basic bud-stem-leaf unit.
4. Increases in thickness.

8 Circle the correct terms in the following sentence:

[**Annual, Biennial**] plants complete their life cycle in one year, whereas [**biennial, perennial**] plants live for three or more years. Wheat is a(n) [**annual, perennial**], whereas Kernza is a(n) [**annual, perennial**]. Kernza is also a [**monocot, dicot**], which has a [**fibrous root, taproot**].

9 Beginning with the stage following spores, place the steps of the plant life cycle in the correct order by numbering them from 1 to 4.

_____ a. zygote develops into an embryo.

_____ b. gametophyte develops.

_____ c. seed germinates.

_____ d. egg is fertilized.

Try Something New

10 Some plants are pollinated by bats, which are active at night. What would you predict about how bat-pollinated flowers differ from bird-pollinated flowers? (*Hint*: Remember that birds are active during the day.)

11 Ethylene is a plant hormone that causes fruits to ripen by converting starches into sugars. It also signals flowers to open, seeds to germinate, and leaves to shed. You may have observed that placing a ripe banana or apple next to another fruit causes it to ripen more quickly. Furthermore, if you place fruit in a paper bag (rather than exposed on the counter), it ripens more quickly. How would you explain this?

12 Several of the plants domesticated by humans have highly modified organs. For example, potatoes are modified stems, and carrots and sweet potatoes are modified roots. From which plant organ would you predict that onions and tomatoes are modified? Why?

Leveling Up

13 **Looking at data** The debate over genetically engineered (GE) crops and genetically modified organisms (GMOs) has been impassioned but not always informed. Genetically engineered crops are those with DNA modified through biotechnology, in order to increase productivity or resistance to disease.

More than 93 percent of the soybeans harvested in the United States, fed almost exclusively to domesticated animals, are genetically engineered for pesticide and herbicide resistance. Examine the graph on the next page that shows the yield in bushels per acre of soybeans from 1980 to 2010. The red circle is the yield for organic soybeans in 2007, which is based on information from 1,331 organic farms.

(a) What does the *x*-axis represent? The *y*-axis? What does the data line on the graph represent?

(b) Describe the trend in soybean productivity from 1980 to 2010.

(c) The yield for organic soybeans in 2007 is only 66 percent of the average yield for that year. What year of conventional soybean production does this production match?

(d) Critics say that the widespread growth and consumption of GE crops may have unexpected health and environmental costs. Proponents argue that genetic engineering is the only way to feed a rapidly growing population. What do you think?

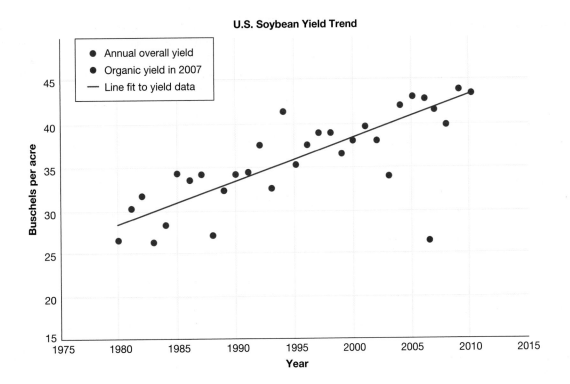

U.S. Soybean Yield Trend

A Critical Choice

A mother peels back the controversy to decide whether to vaccinate her children.

After reading this chapter you should be able to:

- Evaluate a scientific claim, using the process described in the chapter.
- Explain the importance of scientific literacy for making informed decisions.
- Distinguish between secondary and primary literature, and explain the role of peer review in the latter.
- Compare and contrast basic and applied research, and give an example of each.
- Determine whether a scientific claim is based on real science or pseudoscience.

CHAPTER

24

APPLYING THE
SCIENCE

In 2009, Anna Eaton discovered she was pregnant. Several of her friends were also expecting or had recently had children, so the friends' afternoon chats were soon about all things baby: nursery colors, pediatricians, car seats. But every once in a while, one of the mothers broached a subject far more contentious than strollers and baby carriers. One of their hot topics of discussion became vaccination.

Vaccination is the injection of material—typically an inactivated infectious organism or parts of such an organism—that stimulates the immune system to protect against future exposure to that pathogen (review Figure 9.3 on how vaccines work). Over the past 200 years, scientists

Figure 24.1

The first vaccine

In 1796, Dr. Edward Jenner vaccinated his gardener's eight-year-old son against smallpox. At the time, about 60 percent of Europeans became ill with smallpox. Of those infected, 30 percent of adults and 80 percent of infected children died of the disease.

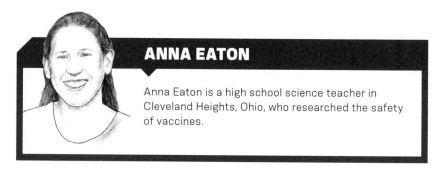

ANNA EATON

Anna Eaton is a high school science teacher in Cleveland Heights, Ohio, who researched the safety of vaccines.

have developed vaccines to protect against dozens of pathogens, starting with smallpox (**Figure 24.1**). Vaccines have also been developed for the Tic Tac–shaped bacteria that cause diphtheria, the round and highly contagious rotavirus, the tiny airborne bacteria that cause pertussis, also known as whooping cough, and many more. Some diseases, like smallpox and polio, have been completely eradicated in the United States through vaccination programs.

Before vaccines, many children died from diseases such as diphtheria, whooping cough, and polio. The infectious pathogens that cause those diseases still exist today in our environment, but because people are protected by vaccines, we almost never see infections. Today, the Centers for Disease Control and Prevention (CDC), the public health branch of the federal government, recommends that children receive 10 vaccines, given over a total of 24 doses, between birth and 15 months of age. Additional vaccinations are recommended between the ages of 18 months and 18 years.

Eaton, pregnant with her first baby, was a microbiologist by training. She worked at the Cleveland Clinic in Ohio, an academic medical center, before moving on to teach biology at a local community college. Some of her friends were also scientists, and one pregnant friend adamantly commented, "Of course we're going to vaccinate. We're scientists."

But Eaton was not so confident. She had seen media stories suggesting that vaccines might cause autism, and when she searched online, she read scary accounts from parents who had a child vaccinated and then watched as their child developed autism. "I'm on message forums, encountering 100 different moms with stories. It was frightening," Eaton recalls.

Stories and anecdotes are powerful. They can change how we feel or think about a subject. But anecdotal evidence is not scientific evidence. Anecdotes are not representative of collected data or collected scientific observations, and therefore they cannot reliably give us the big picture of a subject or phenomenon. Scientific evidence can.

Eaton's scientist friend said there were data backing the safety of vaccines, but Eaton had not seen those data and wasn't about to inject her new baby girl with something she knew nothing about. "Feelings of distrust are in all of us, and possibly the strongest in a new mother," says

Eaton. "Even though she was a friend of mine, it was still not enough for me to just go in the office and have my daughter vaccinated."

True or False?

The statement that vaccines cause autism, or the opposite statement that vaccines are safe, is a **scientific claim**—a statement about how the world works that can be tested using the scientific method. We are exposed to scientific claims every day—dozens of them, in fact. They are not all true. A few years ago, editors at the magazine *Popular Science* asked one of their journalists to write down every scientific claim he heard in a day and evaluate each one. He recorded a whopping 106 claims and evaluated each of them, starting with Cheerios' claim that it reduces cholesterol (supported by scientific evidence) to the claim of a face cream infused with vitamin A that it would revitalize skin (not supported). Of the 106 claims, most were bogus.

The majority of scientific claims come from advertisers. Though companies are legally bound to tell the truth, not all of them do. For example, in 2005 an Italian footwear company named Vibram introduced the $100 FiveFingers running shoes, glove-like shoes that mimic the feel of barefoot running (**Figure 24.2**). In a marketing campaign, the company touted the FiveFingers shoe's health benefits, advertising that the shoes could reduce foot injuries and strengthen foot muscles. In 2013, however, two peer-reviewed studies of over 100 joggers found that the shoes actually increased the likelihood of injury. Soon, Vibram was sued for false advertising, and in 2014 the company settled that lawsuit for $3.75 million, agreeing to refund FiveFingers shoe purchases and remove all health claims about the shoes from its advertisements.

Scientific claims also come from special-interest groups and organizations that exist to advance certain causes, often for political or religious reasons. These claims include statements about global warming, evolution, and medical care. Again, such claims are often untrue. Therefore, it is important to question the "truth" of a claim when you hear it.

We, the public, are not simply consumers of science and technology. By voting on issues that have a scientific underpinning, we shape the course of science and influence which technologies are used, as well as where and how they're used. Although our personal values and political leanings are likely to influence how we vote, the underlying science should also be taken into consideration. **Scientific literacy**, an understanding of the basics of science and the scientific process, enables us to make informed decisions about the world around us and to communicate our knowledge to others. Our hope is that this book has helped you to become scientifically literate.

Eaton didn't know the source of the original claim that vaccines cause autism, but she knew that in order to be scientifically literate and make an informed decision for her daughter, she needed to find it. Scientific claims directly affect our lives because we make decisions based on them. Some of these are small decisions: Should I take a multivitamin every morning? How often should I exercise? Others are larger decisions: Should I vote to support carbon taxes? Does a cell phone cause tumors from radiation? Should I vaccinate my children? The good news is that you can learn to evaluate scientific claims. You

Figure 24.2

Shoes designed to mimic bare feet
FiveFingers running shoes did not live up to the manufacturer's scientific claims.

can be skeptical about claims and, using critical thinking, make scientifically literate decisions for yourself.

Credentials, Please

Caroline was born in January 2010. Eaton had Caroline vaccinated when she was two months old, but then Eaton found herself gripped with fear. "I worried I was doing more harm than good," she says. Because of her rising fears, Eaton did not get Caroline's four-month vaccinations (**Figure 24.3**).

Eaton sought out a pediatrician who would discuss her fears about vaccination. Yet several pediatricians refused to talk with her about a delayed vaccination schedule. Some even refused to treat Caroline if she were not vaccinated. Finally, Eaton found a pediatrician who was willing to listen and discuss. After Eaton shared her concerns about vaccine safety, the pediatrician handed Eaton a book, *Vaccines and Your Child*, by a pediatrician named Paul Offit. In the book, Offit laid out how vaccines work, how they are made, and which risks are real and which are false. In the book, he also provided documentation, with detailed references to scientific papers, showing that there is no scientific evidence that vaccines cause autism.

Eaton found the book to be well written and informative, but she wondered if she could trust Offit. She checked his **credentials**—a first step toward assessing the strength of a person's scientific claim. Does the person making a scientific claim have a PhD or MD? Is the degree in the field in which they are making the claim? PhDs in physics do not have training in germ theory, for example, and medical doctors do not have training in atmospheric science.

Offit had an MD from the University of Maryland and was a professor of vaccinology and pediatrics at the University of Pennsylvania. He was

Figure 24.3

Anna and Caroline Eaton
Anna Eaton had concerns about vaccinations for her first child, Caroline.

also chief of the Division of Infectious Diseases and the director of the Vaccine Education Center at the Children's Hospital of Philadelphia. In other words, Offit had an MD in an appropriate field and held a job at a reputable university. While good credentials alone do not guarantee that a source is trustworthy, scientists practice for many years in their area of expertise, and their scientific claims tend to be based on that expertise and carefully stated.

In addition to someone's credentials, it is important to assess whether the person making a scientific claim has an agenda or **bias**, a prejudice or opinion for or against something. Does the person have an ideological, political, or religious belief that will be supported by the scientific claims being made? Does the person have a conflict of interest? Does he or she stand to make money in any way if others accept those claims?

Doing scientific research almost always requires money, so it is important to take into account where the money comes from. In North America, the vast majority of **basic research** in science is funded by the federal government—that is, by taxpayers. Basic research is intended to expand the fundamental knowledge base of science. In the United States, Capitol Hill appropriates more than $20 billion each year for basic

PAUL OFFIT

Paul Offit is a pediatrician and chief of the Division of Infectious Diseases at the Children's Hospital of Philadelphia.

research in the life sciences, including biomedicine and agriculture. Researchers must compete vigorously for the limited funds, and this competition helps ensure that the public money goes toward supporting high-quality science. Research funded by the government is normally not considered biased, since the funding comes from taxpayers.

But industries and businesses spend a great deal of money funding science as well, often in areas of **applied research**, in which scientific knowledge is applied to human issues and often commercial applications. In some cases, researchers funded by industry may have a bias in favor of whatever that industry is selling. Funding from industry does not necessarily mean that a scientific claim is incorrect, but the claim should be looked at closely to rule out possible bias.

Offit, Eaton read, had worked with other researchers to develop the RotaTeq vaccine against rotavirus, a virus that causes severe diarrhea and in some cases death. The pharmaceutical company Merck had purchased the vaccine, and Offit had received an unspecified amount of money from that transaction. This financial compensation suggested that Offit might have some bias toward the use of the rotavirus vaccine. Though she had enjoyed Offit's book, Eaton decided she did not want to take his word alone as an answer to her questions about vaccines. "In the end, I decided to just read about it, do a lot of research, and talk to a lot of people," she says.

To the Books

Eaton dove into the secondary literature on vaccines. When investigating a scientific claim, your first stop should be the Internet or the library to get a basic overview of the topic from the **secondary literature**, which summarizes and synthesizes an area of research. Textbooks, review articles, and popular science magazines such as *National Geographic*, *Popular Science*, and *Scientific American* are good secondary sources.

Eaton went to her local library and came home with a stack of books, including a thick vaccine textbook so that she could learn the underlying science of how a vaccine works inside the body. She read about how vaccines stimulate cells of the immune system to protect a person from a virus or bacteria. Once, while she was reading, Eaton's husband walked into the room and looked at her with surprise. "What are you doing?" he asked. "I'm looking for answers!" she replied.

For secondary literature on the Internet, try to visit sites that are affiliated with the government, a university, or a respected institution like a major hospital or museum. Wikipedia often has overview articles that link to science blogs and review articles in science journals. Like Eaton did, it's important to check the credentials of the person or people behind a resource, especially on the Internet. Anonymous sources are not to be trusted.

When evaluating a scientific claim, you may need more detailed information than is available in the secondary literature, especially if you're dealing with a particularly important life decision or if the area of science involved is changing rapidly. In that case, you should next review the **primary literature**, where scientific research is first published (**Figure 24.4**). Primary sources include technical reports, conference proceedings, and dissertations, but the most important primary sources are peer-reviewed scientific journals such as *Science*, *Ecology*, and the *Journal of the American Medical Association* (*JAMA*).

Pulling from references she found in the secondary literature, Eaton compiled a pile of primary literature on vaccines. If her library didn't carry a particular journal, she acquired a copy through interlibrary loan or, in some cases, even e-mailed the journal directly and got articles for free. She found dozens and dozens of papers about vaccines.

Correlation or Causation?

One of the first papers Eaton came across in her investigation—one that had made huge waves in the media—was published in *The Lancet* in 1998. *The Lancet* is a well-known peer-reviewed medical journal and therefore a reputable source. The paper was unremarkably titled "Ileal-Lymphoid-Nodular Hyperplasia, Non-specific Colitis, and Pervasive Developmental Disorder in Children." It was a study of 12 children ranging in age from 3 to 10 who had experienced a loss of language skills—a symptom of autism—as well as diarrhea and abdominal pain. Parents of eight

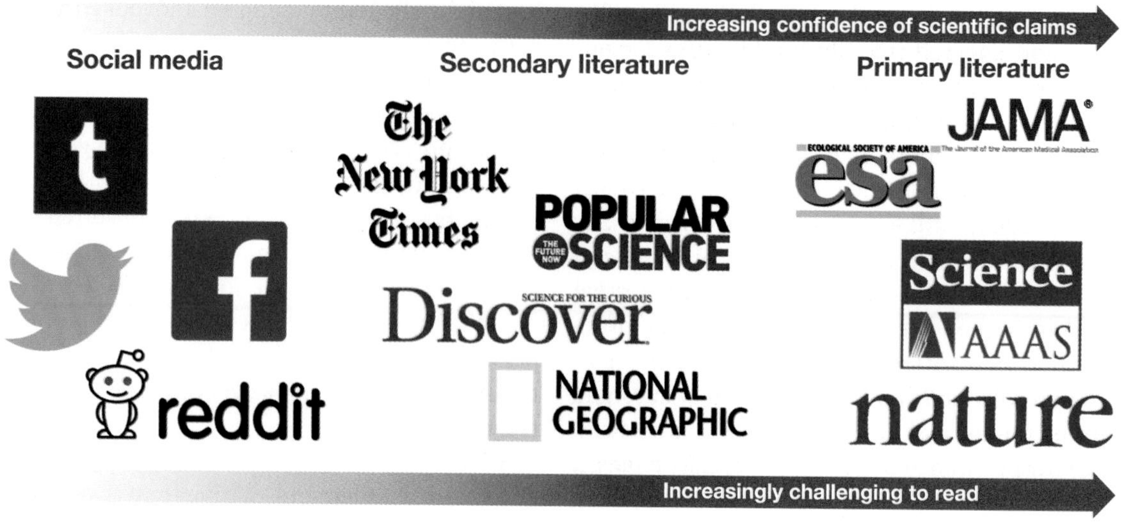

Increasing confidence of scientific claims

Social media | Secondary literature | Primary literature

Increasingly challenging to read

Figure 24.4

Scientific claims in the media and literature

It is easy to find and read scientific claims in social media. However, this is not a good source of scientific information. For help in making important life choices, it is important to go to the secondary literature or even the primary scientific literature for accurate and reliable information.

Q1: Why are we less confident of scientific claims made over social media?

Q2: Where would you place a blog in this figure? Would it matter whether or not it was written by a practicing scientist? Explain your reasoning.

Q3: Give an example of when you would rely on secondary literature to evaluate a scientific claim and an example of when you would go to the primary literature. What is the basis of that decision?

of the 12 children said the onset of symptoms occurred shortly after the child's immunization with the measles, mumps, and rubella (MMR) vaccine. The authors of the paper, a team of 12 researchers, concluded that more research was needed to study a possible relation between the observed brain dysfunction, bowel problems, and the MMR vaccine. In a press conference when the paper was published, one of the authors, a British doctor named Andrew Wakefield, stated that he believed single vaccines, rather than the MMR triple vaccine, were likely to be "safer" for children. The study and press conference sparked widespread fear among parents that the MMR vaccine could cause autism.

The study was published around the same time that officials began documenting an increase in rates of autism. Since the early 1970s, when organizations began counting the number of people with autism for the first time, the incidence of autism has increased 20- to 30-fold in the United States and other countries. In 2002, the estimate was that one in 150 children aged 8 years (the age of peak prevalence) had the disorder. In 2004, that number had risen to one in 125. In 2006, one in 110. By the year 2010, an estimated one in every 68 children 8-years-old had an autism-related disorder (autism spectrum disorder). The CDC has said that the rise in rates is likely due to heightened disease awareness, more screening within schools, and a willingness to label the condition. But after Wakefield suggested that the MMR vaccine might be causing autism, the press and other organizations began to report that the rise in autism rates was caused by the increased use of vaccines.

Linking rising rates of autism with increased use of vaccines is a correlation. As you may recall from Chapter 1, **correlation** means that two or more aspects of the natural world behave in an interrelated manner: if one shows a particular value, we can predict a value for the other aspect. But correlation does not establish **causation**, in which a change in one aspect *causes* a change in another. Correlations may suggest possible causes for a phenomenon, but they do not establish a cause-and-effect relationship. For example, there is also a correlation between organic food sales and the increase in autism. Since 1998, rates of autism have increased hand-in-hand with organic food sales. They are correlated, but there is no scientific evidence that organic food causes autism (**Figure 24.5**).

Another correlation that spurred fears about vaccines is that the onset of autism symptoms occurs at about the same age that children receive vaccinations. Most children receive the MMR vaccine at about 15 months old, which is shortly before the first symptoms of autism are often noticed. Parents of children with autism saw their children begin to exhibit symptoms of the illness after their vaccination and therefore directly observed a correlation between the administration of the vaccine and the onset of autism.

Yet, again, correlation does not prove causation. Only scientific experiments can demonstrate causation. So was Wakefield's claim based on good science? Did the MMR vaccine cause autism?

Real or Pseudo?

Unfortunately, sometimes a claim that superficially looks like science is actually **pseudoscience**. Pseudoscience is characterized by scientific-sounding statements, beliefs, or practices that are not actually based on the scientific method. Asking a few simple questions at each step of the scientific method can help you distinguish science from pseudoscience (**Figure 24.6**).

Using these criteria, Eaton analyzed Wakefield's claim. It was quickly obvious that his study did not live up to the standards of good scientific research (**Figure 24.7**). First, the study was small—only 12 children participated—yet the conclusions were grand: Wakefield suggested that all children should stop receiving the MMR vaccine. Next, the study was not made up of a random sample of children. Instead, the subjects were picked specifically for their symptoms.

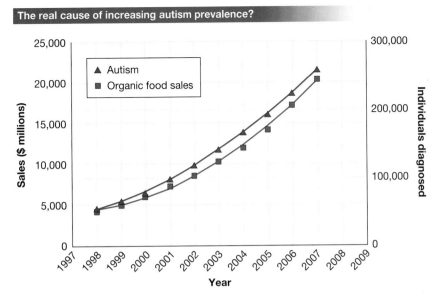

The real cause of increasing autism prevalence?

Figure 24.5

Correlation is not causation: organic food and autism
Reddit user Jasonp55 created a tongue-in-cheek demonstration of why it is important not to assume causation from correlation. Jason used real data on organic food sales and the prevalence of autism from 1997 to 2009. The two are highly correlated, but one does not cause the other.

Q1: How much did organic food sales grow during the period covered in the graph? How much did the incidence of autism grow?

Q2: Why might both organic food sales and autism prevalence have increased during this time period? A Reddit user in the original discussion thread suggested that both might be affected by increasing wealth in the United States. How might increased wealth affect these variables?

Q3: In what way has the vaccine/autism debate been confused by people misinterpreting correlation as causation?

Third, the study did not have a control group, such as children who had been vaccinated but did not show signs of autism, or children who had autism but had not been vaccinated. Finally, and most important, the finding could not be replicated. In fact, as of June 2014, at least 108 published papers, studying millions of children, have found no evidence of a link between vaccines and autism.

In one of the most recent papers, researchers at the University of Sydney in Australia, who stated no conflicts of interest, performed a **meta-analysis**, work combining results from different studies, of five *cohort studies*—observational

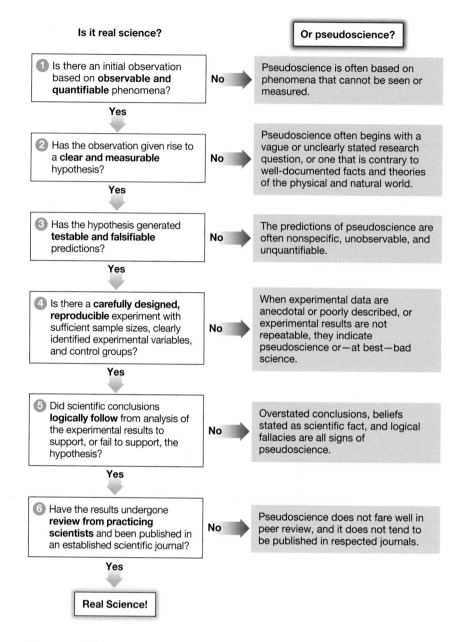

Is it real science?

Or pseudoscience?

① Is there an initial observation based on **observable and quantifiable** phenomena? — **No** → Pseudoscience is often based on phenomena that cannot be seen or measured.

Yes

② Has the observation given rise to a **clear and measurable** hypothesis? — **No** → Pseudoscience often begins with a vague or unclearly stated research question, or one that is contrary to well-documented facts and theories of the physical and natural world.

Yes

③ Has the hypothesis generated **testable and falsifiable** predictions? — **No** → The predictions of pseudoscience are often nonspecific, unobservable, and unquantifiable.

Yes

④ Is there a **carefully designed, reproducible** experiment with sufficient sample sizes, clearly identified experimental variables, and control groups? — **No** → When experimental data are anecdotal or poorly described, or experimental results are not repeatable, they indicate pseudoscience or—at best—bad science.

Yes

⑤ Did scientific conclusions **logically follow** from analysis of the experimental results to support, or fail to support, the hypothesis? — **No** → Overstated conclusions, beliefs stated as scientific fact, and logical fallacies are all signs of pseudoscience.

Yes

⑥ Have the results undergone **review from practicing scientists** and been published in an established scientific journal? — **No** → Pseudoscience does not fare well in peer review, and it does not tend to be published in respected journals.

Yes

Real Science!

Figure 24.6

Science or pseudoscience?

A series of simple questions based on the scientific method can help you determine whether a "scientific" study is real science or pseudoscience.

Q1: State the hypothesis of the people who believe that vaccines cause autism. What is an alternative hypothesis?

Q2: What part(s) of the figure show(s) where Wakefield's study failed to meet the standards of the scientific method?

Q3: Why does only one arrow point to "real science," whereas multiple arrows point to pseudoscience?

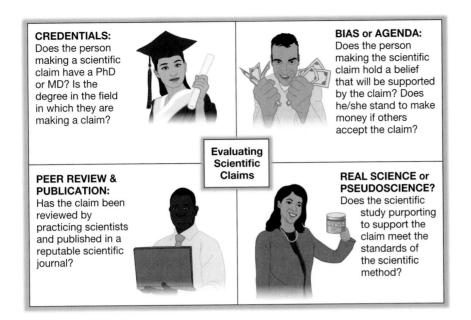

Figure 24.7

Evaluating scientific claims
We are constantly bombarded with scientific claims. A few simple questions will help you evaluate which of these claims are valid and which are not.

Q1: Why is it important to know the education and expertise of a person making a scientific claim?

Q2: List at least five possible biases that people making scientific claims might have.

Q3: Describe a situation in which you might not dismiss the scientific claim of a person who did not have appropriate credentials, or who had a bias toward the claim.

studies of a group of people over a certain period of time—involving 1,256,407 children. The researchers also looked at five *case-control studies*—studies comparing patients with a disease to those without—involving 9,920 children. The data showed no relationship between vaccination and autism, autism and MMR, or autism and thimerosal (a mercury-containing preservative in some vaccines). The work, published in the journal *Vaccine*, included data from the United States, United Kingdom, Japan, and Denmark on vaccines for measles, mumps, diphtheria, tetanus, and pertussis.

Many more peer-reviewed studies with good control groups and sufficient sample sizes, conducted by researchers without a conflict of interest, have shown that the number of vaccines given from birth to 15 months of age under the CDC's recommended vaccination schedule is safe. Researchers have also found that delaying the vaccination schedule increases a child's risk of contracting a disease.

Today it is clear that the paper that started it all, Wakefield's 1998 *Lancet* paper, proposed a claim that is incorrect. All told, it's okay for science to be wrong. There is no expectation that every single one of the thousands of studies published each year will be correct. Yet in the case of the *Lancet* paper, there was bias and wrongdoing. Years after publication of the paper, it came to light that Andrew Wakefield had received large amounts of money as a paid expert for lawyers who were suing vaccine manufacturers. Wakefield had also applied for a patent on a vaccine that would rival the most commonly used MMR vaccine.

Because of these conflicts of interest, in 2010 the General Medical Council, an organization that licenses all medical doctors in the United Kingdom, concluded that Wakefield's conduct had been "irresponsible and dishonest" and revoked his medical license. Eventually, 10 of the paper's 11 other authors retracted their support of the study and its conclusions. Ultimately, in February of 2010, the *Lancet* officially retracted the paper, an extremely rare action for publishers (**Figure 24.8**). A peer-reviewed paper is retracted—withdrawn as untrue or inaccurate—by a publisher or author when its findings are no longer considered trustworthy because of error, plagiarism, a violation of ethical guidelines, or other scientific misconduct.

Fears vs. Facts

Wakefield's paper had not yet been retracted when Eaton was making her decision about whether to vaccinate Caroline. But Eaton followed the guidelines to distinguish science from pseudoscience, and she concluded that Wakefield and his colleagues had looked at too small a sample of children. Therefore, the paper "did not make a lasting impression," recalls Eaton. So she looked beyond that one paper to others. The multitude of papers she found, all with controls and larger sample sizes, assured her that vaccination does not cause autism. In the end, her research and scientific literacy led Eaton to fully vaccinate all of her children on the CDC's recommended schedule.

In making her decision, Eaton was aware that vaccines can have side effects. Most are minor, like a sore arm or a low-grade fever. In some very rare cases, however, vaccination can result in a serious reaction. An allergic reaction to the hepatitis B vaccine, for example, is estimated to occur once in 1.1 million doses. A serious allergic reaction to the MMR vaccine is seen less than once in a million doses. The CDC maintains a list of side effects and constantly monitors the safety of vaccines (**Figure 24.9**).

Weighing the benefits of avoiding disease against the small risk of side effects made the choice to vaccinate her children clear, says Eaton. "We [parents] are often so focused on possible effects of the vaccines that we don't think about our kids getting these diseases" that vaccines prevent, she says.

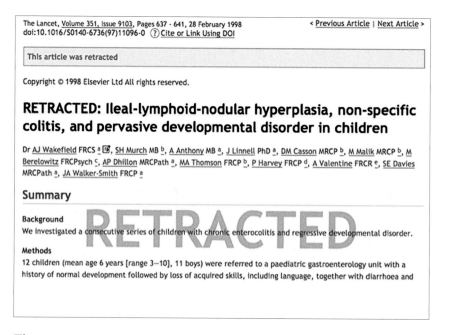

Figure 24.8

The paper that precipitated the vaccine-autism scare is retracted

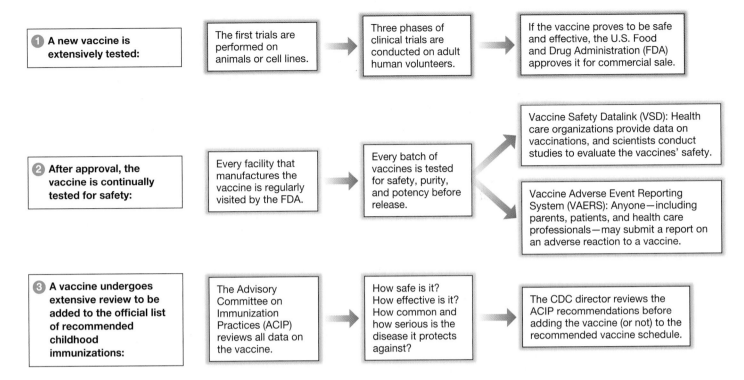

Figure 24.9

Evaluating vaccines: an ongoing process
Vaccines are continually tested and evaluated for effectiveness, safety, and side effects. The severity of the disease a vaccine works against, as well as how likely a child is to be infected with the disease, are also taken into consideration in determining whether to recommend the vaccine.

Q1: Why do vaccine manufacturers begin with tests on animals or cell lines before moving on to adult human subjects?

Q2: What ongoing testing and reporting are vaccines subjected to?

Q3: What do ACIP, FDA, and CDC stand for, and what is the role of each in evaluating vaccines?

Sometimes, when making life choices based on scientific claims, you need to consider nonscientific aspects of your decision. Your values, ethical stances, and religious beliefs will make some choices more acceptable than others. For example, if you believe it is not ethical to eat meat, the scientific finding that leaner meats are healthier than fattier meats won't matter to you.

It is also important to consider how your choices will affect other people. Choosing not to vaccinate a child, for instance, affects the entire community. Rates of vaccination in the United States have been steadily dropping because of misconceptions about the safety of vaccines, and those decreasing rates of immunization put at risk not only the children who are not vaccinated, but also those who cannot be vaccinated, such as infants too young for a vaccine or people who are genetically unable to respond to a vaccine. When a critical portion of a population is vaccinated, then the spread of disease is contained—a concept known as *herd immunity* (**Figure 24.10**). In other words, vaccinating a large number of people keeps germs out of circulation and protects the vulnerable members of the community.

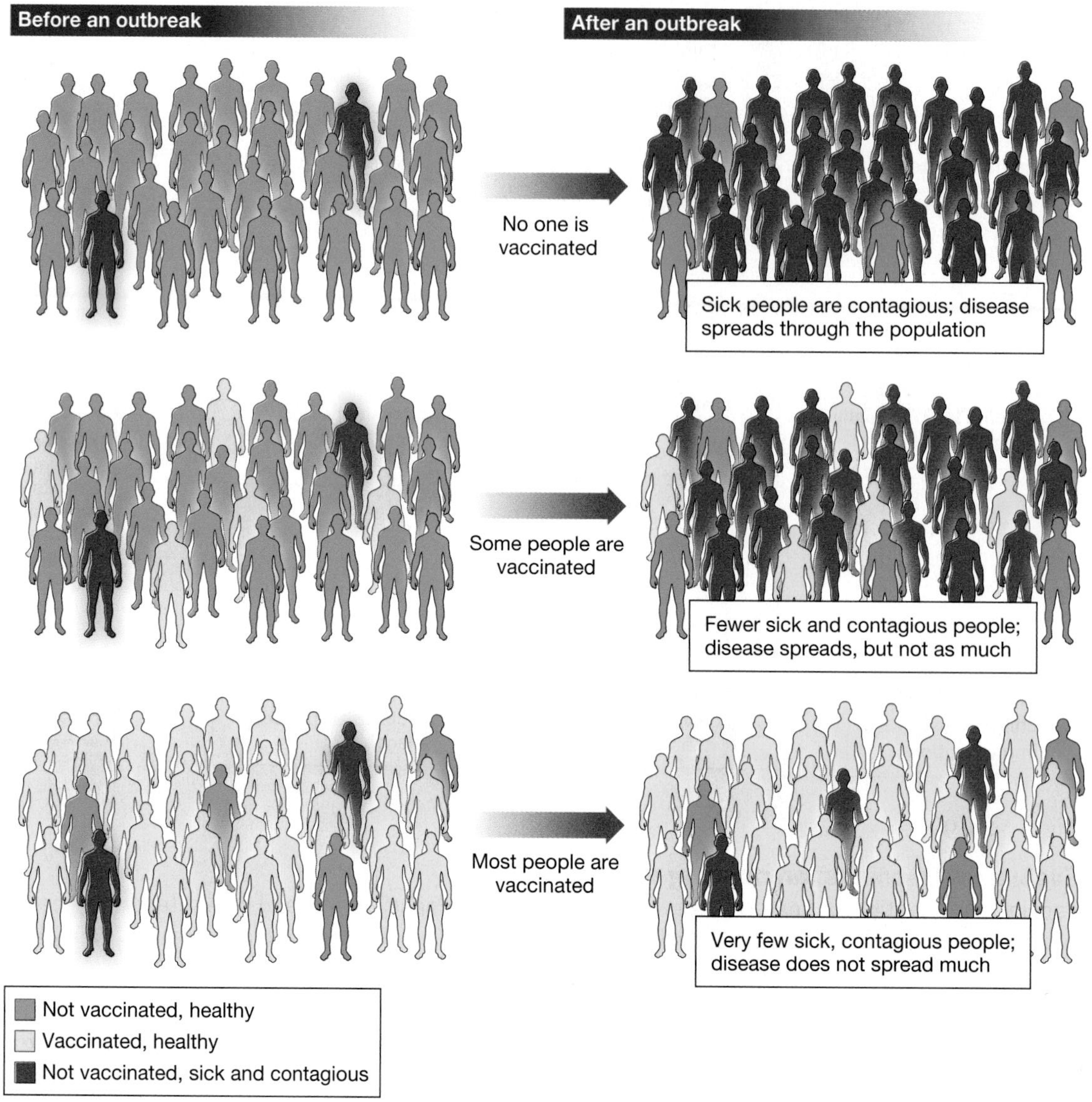

Before an outbreak

After an outbreak

No one is vaccinated

Sick people are contagious; disease spreads through the population

Some people are vaccinated

Fewer sick and contagious people; disease spreads, but not as much

Most people are vaccinated

Very few sick, contagious people; disease does not spread much

- Not vaccinated, healthy
- Vaccinated, healthy
- Not vaccinated, sick and contagious

Figure 24.10

Vaccine prevalence and herd immunity

When most of the population is vaccinated against a contagious disease, the disease will spread little during an outbreak. When the majority of the population is not vaccinated, an outbreak will spread further and may cause disease and death in members of the population who are too young to be vaccinated or have a compromised immune system. Image modified from *The National Institute of Allergy and Infectious Diseases* (NIAID).

Q1: What happens to an immunized person when a disease spreads through a population? (*Hint:* In the graphic, follow an immunized individual before and after a disease spreads.)

Q2: Explain why a disease is less likely to spread to vulnerable members of a population if most people are immunized.

Q3: How does vaccination help an individual person? How does it help that person's community?

Safety in Numbers

In 1999, the United Kingdom began a meningitis vaccination program for children up to the age of 18. Meningitis rates rapidly fell off throughout the population—children and adults alike—thanks to both the effectiveness of the vaccine and the protective effects of herd immunity. Models show that without herd immunity, the number of meningitis cases would have rebounded.

Meningitis cases in England and Wales

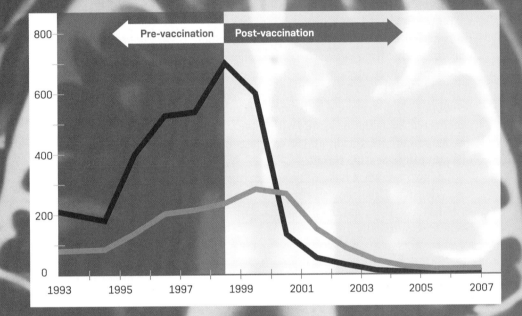

19 years old and younger

20 years old and older

Pre-vaccination   Post-vaccination

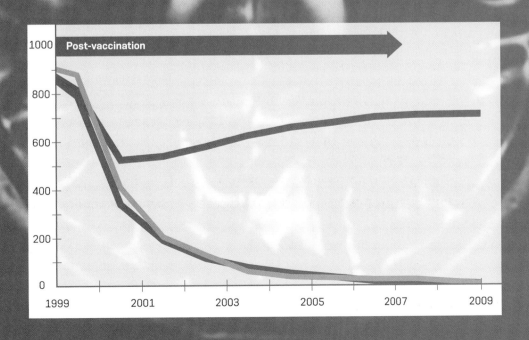

Observed cases

Predicted cases (with herd immunity)

Predicted cases (no herd immunity)

Post-vaccination

When parents opt not to vaccinate their children, herd immunity disappears. And because of that loss of herd immunity, infectious diseases of the past, some thought to be eliminated, are roaring back. In August 2013, a megachurch in Texas whose founder had spoken out against vaccines made headlines after 21 members of its congregation contracted measles. Sixteen of them were unvaccinated. In May 2014, measles cases in the United States hit a 20-year high, with 288 cases reported in a 5-month period. And in the 2013–2014 flu season, 90 percent of the 100 American children who died of complications from the flu had not received the flu vaccine. Decisions based on scientific claims have serious consequences.

Today, Eaton is thankful that all of her children are fully vaccinated (**Figure 24.11**). She encourages her friends, relatives, and colleagues to foster scientific literacy and use their critical-thinking skills to distinguish science from pseudoscience. Not everyone will want to read a vaccine textbook, Eaton admits with a laugh, but anyone can think critically and follow the process of evaluating scientific claims.

"Look to see what has credible backing and what doesn't. This issue transcends vaccines," says Eaton. "We should all have the tools to become scientifically literate citizens."

Figure 24.11

Anna Eaton and her children
Although Eaton initially had concerns about vaccinations for her first child, Caroline, her research gave her confidence that vaccines are a safe and responsible choice. Caroline and her younger brothers are now all up-to-date on their vaccinations.

REVIEWING THE SCIENCE

- A **scientific claim** is a statement about how the world works that can be tested using the scientific method. To evaluate scientific claims, it is important to look at the **credentials** and **bias** of those making the claim.

- **Scientific literacy** requires a basic understanding of scientific facts and theories, and of the process of science. It is important to be scientifically literate in order to make well-informed life decisions.

- Scientific claims can be found in advertising, in social media, in the popular press, and in scientific publications. The best source for a review or overview of a scientific topic is the **secondary literature**, including the popular press, reputable websites, or review articles in scientific journals.

- The actual experimental or observational results related to a scientific claim are found in the **primary literature**, articles published in scientific journals that have undergone peer review.

- With **correlation**, two or more aspects of the natural world behave in an interrelated manner. With **causation**, a change in one aspect *causes* a change in another.

- **Basic research** explores questions about the natural world and expands the fundamental knowledge base of science. **Applied research** seeks to use the knowledge gained from basic research to address human issues and concerns. It may involve developing commercial applications.

- **Pseudoscience** is characterized by scientific-sounding statements, beliefs, or practices that do not meet the standards of the scientific method.

THE QUESTIONS

The Basics

1 Which of the following should receive the least consideration when evaluating a scientific claim?

(a) the scientific credentials of the person making the claim

(b) your personal beliefs and values

(c) whether the study supporting the claim has been published in a peer-reviewed scientific journal

(d) whether the study supporting the claim meets the standards of the scientific method

(e) any possible biases of the person making the claim

2 Scientific literacy means that you

(a) are able to easily read and understand a scientific journal article.

(b) have taken a university-level science course.

(c) understand the process of science and basic scientific facts and theories.

(d) enjoy reading current science news in newspapers and blogs.

(e) are a good critical thinker.

3 An example of basic research is

(a) a study of how hummingbirds learn song.

(b) research on the effects on agriculture of the melting polar ice caps.

(c) looking at possible genetic contributions to autism.

(d) designing more effective vaccines for dangerous infectious diseases.

(e) exploring how agricultural waste can be turned into fuel.

4 "Correlation does not prove causation" means that

(a) if two variables are correlated, one is likely to have caused the other.

(b) if changes in one variable cause changes in another, the variables are not correlated.

(c) only experimental research can answer questions about the natural world.

(d) although two variables are interrelated, changes in one do not necessarily cause changes in the other.

(e) none of the above

5 Link each term with its definition.

SCIENTIFIC LITERACY	1. Uses scientific knowledge to address human issues.
BASIC RESEARCH	2. Helps in making informed life choices.
APPLIED RESEARCH	3. Consists of peer-reviewed scientific journal articles.
SECONDARY LITERATURE	4. Gives an overview of scientific findings on a subject.
PRIMARY LITERATURE	5. Contributes to fundamental science knowledge.

6 Circle the correct terms in the following sentence: Evaluating a scientific claim begins with reviewing the [credentials, fame] of the person making the claim. It is also important to know whether those making a particular claim have a [detachment, bias], a vested interest in whether or not the claim is true. To gain an overview of scientific studies related to the claim, it is helpful to read the [primary literature, secondary literature].

7 You are trying to determining whether a scientific claim is based on real science or pseudoscience. Place the following questions you will address in the correct order by numbering them from 1 to 6.

_____a. Are the study's claims observable and quantifiable?

_____b. Has the study been reviewed by practicing scientists and published in an established scientific journal?

_____c. Are the predictions specific, testable, and falsifiable?

_____d. Is the hypothesis clearly stated, measurable, and aligned with current scientific facts and theories?

_____e. Are the experimental design and analysis well described, well designed, reproducible, and conducted with a large sample size?

_____f. Are the study conclusions logical, based on evidence, and justified, given the study results?

Try Something New

8 Human papillomavirus (HPV) is the most common sexually transmitted disease in the United States; almost all sexually active people are infected with HPV at some point. HPV can be contagious even when an infected person shows no symptoms, and symptoms may not appear until years after infection. Some but not all strains of HPV can cause genital warts and cancer. The HPV vaccine is very effective and is recommended for all young people from the ages of 11 to 26 years. What steps would you go through to decide whether to vaccinate yourself or your child with the HPV vaccine? (*Hint:* Use the method for evaluating scientific claims, including assessing real science versus pseudoscience.)

9 Specify whether each of the following statements is likely to represent real science or pseudoscience. In each case of likely pseudoscience, identify which scientific standard is not met.

(a) Dr. Oz has said that green coffee beans will burn fat, so you can lose weight without dieting.

(b) The *Wall Street Journal* reported that global climate change has been exaggerated by climate scientists.

(c) A group of researchers reported at a national scientific conference that they have found a genetic link to autism.

(d) Many astrologers agree that people born under the sun sign Aquarius are more intelligent than those born under Scorpio.

(e) A study published in the scientific journal *Diabetes* found that sleeping in a cooler room may increase metabolic rate and insulin sensitivity.

10 John Oliver, in a 2014 episode of *Last Week Tonight*, discussed a poll finding that one in four Americans are skeptical of global climate change. He dismissed the poll results and compared it to a poll asking, "Which number is bigger, 15 or 5?" Or "Do owls exist?" Why does Mr. Oliver feel that it is not relevant what the American public believes about climate change? (*Hint:* It could be argued that each of the people polled was making a scientific claim, either in support of or against the scientific consensus on climate change.)

Leveling Up

11 **What do *you* think?** Most American universities require students to have the following immunizations: MMR (measles, mumps, and rubella), varicella (chicken pox), Tdap (tetanus, diphtheria, and pertussis/whooping cough). Most also recommend or require hepatitis A and B, meningococcal conjugate (meningitis), HPV (human papillomavirus), poliovirus, and the annual flu vaccine.

(a) Which of these vaccines have you received?

(b) If you have not received one or more of them, why did you choose not to do so?

(c) Some school districts allow parents to "opt out" of vaccines because of their belief system. Do you think this option should be allowed for university students? Why or why not? What are the possible consequences of allowing people to opt out of vaccinations?

12 **Looking at data** The following graphs show the incidence of pertussis/whooping cough in the United States (a) from 1922 to 2012 and (b) by age group from 1990 to 2012 (the 2013 data are not complete). DTP, Tdap, and DTaP are different formulations of the vaccine that covers tetanus, diphtheria, and pertussis.

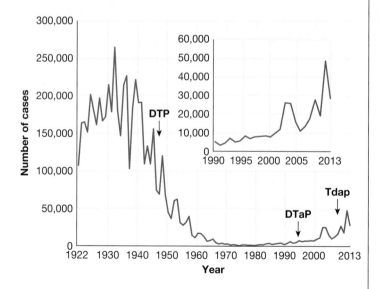

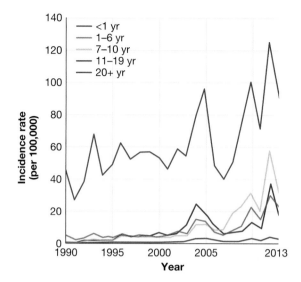

(a) Describe what the first graph shows. What do each of the axes represent? What does one point on the line show? What general trend is seen, if all the data shown in the graph are considered?

(b) Why do you think there was an increase in cases of pertussis in the 2000s?

(c) Describe what the second graph shows. What do each of the axes represent? What does one point on each of the different lines show? What general trend is seen, if all the lines are considered? What differs among the lines?

(d) Compare the incidence of pertussis cases in children under 1 year old and people over the age of 20.

(e) Do a brief Internet search to find the symptoms and possible complications of pertussis, as well as the symptoms and possible complications of vaccination. Which of these seem more concerning to you?

(f) Summarize your reflections on reviewing these graphs and the relative risks of pertussis and the pertussis vaccine. What would you recommend to someone trying to decide whether to vaccinate a child?

Answers

CHAPTER 1

END OF CHAPTER ANSWERS

1. a, c, d

2. a, c, e

3. b

4. c

5. b

6. observation, hypothesis, predictions

7. (a) 1, (b) 4, (c) 5, (d) 3, (e) 6, (f) 2, (g) 7

8. (a) organ, (b) organism, (c) population, (d) ecosystem, (e) organ system, (f) community

9. Experimental variables: e.g. plug it in, check electrical breakers in house (or see if another outlet works), check computer battery, check if screen is not functioning.

    Treatment group & control group: The computer would serve as a treatment group with a sample size of one. Either you would need to have a second computer as a control, or the computer could serve as its own control (under different experimental treatments).

10. If prions are alive, they would need to meet the criteria for life: composed of one or more cells, replicate with DNA, obtain energy from the environment, sense and respond to the environment, maintain an internal environment, and evolve as a group.

11. Observation: Bats are observed with white noses.

    Hypothesis: Bats with white noses are infected with a fungus.

    Experiment: Inject bats with a fungicide and observe whether they are less likely than bats who were given sham injections to develop white nose syndrome.

ANSWERS TO FIGURE QUESTIONS

Figure 1.1

Q1: What were the original observation and question of the scientists studying the sick bats?

A1: They observed many dead bats and many bats with white noses. They questioned whether the high death rates were in some way related to the white noses.

Q2: Where in the scientific method would a scientist decide on the methods she'll use to test her hypothesis?

A2: After predictions are generated from a hypothesis, and before experiments are run to test the predictions of the hypothesis, a scientist would choose the methods for testing the predictions of the hypothesis.

Q3: How might you explain the scientific method to someone who complains that "scientists are always changing their minds; how can we trust what they say?"

A3: Science is a process, and nothing is ever "proved" in science, so we have to expect that our best understanding of nature will change as science proceeds.

Figure 1.2

Q1: Which step(s) in the scientific method does this photograph illustrate?

A1: Observation and testing the predictions of the hypothesis through observational study (either descriptive or analytical).

Q2: What types of environmental data might the researchers have collected?

A2: Examples of environmental data would include temperature and humidity within the cave, and also soil and air samples, to see whether fungal spores are present.

Q3: Why do you think the researchers are wearing protective gear?

A3: So that they do not come into contact with pathogens.

Figure 1.4

Q1: State the hypothesis that this advertisement is claiming was scientifically tested.

A1: Lucky Strike cigarettes taste milder than other leading brands of cigarettes.

Q2: State a prediction that comes from this hypothesis. Is it testable? Why or why not?

A2: Prediction: Subjects who smoke a Lucky Strike cigarette will say it is milder than cigarettes from other brands that they are asked to smoke. Yes, this is testable, because it can be measured and repeated.

Q3: Explain in your own words why the hypothesis cannot be "proved."

A3: Nothing in science can be proved. In this case, only a small number of subjects can be tested; there might always be people (who weren't tested) who think another brand is milder.

Figure 1.5

Q1: Give a possible hypothesis that could be tested by weighing the bats.

A1: Hypothesis: Bats with WNS weigh less than uninfected bats.

Q2: State the hypothesis being tested in the photo on the bottom right.

A2: Hypothesis: Healthy bats injected with a fungicide will have lower rates of infection with WNS than bats that are sham injected (no fungicide).

Q3: Explain in your own words why an experimental study is the only way to show a cause-effect relationship.

A3: Here's one possible: If you find a relationship between two variables in an observational study, you can't know which one causes the other, or even whether a third variable is causing both to occur. In an experimental study, it is possible to manipulate one variable and see whether that manipulation causes a second variable to change.

Figure 1.6

Q1: Which is the control group in this experiment, and what are the three treatment groups?

A1: Control group: Housed alone, with no exposure to WNS. Treatment groups: Applied to wings; housed in physical contact with infected bats; housed in air contact but not physical contact with infected bats.

Q2: What are the hypothesis being tested in this experiment?

A2: Hypothesis: WNS is caused by contact with the fungus *Geomyces destructans*.

Q3: In one or two sentences, state the conclusions you can draw from the experiment. Were the hypotheses supported? Why or why not?

A3: Bats that come into physical contact with *Geomyces destructans* are highly likely to develop WNS, whereas those exposed only through the air do not develop WNS. The hypothesis was supported in part (physical versus air contact).

Figure 1.7

Q1: What is the control group in this experiment, and what are the two treatment groups?

A1: Control group: Sham injection. Treatment groups: Injected with Geomyces destructans from North America; injected with G. destructans from Europe

Q2: At day 40, approximately how many individuals were alive in each treatment group? At day 80? At day 100?

A2: Day 40: 100% of all groups survived. Day 80: 100% of control and North American G. destructans (Gd) groups survived; 14 of 18 in the European Gd group survived. Day 100: 100% of control group survived; 14 of 18 in the North American Gd group survived; only 2 of 18 in the European Gd group survived (although the European Gd experiment was stopped at about day 90).

Q3: In one or two sentences, state the conclusions you can draw from the experiment. Was the hypothesis supported? Why or why not?

A3: Yes, the hypothesis that Geomyces destructans causes WNS and leads to higher mortality was supported.

Figure 1.8

Q1: Give one *fact* about bats that you learned from this chapter.

A1: Here's one possible: Bats can develop fungal infections.

Q2: What is another example of evidence for the *germ theory of disease*? (*Hint:* Think about human diseases.)

A2: One example would be strep throat, caused by *Streptococcus* bacteria and cured by antibiotics.

Q3: Explain in your own words the difference between a fact and a hypothesis, and between a hypothesis and a theory.

A3: Hypotheses are not as certain and are more complex than facts; they are simpler and less well documented than theories.

Figure 1.9

Q1: Give examples of other kinds of organs that mammals such as bats have. (*Hint:* Think of the organs in your own body.)

A1: Examples include kidney, liver, heart, lungs.

Q2: Are bats in California part of the community of bats in upstate New York, if they are of the same species? Why or why not?

A2: No, they are not, because they do not interact with each other.

Q3: Is the soil within a cave in which bats live a part of the bats' population, community, or ecosystem? Explain your reasoning.

A3: Soil is part of the ecosystem; populations and communities are composed only of living things, and soil is part of the physical environment.

CHAPTER 2

END OF CHAPTER ANSWERS

1. c

2. a

3. a

4. d

5. d

6. 3, 4, 1, 2

7. c

8. polymers; sugar; nucleotides; are not; carbon

9. Carbon can be the basis of more complex molecules than hydrogen or oxygen because carbon can form four bonds (vs. one for hydrogen and two for oxygen).

10. The nonpolar end of the detergent will bond to the oil in the salad dressing while the polar end bonds to the water molecules, lifting the oil into the wash water. Vinegar is a polar molecule, so it will dissolve in the wash water; you don't need detergent to remove vinegar.

11. The possibility of contamination cannot be excluded, because some of the amino acids found in the asteroid are synthesized by living organisms on Earth.

    The conclusion that the amino acids in the asteroid were synthesized in space rests on the presence of right-handed amino acids in the asteroid. Right-handed amino acids are not synthesized by living organisms on Earth, so contamination could not account for their presence in the asteroid.

12. Six of the amino acids produced contained sulfur; these could not have formed in the absence of a sulfur-containing reactant.

ANSWERS TO FIGURE QUESTIONS

Figure 2.1

Q1: How many protons, neutrons, and electrons does the hydrogen atom shown here have?

A1: 1 proton, 1 electron, 0 neutrons. The atomic number and atomic mass number are both 1.

Q2: What are the atomic number and the atomic mass number of the carbon isotope shown?

A2: The atomic number is the number of protons: 6. The atomic mass number is the number of protons plus neutrons: 12.

Q3: Nitrogen-11 is an isotope of nitrogen that has 7 protons and 4 neutrons. What are the atomic number and atomic mass number of nitrogen-11?

A3: Nitrogen-11 has an atomic number of 7 (7 protons) and an atomic mass number of 11 (7 protons plus 4 neutrons).

Figure 2.2

Q1: Before the experiment was run, the apparatus was sterilized and then carefully sealed. Why was this an important thing to do?

A1: Having a spotlessly clean apparatus was important to ensure that any amino acids were produced by the experimental conditions, not from contamination.

Q2: Why is inclusion of methane in the gas flasks an essential part of the hypothesis that complex organic molecules were formed in the early atmosphere of Earth? (*Hint:* What makes a molecule organic?)

A2: Amino acids contain carbon, and methane (a simple organic molecule) is the only chemical in the mixture that contains carbon.

Q3: (Answer this question after reading about Miller's "steam injection" experiments.) Where was the steam injected in the experimental apparatus?

A3: The steam was injected directly into the gas chamber, allowing the gases to interact with the water (and the other gases) more directly.

Figure 2.4

Q1: Where are the covalent bonds in this figure?

A1: The covalent bonds are located at the electrons shared between the oxygen atom and the two hydrogen atoms.

Q2: This figure shows a water molecule (H_2O). A hydrogen molecule (H2) consists of two hydrogen nuclei that share two electrons. Draw a simple diagram of a hydrogen molecule indicating the positions of the two electrons.

A2: The electrons are equidistant from the two hydrogen nuclei because there is no difference in the strength of their attraction to electrons.

Q3: When table salt (sodium chloride, NaCl) dissolves in water, it separates into a sodium ion (Na^+) and a chloride ion ($Cl–$). Which portion of a water molecule would attract the sodium ion, and which portion would attract the chloride ion?

A3: The Na^+ ions are attracted to the partial negative charges on the oxygen atoms, and the $Cl–$ ions are attracted to the partial positive charges on the hydrogen atoms.

Figure 2.5

Q1: Describe what will happen to the molecules of olive oil if you shake the bottle and then leave it alone for an hour. What about the molecules of vinegar?

A1: When you shake the bottle, the olive oil will form small beads of oil. As the bottle rests, these beads of oil will float back to the top of the bottle and coalesce into a layer of oil; that is, the bottle will return to its starting condition. Shaking the bottle will distribute the vinegar molecules through the water, and they will remain in solution.

Q2: What would happen if you added another fat to the bottle, such as bacon grease, and shook it?

A2: Bacon grease is a fat and, like the olive oil, would form small beads of fat that would float back to the top of the bottle and coalesce into a layer of grease.

Q3: Given how sugar behaves when it is mixed into coffee or tea, would you predict that it is hydrophobic or hydrophilic?

A3: Sugar dissolves in coffee and tea; therefore, it is hydrophilic.

Figure 2.6

Q1: Identify where in the picture water can be seen in its liquid, solid, and gas states.

A1: The water in the hot springs is in a liquid state. The snow and ice are solid water. Water vapor (steam) in the air is a gas.

Q2: In the gas state, water molecules move too rapidly and are too far apart to form hydrogen bonds. Compare the volumes occupied by an equal number of water molecules in the liquid, solid, and gas states.

A2: The same number of molecules will occupy more volume in both ice (solid) and water vapor (gas) than in liquid water. The volume of ice is defined (9 percent greater than the volume in water) because the water molecules form an ordered array. The volume of water vapor depends on its temperature.

Q3: Explain in your own words how ice floats on water.

A3: Water molecules are farther apart in ice than they are in liquid water. Because the same number of water molecules occupies a greater volume in ice than in water, the density of ice is lower than that of water, and ice floats.

Figure 2.7

Q1: Suppose you were going to repeat Miller's experiments. How would you decide how much of each gas to include in the chamber?

A1: Miller wanted to replicate the atmosphere of early Earth, so he based his mixtures on estimates of the proportions of the gases in the atmosphere at that time.

Q2: Why did the addition of steam to the gases in Miller's second set of experiments increase the yield of amino acids?

A2: The steam (hot water vapor) added heat energy to the mixture, and the water enabled more kinds of chemical reactions.

Q3: Miller used electrical energy in his experiment. What other forms of energy were present in the early atmosphere of Earth that could have led to the formation of complex molecules?

A3: Heat and ultraviolet light were present in early Earth's atmosphere.

Figure 2.8

Q1: How did the NASA scientists protect the fragments of the meteorite from contamination by Earth's amino acids when they collected and transported them to the laboratory for analysis?

A1: As soon as they found a fragment, they picked it up with clean, sterile implements that had no trace chemicals on them and placed them in clean, sterile containers for the trip back to the laboratory. The fragments were allowed to come in contact only with clean, sterile surfaces throughout the analysis.

Q2: What two pieces of evidence suggest that amino acids found in meteorites originated in outer space, rather than being contaminated once they came to Earth?

A2: More amino acids have been found in meteorites (70) than are used in proteins found on Earth (20). In addition, some of the amino acids from meteorites have a "right-handed" orientation, whereas all naturally occurring amino acids on Earth are in a "left-handed" orientation.

Q3: What is the significance of finding extraterrestrial amino acids?

A3: The existence of extraterrestrial amino acids suggests that the first organic molecules may have come from outer space, rather than being formed on Earth.

Figure 2.9

Q1: Which has a higher concentration of free hydrogen ions: vinegar, pH 2.8; or milk, pH 6.5?

A1: A *high* concentration of free hydrogen ions corresponds to a *low* pH. Thus, vinegar has a higher concentration of free hydrogen ions than milk.

Q2: What happens to the concentration of free hydrogen ions in your stomach when you drink a glass of milk?

A2: The concentration of free hydrogen ions decreases (that is, the pH increases).

Q3: Black coffee has a pH of 5. If you use water with a pH of 7 to make coffee, do you increase or decrease the concentration of free hydrogen ions in the liquid?

A3: The concentration of free hydrogen ions increases.

Figure 2.10

Q1: How many different four-unit polymers can be formed by three different monomers?

A1: The formula is 3^4, or 81 different 4-unit polymers.

Q2: Why are monomers so important to living organisms?

A2: Because an enormous number of different biological molecules can be formed from a limited number of monomer building blocks. (A Lego set is an apt analogy.)

Q3: Name three common kinds of organic molecules that are polymers.

A3: Proteins are polymers of amino acids. RNA and DNA are polymers of nucleic acids. Carbohydrates are polymers of monomers containing carbon, hydrogen, and often oxygen.

CHAPTER 3

END OF CHAPTER ANSWERS

1. c
2. b
3. d
4. a
5. b
6. 2, 1, 4, 3
7. 7, 4, 5, 6, 1, 2, 3
8. (See table below)

PROKARYOTES	Eukaryotes	
	Animals	**Plants**
PLASMA MEMBRANE	Plasma membrane	Plasma membrane
		Cellulose cell wall
	Nucleus	Nucleus
	Endoplasmic reticulum	Endoplasmic reticulum
	Golgi apparatus	Golgi apparatus
Ribosomes	Ribosomes	Ribosomes
	Cytoskeleton	Cytoskeleton
	Mitochondria	Mitochondria
		Chloroplasts

9. Water will move from left to right because the concentration of water is higher on the side of the beaker with 5 grams of sugar than it is on the side with 10 grams of sugar. (Note: We have found that students understand osmosis better when it is explained in terms of water moving *down* its concentration gradient than they do with the counterintuitive concept of water moving from an area of low osmotic pressure to an area of high osmotic pressure.)

10. When a liposome makes contact with the plasma membrane of a cell of the nasal passages, the phospholipid bilayer of the liposome fuses with the phospholipid bilayer of the plasma membrane, and the contents of the liposome are released into the cell.

11. Targeting is accomplished by adding molecules to the surface of the liposome that bind only to receptor molecules on the target cells.

ANSWERS TO FIGURE QUESTIONS

Figure 3.3

Q1: What was the purpose of inserting the gene that codes for blue pigment into the synthetic DNA?

A1: The blue color identified the cells that contained the DNA from *M. mycoides*.

Q2: What part of the transformed bacterium is synthetic?

A2: Only the DNA is synthetic; all of the structural components of the cells are from the *M. capricolum* cells into which the synthetic DNA was inserted.

Q3: Did this experiment create life?

A3: Although some articles in the popular press refer to the synthetic bacterium as a new life-form, it is better described as "repackaged life." The DNA is synthetic, but all the intracellular components that enable the DNA to function were already present in the cell.

Figure 3.4

Q1: Why is it important that the phosphate head of a phospholipid is hydrophilic?

A1: The fact that the phosphate head is attracted to water (hydrophilic) and also to other phosphate head means that a bilayer will form.

Q2: What essential component of a cell do liposomes lack, and why is that omission important?

A2: Liposomes lack genetic material (DNA), so the characteristics of a liposome are not transmitted to its descendants.

Q3: Could the tendency of phospholipid bilayers to spontaneously form spheres have played a role in the origin of life? (*Hint:* Refer to "The Characteristics of Living Organisms" on page 6 of Chapter 1.)

A3: Yes. Once phospholipids formed (how that happened is still an open question), they would have formed liposomes spontaneously, trapping substances in their interiors.

Figure 3.5

Q1: In what ways is the plasma membrane a barrier, and in what ways is it a gatekeeper?

A1: It is a barrier in that it keeps out many molecules. It is a gatekeeper in that it selectively allows in other molecules.

Q2: Why can't ions (such as Na^+) cross the plasma membrane without the help of a transport protein?

A2: The electrical charges on ions make them hydrophilic, so they cannot cross the lipid portion of the bilayer, which is hydrophobic, without the help of a transport protein.

Q3: If no energy were available to the cell, what forms of transport would not be able to occur? What forms of transport could occur? (*Hint:* Look ahead at Figures 3.5 and 3.6.)

A3: No form of active transport is possible without the input of energy. Any form of passive transport, including diffusion and osmosis, could occur even in the absence of an energy source.

Figure 3.6

Q1: Is the dye at equilibrium in any of these glasses? Describe how the first glass will look when the dye is at equilibrium with the water.

A1: No, the dye is not at equilibrium. If it were, the first glass would hold uniformly pink or light red water.

Q2: Will diffusion mix the molecules of dye evenly through the water, or is it necessary to shake the container to get a uniform mixture?

A2: Diffusion is sufficient to mix the dye thoroughly, but it is a very slow process and you would probably get tired of waiting.

Q3: Will diffusion mix the dye faster in hot water than in cold water? Why or why not? (*Hint:* Review the discussion of the behavior of water molecules at different temperatures in Chapter 2.)

A3: Diffusion is faster at higher temperatures because the water molecules have more energy, form and break hydrogen bonds at a higher rate, and hence move around more rapidly.

Figure 3.7

Q1: What would the second diagram look like if the pores in the semipermeable membrane were too small to allow water molecules to pass through?

A1: The second diagram would look the same as the first diagram, since neither the sugar molecules nor the water molecules could pass through the membrane.

Q2: What would the second diagram look like if the pores were large enough to let both water molecules and sugar molecules through?

A2: If the pores were large enough for sugar molecules to pass through, some sugar molecules would diffuse down the concentration gradient of sugar from the left side to the right while water molecules were diffusing from right to left down the concentration gradient of water. At equilibrium, the concentration of sugar and water would be the same on both sides of the membrane and the depth of the solution would also be the same on both sides.

Q3: The osmotic concentration of the fluid in an IV bag is the same as the osmotic concentration of blood. What change would you see in the red blood cells of a patient if a bag of distilled water was used in error?

A3: Distilled water is more dilute than blood, so an IV of distilled water would dilute the blood—that is, increase the concentration of water in the blood. Now the red blood cells would have a lower concentration of water than the blood surrounding them, so water would move by osmosis into the red blood cells, causing them to swell.

Figure 3.8

Q1: If endocytosis itself is nonspecific, how does receptor-mediated endocytosis bring only certain molecules into a cell?

A1: The receptor protein embedded in the plasma membrane attracts and holds only specific molecules. When they attach, the plasma membrane bulges inward to engulf the molecule.

Q2: What sorts of molecules could be moved by endocytosis or exocytosis, but not by diffusion?

A2: Endocytosis and exocytosis can move molecules that are too large to pass through the plasma membrane through diffusion.

Q3: How does the fluid that enters a cell via pinocytosis differ from the fluid that enters by osmosis?

A3: The fluid that fills the inward bulge of a cell during pinocytosis may contain molecules that are in solution in the fluid outside the cell, whereas osmosis allows only water molecules to enter a cell.

Figure 3.9

Q1: What structures do prokaryotic and eukaryotic cells have in common?

A1: A plasma membrane, ribosomes, and DNA.

Q2: What cellular processes occur in both prokaryotic and eukaryotic cells?

A2: Both prokaryotic and eukaryotic cells regulate their internal concentration and the movement of substances in and out. Both require a supply of energy and carry out metabolic processes. Both respond to changes in their internal state and to conditions in the environment around them.

Q3: Both plants and animals are eukaryotes, but there are differences in their cellular structure. What are those differences?

A3: Plants have a cell wall and chloroplasts, while animals do not. Most animal cells also do not contain vacuoles.

CHAPTER 4

END OF CHAPTER ANSWERS

1. c

2. a

3. c

4. b

5. d

6. opposite, catabolism, produces

7. (See figure below)

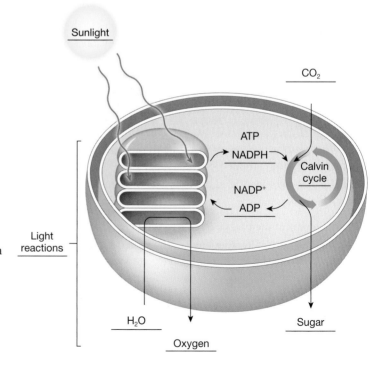

8. (a) 3

   (b) 2

   (c) 1

   (d) 4

9. The Calvin cycle cannot be maintained indefinitely, because it requires a source of energy to produce ATP. This energy source (sugar) is created from sunlight energy during the light reactions.

10. b

11. If metabolic processes occur at a higher rate than can be sustained by cellular respiration, then it is possible to die from lack of usable energy.

ANSWERS TO FIGURE QUESTIONS

Figure 4.2

Q1: Why is photosynthesis called "primary production"?

A1: Because photosynthesis converts two inorganic molecules (carbon dioxide and water) into an organic molecule (glucose).

Q2: How does all animal life depends on photosynthesis?

A2: Only photosynthetic organisms can create organic molecules from inorganic molecules, and animals cannot carry out photosynthesis. (There are a very few exceptions to that generalization.)

Q3: Explain how photosynthesis and cellular respiration are "complementary" processes.

A3: They are complementary processes because they use each others' products and reactants.

Figure 4.4

Q1: Why is it important that enzymes are not permanently altered when they bind with substrate molecules?

A1: The fact that they are not permanently altered means that they can continue to perform their function without having to be present in great numbers or needing be continually produced.

Q2: How would a higher temperature or salt concentration make it more difficult for an enzyme to function effectively?

A2: A higher temperature or salt concentration would change the shape of the enzyme, so it would not be able to bind to its specific molecules.

Q3: If a cell was unable to produce a particular enzyme necessary for a metabolic pathway, describe how the absence of that enzyme would affect the cell.

A3: The metabolic pathway would either proceed very slowly or, possibly, come to a complete halt, so the cell would not be able to function.

Figure 4.5

Q1: What source of energy would algae or plants use for anabolic reactions? Would an animal use the same kind of energy?

A1: Algae and plants would use light energy from the sun for anabolism. Animals would not use light energy, but would instead use energy from food or glucose.

Q2: What source of energy would algae or plants release in catabolic reactions? Would an animal release the same kind of energy?

A2: All living things use energy carriers, especially ATP, for catabolic reactions.

Q3: Create a mnemonic or jingle that helps you remember the difference between anabolism and catabolism.

A3: This question is just for fun and to aid in learning. Here's one possible: *anabolism synthesizes, catabolism cannibalizes* (breaks down molecules).

Figure 4.6

Q1: Define ATP in your own words.

A1: They should answer something like this: "ATP is a molecule that cells use to store and deliver energy."

Q2: How is ATP involved in anabolism and catabolism? (*Hint:* Review Figure 4.6.)

A2: Anabolism requires energy, which is provided most often by ATP. Catabolism releases energy, and would be involved in creating ATP from ADP.

Q3: Arsenic is a chemical compound that disrupts ATP production. Why would this characteristic cause it to be a potent poison?

A3: If ATP stopped being produced, our cells could no longer function and we would quickly die.

Figure 4.7

Q1: Is chlorophyll found only within chloroplasts?

A1: No. In bacteria it is embedded in membranes within the cell.

Q2: What could be an advantage of concentrating chlorophyll molecules in the membranes of chloroplasts?

A2: The location of the chlorophyll molecules maximizes the efficiency of the metabolic pathway of photosynthesis.

Q3: Oil from algae is now being used in beauty products and nutritional supplements. Would you use a moisturizer or eat pills made from algae? Why or why not?

A3: This question asks for an opinion. Some people are squeamish about eating algae, but usually not about applying it as a moisturizer or makeup.

Figure 4.8

Q1: What is the source of the carbon dioxide used for photosynthesis?
A1: Atmospheric CO_2.

Q2: What products of the light reactions of photosynthesis does the Calvin cycle use?
A2: ATP and NADPH.

Q3: What are the two major products of photosynthesis?
A3: Glucose and oxygen

Figure 4.11

Q1: What is produced during fermentation that accounts for the bubbles in beer?
A1: Fermentation releases carbon dioxide (CO_2), which is the cause of bubbles in beer.

Q2: Bakers of yeast breads also rely on fermentation, allowing bread to "rise" before baking. Describe what is occurring with the yeast as the bread rises.
A2: During bread rising, yeasts are performing metabolic functions—and growing and reproducing—using energy created through fermentation. The fermentation produces CO_2 as a waste product, which is trapped within the bread dough and causes it to rise.

Q3: Explain in your own words why lactic acid builds up in your muscles during strenuous physical activity.
A3: They should answer something like this: "During strenuous exercise, my muscles can't get enough oxygen to produce all the needed ATP. So glycolysis produces ATP anaerobically, which produces lactic acid as a by-product."

Figure 4.12

Q1: What are the products of cellular respiration?
A1: Carbon dioxide and water

Q2: Considering the inputs and products of each process, why is cellular respiration considered the reciprocal process to photosynthesis?
A2: Photosynthesis uses carbon dioxide and produces oxygen, whereas cellular respiration does the opposite, using oxygen and producing carbon dioxide.

Q3: Which of the three stages of cellular respiration—glycolysis, the Krebs cycle, or oxidative phosphorylation—could organisms have used 4 billion years ago, before photosynthesis by cyanobacteria released oxygen into the atmosphere?
A3: Glycolysis, because it does not depend on oxygen. As the text says, "glycolysis was probably the earliest means of producing ATP from food molecules, and it is still the primary means of energy production in many prokaryotes."

CHAPTER 5

END OF CHAPTER ANSWERS

1. b
2. c
3. d
4. a
5. b
6. 4, 1, 2, 3
7. Meiosis, binary fission, homologous chromosomes, sister chromatids
8. (a) 1, (b) 2, (c) 5, (d) 3, (e) 4
9. (a) 38, (b) 38, (c) 76, (d) 76.
10. Mitosis separates sister chromatids to opposite poles of the cell; cytokinesis divides the parent cell into two daughter cells.
11. The G_1 checkpoint ensures that the cell is ready to divide—for example, that it is large enough and has enough energy to produce two normal daughter cells. The G2 checkpoint ensures that the cell's DNA has been replicated and packed into pairs of sister chromatids. Bypassing the G1 checkpoint could allow cells to divide before they're ready; bypassing the G2 checkpoint could lead to the production of daughter cells with defective chromosomes.
12. Wound healing is the most visible example of mitosis in progress. Peeling skin following a sunburn and the growth of hair, fingernails, and toenails are additional examples, although in these cases the replicating cells are beneath the surface of the skin.

ANSWERS TO FIGURE QUESTIONS

Figure 5.2

Q1: When is DNA replicated during the cell cycle?
A1: DNA is replicated in interphase.

Q2: When in the cell cycle does DNA separate into two genetically identical daughter cells?
A2: Late in the M phase of cell division.

Q3: If a cell is not destined to separate into daughter cells, what phase does it enter? Is this part of the cell cycle?
A3: It enters the G_0 phase.

Figure 5.3

Q1: In what ways is cell division in prokaryotes and eukaryotes the same?

A1: Both prokaryotes and eukaryotes replicate DNA, separate the DNA to opposite poles of the cell, and then physically split the cell into two parts.

Q2: Why is binary fission referred to as "asexual reproduction"?

A2: It does not involve sharing DNA between individuals, as in sexual reproduction.

Q3: In what ways is cell division in prokaryotes and eukaryotes different?

A3: Prokaryote cells are smaller and simpler than eukaryote cells, so cell division happens more rapidly and is less complex, e.g., prokaryotes do not have organelles, so they do not have to be divided up among daughter cells.

Figure 5.4

Q1: Do all cells in an organism enter each stage of mitosis at the same time? (*Hint:* see image of onion root tip above.)

A1: No.

Q2: What happens between the end of interphase and early prophase that changes the appearance of the chromosomes?

A2: The DNA within the chromosomes is condensed to prepare for division.

Q3: Explain in your own words the role of the mitotic spindle in mitosis.

A3: The mitotic spindle is responsible for accurately separating chromatids into daughter cells.

Figure 5.5

Q1: Why is it important for a chromosome to be copied before mitosis?

A1: There need to be two copies of each chromosome, so that each daughter cell has identical genetic material.

Q2: Are sister chromatids attached at the centromere one or two chromosomes?

A2: Sister chromatids are, when attached, one chromosome. When split apart, they are two separate but identical chromosomes.

Q3: Why is the chromosome's DNA tightly packed for mitosis and cytokinesis? (*Hint:* What would happen if it were unpackaged, as during interphase?)

A3: Tightly packed chromosomes are more easily divided into daughter cells.

Figure 5.6

Q1: Would the cell cycle occur more slowly or more quickly if the cell's checkpoints were disabled?

A1: It would occur more quickly, because the cycle would not stop to ensure that the checkpoint conditions were met.

Q2: What is the advantage of stopping the cell cycle if the cell's DNA is damaged?

A2: If the nutrient supply is inadequate, the daughter cells cannot grow. By stopping the cell cycle, the parent cell can wait until the nutrient supply increases.

Q3: Which part of the cell cycle may have been influenced in Soto and Sonnenschein's breast tumor cells experiments?

A3: The G0/G1 checkpoint

Figure 5.7

Q1: Is a zygote haploid or diploid?

A1: Diploid.

Q2: What cellular process creates a baby from a zygote?

A2: Mitosis.

Q3: If a mother or father is exposed to BPA prior to conceiving a child, how does that explain potential birth defects in the fetus?

A3: If the parent's BPA exposure caused mutation in sex cells (eggs or sperm) that could lead to a birth defect in the fetus.

Figure 5.8

Q1: Is a daughter cell haploid or diploid after the first meiotic division? After the second meiotic division?

A1: After meiosis I, a daughter cell is haploid (has one of each homologous chromosome). After meiosis II, each daughter cell is still haploid (has one of each homologous chromosome), and sister chromatids have split into separate daughter cells.

Q2: What is the difference between homologous chromosomes and sister chromatids?

A2: Sister chromatids are identical DNA molecules, replicated from a single DNA molecule, that remain bound to each other. They exist in a cell only from the S phase until anaphase of mitosis, or anaphase II of meiosis. A homologous chromosome pair consists of the two copies—one maternal, the other paternal—of the same type of chromosome. The pair is present at all times in diploid cells, but a haploid cell has just the paternal or the maternal copy.

Q3: If the skin cells of house cats contain 38 homologous pairs of chromosomes, how many chromosomes are present in the egg cells they produce?

A3: 38.

Figure 5.9

Q1: Why is the term "crossing-over" appropriate for the exchange of DNA segments between homologous chromosomes?

A1: Segments of DNA physically "cross over" between homologous chromosomes.

Q2: At what stage of meiosis (I or II) does crossing-over occur?

A2: Meiosis I.

Q3: What would be the effect of crossing-over between two sister chromatids?

A3: There would be no effect, because sister chromatids are genetically identical.

Figure 5.10

Q1: During meiosis, does random assortment occur before or after crossing-over?

A1: Random assortment occurs after crossing-over.

Q2: What would be the effect on genetic diversity if homologous chromosomes did not randomly separate into the daughter cells during meiosis?

A2: Genetic diversity would decrease.

Q3: With two pairs of homologous chromosomes, four kinds of gametes can be produced. How many kinds of gametes can be produced with three pairs of homologous chromosomes? What does this suggest for the 23 homologous pairs of chromosomes in human cells?

A3: Three pairs of homologous chromosomes can produce eight kinds of gametes. Therefore, 23 homologous pairs could produce humongous variation in gametes. More precisely, it could produce 2^{23} (8,388,608) different combinations of chromosomes in gametes.

CHAPTER 6

END OF CHAPTER ANSWERS

1. heterozygote (1), homozygote (2), recessive (3), genotype (4), Phenotype (5), dominant (6)

2. gene, alleles

3. meiosis, segregation, independent assortment

4. a

5. (a) M, (b) C, (c) C, (d) C, (e) M

6. pleiotropy

7. They would have been a pale purple

8. (a) It suggests that the coat color gene in foxes is Mendelian for red/silver, and the red allele is dominant to the silver allele (recessive).

    (b) Breed true-breeding red foxes (so homozygous) to silver foxes, then breed their offspring to each other

    (c) Predicted results would be 3:1 phenotype ratio (red:silver), and 1:2:1 genotype ratio

9. The first generation result suggests that the round shape allele is dominant to the oval shape allele. The next cross should be offspring, and the proportion of ovals would be 1 in 4 or 25% (3:1 round:oval).

10. Orange color must be dominant to black color. Breed their offspring to test the hypothesis.

11. (a) llff (homozygous recessive for both long hair and no furnishings)

    (b) LlFf, LLFF, LlFF, LLFf

    (c) 9:3:3:1 – short furnished: short unfurnished: long furnished: long unfurnished

ANSWERS TO FIGURE QUESTIONS

Figure 6.2

Q1: What is the physical structure of a gene?

A1: A gene is a strand of DNA within a chromosome.

Q2: How many copies of each gene are found in the diploid cells in a woman's body?

A2: Two copies

Q3: With 46 chromosomes in a human diploid cell, how many chromosomes are from the person's mother and how many are from her father?

A3: 23 chromosomes from the mother, 23 chromosomes from the father

Figure 6.3

Q1: Which can you observe directly: the genotype or the phenotype?

A1: The phenotype may be directly observable

Q2: Which poodle could be heterozygous: the one with the black coat or the one with the brown coat?

A2: Only the black-coated poodle may be heterozygous

Q3: Can you identify with certainty the genotype of a black poodle? A brown poodle?

A3: No for a black poodle, yes for a brown poodle

Figure 6.4

Q1: What would you predict about the color of the F_1 plants' flowers?

A1: They should all be purple

Q2: Why was it important that Mendel begin with pea plants that he knew bred true for flower color? Why couldn't he simply cross a purple-flowered plant and a white-flowered plant?

A2: The purple-flowered plant might be heterozygous OR homozygous

Q3: Over the years, Mendel experimented with over 30,000 pea plants. Why did Mendel collect data on so many plants? Why didn't he study just one cross? *Hint:* Read "What Are the Odds?" on page 100.

A3: With data from more plants, there is a better chance that the results will accurately reflect reality

Figure 6.5

Q1: Why did Mendel's entire F_1 generation look the same?

A1: All of the F1 plants were heterozygous, so they all had a purple phenotype

Q2: The phenotype ratio in the F2 generation is 3:1 purple-to-white flowers. What is the genotype ratio?

A2: 1:2:1

Q3: Draw a Punnett square for a genetic cross of two heterozygous, black-coated dogs. What is the phenotype ratio of their offspring? What is the offspring genotype ratio?

A3: 3:1 black to brown phenotype; genotype is 1:2:1

Figure 6.6

Q1: List all the possible offspring genotypes and phenotypes.

A1: Round yellow (dominant dominant): RRYY, RrYY, RRYy, RrYy
Round green (dominant recessive): RRyy, Rryy
Wrinkled yellow (recessive dominant): rrYY, rrYy
Wrinkled green (recessive recessive): rryy

Q2: What is the offspring phenotype ratio?

A2: 9:3:3:1

Q3: Complete a Punnett square for a genetic cross of two true-breeding Portuguese water dogs—one with a black, wavy coat (homozygous dominant, *BBWW*) and one with a brown, curly coat (homozygous recessive, *bbww*). What is the phenotype ratio of their offspring (F_1)? Now fill out another Punnett square, crossing two of the offspring. What is the phenotype ratio of the F2 generation?

A3: 9:3:3:1 black wavy: black curly: brown wavy: brown curly

Figure 6.7

Q1: Boxers are far more inbred than poodles. Why does that inbreeding make the former a better target for genetic studies of disease than the latter?

A1: Since they are inbred, they are more likely to be homozygous for traits of interest

Q2: Explain why a geneticist interested in finding a gene linked to cancer would want to look at the DNA of senior golden retrievers with *and* without cancer?

A2: To know if there was a genetic difference between goldies with and without cancer, the DNA of cancer-free dogs would have to be known.

Q3: Obsessive-compulsive disorder (OCD) in humans is characterized by obsessive thoughts and compulsive behavior, such as pacing. Canine compulsive disorder (CCD) is characterized by compulsive behavior such as "flank sucking," sometimes seen in Doberman pinschers. Would you predict that the medications given to humans suffering from OCD would decrease compulsive behaviors in CCD dogs? Why or why not?

A3: One would predict that they might, since dogs and humans share many genes and so OCD and CCD may share a common genetic basis and be treatable by the same means.

Figure 6.8

Q1: What are the genotypes of a large and a small dog?

A1: LL and ll

Q2: Is it possible to have a heterozygous large dog? Explain why or why not.

A2: No, because Ll is a medium sized dog

Q3: Crossing a Great Dane and a Chihuahua is likely to be unsuccessful, even though they are members of the same species (and thus have compatible sperm and egg). Why is that? What are some potential risks of such a cross?

A3: It would be difficult for a Great Dane and a Chihuahua to mate, because of the size difference. One risk would be that if a female Chihuahua were to become pregnant this way, the pups would be too large for her to carry safely to term.

Figure 6.9

Q1: What are the possible genotypes (at both genes) of the black dog? The yellow dog? The brown dog?

A1: Black dog: BBEE, BbEE, BBEe, BbEe
Yellow dog: BBee, Bbee, bbee
Brown dog: bbEe, bbEE

Q2: Draw a Punnett square showing possible matings between the black dog (assuming it is heterozygous at both genes) and the yellow dog (assuming it is heterozygous at the *B* gene). List all the possible phenotypes of their offspring. (See Figure 6.6 for an example of a Punnett square made with two traits.)

A2: Black, yellow, brown

Q3: If you wanted the most variable litter possible, what colors of Labrador retrievers would you cross?

A3: If they were true-breeding (homozygous), yellow and brown

Figure 6.10

Q1: The gene that brings about the pale Siamese body fur is also responsible in part for the typical blue eyes of the species. What is the term for this type of inheritance?

A1: Pleiotropy

Q2: Siamese kittens that weigh more tend to have darker fur on their bodies. Why might this be?

A2: It could be that larger kittens have lower core temperatures, thus allowing more melanin to be produced on their bodies

Q3: The Siamese cat pictured is called a "seal point" because it has seal brown extremities. Some Siamese cats show the same color pattern, but the dark areas are of a lighter color or even a different shade—for example, lilac point, red point, blue point. What results would you predict if the experiments described in the text (shaving the cat and then increasing or decreasing temperature) were conducted on cats with these color patterns?

A3: The results should be the same, since the color of the melanin has changed but presumably the mechanism for laying down the melanin (based on temperature) has not.

CHAPTER 7

END OF CHAPTER ANSWERS

1. b, c

2. R D X R

3. a, b

4. chromosomes, genes, loci, alleles

5. A carrier has inherited one copy of a recessive genetic trait, but does not express the trait.

6. c

7. The final pair of chromosomes are sex chromosomes; the individual is a male (one large, one small sex chromosome).

8. CVS, amniocentesis, gene therapy, PGD, IVF

9. XXY, XYY, XXXY – As long as there is at least one copy of the SRY gene, found on the Y chromosome, the individual will be a male.

10. genotype ratio: 1 SS: 2 Ss: 1 ss; phenotype ratio: 3 normal: 1 sickle cell disease; Each child has a 25% chance of having sickle cell disease.

11. (a) from mother, (b) no, (c) yes, (d) two kinds, (e) no, no, yes, female

12. (left) recessive (*d*) on the X chromosome; (right) autosomal recessive (*d*)

ANSWERS TO FIGURE QUESTIONS

Figure 7.2

Q1: How many males and how many females in total does Aldrich's pedigree contain?
A1: 20 males, 22 females

Q2: What proportion of males and what proportion of females were affected by the disorder?
A2: 4/20 males (.20), 0/22 females (.00)

Q3: Why did Aldrich hypothesize that the disease was X-linked? (You will need to read ahead to this question.)
A3: The disease was only observed in males, but their mothers' male relatives were also affected

Figure 7.3

Q1: Why are changes in chromosome *number* almost always more severe than changes in chromosome *structure*?

A1: There are so many genes on an individual chromosome that deleting or adding an entire chromosome will have massive effects on an individual.

Q2: In which part of meiosis would you predict that chromosomal abnormalities are produced? (Refer back to Chapter 6 if necessary.)
A2: During metaphase, when paired chromosomes are separated.

Q3: Create a mnemonic to help remember the four kinds of structural changes (e.g., Doctors Improve Treatment Daily).
A3: One example is "Doctors Improve Treatment Daily"

Figure 7.4

Q1: Is this the karyotype of a male or a female?
A1: Male

Q2: How would the karyotype of a person with Down syndrome differ from this karyotype?
A2: There would be three copies of chromosome 21

Q3: Chromosome size correlates roughly with the number of genes residing on it. Why are an extra copy of chromosome 21 and a missing Y chromosome two of the least damaging chromosomal abnormalities?
A3: These are two of the smallest chromosomes, so fewer genes would be affected by too many or too few of them (vs. with a larger chromosome)

Figure 7.5

Q1: What are the odds that a given *egg* cell will contain an X chromosome? A Y chromosome?
A1: 50/50

Q2: If a couple has two daughters, does that mean that their next two children are more likely to be sons? Explain your reasoning. *Hint: Refer back to "What Are the Odds?" in Chapter 6, on page 100.*
A2: No, each event has an independent probability of prior and future events.

Q3: Sisters share the same X chromosome inherited from their father, but they may inherit different X chromosomes from their mother. What is the probability that brothers share the same Y chromosome? What is the probability that brothers share the same X chromosome?
A3: 100%, 50%

Figure 7.6

Q1: How do we know whether two chromosomes are homologous?
A1: If the chromosomes carry the same genes and align during cell division, they are homologous

Q2: In one sentence, explain how the terms "gene," "locus," and "chromosome" are related.

A2: A gene is found on a particular location – locus – on a chromosome.

Q3: If hair color were determined by a single gene, what would be an example of an allele for this gene?

A3: B for brown hair, b for blonde hair

Figure 7.8

Q1: Which of the children specified in this Punnett square represents Felix? What is his genotype?

A1: Felix would be the "affected son"; his genotype would be XaY

Q2: Explain why Felix is neither homozygous nor heterozygous for the *WAS* gene.

A2: The WAS gene is found on the X chromosome and Felix carries only one X chromosome, since he is a male. Homozygous and heterozygous are only possible with two chromosomes.

Q3: Create a Punnett square to illustrate the offspring if Felix were to have children with a non-carrier woman. What is the probability that a son would have WAS? What is the probability that a daughter would be a carrier of WAS?

A3: 0%, 100%

Figure 7.9

Q1: Which chromosome contains the gene for cystic fibrosis? For Tay-Sachs disease? For sickle-cell disease?

A1: Chromosomes 7, 15, 11

Q2: No known genetic disorders are encoded on the Y chromosome. Why do you think this is?

A2: There are very few genes on the Y chromosome; further, any disorder would always be expressed and therefore selected against.

Q3: In your own words, explain why most single-gene disorders are recessive rather than dominant.

A3: Dominant, single gene disorders would experience heavier selection than recessive disorders because they would always be expressed (no carriers).

Figure 7.11

Q1: Which of the children in this Punnett square represents Zoe? What is her genotype?

A1: Zoe is "affected child", aa

Q2: If Zoe's parents had another child, what is the probability that the child would have cystic fibrosis? That the child would be a CF carrier?

A2: ¼ or 25%, ½ or 50%

Q3: If Zoe is able to have a child of her own someday, and the other parent is not a carrier of cystic fibrosis (he would likely be tested before they chose to have children), what is the probability that the child would have cystic fibrosis? That the child would be a carrier?

A3: 0%, 100%

Figure 7.12

Q1: What is the probability that a child with one parent who has an autosomal dominant disorder will inherit the disease?

A1: 50%

Q2: Why are there no carriers with a dominant genetic disorder?

A2: anyone with the gene would express the disorder

Q3: Because dominant genetic disorders are rare, it is extremely rare for both parents to have the condition (genotype *Aa*). Draw a Punnett square with two *Aa* parents. What proportion of the offspring would have the disorder? What proportion would be normal?

A3: 75%; 25%

Figure 7.13

Q1: What would be the missing or damaged gene" in Felix's case? From what chromosome would a healthy copy be taken?

A1: The WAS gene; from an X chromosome

Q2: Why did Dr. Klein's group first conduct gene therapy on mice rather than humans? What are the advantages and limitations of this approach?

A2: In case there were any unforeseen problems or dangers, a human would not be hurt during the early trials. The limitation is that mice are not humans, so the results are not directly transferable.

Q3: If Felix were to have children of his own someday, would they run the risk of inheriting his disorder, or has gene therapy removed that possibility? Explain your reasoning.

A3: Gene therapy does not replace the damaged gene, it simply produces the gene product that would be made by a healthy gene. Felix remains the same genetically, so there would be a risk that his children would inherit the disorder.

CHAPTER 8

END OF CHAPTER ANSWERS

1. c
2. d
3. a
4. c
5. d
6. (nucleotide) 4, (base pair) 1, (DNA molecule) 3, (base) 2,

7. (See figure below)

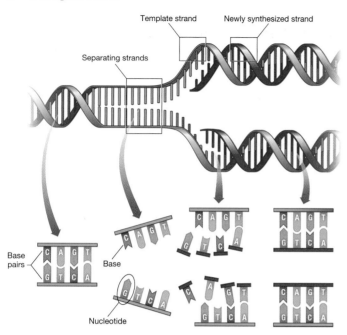

Template strand   Newly synthesized strand

Separating strands

Base pairs   Base

Nucleotide

8. PCR, gene sequencing

9. 1, 5, 3, 2, 4

10. D, S, I

11. E E P P E P E P

12. 20%, 30%, 30%

13. No, because our genome contains more noncoding DNA.

14. Yes, they still perform important functions.

ANSWERS TO FIGURE QUESTIONS

Figure 8.4

Q1: Name two base pairs.

A1: guanine, adenine

Q2: Why is the DNA structure referred to as a ladder? What part of the DNA represents the rungs of the ladder? What part represents the sides of the ladder?

A2: It looks like a ladder; the rungs are hydrogen bonds and the sides of the ladder are the bases

Q3: Is the hydrogen bond that holds the base pairs together a strong or weak chemical bond? Refer to Chapter 2 for more on chemical bonds, if needed.

A3: weak

Figure 8.6

Q1: If all genes are composed of just four nucleotides, how can different genes carry different types of information?

A1: each gene is composed by different numbers of nucleotides in different arrangements, thus allowing for many different types of information to be conveyed

Q2: Do you expect to see more variation in the sequence of DNA bases between two members of the same species (such as humans) or between two individuals of different species (for example, humans and chickens)? Explain your reasoning.

A2: between two individuals of different species; within a species individuals have the same genes but different alleles, but across species there are different genes AND different alleles

Q3: Do different alleles of a gene have the same DNA sequence or different DNA sequences?

A3: different

Figure 8.7

Q1: Which of the DNA strands here are the template strands? Why are they called "template" strands?

A1: The strands from the original double helix are the template strands. They are called this because new DNA is built using the information from them, as a template

Q2: Where in the eukaryotic cell does replication occur? Where in the prokaryotic cell?

A2: eukaryote: nucleus; prokaryote: cytoplasm

Q3: In your own words, explain why replication is referred to as "semiconservative."

A3: It is semi-conservative because each new double helix of DNA is composed of one strand from the original DNA molecule and one new strand.

Figure 8.8

Q1: PCR replicates DNA many times to increase the amount available for analysis. Why is this process called "amplification"?

A1: The quantity of DNA is substantially increased during PCR, aka amplified

Q2: Why are DNA primers necessary for this process?

A2: Primers are necessary to identify the gene of interest, since they bind with the start and end of the gene

Q3: In your own words, explain how PCR is used to identify the strain of *E. coli* that a person is carrying.

A3: A sample from the infected person is taken to the lab, and PCR is done so that there is a larger amount of the E coli DNA to then analyze and identify

Figure 8.9

Q1: Which of the computer screens in this photograph—the one in front of or the one behind the scientist—is displaying a more variable portion of DNA?

A1: the computer in front of her

Q2: Why is it important to know the particular strain of *E. coli* in an outbreak?

A2: If the strain is known, the most appropriate antibiotic can be identified.

Q3: Manning and Whittam did not analyze every base in the O157:H7 *E. coli* genome. Why not?

A3: This would have taken too long and was not necessary

Figure 8.10

Q1: Summarize how DNA repair works and why the repair mechanisms are essential for cells and whole organisms to function normally.

A1: An error is detected and tagged, then the damaged section of DNA is removed and replaced.

Q2: Is DNA repair 100 percent effective?

A2: no

Q3: What would happen to an organism if its DNA repair became less effective?

A3: the cells would have more trouble operating properly as the DNA/genetic instructions became less accurate

Figure 8.11

Q1: What are the three types of point mutations?

A1: substitution, insertion, deletion

Q2: Sickle-cell disease is an autosomal recessive genetic disorder. How many mutated hemoglobin alleles do people with sickle-cell disease have?

A2: two

Q3: Because of improved treatments, individuals with sickle-cell disease are now living into their forties, fifties, or longer. How might this extension of life span affect the prevalence of sickle-cell disease in the population?

A3: it will increase because people will be able to survive long enough to reproduce and pass on the sickle cell allele

Figure 8.12

Q1: Why does horizontal gene transfer enable *E. coli* to evolve more quickly than it could by cell division alone?

A1: horizontal gene transfer can occur across many individuals and within a generation

Q2: In your own words, summarize the major differences between the genomes of prokaryotes and eukaryotes.

A2: prokaryote genomes have less DNA, most of which codes for proteins, in a single chromosome. Eukaryote genomes have more DNA, much of which is non-coding, in multiple chromosomes.

Q3: What is the advantage of having a large genome, with DNA spread out among many chromosomes, as in eukaryotes? What is the advantage of having a small genome on a single chromosome, as in prokaryotes?

A3: Eukaryotes can be much more complex, whereas prokaryotes can reproduce much more quickly

CHAPTER 9

END OF CHAPTER ANSWERS

1. transcription, gene expression, gene regulation, translation

2. tRNA, rRNA, mRNA, tRNA, rRNA

3. Asn, STOP, Ile, Gly, Pro

4. CGU, GCU, AUG, GGU

5. redundancy, ambiguity

6. a, b, c, d

7. (See figure below)

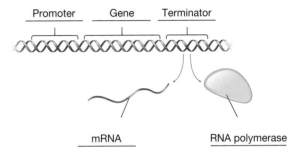

8. (a) 2, (b) 4, (c) 6, (d) 8, (e) 1, (f) 9, (g) 5, (h) 3, (i) 7

9. similarity: both involve building one molecule from another

   different: the first is DNA to RNA, the second is RNA to protein

10. This is too high a level to write an answer to – perhaps it belongs in Leveling Up?

11. Control point 2: regulation of transcription, might have an error in up-regulation in these cases; Control point 3: breakdown of mRNA, might have an error in down-regulation in these cases.

ANSWERS TO FIGURE QUESTIONS

Figure 9.3

Q1: Describe in one sentence how a vaccine creates immunity to a virus.

A1: A vaccine causes the immune system to create antibodies in response to an inactive/harmless virus, those antibodies then immediately identify and destroy any live virus (of that type) that enters the body.

Q2: Why is it impossible to become infected with a virus from a vaccine composed of viral proteins?

A2: Proteins don't replicate, DNA/RNA are needed for a virus to replicate

Q3: Natural immunity occurs without a vaccine, just by being exposed to a particular stimulus, like the chicken pox virus. Explain why people don't get chicken pox twice.

A3: Your body creates antibodies when you are exposed to eg chicken pox virus, therefore if you are exposed a second time those antibodies immediately identify and destroy the chicken pox virus (just as in Q1 above).

Figure 9.4

Q1: Why do you think biopharming with plants is so much faster than biopharming with eggs?

A1: Greater quantities of vaccine can be produced more quickly with plants than with eggs

Q2: Why is speed of production so important for vaccines?

A2: The sooner a good vaccine can be produced, the sooner people can be protected from a disease.

Q3: Why must tobacco-derived vaccines, or any new medications for that matter, be approved by the FDA?

A3: The FDA is responsible for ensuring the safety of drugs (and food), and vaccines are drugs, therefore FDA approval is necessary.

Figure 9.6

Q1: In which of the steps illustrated here does DNA replication occur? In which steps does gene expression occur?

A1: replication: 4, 5; expression: 6

Q2: Why do vaccine producers not simply replicate the entire viral genome? Why do they instead isolate the gene for one protein and replicate only that gene?

A2: Two reasons: it would take more time to replicate the entire genome, and more importantly, that would make the vaccine more dangerous

Q3: What role do the bacteria play in this process? Why are they needed?

A3: Bacteria rapidly produce the proteins needed for the vaccine; bacteria can produce proteins much more quickly than eukaryotes

Figure 9.7

Q1: Why is only one strand of DNA used as a template?

A1: Only one strand is used, because the other would code for the OPPOSITE mRNA

Q2: If a mutation occurred within the promoter or terminator region, do you think it would affect the mRNA transcribed? Why or why not?

A2: yes, because it wouldn't be clear where the gene started or stopped, so it might not be transcribed at all, or it would grow too long

Q3: The template strand of a gene has the base sequence TGAGAAGACCAGGGTTGT. What is the sequence of RNA transcribed from this DNA, assuming that RNA polymerase travels from left to right on this strand?

A3: ACUCUUCUGGUCCCAACA

Figure 9.8

Q1: In your own words, define RNA splicing. When during gene expression does it occur?

A1: RNA splicing occurs after transcription, before the mRNA transcript is translated into a protein

Q2: What do you predict would happen if the introns were not removed from RNA before translation? Why would it be a problem if the introns were not removed?

A2: if the introns were not removed, they would be translated, and the protein would be much larger and presumably non-functional

Q3: Where is the mRNA destined to go once it has been transported out of the nucleus?

A3: into the cytoplasm

Figure 9.9

You will need to finish reading the section on translation to the following three questions.

Q1: Which amino acid always begins an amino acid chain? Which codon and anticodon are associated with that amino acid?

A1: methionine

Q2: Each of the codons for stopping translation binds to a tRNA molecule that does not carry an amino acid. How would the binding of a stop codon cause the completed amino acid chain to be released?

A2: because there is no amino acid, the growing protein becomes detached from the mRNA

Q3: Given the mRNA sequence that you transcribed from the DNA template strand in Figure 9.7, what is the amino acid sequence that would be translated?

A3: ACUCUUCUGGUCCCAACA: Threonine, Leucine, Leucine, Valine, Proline, Threonine

Figure 9.10

Q1: How many codons code for isoleucine? For tryptophan? For leucine?

A1: 3, 1, 4

Q2: What codons are associated with asparagine? With serine?

A2: asparagine: AAU, AAC; serine: AGU, AGC

Q3: From the mRNA sequence that you transcribed from the DNA template strand in Figure 9.7, remove only the first A. What amino acid sequence would be translated as a result of this change? How does that sequence compare to the amino acid sequence you translated from the original mRNA sequence? *Bonus:* What kind of mutation is this? (See Chapter 8.)

A3: CUCUUCUGGUCCCAACA Leucine, Phenylalanine, Tryptophan, Serine, Glutomine; there is one less amino acid, and each of the amino acids is different; Bonus: deletion

Figure 9.11

Q1: Why is an insertion or a deletion in a gene more likely to alter the protein product than a substitution, such as A for C, would?

A1: Because an insertion or deletion "frameshifts", so every single amino acid is likely to be different, versus in a substitution, where at most one amino acid changes

Q2: Which would you expect to have more impact on an organism—a point mutation as shown here, or the insertion or deletion of a whole chromosome, as discussed in Chapter 8?

A2: The loss or addition of an entire chromosome, with all the genes on it, is likely to have far more impact on an organism than a mutation within a single gene.

Q3: Which mechanisms in a cell prevent mutations? (*Hint:* Refer back to Chapter 8 if needed.)

A3: checkpoints in the cell cycle prevent (or at least repair) mutations

Figure 9.12

Q1: As illustrated here, at what control point is transcription regulated?

A1: control point 2

Q2: What is a possible advantage of regulating gene expression before versus after transcription?

A2: then the cell does not waste time and energy producing mRNA transcripts that it will not use

Q3: If you wanted to up-regulate the production of the hemagglutinin protein in a tobacco plant carrying the hemagglutinin gene, at which control point(s) would that be possible? Justify your reasoning.

A3: control points 1 through 4 all could have an impact on the amount of humagglutinin being produced by a cell

CHAPTER 10

END OF CHAPTER ANSWERS

1. b
2. d
3. b
4. c
5. 4, 1, 2, 5, 3
6. d
7. b
8. d
9. a

ANSWERS TO FIGURE QUESTIONS

Figure 10.3

Q1: What is selective breeding, and how does it work?

A1: Selective breeding is the process by which humans allow only individuals with certain inherited characteristics to breed. Generation after generation, the selective breeder chooses which individuals mate and pass their traits to the offspring.

Q2: Describe how selective breeding leads to artificial selection.

A2: Over time, the population resulting from generations of selective breeding can change significantly, and we can then say that artificial selection has occurred.

Q3: Name as many organisms as you can whose current phenotype is due to artificial selection.

A3: Examples include any domesticated animal, including common pets and agriculturally important animals. All agricultural plants are also products of artificial selection.

Figure 10.4

Q1: What is natural selection?

A1: Natural selection is the process by which individuals with genetic characteristics that are advantageous for a particular environment survive and reproduce at a higher rate than do individuals that have other, less useful characteristics. In other words, whoever has the most kids wins!

Q2: If humans are the selective force in artificial selection, what is the selective force in natural selection?

A2: Nature, or the environment.

Q3: Compare and contrast artificial selection and natural selection. Name two ways in which they are similar. How are they different?

A3: They are similar in that both occur in populations, not individuals; that there must be a change or changes in the population; and that they occur over time—usually many, many generations. They are different in that artificial selection results from selective breeding performed by human beings, while natural selection occurs by the breeding of individuals in a population that survive in a particular environment.

Figure 10.6

Q1: What is the general definition of a fossil?

A1: Fossils are the mineralized remains of formerly living organisms or the impressions of formerly living organisms.

Q2: Describe how the fossil record provides strong evidence for evolution.

A2: The fossil record can reveal homologous traits, vestigial traits, DNA sequence similarities, and biogeographic evidence of evolution. The fossil record shows changes in all of these categories, which can be organized by time and location to re-create the history of evolutionary change on Earth.

Q3: What is meant by the term "intermediate fossil" when referring to the fossil record?

A3: Intermediate forms or fossils are fossils of species with some similarities to the known extinct ancestral group and some similarities to the descendant or currently living species. They can be thought of as "missing links" in evolution.

Figure 10.9

Q1: Why do water-dwelling animals have thicker bones than land-dwelling animals?

A1: Thick bone is an adaptation to living in water. Thick bones are heavier and help water-dwelling animals to control their buoyancy (ability to float).

Q2: Why does this thick-bone adaptation suggest a water-dwelling lifestyle?

A2: Fossil animals with thick bones are presumed to have been water dwellers because almost all currently living water-dwelling animals have thick bones.

Q3: How did this adaptation likely increase survival or reproduction in *Indohyus*?

A3: Thick bones probably enabled *Indohyus* to forage on the bottoms of lakes or ponds more efficiently than other species with lighter bones that had to work harder to stay submerged. A feeding advantage could have enabled *Indohyus* to eat more, live longer, and have a higher reproductive rate (producing more babies that survived) than those with lighter bones.

Figure 10.11

Q1: What is meant by the term "common ancestor"? Give an example.

A1: A common ancestor is the species from which at least two currently living species both descended, ancestral species that two or more new species arose from through a change in the traits of a population over time.

Q2: How are homologous structures among organisms evidence for evolution?

A2: Homologous structures are parts of an organism that have changed in size or specific form over time but are easily determined to be the same structure in the ancestral species from which the organism evolved. Species with homologous traits are all related by originally coming from an ancestor with those specific structures.

Q3: Aside from skeletal structural similarities, what other commonalities among organisms are considered homologous?

A3: Any structures that are shared by related organisms and also shared in an ancestor are homologous traits. Some examples include mammary glands, egg laying, structures to extract oxygen from air or water, and the use of DNA as the genetic material.

Figure 10.12

Q1: How are vestigial structures among organisms evidence for evolution? Give examples.

A1: Vestigial structures are evidence for evolution because they are shared among related species that all have a common ancestor. Goose bumps in humans are an example. In our furry ancestors, goose bumps fluffed the fur, thereby increasing its insulating effects and helping the animals to keep warm.

Q2: Are vestigial structures also homologous structures? Explain.

A2: Yes. Vestigial traits are shared in organisms that have a common ancestor. All organisms that descended from a furry ancestor have goose bumps when they are chilled or cold.

Q3: Why do vestigial structures still exist if they are no longer useful?

A3: Only traits that harm an organism's ability to survive and reproduce disappear from the fossil record, because organisms having these traits die and do not reproduce. Traits that are merely useless and not harmful will persist in the organisms that have them. They may diminish because these structures are no longer needed and do not give organisms a selective advantage, but the organisms survive and reproduce just as well with or without them.

Figure 10.13

Q1: If a sequence from another species were compared and showed a 96 percent sequence similarity to humans, would that species be more- closely related to humans than chimpanzees are?

A1: The hypothetical species would be less similar to humans than chimpanzees are (96 percent similar versus 98.4 percent in chimps). This species would be more similar to humans than are mice and chickens (96 percent similar versus 83 percent and 72 percent, respectively).

Q2: Are similarities in the DNA sequences of genes considered evolutionary homology? Explain.

A2: All living organisms use DNA as their genetic material, suggesting that the first true ancestral cell used DNA as its genetic material. DNA sequence similarity is a homologous trait because all descendant cells use DNA; the exact same four nucleotide molecules in different orders make up the DNA in all cells on Earth.

Q3: How is the increased similarity in the DNA sequences of genes between more-related organisms—and the decreased similarity between less-related organisms—evidence for evolution? Use the examples in this figure to support your.

A3: DNA sequence similarity is the gold standard for determining species relatedness because it is the genetic material in all cells on Earth—the best homologous trait that exists. The changes in DNA sequences in populations over time create the changes in traits that drive evolution. We can map the changes that occur in populations or species by looking at sequence similarity and re-creating a family tree. All other evidence for evolution supports the theory that all organisms derived from an ancestral cell that used DNA as its genetic material. The more related a species is to another species, the more similar the DNA sequences are. Humans and chimps are primates, mammals, and vertebrates, and they are more similar in DNA sequence than are humans and mice, which are only both vertebrates and mammals, but not both primates. Of the examples in this figure, chickens are the least similar to humans because although they are birds, which are vertebrates, they are not mammals or primates. A nonvertebrate animal like a jellyfish or worm would be even less similar to humans than the species named here, but it would likely still show some similarity.

Figure 10.14

Q1: Why should we expect to find *N. fosteri* fossils all over the world, given that it first evolved in Pangaea?

A1: We expect to find *Neoceratodus fosteri* fossils all over the world because these organisms existed before the breakup of the mass continent Pangaea. During the breakup, these organisms traveled with their continent to the current locations.

Q2: Can we use biogeographic evidence to support evolution without using fossil evidence? Explain and give examples.

A2: Yes. We can use the current locations of living organisms that are related to support evolution by biogeography. For example, members of the primate family live on almost all the continents on Earth, suggesting that their common ancestor lived at the time of Pangaea.

Q3: Can we use DNA sequence similarities together with biogeography as evidence for evolution? Explain, using examples.

A3: Yes. We can couple DNA sequencing with the locations of fossil or living organisms to support evolution by common descent. We can perform DNA sequence analysis on all the primates on Earth, coupled with their current locations, to reenact the history of primate evolution.

Figure 10.15

Q1: How are the similarities among organisms during early development evidence for evolution? Give examples.

A1: Similarities between organisms during early development suggest that they have a common ancestor whose early development occurred in the same or similar manner. All vertebrates go through similar stages of development in the early embryo. Many invertebrate organisms also share the same steps in embryonic development.

Q2: Are the similar structures among vertebrate species during embryogenesis homologous structures? Explain.

A2: Yes. The similar structures are homologous traits because they are shared with a common ancestor.

Q3: Why do embryonic structures still exist at points during embryogenesis if they are not used after birth?

A3: These structures can be considered vestigial traits, since they are now useless to the organism in which they still exist

embryonically. Remember, vestigial traits still exist because they do not harm the organisms' ability to survive and reproduce.

CHAPTER 11

END OF CHAPTER ANSWERS

1. a
2. b
3. b
4. b
5. c
6. b
7. a
8. a
9. b
10. b

ANSWERS TO FIGURE QUESTIONS

Figure 11.2

Q1: What is natural selection selecting for here?

A1: Methicillin-resistant S. aureus (MRSA).

Q2: Why do bacteria that are not randomly resistant to antibiotics die out when exposed to antibiotics?

A2: After entering a bacterium, an antibiotic generally blocks or poisons one or more processes of the bacterium's life cycle so that it cannot survive or reproduce. Bacteria that have a mechanism to pump out the poison generally survive the poison and live to reproduce; they are termed "resistant."

Q3: Why is the antibiotic represented by a kitchen strainer in this figure?

A3: The antibiotic is depicted as a kitchen strainer because antibiotics act like strainers: they can "catch" or kill most bacteria in a population, but there will always be at least one bacterium that can survive the antibiotic assault and slip through the strainer.

Figure 11.4

Q1: What is the difference between MRSA and VRSA?

A1: Both of these bacteria are members of the species *Staphylococcus aureus*. The species name is the "SA" part of both names. The "MR" in MRSA stands for "methicillin-resistant," and the "VR" in VRSA stands for "vancomycin-resistant." The only difference between these two populations is that MRSA survives in the presence of the antibiotic methicillin but can be killed by vancomycin, while VRSA survives in the presence of both methicillin and vancomycin.

Q2: Why is there a clear zone around the paper disk in the top dish and not the bottom dish?

A2: The clear zone represents the area where the antibiotic has seeped into the medium from the antibiotic-soaked paper disk and killed off the bacteria. The bacteria cannot grow here, so all you see is the growth medium in the dish, with no bacteria growing on it. The rest of the dish is covered by bacteria and appears opaque.

Q3: Why is the lack of a clear zone around the paper disk in the bottom dish so alarming?

A3: The lack of a clear zone in the bottom dish suggests that the antibiotic of last resort, vancomycin, cannot kill the bacteria and they grow just as well in the antibiotic area as in the areas away from the vancomycin-soaked paper disk. If the last-resort antibiotic doesn't kill these bacteria, there is no current antibiotic that will. VRSA is a deadly bacterial infection against which we have no good defense.

Figure 11.5

Q1: What would the white fur pigment allele frequency be if three of the homozygous black allele mice (having two black alleles) were heterozygous (having one white and one black allele) instead?
A1: 16/30 = 53%.

Q2: What would the white fur pigment allele frequency be if all of the white mice died and were therefore removed from the population? Would the black fur pigment allele frequency be affected? If so, how?
A2: The white fur pigment allele frequency would be 3/20 = 15%. Yes, the black fur pigment allele frequency would be affected; there would then be 17 black alleles out of a total of only 20 alleles: 17/20 = 85%.

Q3: What would the white fur pigment allele frequency be if all of the gray mice died and were therefore removed from the population?
A3: 10/10 = 100%.

Figure 11.6

Q1: Why does the population of *S. aureus* bacteria *not* pose a life-or-death health threat outright?

A1: These are the bacteria that normally live on our skin and do not harm us unless there is a major skin disturbance like a burn or a large scrape that is not cleaned and kept protected.

Q2: Why do the vancomycin-resistant bacteria have a higher frequency in the population after treatment with vancomycin?

A2: All of the bacteria that do not contain the resistance allele are killed by the vancomycin and therefore no longer exist. The only bacteria left are the vancomycin-resistant bacteria (VRSA).

Q3: If this figure used the mouse example of allele frequency from Figure 11.5 and the white mice increased in numbers like the vancomycin-resistant bacteria here did, what would happen to the allele frequency of the white fur pigment allele? What would happen to the black fur pigment allele frequency?

A3: The frequency of the white fur pigment allele would increase, and the black fur pigment allele would decrease in frequency.

Figure 11.7

Q1: If one extreme phenotype makes up most of a population after directional selection, what happened to the individuals with the other phenotypes?

A1: They were killed and eaten by predators.

Q2: What do you think would happen to the phenotypes of the peppered moth if the tree bark was significantly darkened again by disease or pollution?

A2: The moths that survived would be those that were more similar to the color of the bark because the birds would not as easily see them to kill and eat them. The phenotype of the population of peppered moths would become darker like the trees.

Q3: What do you think would happen to the phenotypes of the peppered moth if the tree bark became a medium color, neither light nor dark?

A3: Stabilizing selection would likely occur, and only medium-colored moths would not be killed and eaten by birds.

Figure 11.8

Q1: Think of another example of stabilizing selection in human biology. Has modern technology or medicine changed its impact on the resulting phenotypes?

A1: Stabilizing selection probably affected many human traits before modern technology/medicine played a major role in survival and quality of life. Examples include adult height and weight, which would be affected by many hormone levels and overall metabolism.

Q2: How do you think a graph of birth weight versus survival for a developing country with little health care and homebirths would compare to the graph shown here?

A2: This graph would be even sharper, with less survival at either end. Evolution would be more stabilizing than in the example shown here.

Q3: How do you think a graph of birth weight versus survival for an affluent city in the United States today would compare to the graph shown here?

A3: This graph would be much wider, with more survival at both ends. Evolution would be much less stabilizing than in the example shown here.

Figure 11.9

Q1: Almost all birds starved during the dry season depicted here. What type of selection would have been present if only the intermediate-beaked birds had survived (instead of the small- and large-beaked birds surviving)?

A1: Stabilizing selection.

Q2: Think of another example of disruptive selection. Now change the parameters so that your example illustrates directional selection instead. Which individuals survive? Which individuals die?

A2: For directional selection, only the large-beaked or small-beaked birds survive, but not both.

Q3: Of the three patterns of natural selection presented in this discussion, which one always results in two different phenotypes left standing?

A3: Disruptive selection.

Figure 11.10

Q1: How is convergent evolution different from evolution by common descent?

A1: Convergent evolution is essentially the opposite of evolution by common descent. Convergent evolution begins with two distantly related organisms that, over many generations, end up with similar phenotypes because they have adapted to similar environments. Evolution by common descent begins with an original common ancestor and, over many generations, may split into many different populations that are phenotypically different.

Q2: What is the main difference between a homologous structure and an analogous structure?

A2: A homologous structure is shared between organisms because a common ancestor had that structure; an analogous structure performs a similar function in different organisms but is not shared by a common ancestor. Analogous structures form through convergent evolution.

Q3: Why is convergent evolution considered evidence for evolution (see Chapter 10)?

A3: Convergent evolution occurs through a change in allele frequencies or traits over time—essentially the definition of evolution. Convergent evolution results in an organism that is better adapted to its environment—again, evolution.

Figure 11.13

Q1: If a goose with genotype *AA* had migrated instead of the goose with genotype *aa*, would this still be considered gene flow? Why or why not?

A1: No, this is technically not gene flow. Although alleles are being exchanged, they are the same as the existing alleles in the population and will not change allele frequencies over time and many generations.

Q2: If a goose with genotype *Aa* had migrated instead of the goose with genotype *aa*, would this still be considered gene flow? Why or why not?

A2: Yes, this is gene flow. Although the effect is not as extreme as with the *aa* genotype, the *Aa* genotype introduces a new allele into an existing population, creating offspring that can be *Aa*,

and thereby changing allele frequencies over time and many generations.

Q3: If the goose with genotype *aa* migrated to population 2 as shown but failed to mate with any of the *AA* individuals, would this still be considered gene flow? Why or why not?

A3: No, this is not gene flow. Just adding an individual with different alleles to a population does not count as gene flow. There must be an exchange of alleles between the newcomer and an existing individual.

Figure 11.14

Q1: Why do you think a genetic bottleneck is more likely to occur in a small population than in a very large population?

A1: In a large population, it is less likely that a chance event can kill off almost all of the individuals, leaving only a few behind that randomly represent only one of multiple phenotypes. In a small population a tsunami, hurricane, volcanic eruption, or other natural disaster could easily kill off all but a few individuals. All subsequent offspring would arise from these few individuals, whatever phenotype they might be, regardless of which phenotypes in the original population were best adapted.

Q2: Genetic drift is often described as a "chance event." Name several chance events that could cause a genetic bottleneck.

A2: Examples that could cause a genetic bottleneck include deadly viruses, famine, drought, immigration of many predators, habitat loss, tsunami, or other natural disaster.

Q3: Which resulting population has the most genetic diversity?

A3: The wide mouth jar population.

CHAPTER 12

END OF CHAPTER ANSWERS

1. b
2. a
3. c
4. c
5. a
6. a
7. c
8. d
9. a
10. b
11. d
12. b

ANSWERS TO FIGURE QUESTIONS

Figure 12.4

Q1: What are the three requisite parts of the biological species concept?

A1: The biological species concept requires that individuals mate, that they have live offspring, and that these offspring are fertile.

Q2: How would you design an experiment to determine whether two populations are distinct species according to the biological species concept?

A2: Your experiment would require mixing individuals from the two populations under conditions conducive to sexual reproduction. If the individuals did mate, the offspring would then need to be raised to maturity and be set up to also mate and produce live offspring.

Q3: For which types of populations does the biological species concept not work as a way of determining how they're related?

A3: The biological species concept cannot be applied to populations that reproduce asexually (such as bacteria), because it requires sexual reproduction.

Figure 12.5

Q1: What is the definition of a morphological species?

A1: The morphological species concept is based on the notion that species can be identified as a separate and distinct group of organisms solely by their morphology (their physical characteristics).

Q2: Using the morphological species concept, how would you determine whether two populations are distinct?

A2: A scientist would look at anatomical and physical characteristics and make a determination about whether two populations are different enough to be considered two separate species.

Q3: How is genetic divergence between populations determined?

A3: Genetic divergence between two populations is determined by examination of the DNA sequences of many individuals in the two populations. A lot of similarity between DNA sequences suggests little genetic divergence, while many changes between the gene sequences suggest much greater genetic divergence.

Figure 12.6

Q1: What is the definition of gene flow? How is gene flow blocked by geographic barriers?

A1: Gene flow is defined as the passing of alleles between different populations of the same species. Alleles are passed between populations only by sexual reproduction, so a geographic barrier that prevents mating automatically prevents gene flow.

Q2: Name as many types of geographic barriers as you can. Which do you think would be the best at blocking gene flow?

A2: Examples of geographic barriers include but are not limited to rivers, lakes, oceans, glaciers, mountains, canyons, brick walls, freeways, fences. The larger the barrier, the better it blocks two individuals from finding each other and mating.

Q3: Are geographic barriers universal for all species? If not, name a geographic barrier that would block gene flow for one species but not another.

A3: No, they are not universal. A river, for example, would block gene flow between two lizard populations but not two bird populations.

Figure 12.7

Q1: What factors must be present for allopatric speciation to occur?

A1: A geographic barrier.

Q2: If a geographic barrier is removed and the two reunited populations intermingle and breed, what attributes must the offspring have in order to conclude by the biological species concept that the two populations are still the same species?

A2: The offspring must be viable (alive) and fertile (be able to reproduce).

Q3: If the two populations in question 2 are determined to still be the same species, did allopatric speciation occur?

A3: If they are still the same species, no speciation occurred.

Figure 12.11

Q1: Describe how coevolution is distinct from evolution as described in Chapters 10 and 11.

A1: In coevolution, a species evolves directly to interact better with another species. Coevolution can be both species evolving to interact better with each other, or it can be just one of the species adapting to the other species.

Q2: Is coevolution the same thing as convergent evolution, described in Chapter 11?

A2: No. In convergent evolution, two genetically different species look more alike over time because they are adapting to similar environments. In coevolution, two different species adapt to each other's adaptations over time.

Q3: Do you think one species' adapting over time to feed specifically and extremely successfully on another species is an example of coevolution? Why or why not?

A3: Yes. Coevolution does not have to be reciprocal.

Figure 12.12

Q1: What is the main difference between allopatric and sympatric speciation?

A1: Allopatric speciation requires a geographic barrier, and sympatric speciation cannot include a geographic barrier.

Q2: Name two events that must happen for both allopatric speciation and sympatric speciation to occur.

A2: The two populations must be reproductively isolated, and genetic change must occur.

Q3: Do you think all of the 500 species in Lake Victoria arose through sympatric speciation? Why or why not?

A3: Yes, because there are no geographic barriers to separate the populations in the lake. There would have to be a human-made wall or fence for allopatric speciation to occur.

Figure 12.13

Q1: What does "prezygotic" mean?

A1: "Prezygotic" means "before a zygote" or "before fertilization of an egg by a sperm"—in other words, no fusion of egg and sperm.

Q2: How is the ritual dance a prezygotic reproductive barrier?

A2: This dance happens before mating. If the dance is not correct, no mating happens. No mating means no fusion of egg and sperm.

Q3: What are some other prezygotic reproductive barriers besides a ritualistic mating dance?

A3: Examples include all geographic barriers, inability of egg and sperm to fuse for genetic reasons (gamete incompatibility and isolation), inability to mate because the genitalia are physically incompatible (mechanical incompatibility and isolation), and ecological isolation in which two species breed in different portions of their habitat, at different seasons, or at different times of day.

Figure 12.14

Q1: Which species is/are sympatric with *E. lucunter*?

A1: *E. viridis*, because it is found on the same side of the Panama landmass as *E. lucunter*.

Q2: Which species is/are allopatric with *E. lucunter*?

A2: *E. vanbrunti*, because it is found on the opposite side of the Panama landmass as *E. lucunter*.

Q3: If gametes are incompatible, what will be the result of a mating event between them?

A3: No zygote will form. Egg and sperm will not fuse.

CHAPTER 13

END OF CHAPTER ANSWERS

1. e

2. (a) Prokarya; Bacteria, (b) Eukarya; Fungi, (c) Eukarya; Plantae, (d) Eukarya; Plantae, (e) Eukarya; Animalia

3. (a) 3, (b) 1, (c) 4, (d) 2, (e) 5

4. (See figure below)

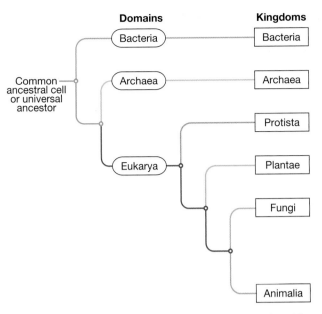

5. (clade) 2, (node) 3, (lineage) 5, (evolutionary tree) 4, (shared derived trait) 1

6. prokaryotes, Eukarya, Plantae, Animalia

7. d

8. (See one example below)

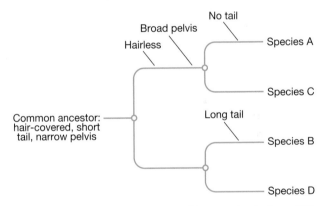

9. (a) Bacteria, Archae,Protista; (b) Bacteria, Archae; (c)Bacteria, Archae; (d)Animalia; (e)Plantae, Bacteria

10. Xu: We hypothesize that *Archaeopteryx* and *Xiaotingia* are dinosaurs, closely related to the deinonychosaurs.

    Godefroit: We hypothesize that *Archaeopteryx* and *Xiaotingia* are early birds, and Aurornis is an earlier bird.

11. (a) mycorrhizae – A symbiotic relationship between a fungus (Domain Eukarya, Kingdom Fungi) and a plant (Domain Eukarya, Kingdom Plantae) that is mutualistic.

    (b) lichens – A symbiotic relationship between an algae or cyanobacteria (Domain Bacteria & Kingdom Eubacteria) with a fungus (Domain Eukarya, Kingdom Fungi) that is mutualistic.

(c) microbiome – One example is in humans (Domain Eukarya, Kingdom Animalia) with its resident bacteria (Domain Bacteria & Kingdom Eubacteria); these can be commensal, mutualist or parasitic).

(d) hermit crabs/shells – A symbiotic relationship between a crab (Domain Eukarya, Kingdom Animalia) and the shell of a dead snail (Domain Eukarya, Kingdom Animalia). It is commensal, since the snail is unaffected.

(e) malaria - A symbiotic relationship between an animal (Domain Eukarya, Kingdom Animalia) and a virus (no domain/kingdom). It is parasitic, as the host is harmed, to the benefit of the virus.

ANSWERS TO FIGURE QUESTIONS

Figure 13.2

Q1: Why is there a shared line from the universal ancestor for Archaea and Eukarya?
A1: Archaea and Eukarya share a common ancestor more recently than either one does with Bacteria.

Q2: Where would birds be found within this figure? What about humans?
A2: Birds and humans would both be found as branches of Eukarya.

Q3: To which domain would you expect a disease-causing organism to belong? What if the organism were multicellular?
A3: Disease-causing organisms could be found within Bacteria, but also within Eukarya (kingdom Fungi). Multicellular organisms would be within Eukarya.

Figure 13.3

Q1: During what geologic period did life on Earth begin?
A1: The Precambrian.

Q2: How long ago did species begin to move from water to land? What period was this?
A2: About 480 mya, in the Ordovician.

Q3: In what period would *Archaeopteryx* have been alive?
A3: The Jurassic.

Figure 13.4

Q1: In what ways were theropods the same as modern birds? Give at least two similarities.
A1: They ran on two legs and had hollow, thin-walled bones.

Q2: In what ways did theropods differ from modern birds? Give at least two differences.
A2: They were more variable in size and in skin covering.

Q3: Birds are often referred to as "living dinosaurs." Is this accurate? Why or why not?
A3: Birds are direct descendants of dinosaurs, so they could be argued to *be* dinosaurs.

Figure 13.5

Q1: In the traditional tree, identify the node showing the common ancestor for early birds and dinosaurs.
A1: The common ancestor was after the split from the deinonychosaurs.

Q2: What do both the traditional tree and Xu's tree suggest about troodontids and dromaeosaurids?
A2: Both trees suggest that these two groups were closely related.

Q3: In both trees, identify the node for the common ancestor of *Archaeopteryx* and other birds. In what way are the nodes different in the two trees?
A3: The traditional tree shows *Archaeopteryx* on the bird side of the split between birds and dinosaurs. Xu's tree shows *Archaeopteryx* on the dinosaur side of the split.

Figure 13.7

Q1: What group of organisms shares a common ancestor with plants?
A1: Green algae.

Q2: Are fungi more closely related to plants or to animals? Does this surprise you? Why or why not?
A2: Fungi are more closely related to animals. The remainder of this question asks for an opinion.

Q3: If you were to create an evolutionary tree in which amoebas were included within the kingdom of organisms to which they were the most closely related (rather than with protists, where they are currently placed), where would you put them?
A3: This question asks for an opinion.

Figure 13.12

Q1: Is *Xiaotingia* an earlier or later bird than *Archaeopteryx* in this tree?
A1: *Xiaotingia* is later than *Archaeopteryx*.

Q2: If a future study, based on more fossils or new measurements, placed *Archaeopteryx* back with dinosaurs, would this suggest that birds are not related to dinosaurs? Why or why not?
A2: No. What is in question is not *whether* birds are related to dinosaurs, but when exactly they split off from related dinosaur species and what species is the first example of that split.

Q3: If you were to create an evolutionary tree of modern birds, where would you expect to place the roadrunner (based on its appearance in the above figure) as compared to a house sparrow or pigeon?

A3: The roadrunner looks more like a dinosaur, so it might be argued that it belongs closer to the base of the tree.

Figure 13.13

Q1: What extinction event occurred about 200 mya? What animal groups were most affected by this event?
A1: Triassic; reptiles.

Q2: Which of the mass extinctions appears to have removed the most animal groups? How long ago did this extinction occur?
A2: Permian; about 250 mya.

Q3: The best studied of the mass extinctions is the Cretaceous extinction. Why do you think it has been better studied than the other extinctions?
A3: Because it is more recent, more fossils may be available from this extinction than from earlier ones.

CHAPTER 14

END OF CHAPTER ANSWERS

1. a
2. a
3. a
4. c
5. c
6. b
7. b
8. d
9. a

ANSWERS TO FIGURE QUESTIONS

Figure 14.2

Q1: Within which category are individuals most closely related to one another?
A1: Species.

Q2: Within which category are individual species most distantly related?
A2: Kingdom.

Q3: Are individual species more closely related within the same order or within the same family?
A3: Family.

Figure 14.3

Q1: According to this evolutionary tree, which primate group is most closely related to humans?

A1: Chimpanzee.

Q2: According to this evolutionary tree, which primate group is most distantly related to humans?
A2: Lemurs, and others in this branch.

Q3: What are the common characteristics among all the primates, including humans?
A3: Use of tools, a capacity for symbolic language, and the performance of deliberate acts of deception.

Figure 14.4

Q1: According to natural selection, deleterious traits disappear from a population over time. What traits must have been deleterious for ground-dwelling early hominins?
A1: Opposable toes; walking on four legs.

Q2: According to natural selection, advantageous traits persist in a population over time. What traits must have been advantageous for ground-dwelling early hominins?
A2: Upright posture; thick tooth enamel.

Q3: What traits may have been neither deleterious nor advantageous for ground-dwelling early hominins?
A3: Hair; opposable thumbs; use of tools; a capacity for symbolic language; the performance of deliberate acts of deception.

Figure 14.5

Q1: What other reason besides continuing to use trees might explain why early hominins had partially opposable big toes?
A1: This trait may have taken a long, long time to lose. A major change in structure cannot occur within a few generations. It was likely a gradual change over thousands of generations.

Q2: In what way does the pattern of these footprints suggest that the print makers were walking upright?
A2: The lack of hand or knuckle prints accompanying the footprints.

Q3: Why do you think we no longer have partially opposable big toes?
A3: The opposable big toes would have made walking upright or running more difficult. If we no longer returned to the trees, there was no selective advantage to having them. In fact, those with fully opposable toes were at a disadvantage.

Figure 14.6

Q1: Where would the Neanderthal species branch be on this tree?
A1: The *Homo neanderthalensis* lineage would branch either from the *H. erectus* and *H. sapiens* common ancestor or from *H. erectus* with *H. sapiens*. The exact hereditary line for these two closely related species is still unclear.

Q2: How would the Neanderthal skull differ from the *Homo erectus* skull?

A2: The skull would be larger, more like the size of *Homo sapiens* or larger.

Q3: How would the Neanderthal skull differ from the *Homo sapiens* skull?

A3: The skull would be longer, the forehead would slope more, and there would be no chin.

Figure 14.7

Q1: Why does mitochondrial DNA come only from your mother?

A1: Mitochondria are found in the egg but not the sperm. Eggs are from your mom, and sperm are from your dad.

Q2: If a Neanderthal-human hybrid was born to a human mother and a Neanderthal father, could you tell by mitochondrial-DNA sequencing?

A2: No. All the mitochondrial DNA would be from *Homo sapiens*.

Q3: If a Neanderthal-human hybrid was born to a human father and a Neanderthal mother, could you tell by mitochondrial-DNA sequencing?

A3: Yes. In this case, all the mitochondrial DNA would be Neanderthal.

Figure 14.8

Q1: If a human-Neanderthal hybrid was born to a human mother and a Neanderthal father, could you tell by whole-genome DNA sequencing?

A1: Yes. The offspring would have nuclear DNA from both the egg and the sperm.

Q2: If a human-Neanderthal hybrid was born to a Neanderthal mother and a human father, could you tell by whole-genome DNA sequencing?

A2: Yes. The offspring would have nuclear DNA from both the egg and the sperm.

Q3: Under what circumstances are scientists able to do whole-genome sequencing, and when are they restricted to mitochondrial-DNA sequencing?

A3: Whole-genome sequencing requires well-preserved cells or tissues with fully intact DNA. Mitochondrial DNA can be isolated from less well preserved cells and tissues, and more damaged DNA.

Figure 14.9

Q1: Are you surprised by the interpretations of the hominins in this picture? If you are, explain why. If you're not surprised, explain why not.

A1: This question asks for an opinion. Here's one possible: *Homo sapiens* looks more primitive than expected, and the other species look strikingly like we do.

Q2: Describe the main differences between the hominin species.

A2: Height, musculature, size of skull, slope of forehead, amount of hair and its location.

Q3: From what you've learned about these species, do you think these representations are accurate? How can you find more information about each species to help you this question?

A3: The first part of the question asks for an opinion. Here's one possible for the second part: I can do much more extensive research about all of our family members online and at museums of natural history.

Figure 14.10

Q1: Which major branch represents the oldest known group of hominins?

A1: *Ardipithecus*.

Q2: With what other species of *Homo* did modern humans overlap in time?

A2: *H. neanderthalensis* and *H. erectus*.

Q3: According to this tree, many hominin groups overlapped in time with other groups. Do you think these species intermingled?

A3: This question asks for an opinion.

Figure 14.11

Q1: What evidence suggests that Neanderthals never lived in Africa?

A1: Modern humans of African descent have neither mitochondrial nor nuclear Neanderthal DNA sequences in their genomes.

Q2: How does the hypothesized origin of modern humans (*Homo sapiens*) differ from the hypothesized origin of Neanderthals (*Homo neanderthalensis*)?

A2: Modern humans evolved from archaic humans in Africa and spread to the rest of the world. Neanderthals are thought to have evolved from archaic humans living in the Middle East.

Q3: What species of hominins other than the Neanderthals may have co-mingled with modern humans?

A3: *Homo erectus*.

Figure 14.12

Q1: Which skull features are distinctly modern human?

A1: Prominent chin; high, straight forehead.

Q2: Which skull features are distinctly Neanderthal?

A2: Distinct eyebrow ridge; no chin; sloping forehead.

Q3: Why would you expect a hybrid of the two species to have intermediate features?

A3: The hybrids would have half their DNA from *Homo sapiens* and half from *Homo neanderthalensis*. Hybrids have features of both parents. Some might have intermediate features, while others could have mixed features (such as a sloping forehead but a prominent chin).

CHAPTER 15

END OF CHAPTER ANSWERS

1. e
2. e
3. a
4. b
5. a
6. d
7. b
8. c
9. a
10. d

ANSWERS TO FIGURE QUESTIONS

Figure 15.2

Q1: Clouds cover much of Earth. Are these part of the biosphere? Explain.

A1: Yes; clouds are part of the biosphere because they are in our atmosphere and contribute to humidity, precipitation, and overall climate of a given area on Earth. These factors interact with the living organisms on Earth and determine what types of organisms can survive in different locations on Earth.

Q2: Polar ice caps cover part of Earth. Are these part of the biosphere? Explain.

A2: Any areas on Earth that are devoid of life are not part of the biosphere. Glacial areas where no living things exist are not part of the biosphere. Areas of the polar ice caps where organisms such as polar bears and penguins survive are definitely part of the biosphere.

Q3: The photo shows Earth as it is found in our solar system surrounded by outer space. Is outer space part of the biosphere? Explain.

A3: Outer space is not part of the biosphere, because it does not interact with the organisms on our planet. Yet the sun does provide the energy and heat necessary for life on Earth, and our biosphere would not exist without it.

Figure 15.3

Q1: List as many biotic and abiotic factors in this photograph as you can.

A1: Biotic = all living things, such as plants, as well as all animals, including microscopic bacteria and algae. Abiotic = all nonliving things, such as rocks and water.

Q2: Is the forest part of the biotic or abiotic environment? Explain.

A2: It can be considered both. The plants and microorganisms are biotic, while the dirt and minerals are abiotic.

Q3: Is the river part of the biotic or abiotic environment? Explain.

A3: It can be considered both. The algae, plants, and aquatic organisms like fish living in the river are biotic, while the water and rocks of the river are abiotic.

Figure 15.5

Q1: Name two ways in which climate change affects plants, animals, and humans.

A1: Extinctions and changes in where species can survive resulting from habitat loss or change.

Q2: Name two ways in which climate change affects the frequency and severity of floods.

A2: Changes in rainfall patterns and melting ice can cause rivers and lakes to overflow.

Q3: How will climate change cause a rise in sea level?

A3: Melting of glaciers and polar ice will cause sea levels to rise.

Figure 15.6

Q1: What is a main determinant of temperature in different areas of Earth?

A1: The angle at which the sun's rays strike Earth. More direct = warmer.

Q2: Why is it colder at the poles than at the equator?

A2: The sun's rays are spread wider at the poles and strike Earth less directly than at the poles. Less direct = cooler.

Q3: Why is it warmer at the equator than at the poles?

A3: All of the sun's energy is directed at the equator, and the rays strike Earth at a direct angle there. The result is more intense heat at the equator.

Figure 15.7

Q1: How much of the incoming solar energy is reflected back to outer space?

A1: About a third.

Q2: What kind of energy is reemitted to the atmosphere after being absorbed by Earth's surface?

A2: Infrared radiation.

Q3: How are greenhouse gases like a blanket on your bed at night?

A3: Greenhouse gases trap the heat around Earth and hold it near the surface just as a blanket traps body heat from escaping and holds it near your body.

Figure 15.8

Q1: What measurements do the green circles represent?
A1: The green circles indicate CO_2 levels measured from bubbles of air trapped in ice that formed many hundreds of years ago.

Q2: What measurements do the red circles represent?
A2: The red circles are direct measurements of CO_2 levels at the Mauna Loa Observatory in Hawaii.

Q3: For approximately how many years has the Mauna Loa Observatory been recording CO_2 levels?
A3: Directly for about 50 years.

Figure 15.9

Q1: In what years were global temperatures the coolest?
A1: The years round 1910.

Q2: In what years were global temperatures the warmest?
A2: The years around the late 1990s through the present.

Q3: What trend is apparent in this graph of actual global temperatures?
A3: Average global temperatures are rising.

Figure 15.11

Q1: Where will fire most seriously affect the Amazon rainforest?
A1: Fires will most likely affect areas labeled "Dry and/or logged forest" (shown in orange).

Q2: Where will fire be the least damaging to the Amazon rainforest?
A2: Fires will least affect the old-growth forest (shown in green).

Q3: This map does not include an increase in pasturelands for grazing animals. Do you think more or less pastureland will be needed in 2030? Explain.
A3: As populations increase, the need for more pastureland to raise cattle for human consumption will increase as well.

Figure 15.12

Q1: What is transpiration?
A1: Transpiration is the process of plants absorbing water through their roots and releasing this water to the atmosphere through their leaves.

Q2: Why is transpiration important to the water cycle?
A2: Transpiration returns water from the soil back to the atmosphere to form clouds and eventually precipitation.

Q3: If there are fewer plants and therefore less transpiration in a given area, what will happen to the humidity or cloud cover in this area?

A3: The humidity and cloud cover will decrease where there is a substantial decrease in transpiration.

Figure 15.13

Q1: What pattern emerges when you compare rainfall in the Northern and Southern Hemispheres?
A1: The same patterns emerge as you move away from the equator either northward or southward.

Q2: What pattern emerges when you compare the major biomes of the Northern and Southern Hemispheres?
A2: Again, as you move away from the equator, the major biomes are equivalent distances from the equator either north or south.

Q3: What happens at the equator to make this region so wet?
A3: The density of plants is very high, and therefore the amount of transpiration is very high, resulting in cloud cover and high precipitation. High precipitation results in high plant growth and high transpiration.

Figure 15.14

Q1: What are three ways that carbon is released into the atmosphere?
A1: Respiration from animals; the burning of organic matter, including fossil fuels and wood; and the decomposition of dead organic material.

Q2: Are all of the pathways you listed for question 1 affected by human activity?
A2: Almost everything on our planet is affected by human activity in some way. Of the three pathways of carbon release, the one most affected by humans is the burning of organic materials for energy.

Q3: What are two biotic reservoirs of carbon?
A3: Plants and animals.

Figure 15.15

Q1: How does a carbon source contribute to global warming?
A1: A carbon source is a reservoir of carbon that releases more than it absorbs, thereby dumping CO_2 into the environment, increasing greenhouse gases, and causing global warming.

Q2: How does a carbon sink protect against global warming?
A2: A carbon sink absorbs more carbon than it releases, thereby removing CO_2 from the environment, decreasing greenhouse gases, and blocking global warming.

Q3: How can trees act as both a source and a sink?
A3: Trees act as a carbon sink when photosynthesizing and absorbing CO_2 and as a carbon source when they are burned for fuel or in a wildfire.

CHAPTER 16

END OF CHAPTER ANSWERS

1. d
2. b
3. c
4. d
5. a
6. d
7. b
8. b
9. (a) logistic growth
10. b

ANSWERS TO FIGURE QUESTIONS

Figure 16.1

Q1: This map does not specify the population densities of individual cities. How do the rural areas around Shanghai compare to the rural areas around your city?

A1: Somis, California, is a small farming community near Los Angeles and has only 74 people per square mile, compared to 1,035 people per square mile in rural areas around Shanghai.

Q2: Look up the population density of your state. How does it compare to the most and least populated areas of China?

A2: The most populated rural areas in China have 1,035 people per square mile, and the least have 2.6 people per square mile.

Q3: Look up the population density of the United States. How do its most and least populated areas compare to China's?

A3: The most populated rural areas in China have 1,035 people per square mile, and the least have 2.6 people per square mile.

Figure 16.3

Q1: What features in an environment define its carrying capacity?

A1: Everything an organism living in that environment needs to survive and reproduce, such as food, shelter, water, sunlight, building materials, and so on.

Q2: Why is the carrying capacity different for different environments?

A2: Each environment has different amounts of required items for survival and reproduction of the organisms living there. For example, a desert has much less available water for organisms than does a tropical rainforest.

Q3: Why is the carrying capacity different for individual species within the same environment?

A3: Each species has specific requirements for survival and reproduction that may be drastically different from those of other species. For example in a desert environment the carrying capacity for a desert tortoise is much higher than the carrying capacity for a water-loving frog. Essentially, frogs cannot survive in deserts at all, so they would have a carrying capacity of zero in a desert.

Figure 16.4

Q1: What is the carrying capacity in the logistic-growth environment?

A1: About 450 willow trees.

Q2: What factor(s) determined the carrying capacity of willow trees in this environment before 1954?

A2: The number of grazing rabbits.

Q3: The exponential-growth graph does not include a carrying-capacity line. Why not?

A3: Nowhere on the exponential-growth graph does the population size level off or seem to be slowing down; therefore, the population has not yet reached the carrying capacity of the environment.

Figure 16.5

Q1: According to this graph, approximately when did exponential growth begin?

A1: Around the year 1800 CE.

Q2: What milestone corresponds to the transition from logistic to exponential population growth?

A2: The onset of the industrial revolution.

Q3: What is the UN's projected carrying capacity of the Earth, and when will we reach it?

A3: About 8 billion people, which we will reach in about AD 2030.

Figure 16.7

Q1: Within the first 12 years of the one-child policy, did the fertility rates of women in China decrease substantially?

A1: No. The fertility rates of women in China remained stable, at about 2.6 children per woman, until about 1990.

Q2: What is the current fertility rate of women in China?

A2: 1.7 births per woman.

Q3: From the data plotted on this graph, why do you think it took a while for fertility in China to decline after the one-child policy was enacted, and what, ultimately, do you think caused the decline?

A3: There could be many reasons why the decrease in fertility was delayed after the start of the one-child policy in China. These examples are likely dependent on each other and may include a difficulty in enforcing the new rule and therefore a lag in reduction of births per women. Ultimately, though, fertility may have declined because many couples decided that having more than one child was too expensive anyway; more and more couples moved to urban areas, where policy enforcement was more strict than in rural areas; and so on.

Figure 16.8

Q1: How many fewer children would women have had on average without the one-child policy in effect, according to Cai and Wang?
A1: 0.5 less children

Q2: What factors likely contributed to lower fertility rates of women in China in the last 10 years?
A2: Better access to health care, birth control, and education

Q3: What do Cai and Wang project will happen if the one-child policy is lifted?
A3: Birth rates will stabilize or continue to decline.

Figure 16.9

Q1: What factors may be limiting growth and reproduction in these crowded conditions?
A1: Possible factors include availability of nutrients, water, sunlight, room for root and shoot growth.

Q2: Why are overcrowded conditions considered density-dependent population changes?
A2: Because the more organisms there are in the environment or the more densely packed they are, the more competition there is for resources and the less each individual will likely receive.

Q3: Relate this example of overcrowded conditions to China's one-child policy. How do you think the situations are similar? How are they different?
A3: The overcrowded conditions in China seem to have led to the increase in competition for jobs, housing, and food. Unlike in the plant example, the resources actually seem to be available to the Chinese people, but they are too expensive to attain in the current economic situation. The plants seem to struggle because they do not have enough resources, while the people in China have the resources but just cannot attain them.

Figure 16.10

Q1: In what year did the reindeer's numbers begin to rise exponentially?
A1: In about 1933.

Q2: In what years was the reindeer's population growth logistic?
A2: Between about 1911 and 1932.

Q3: What do you think happened to this population of reindeer? What environmental conditions might support your hypothesis?
A3: The population probably depleted its food resources, and many individuals probably starved to death beginning about 1938. Their

winter food sources were overgrazed, leaving not enough food to support the high population of reindeer in this environment.

Figure 16.11

Q1: In what year did the bald eagle population rise to more than 2,000 breeding pairs?
A1: In about 1986 or 1987.

Q2: Give some examples of possible density-dependent limits on bald eagle populations.
A2: Examples include numbers of prey available, adequate habitat for nesting and hunting, availability of water.

Q3: Is the population growth of bald eagles more like logistic or exponential growth? Explain why you think so.
A3: The population growth of bald eagles is more like logistic growth in that it is more of an S-shaped curve—at least between 1960 and 2005, where it seemed to level off for a few years. On the other hand, between 2005 and 2010 the growth appears to be exponential.

Figure 16.12

Q1: During which years did the hare likely have the greatest food supply?
A1: The years 1865 and 1888.

Q2: Besides the number of hare, what other factors might contribute to the number of lynx besides the number of hare?
A2: Other possible factors include the quality of habitat for building dens and raising young, the availability of freshwater, the competition with other lynx and other predators of the hare.

Q3: Can you draw an average carrying-capacity line on these graphs? Why or why not?
A3: It would be difficult to draw a line representing a carrying capacity for both of these animals because of the way they go through cycles of "boom and bust" in population size. They do not show logistic growth, where the leveling off of the S curve provides an obvious carrying capacity.

Figure 16.14

Q1: What is China's current ratio of children to elderly people?
A1: about 70% to 30%

Q2: What is China's projected ratio of children to elderly people in 2050?
A2: about 40% to 60%

Q3: Why will there be more elderly people than children in the future in China?
A3: increased healthcare for elderly

CHAPTER 17

END OF CHAPTER ANSWERS

1. d
2. b
3. c
4. c
5. b
6. c.
7. a
8. c
9. a
10. b
11. b
12. (a) humans and killer whales, (b) phytoplankton, (c) Baleen whale and sperm whale, (d) other herbivorous plankton

ANSWERS TO FIGURE QUESTIONS

Figure 17.2

Q1: What is an ecological community?
A1: The associations and interactions of all the species that live in a given area.

Q2: Of what community could this aspen woodland be a smaller part?
A2: A larger deciduous forest community.

Q3: What other small communities could be found within this larger community?
A3: Many other communities could exist, including soil communities (dwelling insects, other invertebrates, and microbes), communities of plants and animals residing in the undergrowth of the forest, canopy communities of animals that live in the treetops, among others.

Figure 17.3

Q1: How does relative species abundance compare between the two communities in this figure?
A1: The abundance of one species is much higher in the first community than in the second.

Q2: How does species richness compare between the two communities?
A2: The richness of species (how many are represented) is higher in the second community than in the first.

Q3: How does relative species abundance and species richness define the species diversity of a forest community?

A3: High species diversity relies on having each species in decent abundance and not one species taking up the majority of space in an area (richness), and on having many different species in a given area (diversity).

Figure 17.4

Q1: How many species were left in 1966 when sea stars were *not* removed from a community?
A1: About 17 or 18.

Q2: How many species were left in 1966 when sea stars were removed from a community?
A2: Only 2 or 3.

Q3: How do yours to questions 1 and 2 demonstrate the importance of a keystone species for the maintenance of diversity in a community?
A3: The species diversity plummeted as a result of the loss of this sea star. Without this species, the entire mussel community changed from many species to only a few.

Figure 17.5

Q1: What is the difference between a food chain and a food web?
A1: A food chain is a single linear feeding relationship of who eats whom. A food web incorporates all the individual food chains and how they overlap in a community.

Q2: What species eat the coyote?
A2: The gray wolf eats the coyote.

Q3: What species does the coyote eat?
A3: The coyote eats the vole, the snowshoe hare, berries, the pronghorn, and the elk.

Figure 17.6

Q1: What do producers produce?
A1: Producers produce the food (energy) for primary consumers.

Q2: Where do producers acquire the energy they need to perform their function in the food chain?
A2: From the sun. In combination with CO_2 and water, they produce glucose through photosynthesis.

Q3: Why are producers necessary for life on Earth?
A3: Without producers, there would be no influx of energy into the Earth's biosphere—no energy source for consumers to acquire.

Figure 17.8

Q1: How does the clownfish help the anemone?
A1: The clownfish provides nutrients and protection from predators.

Q2: How does the anemone help the clownfish?
A2: The anemone provides shelter from predators.

Q3: Describe what might happen to an anemone without a resident clownfish?

A3: A clownfish-free anemone unprotected from anemone-eating fish could be grazed extensively and be unable to support its growth needs without the nutrients in the clownfish excrement.

Figure 17.9

Q1: How is commensalism different from mutualism?

A1: In commensalism, only one member of the association benefits; in mutualism, both members benefit from the relationship.

Q2: How does living on a whale help barnacles?

A2: The whale provides a home and a constant stream of water passing over the barnacles to bring them the tiny particles of food they filter from the sea.

Q3: Do you think a whale could avoid being colonized by barnacles? Why or why not?

A3: It is probably unlikely that a whale could avoid being covered by barnacles, although since the barnacles do not help or hurt the whale, there would be no reason for a whale to try to avoid them.

Figure 17.10

Q1: What type of harm usually comes to prey species?

A1: Prey species are actually killed and eaten by predators; parasite hosts are usually only hurt, but not killed.

Q2: What type of harm usually comes to the hosts of parasites?

A2: Parasite hosts are usually only harmed by the parasite. If they are killed by a parasite, it usually takes a significant amount of time. Another difference is that when they do die from a parasitic infection, the parasite usually dies along with the host, because their food source is now gone. Unlike a predator, the parasite does not eat the entire host organism.

Q3: Are the elk that graze on aspen tree saplings predators or parasites? Why?

A3: This depends on what happens to the sapling. If it is small enough that one elk eats the whole thing and kills it, we would consider the elk a predator. But if the elk nibbles on small sections of multiple saplings and does not kill them immediately but eventually causes their demise, we would consider the elk a parasite.

Figure 17.11

Q1: How do predators know that brightly colored prey are usually toxic?

A1: Most predators have experienced through the trial and error of tasting prey and becoming sick that brightly colored prey are toxic. Predators must learn by trying one to avoid others later. If the organism were so toxic that the predator died from eating one, then bright displays would not help prey avoid predation.

Q2: Do you think mimicry works if the toxic species is in low abundance? Why or why not?

A2: Mimicry is similar to warning coloration in that its benefit is usually accomplished through trial and error. If there are many more nontoxic mimics than real toxic individuals, predators that successfully eat the mimics will not learn to avoid them. Only if it is more likely that a predator will encounter an actual toxic and bad-tasting prey will it learn to avoid anything that looks similar to the prey it encountered

Q3: Why is camouflage considered an adaptive response to predation?

A3: Camouflage enables prey species to blend in with their surroundings so that they're difficult for predators to detect. Random variation in coloration of a population enables those that are better hidden from predators to survive and reproduce, while those that do not match the surroundings well are easily seen and eaten. Over time, only well-camouflaged prey survive to reproduce, passing the camouflage coloration on to their offspring.

Figure 17.12

Q1: What percentage of pigeons are caught when they are alone and not in a flock?

A1: Very close to 80 percent.

Q2: For wood pigeons, what is the minimum number of individuals that provide protection from goshawks?

A2: 11 individuals.

Q3: Why do you think a group of musk oxen versus a lone musk ox would be safer from a pack of wolves?

A3: The group of musk oxen could form a circle in which they faced out and use their large horns to impale the wolves, keeping their sides and rear ends protected from attack. A lone ox would be completely unprotected from a pack of wolves working together.

Figure 17.13

Q1: Which species is the superior competitor?

A1: *Aphytis lingnanensis.*

Q2: In what ecological niche do these wasp species compete?

A2: The citrus groves of southern California.

Q3: Why is this example considered exploitative competition? How do you know?

A3: These wasps feed on the same foods in the same place but never physically come in contact with each other. They are in direct competition but do not physically interact.

Figure 17.14

Q1: How does interference competition work?

A1: Two different species are in direct competition for a resource or resources, and they physically interact by fighting or intimidating each other.

Q2: Does interference competition happen in the higher portions of the shoreline or the lower?

A2: The two species physically interact only on the lower portions of the shoreline, so this is where interference competition happens. On the higher portions, competition is less important because *Semibalanus* cannot tolerate the environment regardless of the presence of *Chthamalus*.

Q3: Which species is the better competitor?

A3: *Semibalanus* is the better competitor because in areas where the young of both *Semibalanus* and *Chthamalus* are deposited, *Semibalanus* survives by being better at using the resources available to both.

Figure 17.15

Q1: What species represents the first colonizers of the sand dunes?

A1: Dune-building grasses like marram grass.

Q2: What species is the intermediate species, and how does it become the dominant species?

A2: The pine is the intermediate species, and pine trees likely become dominant by outcompeting the grasses for water, nutrients, and sunlight. Once established, their shade further inhibits the growth of grasses.

Q3: What species is the mature, climax community species and how does it become the dominant species?

A3: The black oak is the mature, climax community species, and it likely becomes the dominant species because it outcompetes the pines for water, nutrients, and sunlight.

Figure 17.16

Q1: What other types of disturbances could you imagine destroying a forest?

A1: Other types of disturbances could include clear-cutting or heavy logging, abandonment of agricultural fields, flooding, clearing by tornado or hurricanes, the creation of empty lots in a newly built subdivision, among others.

Q2: How is secondary succession different from primary succession?

A2: Secondary succession starts with some existing producers left over from the disturbance, whereas primary succession starts from nothing—pure rock or sand. A disturbance that would result in primary succession would be a lava flow that covers an entire area; succession would have to start from pure lava rock and no existing producers.

Q3: What is a climax community?

A3: A climax community is the final step in succession, in which the species in the community remain and are not replaced by any other species as time goes by.

CHAPTER 18

END OF CHAPTER ANSWERS

1. a
2. c
3. a
4. d
5. c
6. d
7. c
8. b
9. c
10. c.

ANSWERS TO FIGURE QUESTIONS

Figure 18.2

Q1: What process do producers use to capture energy from the sun?

A1: Producers use photosynthesis to capture energy from the sun and, with water and CO_2, convert it to chemical energy—sugar.

Q2: What class of organisms breaks down the dead bodies of other organisms?

A2: Decomposers break down the dead bodies of other organisms, use the nutrients, and return the excess to the environment.

Q3: What happens to most of the energy in an ecosystem?

A3: It is used for the growth and survival of organisms and therefore is called metabolic heat.

Figure 18.4

Q1: What percentage of the original 10,000 calories is available to a shark that ate the tuna?

A1: 1 percent.

Q2: At what trophic level would we categorize the tuna?

A2: The tuna is at the fourth trophic level and is a tertiary consumer.

Q3: Why is energy "lost" at each trophic level?

A3: Much of the energy consumed is used for survival and growth (metabolism) of the consuming organism. As energy is used by an organism, it is either incorporated into that organism's tissues or is lost as metabolic heat.

Figure 18.5

Q1: Which organisms are the producers in this ecosystem?

A1: Phytoplankton.

Q2: Which nutrient is captured through photosynthesis?

A2: Carbon is captured as CO_2 and converted into sugar through photosynthesis.

Q3: Name the biotic factors and the abiotic factors shown in this figure.

A3: The biotic factors include the phytoplankton, the zooplankton, and any other organisms living in the water. The abiotic factors include the free nutrients in the water, the water itself, the sunlight, and CO_2.

Figure 18.6

Q1: What is the main difference between a consumer and a decomposer?

A1: All decomposers are consumers, but not all consumers are decomposers. Decomposers are a specific type of consumer that feeds off of only dead organic matter (dead plants or animals). All other types of consumers feed off of live plants and animals.

Q2: Which of the things shown in the figure are abiotic?

A2: Abiotic parts of the figure include the soil and nutrients within the soil, any water in the environment, and sunlight energy.

Q3: Describe all the points at which heat is lost in this figure.

A3: During the survival and reproduction of producers, consumers, and decomposers.

Figure 18.10

Q1: Which terrestrial biome is represented by the red color?

A1: Tropical forest has the highest NPP, so it is represented by red.

Q2: Which biome is represented by the blue color?

A2: The estuaries, as well as coastal and continental shelf ecosystems of the marine biome, are shown in blue.

Q3: Which biomes are represented by the lowest NPP in each panel?

A3: The lowest terrestrial NPP appears to occur in the desert and tundra biomes in the terrestrial map, and in the open ocean in the aquatic map.

Figure 18.11

Q1: In what years were chlorophyll levels the highest?

A1: 1999 and 2008.

Q2: In what years were the temperature changes from the average the greatest?

A2: 2000 (low), 2003–4 (high), and 2008 (low).

Q3: Are there any points where a temperature increase did not cause a chlorophyll decrease?

A3: Yes. In 2007, both temperatures and chlorophyll levels appear to have increased slightly from the average.

CHAPTER 19

END OF CHAPTER ANSWERS

1. d
2. a
3. c
4. d
5. a
6. (a) 2, (b) 3, (c) 4, (d) 1
7. homeostatic, set point, negative, thermoregulation
8. (a) 4, (b) 1, (c) 3, (d) 5, (e) 2
9. Yes, there is a difference, and there are two possible causes. First, if intercourse occurs too soon after the surgery, there may be an egg past the point of ligation or (more commonly) sperm still in the vas deferens. Second, occasionally the tubes will grow back together, which happens in about one out of 2,000 surgeries for men (0.05% failure) and one out of 200 surgeries for women (0.5% failure).
10. Because newborns have a higher surface area–to–volume (SA-to-V) ratio, they can chill or overheat more quickly than older and larger people do. They also need to be fed more often, and they become dehydrated more quickly because of their high SA-to-V ratio.

ANSWERS TO FIGURE QUESTIONS

Figure 19.2

Q1: In which of the countries shown in the graph did women want the most children, on average? The least?

A1: Most: Finland. Least: Austria.

Q2: Where is the largest gap between the number of children women would like to have and the number they actually have?

A2: Sweden.

Q3: Which country had the highest average number of children? The lowest?

A3: Highest: Poland. Lowest: Italy.

Figure 19.3

Q1: Which tissue type is skin primarily composed of?

A1: Epithelial.

Q2: Which tissue type is bone primarily composed of?

A2: Connective.

Q3: Looking at your hand in its entirety, which tissue types are part of it?

A3: All four tissues are found in the hand.

Figure 19.4

Q1: Which organ system defends the body from infectious diseases such as the common cold or flu?

A1: The immune system.

Q2: Which organ systems transport oxygen to cells?

A2: The respiratory and circulatory systems.

Q3: Which organ systems regulate the activities of the other organ systems?

A3: The nervous and endocrine systems.

Figure 19.5

Q1: Why is it important to maintain a stable body temperature?

A1: All of the chemical processes that occur in living things have an optimal temperature range, and will be less efficient (or unable to occur) when temperature deviates too much from that range.

Q2: Which organ system is involved as a temperature sensor? (See Figure 19.4 for an overview of organ systems.)

A2: Nervous system.

Q3: Give another example of a homeostatic pathway in humans.

A3: Systems for control of internal water content; systems for control of internal pH levels. Other answers are possible.

Figure 19.7

Q1: Identify one way in which spermatogenesis and oogenesis are the same.

A1: Here's one possible answer: They each produce haploid gametes through meiosis.

Q2: How many sperm are produced from each precursor cell? How many eggs are produced from each precursor cell?

A2: Sperm: four. Eggs: one.

Q3: How much time elapses between the appearance of a precursor cell and the formation of a sperm? How does this process differ for an egg? (You will need to read ahead to answer this question.)

A3: New sperm are produced daily, whereas each egg takes approximately a month to develop.

Figure 19.8

Q1: Which hormones important for the menstrual cycle are produced in the pituitary gland?

A1: Follicle stimulating hormone (FSH) and luteinizing hormone (LH).

Q2: Which hormone is involved in producing the uterine lining?

A2: Progesterone.

Q3: How is the egg follicle involved in producing hormones?

A3: The egg is carried within a follicle, which produces estradiol. After the egg is released from the follicle, the remaining corpus luteum produces progesterone.

Figure 19.9

Q1: If an egg is released but no sperm enter the oviduct, what is likely to occur?

A1: The unfertilized egg will travel into the uterus and then out of the body, and the menstrual cycle will continue.

Q2: If sperm enter the oviduct but no egg is present, what is likely to occur?

A2: The sperm will eventually die, and no pregnancy will occur.

Q3: Sperm can live for up to five days inside a woman's body. Furthermore, eggs may be released at any point in the cycle, although mid-cycle ovulation is the norm. If you are trying *not* to become pregnant, when is it safe to have unprotected intercourse?

A3: Essentially never, since sperm can live for so long and eggs may be released at any point (and, in fact, may be released by the act of intercourse).

Figure 19.10

Q1: Place these terms in the correct order of development: embryo, fetus, infant, zygote.

A1: Zygote, embryo, fetus, infant.

Q2: In what trimester is the fetus most likely to survive outside its mother's body?

A2: The third trimester.

Q3: The first trimester is the most sensitive time for exposure to mutagens. Why might that be?

A3: This is when most organ systems develop, so any mutations could have profound effects on the viability of the developing fetus.

Figure 19.11

Q1: What is the role of estrogen in childbirth?

A1: It makes uterine muscles more sensitive to oxytocin.

Q2: Explain how the involvement of hormones in childbirth is an example of a positive feedback loop.

A2: Contractions of the uterus are caused by oxytocin, and increasing contractions cause more oxytocin to be produced.

Q3: If a woman has been pregnant for more than 40 weeks, her doctor might give her an injection of oxytocin to precipitate labor. How would that bring about labor?

A3: Because of the positive feedback loop with oxytocin and uterine contractions, the initial injection of oxytocin should cause contractions—which would then kick off the feedback loop, with the woman's body now releasing more oxytocin.

CHAPTER 20

END OF CHAPTER ANSWERS

1. c

2. a

3. b

4. d

5. c

6. (a) 3, (b) 4, (c) 2, (d) 1, (e) 5

7. muscle, myosin, actin, Z, sarcomere

8. (a) 4, (b) 1, (c) 3, (d) 5, (e) 2

9. d: Swimming. It is not weight bearing and does not directly build muscle (as in weight lifting), so it is the least effective for increasing bone strength.

10. "Cavity": mouth; "32 protruding bones": teeth; "meat tentacle": tongue; "vat of acid": stomach; "absorb its essence and transform it into energy": small intestine (mainly).

ANSWERS TO FIGURE QUESTIONS

Figure 20.2

Q1: Which vitamins described in the figure are important for healthy bones?
A1: Vitamins A, B_{12}, C, and D.

Q2: Which vitamins are you more likely to overaccumulate?
A2: The fat soluble vitamins: A, D, E, and K.

Q3: In your own diet, are there any vitamins that you may not be eating enough of?
A3: Possible answers include A if no yellow vegetables are eaten, and B_{12} if no meat is consumed.

Figure 20.3

Q1: List, in order, the structures of the digestive system that a piece of swallowed food would pass through, beginning with the mouth.
A1: Mouth, esophagus, stomach, small intestine, colon, anus.

Q2: What part of the digestive system hosts bacteria that produce vitamins?
A2: Large intestine.

Q3: What is the shared function of the liver and gallbladder?
A3: They produce, store, and release bile to digest fats.

Figure 20.4

Q1: Why is a larger surface area important for absorption?
A1: With a larger surface area, a higher proportion of nutrients traveling through the small intestine will be absorbed into the body, since they are more likely to come into contact with the cells lining the small intestine.

Q2: In what way are the epithelial cells lining the villi modified to increase absorption?
A2: The microvilli of the cells increase surface area even further.

Q3: Explain the role of the capillaries within each villus.
A3: The capillaries are important for moving nutrients from the villi and out to the rest of the body.

Figure 20.5

Q1: What organs of the digestive system are involved in processing vitamin D?
A1: The liver and kidneys.

Q2: Do you get the majority of the vitamin D you need from the sun, from natural foods, or from dietary supplements?
A2: Answers will vary.

Q3: It is true that you can increase your vitamin D levels by visiting a tanning booth. Why is it considered a bad idea to increase vitamin D in this way?
A3: There is a much higher risk of skin cancer from tanning booths compared to natural sunlight.

Figure 20.6

Q1: The collarbone is part of which skeleton: axial or appendicular?
A1: Appendicular.

Q2: Which parts of the skeleton are made of cartilage?
A2: Joints, such as elbows, knees, and shoulders, as well as parts of most ribs.

Q3: Which part of the skeleton protects the central nervous system (the brain and spinal cord)?
A3: Appendicular.

Figure 20.9

Q1: What is the function of the synovial sac?
A1: It produces lubricating fluid, to allow the knee joint to move more easily.

Q2: Compare and contrast ligaments and tendons.
A2: Both connect a bone to something else. Ligaments connect bone to another bone, while tendons connect bone to muscle.

Q3: Knee injuries are some of the most common sports injuries. Why do you think that is?
A3: Knees are critical for running, and many sports involve running. In addition, so many components of the knee must function together that damage to one affects functioning of the entire knee.

Figure 20.10

Q1: Which types of muscles can you voluntarily contract?
A1: Skeletal muscles.

Q2: Do the muscles in the heart contract voluntarily or involuntarily?
A2: Involuntarily.

Q3: You do not have to think about breathing (otherwise, sleeping would be dangerous!), but you *can* increase or decrease your rate of breathing. Are the muscles involved in breathing, then, voluntary or involuntary?
A3: Involuntary, but we are able to override them for conscious control of breathing.

Figure 20.11

Q1: List muscle structures from smallest to largest, beginning with sarcomeres.
A1: Sarcomere, myofibril, muscle fiber, muscle fiber bundle, muscle.

Q2: What are the components of the sarcomere?
A2: Myosin, actin, Z disc.

Q3: Across animal species, the microscopic structure of muscles is the same. Why, then, are there differences in strength among animals?
A3: Animals are of different sizes and shapes, so their muscles are too. If a mouse were as large as an elephant, it would likely be as strong.

CHAPTER 21

END OF CHAPTER ANSWERS

1. d
2. c
3. a
4. b
5. b
6. (a) 3, (b) 2, (c) 1, (d) 4
7. nephron; reabsorb; secrete; ureter; urethra
8. (a) 4, (b) 2, (c) 1, (d) 5, (e) 3
9. The net transport of CO_2 will be from the lung air space to the alveolar capillaries. Gas exchange occurs through diffusion from an area of higher concentration to an area of lower concentration.
10. (a) With age, it is easier to sustain damage to the cardiovascular system with increased blood pressure or heart rate.

(b) Short term: increased heart rate. Long term: decreased resting heart rate (because the heart is stronger from exercise).

(c) Anxiety can increase blood pressure, and some people are afraid of the doctor's office.

ANSWERS TO FIGURE QUESTIONS

Figure 21.3

Q1: Where is the majority of carbon dioxide carried in the blood?
A1: In the plasma.

Q2: Where is the majority of oxygen carried in the blood?
A2: In the red blood cells, attached to hemoglobin molecules.

Q3: What would happen if your red blood cells carried a mutation that made the hemoglobin less effective at binding to oxygen (as in sickle-cell disease)?
A3: Less oxygen would be available to the body, which could have many negative effects; for example, it would be hard to exercise.

Figure 21.4

Q1: Why are arteries more muscular than veins?
A1: They need to sustain higher pressures to move blood throughout the body.

Q2: What structural feature(s) of capillaries enable easier diffusion into and out of surrounding tissues?
A2: They are very narrow and have thin, porous walls.

Q3: Why do you think that capillaries are not typically transplanted?
A3: The small size of capillaries makes them harder to transplant than larger veins and arteries.

Figure 21.5

Q1: Beginning with the left atrium, list the locations of a drop of blood as it moves through the circulatory system.
A1: Left atrium, left ventricle, arteries, capillaries, veins, right atrium, right ventricle, lungs, left atrium.

Q2: Why is the left ventricle larger than the right ventricle, and why are its walls thicker and more muscular?
A2: It must generate higher pressures to drive blood through the long systemic circuit. The right ventricle pushes blood through the much shorter pulmonary circuit.

Q3: Some people have an artificial pacemaker implanted in their heart. What is its function?
A3: If someone's natural pacemaker (SA node) is not working, then the heart will not receive the necessary signal for it to contract in unison. An artificial pacemaker ensures that the heart receives the signal it needs.

Figure 21.6

Q1: When is air richer in oxygen—as it enters the body or as it exits the body?

A1: As it enters.

Q2: The figure shows air entering through the nose. Where else can air enter the respiratory system?

A2: Through the mouth.

Q3: Explain how the body creates a change in air pressure during breathing.

A3: During inhalation, the diaphragm pulls down and rib muscles push out to create a larger volume and thus decreased air pressure in the lungs, bringing air into the lungs from outside the body. During exhalation, the diaphragm moves up and rib muscles pull in, decreasing volume and increasing pressure so that air moves out.

Figure 21.7

Q1: Beginning with the nose, list the locations of a molecule of oxygen as it moves through the respiratory system.

A1: Nose, nasal cavity, pharynx, larynx, trachea, bronchi, lung.

Q2: From the lung, where would the oxygen molecule move?

A2: From alveoli in the lungs into veins in the circulatory system and then into cells through diffusion.

Q3: How might an upper respiratory infection (URI) like the common cold affect your respiratory system and thus your breathing?

A3: Tissues would become inflamed and mucus production would increase, making breathing more difficult.

Figure 21.8

Q1: Why is a large surface area important for gas exchange?

A1: Gas exchange occurs through diffusion, and the rate of diffusion is directly related to the surface area available for it to occur.

Q2: Does carbon dioxide move into or out of the alveoli? Into or out of capillaries at the surface of the alveoli?

A2: Into alveoli, out of capillaries.

Q3: When a person has pneumonia, the alveoli may fill with fluids. Why would this be a problem?

A3: Gas exchange—oxygen and carbon dioxide—would be more difficult.

Figure 21.10

Q1: What is released from the kidney, and where does it go after release?

A1: Urine is released, and it enters the bladder and is then excreted from the body.

Q2: What is the difference between reabsorption and secretion?

A2: Reabsorption is the process of bringing substances back into the body; secretion is removing substances from the body (in urine).

Q3: Alcohol suppresses the kidney's ability to reabsorb water. What common consequence of drinking alcohol is related to this fact, and how might you alleviate this problem?

A3: Decreased reabsorption can lead to dehydration, a major cause of hangovers. Increased intake of (nonalcoholic) fluids can alleviate this problem.

Figure 21.12

Q1: Are sensory neurons in your eye part of the CNS or PNS?

A1: PNS; all sensory neurons are part of the peripheral nervous system.

Q2: Is the spinal cord part of the CNS or PNS?

A2: CNS

Q3: In a *reflex arc*, sensory input is processed in the spinal cord and an immediate signal is sent to activate motor output, without first routing the input signal to the brain. Examples of this include the "knee jerk" and other reflexes. Give another example of this kind of reflexive reaction to sensory input.

A3: Examples include squinting in bright light and/or flinching at a loud sound or rapidly moving object.

Figure 21.13

Q1: How do neurons look different from other cells you have learned about in this book?

A1: They have long extensions—dendrites and axons.

Q2: What is the function of these differences?

A2: Dendrites and axons are used to receive and send information rapidly from cell to cell.

Q3: Some people are born without the capacity to feel pain. Although it might initially sound nice not to feel pain, it is in fact quite dangerous. Describe a situation in which it would be dangerous not to feel pain.

A3: One possible answer: If you were going to have a bath or shower but couldn't tell that the water was too hot, you could be badly burned.

CHAPTER 22

END OF CHAPTER ANSWERS

1. c
2. b
3. d

4. a

5. b

6. (a) 4, (b) 1, (c) 2, (d) 3

7. primary, secondary, active, passive, active

8. (a) 4, (b) 5, (c) 3, (d) 1, (e) 2

9. (a) A, (b) A, (c) C, (d) A, (e) C

10. Higher levels of cortisol could be hard on the body over the longer term (since they increase both liver activity and heart rate). Decreased immune function could make a person more vulnerable to pathogens such as cold and flu viruses.

11. A B cell is a lymphocyte, a particular kind of white blood cell. It is produced and matures in the bone marrow. It is an important component of the adaptive immune response.

12. If you "turn off" one of the body's immune response mechanisms, it may not be able to fight a pathogen/infection as well. In addition, some people are allergic to one or more of these medications.

ANSWERS TO FIGURE QUESTIONS

Figure 22.3

Q1: What organ coordinates the endocrine system?
A1: Hypothalamus.

Q2: How does an endocrine gland differ from an exocrine gland?
A2: Endocrine glands are ductless; exocrine glands are ducted.

Q3: How do male and female endocrine systems differ?
A3: Different sex organs contribute to the endocrine system: in males the testes, in females the ovaries.

Figure 22.4

Q1: How do hormones travel to target cells?
A1: Through body fluids, especially the blood.

Q2: Distinguish between an endocrine cell and a target cell.
A2: Endocrine cells produce and secrete hormones; target cells change in response to the presence of hormones.

Q3: Why is a hormone called a signaling molecule?
A3: It signals that something has changed in the organism or that its environment requires a cellular response.

Figure 22.5

Q1: Describe the two ways that a hormone outside a cell can exert its effect on a cell.
A1: The hormone may cross the plasma membrane, or it may bind to a receptor embedded in the membrane.

Q2: Within the cell, how does a hormone bring about a change in cell activity?
A2: By changing gene expression, metabolism, cytoskeletal organization, or membrane transport.

Q3: It takes very little of the hormone cortisol to have large effects throughout the body. Explain why.
A3: The binding of cortisol to specific cells (with the appropriate receptors) activates many proteins within those cells.

Figure 22.6

Q1: Describe an event (other than the one illustrated in the figure) that might cause the release of adrenaline.
A1: Here's one possible answer: Taking an exam.

Q2: What organs does adrenaline affect?
A2: The liver and heart (and also the immune system).

Q3: What do you think would happen if your adrenal glands were constantly releasing adrenaline?
A3: The text mentions the breakdown of glycogen to glucose, the increase in heart rate, and suppression of the immune system. All of these may be damaging over a long time period (hence the health problems associated with constant stress.)

Figure 22.7

Q1: What is the main physical barrier that animals use to keep out pathogens?
A1: Skin.

Q2: Give an example of a chemical defense within the digestive system.
A2: Low pH in the stomach.

Q3: Explain why rubbing your eyes and nose during flu and cold season is not recommended.
A3: Viruses from your hands can move into your eyes and nose, where it is easier for them to gain access and infect you.

Figure 22.8

Q1: Place these terms in order from most to least inclusive: Neutrophil, white blood cell, phagocyte, innate immune system.
A1: Innate immune system, white blood cell, phagocyte, neutrophil.

Q2: Compare and contrast macrophages and neutrophils.
A2: Both are phagocytes, a type of white blood cell, and both engulf pathogens. Macrophages are larger than neutrophils, and neutrophils also differ from macrophages by using chemicals to attack.

Q3: Why would it be a problem if your innate immune system identified the insulin-producing cells in your pancreas as "nonself?"

A3: If your immune system thought that your insulin producing cells were pathogens, it would attack and destroy them. This is, in fact, what occurs in type 1 diabetes.

Figure 22.9

Q1: What is the role of white blood cells in inflammation?
A1: They destroy pathogens and engulf cellular debris.

Q2: What would happen if histamines were not produced during inflammation?
A2: Macrophages could not leave the bloodstream to attack invading pathogens at the site of cellular damage.

Q3: Why is inflammation called a "nonspecific" immune response?
A3: It responds in the same way to any invading pathogen (and, in fact, to any cellular damage).

Figure 22.10

Q1: Why is blood clotting an important immune response?
A1: It seals a potential point of entry for pathogens, and also reduces blood loss from the wound.

Q2: How are the inflammatory response and blood clotting similar?
A2: Both are components of the innate immune system, both are rapid responses to a wound, and both seal off the site of damage.

Q3: Some people have a genetic disorder in which their blood cannot clot. Why would this be a problem?
A3: If your blood can't clot, even the smallest wound can cause enormous loss of blood.

Figure 22.11

Q1: Why are B and T cells so named?
A1: B cells mature in the bone marrow; T cells mature in the thymus.

Q2: In what way is this immune system "adaptive"?
A2: It adapts to specific invading pathogens, rather than having a generalized response to all pathogens as in the innate immune system.

Q3: Why is the adaptive immune response considered the third layer of the immune system?
A3: It responds more slowly to pathogens than do either the external defenses or the innate immune system. It also evolved later and is found only in vertebrates.

CHAPTER 23

END OF CHAPTER ANSWERS

1. a

2. c

3. e

4. b

5. c

6. a

7. (a) 3, (b) 2, (c) 1, (d) 4

8. Annual; perennial; annual; perennial; monocot; fibrous root

9. (a) 4, (b) 2, (c) 1, (d) 5, (e) 3

10. Color will not be as important at night, and smell should be more important, to help the bat find the plant. Since bats are generally larger than birds, bat-pollinated flowers need to be sturdier and have larger nectar rewards too.

11. Ethylene is clearly released as a gas into the air, because it doesn't require physical contact to act. This would explain why it works more effectively in an enclosed space like a bag, which would prevent the ethylene gas from dispersing through the air as quickly.

12. Onions are modified leaves, which is clear from their structure if you look at the whole onion plant. Tomatoes are a fruit; you can see this because they are full of seeds and are sweet when ripe.

ANSWERS TO FIGURE QUESTIONS

Figure 23.3

Q1: What is the function of the vascular tissue?
A1: It moves water and nutrients through the plant.

Q2: A plant organ is green if the cells within it contain chloroplasts. In the figure, which plant organ does not contain chloroplasts, and why do you think that is?
A2: The roots, because they are not exposed to sunlight and so cannot photosynthesize.

Q3: Which tissue type has chloroplast-containing cells? Why?
A3: Ground tissue, because this is where photosynthesis occurs.

Figure 23.4

Q1: How do root hairs increase the amount of water and nutrients that a plant can absorb?
A1: Root hairs increase the plant surface area exposed to the soil, enabling more water and nutrients to be absorbed.

Q2: How do roots make a plant more stable?
A2: Plants without broad and deep roots are more easily knocked over by wind or rain.

Q3: Given question 2, which do you predict would be more stable in harsh weather conditions—annuals or perennials?
A3: Perennials should be more stable because their roots are longer (since they grow over multiple years).

Figure 23.5

Q1: Which part of the shoot system—the bud, stem, or leaf unit—produces flowers?

A1: Buds produce either leaves or flowers.

Q2: Find the stoma in the figure. How is its location on the leaf important for its function?

A2: Stomata take in carbon dioxide and release water and oxygen, so they need to be on the surface of the leaf.

Q3: What nutrients does phloem move from the leaves to other parts of the plant?

A3: Sugars created from photosynthesis.

Figure 23.7

Q1: Give an example of how plants use chemicals to defend themselves.

A1: Here's one possible answer: Plants store poisons in their leaves so that herbivores won't eat them.

Q2: Of the plant hormones introduced in this figure, which promote growth?

A2: Auxins and cytokinins.

Q3: What is the first line of plant defense? Is this a chemical defense?

A3: A tough outer surface and organs modified for defense (for example, thorns); these are mainly physical defenses, not chemical.

Figure 23.9

Q1: Are eggs and sperm haploid or diploid? Are they part of the sporophyte or gametophyte generation?

A1: Haploid; gametophyte.

Q2: Why is the plant life cycle, but not the animal life cycle, referred to as an "alternation of generations"?

A2: Because plants have a complete haploid organism, the gametophyte, while animals have only haploid gametes.

Q3: How does asexual reproduction differ from the plant life cycle diagrammed here?

A3: In asexual reproduction, no haploid gamete is produced, so the entire gametophyte portion of the life cycle (bottom half of the figure) does not occur.

Figure 23.10

Q1: What kinds of seeds would you expect to be contained within a sugary fruit: seeds spread by water or by animals? Why?

A1: Seeds spread by animals, since animals must be attracted to the seeds/fruit as food.

Q2: What kinds of seeds would you expect to weigh less: those spread by air or those artificially selected by humans? Why?

A2: Those spread by air, because lighter seeds will travel farther, and humans want larger seeds or fruit because they are more edible.

Q3: How does the height of a tree affect its ability to spread its seeds widely?

A3: Seeds from taller trees should be able to disperse farther because they are higher up and better able to be lifted and carried long distances by wind.

Figure 23.11

Q1: Many crops, including apples, peppers, and tomatoes, depend on insect pollinators. What would happen if no pollinator visited a plant?

A1: If a plant is not pollinated, the fruit won't develop.

Q2: Although they look very different, flower petals are actually modified leaves. Do flowers still perform the main function of leaves? Explain your reasoning.

A2: Most flowers aren't green, suggesting that they don't have chloroplasts and therefore cannot produce food via photosynthesis.

Q3: Some flowers look like an insect. How might this resemblance attract pollinators?

A3: Male insects in search of a mate might be attracted to the flower, thinking it is a female of its own species. Alternatively, insect predators might be attracted to the flower, thinking it was food.

Figure 23.12

Q1: What traits are scientists selecting for in each new generation of Kernza?

A1: Larger seeds, better flavor, and deeper roots.

Q2: What would happen if the Kernza plants were allowed to freely pollinate, rather than being selectively pollinated by hand?

A2: Pollination would be random, so the desirable traits could be lost in the next generation of plants.

Q3: After the flowers have been hand-pollinated, the scientist covers the flower stalk with the bag seen in the photo. Why?

A3: To ensure no inadvertent pollination by wind or by an insect pollinator.

CHAPTER 24

END OF CHAPTER ANSWERS

1. b
2. c
3. a
4. d
5. (a) 2, (b) 5, (c) 1, (d) 4, (e) 3

6. credentials; bias; secondary literature

7. (a) 1, (b) 6, (c) 3, (d) 2, (e) 4, (f) 5

8. (1) Evaluate the credentials of the group members recommending the vaccine (in this case it would be members of the ACIP, FDA, and CDC).
(2) Assess any possible biases of the people in step 1.
(3) Read the secondary literature (for example, recent articles in *Popular Science* or the *New York Times*) for an overview of the issue.
(4) Read recent primary literature that was cited in the secondary literature (for example, articles in JAMA or the New England Journal of Medicine or Vaccine).
(5) Review papers from the primary literature for credentials and biases of the authors, good research design, reasonable sample size, and conclusions.

9. (a) Pseudoscience. Dr. Oz has relevant credentials but appears to have an agenda based on the success of his TV show. In addition, the idea of a "fat-burning" dietary aid does not align with current scientific understanding.

(b) Pseudoscience. The *Wall Street Journal* is a respected newspaper but tends to how a conservative bias, especially since being bought by Rupert Murdoch. The idea of a scientific conspiracy is also a red flag, as is the rejection of scientific consensus.

(c) Real science. Practicing scientists have reported experimental findings to their peers, but they do not yet appear to have published in a peer-reviewed journal, which will need to be the next step.

(d) Pseudoscience. Astrologists are not scientists. There is no scientific evidence that date of birth has any effect on personality.

(e) Real science. The journal *Diabetes* is a peer-reviewed and well-established scientific journal. We do not know, however, the design of the study or the sample size.

10. Mr. Oliver feels that people's opinions about basic scientific understanding are not relevant; what is relevant is the scientific consensus on the matter. The people who claim to agree or not agree with the scientific consensus likely do not have expertise relevant to evaluating the global climate.

ANSWERS TO FIGURE QUESTIONS

Figure 24.4

Q1: Why are we less confident of scientific claims made over social media?

A1: Claims made over social media are not subject to peer review before being "published." Therefore, people can—and do—make ridiculous scientific claims in social media.

Q2: Where would you place a blog in this figure? Would it matter whether or not it was a science blog and/or written by a practicing scientist? Explain your reasoning.

A2: A blog could be listed either under social media or under secondary literature, depending on the quality of the content. In the case of a science blog written by a practicing scientist, the choice would depend on the subject matter, but it would likely be listed under secondary literature. (Counterargument: the physicist who thinks global climate change is a scientific conspiracy.)

Q3: Give an example of when you would rely on secondary literature to evaluate a scientific claim and an example of when you would go to the primary literature. What is the basis of that decision?

A3: You might go to the secondary literature only for something that is not life-threatening—for example, what kind of exercise to do, or whether to turn down the thermostat in your bedroom. Primary literature is challenging reading, so it is usually left for life-critical choices—for example, whether to vaccinate your children, or possibly whether to eat a vegetarian or vegan diet.

Figure 24.5

Q1: How much did organic food sales grow during the period covered in the graph? How much did the incidence of autism grow?

A1: Organic food sales grew from $5 billion to $25 billion (a fivefold increase). Autism diagnoses grew from about 50,000 individuals to over 250,000 (also a fivefold increase).

Q2: Why might both organic food sales and autism prevalence have increased during this time period? A Reddit user in the original discussion thread suggested that both might be affected by increasing wealth in the United States. How might increased wealth affect these variables?

A2: People with more disposable income are able to spend more on food (hence the rise in organic food sales), and they are also better able to take their children in for advanced medical care (possibly the reason for increased identification of autism disorder).

Q3: In what way has the vaccine/autism debate been confused by people misinterpreting correlation as causation?

A3: Because vaccination was increasing at the same time that autism rates were rising (correlation), people suggested that the former caused the latter (causation). In addition, the time at which children are typically vaccinated is about the same age at which autism symptoms typically appear.

Figure 24.6

Q1: State the hypothesis of the people who believe that vaccines cause autism. What is an alternative hypothesis?

A1: People who believe that vaccines cause autism hypothesize the following: Vaccination harms a child's immune system and stimulates the development of autism. An alternative hypothesis might say this: Autism is caused by a genetic predisposition, and symptoms begin to show up in the second year of a child's life.

Q2: What part(s) of the figure show(s) where Wakefield's study failed to meet the standards of the scientific method?

A2: Wakefield's study does not meet really any expectations of the scientific method.

Q3: Why does only one arrow point to "real science," whereas multiple arrows point to pseudoscience?

A3: All of the issues shown in the figure must be addressed for a study to meet the expectations of the scientific method.

Figure 24.7

Q1: Why is it important to know the education and expertise of a person making a scientific claim?

A1: The opinion of a person who doesn't really understand the science behind a claim is not valid.

Q2: List at least five possible biases that people making scientific claims might have.

A2: (1) They could make money if you buy a product related to the claim. (2) They might win a lawsuit if the judge believes the claim. (3) They could become famous if people accept the claim. (4) Their religious beliefs might be supported if the claim is true. (5) Their political beliefs might be supported if the claim is true.

Q3: Describe a situation in which you might not dismiss the scientific claim of a person who did not have appropriate credentials, or who had a bias toward the claim.

A3: In the case of Offit, he made a great deal of money by selling a rotavirus vaccine he created. However, his expertise is extremely strong and he does not have an ongoing financial interest in supporting vaccination.

Figure 24.9

Q1: Why do vaccine manufacturers begin with tests on animals or cell lines before moving on to adult human subjects?

A1: They need to be sure of the safety of the vaccine before exposing people to it.

Q2: What ongoing testing and reporting are vaccines subjected to?

A2: Manufacturers test all vaccine lots, the FDA regularly inspects manufacturing facilities, the Advisory Committee on Immunization Practices (ACIP) and the CDC director review all test results before approving vaccines, and vaccine safety is continuously monitored through the Vaccine Adverse Event Recording System (VAERS) and the Vaccine Safety Datalink (VSD).

Q3: What do ACIP, FDA, and CDC stand for, and what is the role of each in evaluating vaccines?

A3: ACIP = Advisory Committee on Immunization Practices; FDA = Food and Drug Administration; CDC = Centers for Disease Control and Prevention. All are involved with initial approval and ongoing monitoring of vaccines.

Figure 24.10

Q1: What happens to an immunized person when a disease spreads through a population? (*Hint:* In the graphic, follow an immunized individual before and after a disease spreads.)

A1: The immunized person does not contract the disease.

Q2: Explain why a disease is less likely to spread to vulnerable members of a population if most people are immunized.

A2: "Herd immunity" means that fewer people contract the disease, and therefore vulnerable people are less likely to be exposed to a contagious person.

Q3: How does vaccination help an individual person? How does it help that person's community?

A3: An individual who is vaccinated is much less likely to become ill and is therefore less likely to pass on a disease to others in the community.

Glossary

A

ABA See **abscisic acid**.

abiotic Nonliving. Compare *biotic*.

abscisic acid (**ABA**) A plant hormone that mediates adaptive responses to drought, cold, heat, and other stresses.

absorption The uptake of mineral ions and small molecules by cells lining the cavity of the digestive tract.

acid A chemical compound that loses hydrogen ions (H^+) in aqueous surroundings. Compare *base*.

active carrier protein A transport protein that facilitates active transport. Compare *passive carrier protein*.

active immunity Immunity to a particular pathogen that is conferred by antibodies made by the body itself. Compare *passive immunity*.

active site The location within an enzyme where substrates are bound.

active transport The movement of a substance in response to an input of energy. Compare *passive transport*.

adaptation 1. An evolutionary process by which a population becomes better matched to its environment over time. 2. See **adaptive trait**.

adaptive immune system The immune system's third line of defense against pathogens, which mounts responses against specific invaders via antibodies (antibody-mediated immunity) and phagocytes and other specialized cells (cell-mediated immunity). Compare *external defenses* and *innate immune system*.

adaptive radiation The expansion of a group of organisms to take on new ecological roles and to form new species and higher taxonomic groups.

adaptive trait Also called an *adaptation*. A feature that gives an individual improved function in a competitive environment.

adenine (**A**) One of the four nucleotides that make up DNA. The other three are thymine (T), guanine (G), and cytosine (C).

ADP Adenosine diphosphate.

adrenal glands The paired glands, located atop the kidneys, that release the hormones cortisol, adrenaline (epinephrine), and noradrenaline (norepinephrine), which launch a number of rapid physiological responses, including boosting blood glucose levels.

aerobic Requiring oxygen. Compare *anaerobic*.

allele frequency The percentage of a specific allele in a population.

alleles Different versions of a given gene.

allopatric speciation The formation of new species from geographically isolated populations. Compare *sympatric speciation*.

alternation of generations The life cycle of flowering plants, in which haploid stages (gametophytes) alternate with diploid stages (sporophytes).

alveoli (**sing. alveolus**) Small clusters of minute sacs resembling a bunch of grapes where gases are exchanged in the lungs.

amino acid Any of a class of small molecules that are the building blocks of proteins.

amniocentesis A prenatal genetic screening technique in which amniotic fluid is extracted from the pregnancy sac that surrounds a fetus by means of a needle that is inserted through the abdomen into the uterus. Compare *chorionic villus sampling*.

anabolism Metabolic pathways that create complex biomolecules from smaller organic compounds. Compare *catabolism*.

anaerobic Not requiring oxygen. Compare *aerobic*.

analogous trait A feature that is shared across species because of convergent evolution, not because of modification by descent from a recent common ancestor. Compare *homologous trait*.

analytical study An observational study that looks for patterns in the information collected and addresses how or why those patterns came to exist. Compare *descriptive study*.

anatomy The study of the structures that make up a complex multicellular body. Compare *physiology*.

androgens Any of the hormones produced in the testes that stimulate cells to develop characteristics of maleness, such as beard growth and the production of sperm. Compare *estrogens* and *progestogens*.

Animalia One of the six kingdoms of life, encompassing all animals, including humans, birds, and dinosaurs.

annual A plant that completes its life cycle in one year. Compare *biennial* and *perennial*.

antibody Any of various Y-shaped proteins that recognize and attack invaders in the antibody-mediated response of the adaptive immune system.

anticodon A sequence of three nitrogenous bases at one end of a tRNA molecule that binds the corresponding *codon* on an mRNA molecule.

anus The muscle-lined opening through which solid waste is eliminated by the body.

appendicular skeleton The part of the skeleton that has to do with motion. It is made up of the arms, legs, and pelvis. Compare *axial skeleton*.

applied research Research in which scientific knowledge is applied to human issues and often commercial applications. Compare *basic research*.

Archaea One of the three domains of life (compare *Bacteria* and *Eukarya*) and also one of the six kingdoms of life. The domain and kingdom Archaea consists of single-celled organisms best known for living in extremely harsh environments

artery Any of the large vessels that transport blood away from the heart. Compare *capillary* and *vein*.

artificial selection The process by which individuals display specific traits through selective breeding. Compare *natural selection*.

asexual reproduction The process by which clones, offspring that are genetically identical to the parent, are generated. Compare *sexual reproduction*.

atom The smallest unit of an element that retains the element's distinctive properties. Atoms are the building blocks of all matter.

atomic mass number The sum of the number of protons and the number of neutrons in an atom's nucleus. Compare *atomic number*.

atomic number The number of protons in an atom's nucleus. Compare *atomic mass number*.

ATP Adenosine triphosphate, a small, energy-rich organic molecule that is used to store energy and to move it from one part of a cell to another. Every living cell uses ATP.

atrioventricular (AV) node The part of the heart that passes a signal to contract on from the atria to the ventricles. Compare *sinoatrial node*.

atrium (pl. atria) Either of the two upper chambers of the heart. Compare *ventricle*.

autosome Any chromosome that is not one of the *sex chromosomes*.

auxin Any of a group of plant hormones that are necessary for cell division and promote root formation. Compare *cytokinin*.

AV node See **atrioventricular node**.

axial skeleton The part of the skeleton that supports and protects the long axis of the body. It includes the skull, the ribs, and a long, bony spinal column. Compare *appendicular skeleton*.

axon The part of a neuron that sends information to other cells. Compare *dendrite*.

B

B cell A type of lymphocyte that matures in the bone marrow. B cells are involved in antibody-mediated immunity. Compare *T cell*.

Bacteria One of the three domains of life (compare *Archaea* and *Eukarya*) and also one of the six kingdoms of life. The domain and kingdom Bacteria includes familiar disease-causing bacteria.

base A chemical compound that accepts hydrogen ions (H⁺) in aqueous surroundings. Compare *acid*.

base pair Also called *nucleotide pair*. Two nucleotides that form one rung of the DNA ladder.

base-pairing rules The rules that govern the pairing of nucleotides in DNA. Adenine (A) on one strand can pair only with thymine (T) on the other strand, and cytosine (C) can pair only with guanine (G).

basic research Research that is intended to expand the fundamental knowledge base of science. Compare *applied research*.

behavioral trait A characteristic of an individual's behavior, such as shyness or extroversion. Compare *biochemical trait* and *physical trait*.

bias A prejudice or opinion for or against something.

biennial A plant that takes two years to complete its life cycle, growing and maturing in the first year and reproducing in the second year. Compare *annual* and *perennial*.

bile A fluid, produced by the liver, that helps digest fats by creating a coating that enables the fat globules to interact with water molecules to partially dissolve them.

binary fission A type of cell division in which a cell simply divides into two equal halves, resulting in daughter cells that are genetically identical to each other and to the parent cell.

biochemical trait A characteristic due to specific chemical processes of an individual, such as the level of a particular enzyme. Compare *behavioral trait* and *physical trait*.

biofuel A fuel produced by the conversion of organic matter.

biogeography The geographic locations where organisms or the fossils of a particular species are found.

biological hierarchy A way to visualize the breadth and scope of life, from the smallest structures to the broadest interactions between living and nonliving systems that we can comprehend.

biological species concept The idea that a species is defined as a group of natural populations that can interbreed to produce fertile offspring and cannot breed with other such groups. Compare *morphological species concept*.

biome A large region of the world defined by shared physical characteristics, especially climate, and a distinctive community of organisms.

biosphere All the world's living organisms and the physical spaces where they live.

biotic Living. Compare *abiotic*.

bipedal Walking upright on two legs.

blood The cells, cell fragments, and plasma that circulate through the heart and blood vessels.

blood pressure The force of blood pushing from the heart through blood vessels.

blood vessel A vessel that transports blood throughout the body.

boreal forest A terrestrial biome that is found in northern or high-latitude regions, has cold, dry winters and mild summers, and is dominated by coniferous trees.

brain The part of the central nervous system that controls and coordinates nerve signals throughout the body and has a large capacity for processing diverse types of sensory information.

breathing The process of taking air into the lungs (inhaling) and expelling air from them (exhaling).

bronchioles A series of ever-smaller tubes that branch from the bronchi and open into the alveoli.

bronchus (pl. bronchi) Either of two small tubes that the trachea branches into in the chest and that lead to the lungs.

C

Calvin cycle Also called *carbon fixation*. The second of two principal stages of photosynthesis, in which a series of enzyme-catalyzed chemical reactions converts carbon dioxide (CO_2) into sugar, using energy delivered by ATP and electrons and hydrogen ions donated by NADPH. Compare *light reactions*.

Cambrian explosion The burst of evolutionary activity, occurring about 540 million years ago, that resulted in a dramatic increase in the diversity of life. Most of the major living animal groups first appear in the fossil record during this time.

camouflage Any type of coloration or appearance that makes an organism hard to find or hard to catch. Compare *mimicry* and *warning coloration*.

capillary Any of the tiny blood vessels that exchange materials by diffusion with nearby cells. Compare *artery* and *vein*.

carbohydrate Any of a major class of macromolecules, including sugars and starches, built of repeating units of carbon, hydrogen, and oxygen.

carbon cycle The transfer of carbon within biotic communities, between living organisms and their physical surroundings, and within the abiotic world.

carbon dioxide (CO_2) The most abundant and consequential of the greenhouse gases.

carbon fixation See **Calvin cycle**.

carbon sink A natural or artificial reservoir that absorbs more carbon than it releases. Compare *carbon source*.

carbon source A natural or artificial reservoir that releases more carbon than it absorbs. Compare *carbon sink*.

cardiac muscle The specialized muscle that helps produce the coordinated contractions known as heartbeats. It has a banded appearance and branched muscle fibers. Compare *skeletal muscle* and *smooth muscle*.

cardiovascular system A closed circulatory system consisting of a muscular heart, a complex network of blood vessels that collectively form a closed loop, and blood that circulates through the heart and blood vessels.

carnivore An animal (or, rarely, plant) that kills other animals for food. Compare *herbivore*.

carrying capacity The maximum population size that can be sustained in a given environment.

cartilage A dense tissue of the skeleton that combines strength with flexibility. It is found almost everywhere two bones meet and prevents them from grinding together.

catabolism Metabolic pathways that release chemical energy in the process of breaking down complex biomolecules. Compare *anabolism*.

causation A statistical relation indicating that a change in one aspect of the natural world causes a change in another aspect. Compare *correlation*.

cell The smallest and most basic unit of life—a microscopic, self-contained unit enclosed by a water-repelling membrane.

cell cycle The sequence of events that make up the life of a typical eukaryotic cell, from the moment of its origin to the time it divides to produce two daughter cells.

cell division Also called *M (mitotic) phase*. The final stage of the cell cycle. Cell division includes the transfer of DNA from the parent cell to the daughter cells.

cell theory One of the unifying principles of biology, a theory stating that every living organism is composed of one or more cells, and that all cells living today came from a pre-existing cell.

cellular respiration The reciprocal process to *photosynthesis* in which sugars are broken down into energy usable by the cell.

central nervous system (**CNS**) One of two main pars of the nervous system, consisting of the brain and spinal cord. Compare *peripheral nervous system*.

centromere The central region of a chromosome that attaches sister chromatids together.

cervix The lower portion of the uterus, which narrows and connects the uterus to the vagina.

channel protein A transport protein that helps specific substances move passively across the plasma membrane. See also **passive carrier protein**.

chaparral A terrestrial shrubland biome characterized by cool, rainy winters and hot, dry summers, and dominated by drought-resistant plants.

chemical bond A force that holds two atoms together.

chemical compound A molecule that contains atoms from two or more different elements.

chemical reaction The process of breaking existing chemical bonds and creating new ones.

chlorophyll A green pigment that is specialized for absorbing light energy.

chloroplast An organelle of plant cells and some protist cells that captures energy from sunlight and uses it to manufacture food molecules via photosynthesis.

chorionic villus sampling (**CVS**) A prenatal genetic screening technique in which cells are extracted by gentle suction from the villi (a cluster of cells that attaches the pregnancy sac to the wall of the uterus). Ultrasound is used to guide the narrow, flexible suction tube through a woman's vagina and into her uterus. Compare *amniocentesis*.

chromatid Either of two identical DNA molecules produced by the replication of a chromosome.

chromosomal abnormality Any change in the chromosome number or structure, compared to what is typical for a species.

chromosome A DNA double helix wrapped around spools of proteins.

chromosome theory of inheritance The theory that genes are located on chromosomes, and that these chromosomes are the basis for all inheritance.

circulatory system The organ system that moves oxygen from the respiratory system to the heart, which then pumps oxygen-rich blood to the rest of the body.

citric acid cycle See Krebs cycle.

clade A branch of an evolutionary tree, consisting of an ancestor and *all* its descendants.

class The unit of classification in the Linnaean hierarchy above order and below phylum.

climate The prevailing *weather* of a specific place over relatively long periods of time (30 years or more).

climate change A large-scale and long-term alteration in Earth's climate, including such phenomena as global warming, change in rainfall patterns, and increased frequency of violent storms.

climax community A mature community whose species composition remains stable over long periods of time.

CNS See **central nervous system**.

codominance Interaction between two alleles of a gene that causes a heterozygote to display a phenotype that clearly displays the effects of both alleles. Compare *incomplete dominance*.

codon A unique sequence of three mRNA bases that either specifies a particular amino acid during translation or signals the ribosomes where to start or stop translation. Compare *anticodon*.

coevolution The tandem evolution of two species that results because interaction between the two so strongly influences their survival.

collagen A tough but pliable protein that, in addition to being the main component of cartilage, is found in a great variety of tissues.

colon See **large intestine**.

commensalism A species interaction in one species benefits at no cost to the other species. Compare *competition*, *exploitation*, and *mutualism*.

common ancestor An organism from which many species have evolved.

common descent The sharing of a common ancestor by two or more different species.

community The populations of different species that live and interact with one another in a particular place.

compact bone One of the two major types of bone tissue. It forms the hard, white outer region of bones. Compare *spongy bone*.

competition A species interaction in which both species may be harmed. Compare *commensalism*, *exploitation*, and *mutualism*.

complex trait A genetic trait whose pattern of inheritance cannot be predicted by Mendel's laws of inheritance.

connective tissue A tissue that binds and supports tissues and organs.

consumer An organism that obtains energy by eating all or parts of other organisms or their remains. Compare *producer*.

contraceptive Any means of preventing pregnancy.

control group The group of subjects in an experiment that is maintained under a standard set of conditions with no change in the independent variable. Compare *treatment group*.

controlled experiment An experiment that measures the value of a dependent variable for two groups of subjects that are comparable in all respects except that one group (the treatment group) is exposed to a change in the independent variable and the other group (the control group) is not.

convection cell A large and consistent atmospheric circulation pattern in which warm, moist air rises and cool, dry air sinks. Earth has four giant convection cells.

convergent evolution Evolution that results in organisms that have different genetics but appear very much alike.

corpus luteum The cells of the ruptured follicle that remain behind in the ovary after ovulation to produce the hormone progesterone.

correlation A statistical relation indicating that two or more aspects of the natural world behave in an interrelated manner: if one shows a particular value, we can predict a particular value for the other aspect. Compare *causation*.

cotyledon A food-storing organ that is part of the tiny, embryonic seedling that lies inside a plant seed.

covalent bond The sharing of electrons between two atoms. Compare *hydrogen bond* and *ionic bond*.

credentials Evidence of qualifications and competence to be recognized as an authority on a subject. Such evidence would include education and accomplishments.

cross See **genetic cross**.

crossing-over The physical exchange of chromosomal segments between homologous chromosomes. Compare *genetic recombination*.

CVS See **chorionic villus sampling**.

cyclical fluctuation A relatively predictable pattern of change in population size that occurs when at least one of two species is strongly influenced by the other. Compare *irregular fluctuation*.

cytokine Any of a group of signaling proteins that immune system cells use to communicate when an invader is present.

cytokinesis Division of the cytoplasm, the second step of mitotic division, resulting in two self-contained daughter cells. Compare *mitosis*.

cytokinin Any of a group of plant hormones that are necessary for cell division and promote shoot formation. Compare *auxin*.

cytoskeleton The network of protein cylinders and filaments that forms the framework of a cell.

cytosine (C) One of the four nucleotides that make up DNA. The other three are adenine (A), thymine (T), and guanine (G).

D

data (sing. datum) Information collected in a scientific study.

decomposer An organism that breaks down the dead bodies of other organisms.

deletion A mutation in which a base is deleted from the DNA sequence of a gene. Compare *insertion* and *substitution*.

dendrite The part of a neuron that receives information from other cells. Compare *axon*.

density-dependent population change A change in population size that occurs when birth and death rates change as the population density changes. Compare *density-independent population change*.

density-independent population change A change in population size that occurs when populations are held in check by factors that are not related to the density of the population. Compare *density-dependent population change*.

deoxyribonucleic acid See **DNA**.

dependent variable Any variable that responds, or could potentially respond, to changes in an *independent variable*.

dermal tissue One of three types of plant tissue. Forming the outermost layer of the plant, dermal tissue protects the plant from the outside environment and controls the flow of materials into and out of the plant. Compare *ground tissue* and *vascular tissue*.

descriptive study An observational study that reports information about what is found in nature. Compare *analytical study*.

desert A terrestrial biome characterized by a lack of moisture and temperature fluctuation from very hot during the day to very cold at night, and dominated by succulent plants such as cacti.

diaphragm A thick sheet of muscle that forms the floor of the chest cavity and plays an important role in controlling breathing.

dicot Any plant that has two cotyledons in each of its seeds. Dicots are characterized by parallel veins in the leaves, vascular bundles scattered, fibrous roots, and floral parts in multiples of three. Examples include beans, squash, oak trees, and roses. Compare *monocot*.

diffusion The movement of a substance from a region of higher concentration to a region of lower concentration.

digestion The chemical breakdown of food.

digestive system The organ system that breaks down food in the mouth, stomach, and small intestine and absorbs nutrients in the small and large intestine.

dihybrid cross A controlled mating experiment involving organisms that are heterozygous for two traits.

diploid Possessing a double set of genetic information, represented by $2n$. Somatic cells are diploid. Compare *haploid*.

directional selection The most common pattern of natural selection, in which individuals at one extreme of an inherited phenotypic trait have an advantage over other individuals in the population. Compare *disruptive selection* and *stabilizing selection*.

disruptive selection The least common pattern of natural selection, in which individuals with either extreme of an inherited trait have an advantage over individuals with an intermediate phenotype. Compare *directional selection* and *stabilizing selection*.

DNA Deoxyribonucleic acid, the genetic code of life, consisting of two parallel strands of nucleotides twisted into a double helix. DNA is the genetic material that transfers information from parents to offspring.

DNA polymerase The enzyme that builds new strands of DNA in DNA replication.

DNA replication The duplication of a DNA molecule.

DNA sequence similarity The degree to which the sequences of two different DNA molecules are the same, a measure of how closely related two DNA molecules are to each other.

domain The highest hierarchical level in the organization of life, describing the most basic and ancient divisions among living organisms. The three domains of life are Bacteria, Archaea, and Eukarya.

dominant allele An allele that prevents a second allele from affecting the phenotype when the two alleles are paired together. Compare *recessive allele*.

dominant genetic disorder A genetic disorder that is inherited as a dominant trait on an autosome. Compare *recessive genetic disorder*.

double helix The spiral formed by two complementary strands of nucleotides that is the backbone of DNA.

down-regulation The slowing down of gene expression. Compare *up-regulation*.

E

Earth equivalent The number of planet Earths needed to provide the resources we use and absorb the wastes we produce.

ecological community An association of species that live in the same area.

ecological footprint The area of biologically productive land and water that an individual or a population requires to produce the resources it consumes and to absorb the waste it produces.

ecological isolation The condition in which closely related species in the same territory are reproductively isolated by slight differences in habitat. Compare *geographic isolation* and *reproductive isolation*.

ecological niche The set of conditions and resources that a population needs in order to survive and reproduce in its habitat.

ecology The scientific study of interactions between organisms and their environment, where the environment of an organism includes both biotic factors (other living organisms) and abiotic (nonliving) factors.

ecosystem A particular physical environment and all the communities in it.

ecosystem process Any of four processes—nutrient cycling, energy flow, water cycling, and succession—that link the biotic and abiotic worlds in an ecosystem.

egg Also called *ovum*. The female gamete. Compare *sperm*.

electron A negatively charged particle found outside the nucleus of an atom. Compare *neutron* and *proton*.

electron transport chain An elaborate chain of chemical events that ultimately generates ATP and NADPH.

element A pure substance that has distinctive physical and chemical properties, and that cannot be broken down into other substances by ordinary chemical methods.

elimination The removal from the body of solid waste, consisting mostly of indigestible material and bacteria that inhabit the digestive tract.

embryo The earliest stage of development of an individual after fertilization, up to 2 months of age in humans. Compare *fetus*.

embryonic development The process by which an embryo develops. Common patterns of embryonic development across species provide evidence of evolution.

endocrine gland A gland that releases hormones into body fluids, such as the bloodstream, for transport to target cells throughout the body.

endocrine system The organ system, consisting of a number of glands and secretory tissues that produce and secrete hormones, that works closely with the nervous system to regulate all other organ systems.

endocytosis The process by which materials are transported into a cell via vesicles. Compare *exocytosis*.

endoplasmic reticulum (ER) An extensive and interconnected network of sacs made of a single membrane that is continuous with the outer membrane of the nuclear envelope. See also **rough ER** and **smooth ER**.

endoskeleton An internal skeleton. Compare *exoskeleton*.

energy The capacity of any object to do work, which is the capacity to bring about a change in a defined system.

energy capture The trapping and storing of solar energy by the producers at the base of an ecosystem's energy pyramid.

energy carrier A molecule that can store and deliver usable energy.

energy pyramid A pyramid-shaped representation of the amount of energy available to organisms in a food chain.

enzyme Any of a class of small molecules that speed up chemical reactions.

epistasis A form of inheritance in which the phenotypic effect of the alleles of one gene depends on the presence of alleles for another, independently inherited gene. Compare *pleiotropy*.

epithelial tissue A tissue that covers organs and lines body cavities.

ER See **endoplasmic reticulum**.

esophagus The food tube that connects the pharynx to the stomach.

essential amino acid Any of the eight amino acids that can be obtained only from food.

estradiol The primary estrogen. Compare *progesterone* and *testosterone*.

estrogens Any of the hormones produced in the ovaries that play a role in determining female characteristics such as wide hips, a voice that is pitched higher than that of males, and the development of breast tissues. Compare *progestogens* and *androgens*.

estuary An aquatic biome characterized by shallow depth and high productivity. An estuary is the tidal ecosystem where a river flows into the ocean and consequently has a constant ebb and flow of fresh and salt water. It is dominated by grasses and sedges.

ethylene A plant hormone that stimulates the ripening of some fruits, including apples, bananas, avocadoes, and tomatoes.

Eukarya One of the three domains of life, including all the living organisms that do not fit into the domains *Archaea* or *Bacteria*, from amoebas to plants to fungi to animals.

eukaryote Any of one of the two major groups of living organisms. Animals, plants, fungi, and protists are all eukaryotes. Compare *prokaryote*.

evaporation The transition from liquid to gas.

evolution A change in the overall inherited characteristics of a group of organisms over multiple generations.

evolutionary tree A model of evolutionary relationships among groups of organisms that is based on similarities and differences in their DNA, physical features, biochemical characteristics, or some combination of these. It maps the relationships between ancestral groups and their descendants, and it clusters the most closely related groups on neighboring branches.

exocytosis The process by which materials are exported out of a cell via vesicles. Compare *endocytosis*.

exon A stretch of DNA that carries instructions for building a protein. Compare *intron*.

exoskeleton An external skeleton. Compare *endoskeleton*.

experiment A repeatable manipulation of one or more aspects of the natural world.

experimental group See **treatment group**.

exploitation A species interaction in which one species benefits and the other species is harmed. Compare *commensalism*, *competition*, and *mutualism*.

exploitative competition Competition between species in which the two species indirectly compete for shared resources, such as food. Compare *interference competition*.

exponential growth A pattern of population growth in which the population increases by a constant proportion over a constant time interval, such as one year. Exponential growth occurs when there are no constraints on resources and is represented by a J-shaped curve. Compare *logistic growth*.

external defenses The immune system's first line of defense against pathogens, consisting of physical and chemical barriers that reduce the likelihood of harmful organisms or viruses gaining access to internal tissues. Compare *adaptive immune system* and *innate immune system*.

F

F_1 generation The first generation of offspring in a series of genetic crosses. Compare *F_2 generation* and *P generation*.

F_2 generation The second generation of offspring in a series of genetic crosses. Compare *F_1 generation* and *P generation*.

facilitated diffusion Diffusion that requires transport proteins. Compare *simple diffusion*.

fact A direct and repeatable observation of any aspect of the natural world. Compare *theory*.

fallopian tube See **oviduct**.

falsifiable Able to be refuted.

family The unit of classification in the Linnaean hierarchy above genus and below order.

fat-soluble vitamin A vitamin that cannot dissolve in water and therefore tends to accumulate in body fat because it cannot be so easily excreted in urine. Compare *water-soluble vitamin*.

feces The solid waste produced by digestion.

feedback loop The steps of a process that either decreases (*negative feedback*) or increases (*positive feedback*) the output of a process.

fermentation A metabolic pathway by which most anaerobic organisms extract energy from organic molecules. It begins with glycolysis and is followed by a special set of reactions whose only role is to help perpetuate glycolysis. Fermentation enables organisms to generate ATP anaerobically.

fertilization The fusion of two gametes, results in a zygote.

fetus The second stage of development of an individual, from 2 months to birth in humans. Compare *embryo*.

filtration The blood-cleansing work of the kidneys. Compare *reabsorption* and *secretion*.

follicle-stimulating hormone (FSH) A hormone, produced by the pituitary gland, that launches each menstrual cycle by stimulating the growth of ovarian follicles.

food chain A single direct line of who eats whom among species in a community. Compare *food web*.

food web The way in which various and often overlapping *food chains* of a community are connected.

fossil The mineralized remains of a formerly living organism or the impression of a formerly living organism.

founder effect A form of genetic drift that occurs when a small group of individuals establishes a new population isolated from its original, larger population.

freshwater biome An aquatic biome whose character is defined by the terrestrial biomes that it borders or through which its water flows. Lakes, rivers, and wetlands are all part of the freshwater biome.

FSH See **follicle-stimulating hormone.**

Fungi One of the six kingdoms of life, including mushrooms, molds, and yeasts.

G

G_0 phase A nondividing state of the cell.

G_1 phase "Gap 1," the first phase in the life of a newborn cell.

G_2 phase "Gap 2," the phase of the cell cycle between the S phase and cell division.

gallbladder An organ of the digestive system that stores bile made by the liver and dispenses the bile into the small intestine as needed.

gamete A sex cell. Male gametes are sperm; female gametes are eggs. Compare *somatic cell*.

gametophyte A haploid, multicellular structure produced by mitotic division of a plant's spore. Cells in the gametophyte differentiate into egg or sperm. Compare *sporophyte*.

gene The basic unit of information, consisting of a stretch of DNA, that codes for a distinct genetic characteristic.

gene expression The process by which genes are transcribed into RNA and then translated to make proteins.

gene flow The exchange of alleles between populations.

gene regulation The changing of which genes are expressed in response to internal signals or external cues that allows organisms to adapt to their surroundings by producing different proteins as needed.

gene therapy A genetic engineering technique for correcting defective genes responsible for disease development.

genetic bottleneck A form of genetic drift that occurs when a drop in the size of a population causes a loss of genetic variation.

genetic carrier An individual who has only one copy of a recessive allele for a particular disease and therefore can pass on the disorder allele but does not have the disease.

genetic code The information specified by each of the 64 possible codons.

genetic cross A controlled mating experiment performed to examine how a particular trait may be inherited.

genetic disorder A disease caused by an inherited mutation, passed down from a parent to a child.

genetic divergence The presence of differences in the DNA sequences of genes.

genetic drift A change in allele frequencies produced by random differences in survival and reproduction among the individuals in a population.

genetic engineering The permanent introduction of one or more genes into a cell, tissue, or organism.

genetic recombination The exchange of DNA between homologous chromosomes brought about by *crossing-over*, contributing to variation in gametes.

genetic trait Any inherited characteristic of an organism that can be observed or detected in some manner.

genome The complete set of genes of an organism.

genotype The allelic makeup of a specific individual with respect to a specific genetic trait. Compare *phenotype*.

genus (pl. genera) The unit of classification in the Linnaean hierarchy above species and below family.

geographic isolation The condition in which populations are separated by physical barriers. Compare *ecological isolation* and *reproductive isolation*.

global warming A significant increase in the average surface temperature of Earth over decades or more.

glycolysis The first of three stages of cellular respiration. During glycolysis, sugars (mainly glucose) are split to make the three-carbon compound pyruvate. For each glucose molecule that is split, two molecules of ATP and two molecules of NADH are released. Compare *Krebs cycle* and *oxidative phosphorylation*.

Golgi apparatus A collection of flattened membranes that packages and directs proteins and lipids produced by the ER to their final destinations either inside or outside the cell.

grassland A terrestrial biome characterized by low moisture levels (but not as low as in deserts), and dominated by grasses and herbaceous plants.

greenhouse effect The process by which greenhouse gases let in sunlight and trap heat.

greenhouse gas A gas in Earth's atmosphere that absorbs heat that radiates away from Earth's surface. Examples include carbon dioxide (CO_2), water vapor (H_2O), methane (CH_4), and nitrous oxide (N_2O).

ground tissue One of three types of plant tissue. Forming the intermediate layer of the plant, ground makes up the bulk of the plant body and performs a wide range of functions, including support, wound repair, and photosynthesis. Compare *dermal tissue* and *vascular tissue*.

guanine (G) One of the four nucleotides that make up DNA. The other three are adenine (A), thymine (T), and cytosine (**C**).

H

haploid Possessing a single set of genetic information, represented by 2*n*. Gametes are haploid. Compare *diploid*.

heart A muscular organ the size of a fist that works as the body's circulatory pump.

heart rate The number of times a heart beats per minute.

hemoglobin An oxygen-binding protein found in red blood cells.

herbivore An animal that kills plants for food. Compare *carnivore*.

hermaphrodite An individual that produces both functional testes and functional ovaries and is therefore both male and female.

heterozygous Carrying two different alleles for a given phenotype (*Bb*). Compare *homozygous*.

homeostasis The process of maintaining constant internal conditions.

homeostatic pathway The sequence of steps that reestablishes homeostasis if there is any departure from the genetically determined normal state of a particular internal characteristic.

hominids The ape family, which includes humans and chimpanzees. All hominids are capable of tool use, symbolic language, and deliberate acts of deception.

hominins The "human" branch of the hominids, including modern humans and extinct relatives such as Neanderthals.

homologous pair A pair of chromosomes consisting of one chromosome received from the father and one from the mother.

homologous trait A feature that is similar across species because of common descent. Homologous traits may begin to look different from one another over time. Compare *analogous trait*.

homozygous Carrying two copies of the same allele (such as *BB* or *bb*) for a particular gene. Compare *heterozygous*.

horizontal gene transfer The transfer of genes on plasmids from one bacterium to another.

hormone A chemical that coordinates internal activities necessary for growth and reproduction.

host 1. An organism in which a *parasite* lives. 2. An individual that becomes infected by a *pathogen*.

hydrogen bond The weak electrical attraction between a hydrogen atom with a partial positive charge and a neighboring atom with a partial negative charge. Compare *covalent bond* and *ionic bond*.

hydrologic cycle The movement of water as it circulates from the land to the sky and back again.

hydrophilic Literally, "water-loving." Soluble in water. Compare *hydrophobic*.

hydrophobic Literally, "water-fearing." Excluded from water. Compare *hydrophilic*.

hypertonic Describes a fluid having a solute concentration higher than that of the cell it surrounds. Compare *hypotonic* and *isotonic*.

hypothalamus (**pl. hypothalami**) A small organ at the base of the vertebrate brain that coordinates the endocrine system and integrates it with the nervous system.

hypothesis (pl. **hypotheses**) An informed, logical, and plausible explanation for observations of the natural world.

hypotonic Describes a fluid having a solute concentration lower than that of the cell it surrounds. Compare *hypertonic* and *isotonic*.

I

immune memory The capacity of the adaptive immune system to remember a first encounter with a specific pathogen and to mobilize a speedy and targeted response to future infection by the same strain.

immune system The organ system that defends the body from invaders such as viruses, bacteria, and fungi.

in vitro fertilization (**IVF**) Fertilization of an egg by a sperm in a petri dish, followed by implantation of one or more embryos into a woman's uterus.

incomplete dominance Interaction between two alleles of a gene in which neither one can exert its full effect, causing a heterozygote to display an intermediate phenotype. Compare *codominance*.

independent assortment The random distribution of the homologous chromosomes into daughter cells during meiosis I.

independent variable The variable that is manipulated by the researcher in a scientific experiment. Compare *dependent variable*.

indeterminate growth The general growth pattern of plants, in which they grow throughout their lives.

induced fit A process in which an enzyme changes shape when molecules bind to its active site.

inflammation The immediate and coordinated sequence of events mounted by cytokines in the innate immune system in response to tissue damage from a pathogen invasion or wound.

ingestion The taking in of food, the first stage in the processing of food by the digestive system.

innate immune system The immune system's second line of defense against pathogens, consisting of cells and proteins that recognize the presence of an invader and mount an internal defense to kill, disable, or isolate it. Compare *adaptive immune system* and *external defenses*.

insertion A mutation in which a base is inserted into the DNA sequence of a gene. Compare *deletion* and *substitution*.

integumentary system The largest organ system in the human body, covering and protecting the surface of the body.

interference competition Competition between species in which one organism directly excludes another from the use of a resource. Compare *exploitative competition*.

intermediate fossil A fossil that displays physical characteristics in between those of two known fossils in a family tree.

interphase The longest stage of the cell cycle. Most cells spend 90 percent or more of their life span in interphase.

intron A stretch of DNA that does not code for anything. Compare *exon*.

invariant trait A trait that is the same in all individuals of a species. Compare *variable trait*.

ion An atom that has lost or gained electrons and therefore is either negatively or positively charged.

ionic bond The chemical attraction between a negatively charged ion and a positively charged ion. Compare *covalent bond* and *hydrogen bond*.

irregular fluctuation An unpredictable pattern of change in population size. Compare *cyclical fluctuation*.

isotonic Describes a fluid having a solute concentration equal to that of the cell it surrounds. Compare *hypertonic* and *hypotonic*.

isotopes Two or more forms of an element that have the same number of protons but different numbers of neutrons.

IVF See **in vitro fertilization**.

J

J-shaped growth curve The type of graphical curve that represents exponential growth. Compare *S-shaped growth curve*.

joint A junction in the skeletal system that lets the skeleton move in specific ways.

jumping gene See **transposon**.

K

karyotype A depiction showing all the chromosomes of a particular individual or species arranged in homologous pairs.

keystone species A species that has a disproportionately large effect on a community, relative to the species' abundance.

kidneys The paired organs that maintain water and solute homeostasis.

kingdom The second-highest hierarchical level in the organization of life; the unit of classification in the Linnaean hierarchy above phylum and below domain. The six kingdoms of life are Bacteria, Archaea, Protista, Plantae, Fungi, and Animalia.

Krebs cycle The second of three stages of cellular respiration. In this sequence of enzyme-driven reactions, the pyruvate made in glycolysis is broken down, releasing CO_2 and producing large amounts of energy carriers, including ATP, NADH, and $FADH_2$. Compare *glycolysis* and *oxidative phosphorylation*.

L

large intestine Also called *colon*. The final portion of the digestive system, where water and nutrients are absorbed before the remaining solid waste is expelled.

larynx Also called *voice box*. The breathing and sound-producing structure that forms the entryway to the trachea.

law of independent assortment The law, proposed by Gregor Mendel, stating (in modern terms) that when gametes form, the two alleles of any given gene segregate during meiosis independently of any two alleles of other genes. Compare *law of segregation*.

law of segregation The law, proposed by Gregor Mendel, stating (in modern terms) that the two alleles of a gene are separated during meiosis and end up in different gametes. Compare *law of independent assortment*.

LH See **luteinizing hormone**.

ligament A specialized, flexible band of tissue that joins bone to bone. Compare *tendon*.

light reactions The first of two principal stages of photosynthesis, in which chlorophyll molecules absorb energy from sunlight and use that energy for the splitting of water, which in turn produces oxygen gas (O_2) as a by-product that is released into the atmosphere. Compare *Calvin cycle*.

light independent-reactions See **Calvin cycle**.

lineage A single line of descent.

Linnaean hierarchy A system of biological classification devised by the Swedish naturalist Carolus Linnaeus in the eighteenth century.

lipid Any of a major class of molecules built of fatty acids and insoluble in water.

liposome A sphere formed by a phospholipid bilayer.

liver A large organ of the digestive system that produces bile, stores glycogen, and detoxifies dangerous chemicals in the body.

locus (pl. loci) The physical location of a gene on a chromosome.

logistic growth A pattern of population growth in which the population grows nearly exponentially at first but then stabilizes at the maximum population size that can be supported indefinitely by the environment. Logistic growth is represented by an S-shaped curve. Compare *exponential growth*.

lower respiratory system The part of the respiratory system made up of the trachea, bronchi, and lungs. Compare *upper respiratory system*.

lungs The paired organs in which gases (oxygen and carbon dioxide) are exchanged.

luteinizing hormone (LH) A hormone, produced by the pituitary gland, that triggers ovulation.

lymphatic system The network of ducts, lymph nodes, and associated organs that are the primary sites for action by the adaptive immune system.

lymphocyte A type of white blood cell that confers specific immunity as part of the adaptive immune system. Immature lymphocytes differentiate into B cells and T cells.

lysosome An organelle in animal cells that acts as a garbage or recycling center. Compare *vacuole*.

M

M (mitotic) phase See **cell division**.

macromolecule A large organic molecule.

macronutrient Any of nine nutrients—carbon, oxygen, hydrogen, nitrogen, phosphorus, potassium, calcium, sulfur, and magnesium—that plants need in relatively large amounts. Compare *micronutrient*.

marine biome An aquatic biome characterized by salt water and encompassing both the coastal regions of all continents and the open ocean.

marrow A tissue found in the cavities of hollow bones that, depending on the type of bone, stores fat or produces blood cells.

mass extinction A period of time during which more than 50 percent of Earth's species has gone extinct. The fossil record shows that there have been five mass extinctions in the history of Earth.

matter Anything that has mass and occupies a volume of space.

Mendelian trait A trait that is controlled by a single gene and unaffected by environmental conditions.

menstrual cycle The process in human females by which individual eggs mature and are released in a hormone-driven sequence of events approximately monthly.

messenger RNA (**mRNA**) A strand of RNA that is complementary to a DNA template strand. Compare *ribosomal RNA* and *transfer RNA*.

meta-analysis Work combining results from different studies.

metabolic heat The heat released as a by-product of chemical reactions within a cell, typically during cellular respiration.

metabolic pathway Any of various chains of linked events that produce key biological molecules in a cell, including important chemical building blocks like amino acids and nucleotides.

metabolism All the chemical reactions that occur inside living cells, including those that release and store energy.

meiosis A specialized type of cell division that kicks off sexual reproduction. It occurs in two stages: meiosis I and meiosis II, each involving one round of nuclear division followed by cytokinesis. Compare *mitosis*.

meiosis I The first stage of meiosis, in which the chromosome set is reduced by the separation of each homologous pair into two different daughter cells. Each homologous chromosome lines up with its partner and then separates to the two ends of the cells. Compare *meiosis II*.

meiosis II The second stage of meiosis, in which sister chromatids are separated into two new daughter cells. Compare *meiosis I*.

micronutrient Any of a variety of nutrients (including iron, zinc, and copper) that plants need in relatively small amounts. Compare *macronutrient*.

mimicry Coloration of a nonpoisonous animal that resembles the coloration of a toxic species. Compare *camouflage* and *warning coloration*.

mineral An inorganic chemical that has a critical biological function.

mitochondrial-DNA inheritance The passing down of DNA from the mitochondria in an egg cell to a new generation. Mitochondrial DNA passes virtually unchanged from mother to child, so it can be tracked from one generation, or one species, to another. Sequencing of mitochondrial DNA can determine how related an individual is to its female ancestors on its mother's side. Compare *nuclear-DNA inheritance*.

mitochondrion (**pl. mitochondria**) An organelle that is a tiny power plant fueling cellular activities. Mitochondria are the main source of energy in eukaryotic cells.

mitosis Division of the nucleus, the first step of mitotic division. Mitosis is divided into four main phases: prophase, metaphase, anaphase, and telophase. Compare *cytokinesis* and *meiosis*.

mitotic division A type of cell division that generates two genetically identical daughter cells from a single parent cell in eukaryotes. It consists of two steps: mitosis and cytokinesis.

mitotic phase See **cell division**.

molecule An association of atoms held together by chemical bonds.

monocot Any plant that has one cotyledon in each of its seeds. Monocots are characterized by netlike veins in the leaves, vascular bundles arranged in rings, taproots, and floral parts in multiples of four or five. Examples include all the grasses, members of the lily family, palm trees, and banana plants. Compare *dicot*.

monomer A small molecule that is the repeating unit of a *polymer*. For example, amino acids are the monomers that make up protein polymers.

morphological species concept The idea that most species can be identified as separate and distinct groups of organisms solely on the basis of their physical characteristics. Compare *biological species concept*.

morphology An organism's physical characteristics.

most recent common ancestor The most immediate ancestor that two lineages share.

mRNA See **messenger RNA**.

muscle fiber A long, narrow cell that can span the length of an entire muscle because it is made up of several muscle cells that fused together during development.

muscle tissue A tissue that generates force by contracting.

muscular system The organ system that, working closely with the skeletal system, produces the force that moves structures within the body.

mutation A change to the sequence of bases in an organism's DNA.

mutualism A species interaction in which both species benefit. Compare *commensalism, competition*, and *exploitation*.

myofibril Any of the cylindrical structures packed inside of muscle fibers containing proteins that contract by bracing against each other.

N

natural selection The process by which individuals with advantageous genetic characteristics for a particular environment survive and reproduce at a higher rate than do individuals with other, less useful characteristics. Compare *artificial selection*.

negative feedback The steps of a process that decrease its output. Compare *positive feedback*. See also **feedback loop**.

nephron The basic functional unit of the kidney.

nervous system The organ system that directs the rapid contractions of muscles and processes information received by the senses, such as touch, sound, and sight.

nervous tissue A tissue that communicates and processes information.

net primary productivity (NPP) The energy acquired through photosynthesis that is available for the growth and reproduction of producers within an ecosystem. It is the amount of energy captured by photosynthetic organisms, minus the amount they expend on cellular respiration and other maintenance processes.

neuron A specialized cell of the nervous system that transmits signals from one part of the body to another in a fraction of a second.

neutron An electrically neutral particle found in the nucleus of an atom. Compare *electron* and *proton*.

node The point on an evolutionary tree indicating the moment in time when an ancestral group split, or diverged, into two separate lineages. The node represents the most recent common ancestor of the two lineages in question.

noncoding DNA DNA that does not code for any kind of functional RNA.

nonspecific response A response mounted by the immune system against pathogens that is indiscriminate as to the invaders it repels. External defenses and the innate immune system are both nonspecific responses. Compare *specific response*.

NPP See **net primary productivity**.

nuclear-DNA inheritance The passing down of DNA from the nucleus in an egg or sperm cell to a new generation. Sequencing of nuclear DNA can determine how related an individual is to all of its ancestors, both male and female. Compare *mitochondrial-DNA inheritance*.

nuclear envelope The boundary of a cell's nucleus, consisting of two concentric phospholipid bilayers.

nuclear pore Any of many small openings in the nuclear envelope that allow chemical messages to enter and exit the nucleus.

nucleic acid Any of a major class of molecules, including DNA and RNA, built of chains of nucleotides.

nucleotide The basic repeating subunit of DNA, composed of the sugar deoxyribose, a phosphate group, and one of four bases: adenine (A), cytosine (C), guanine (G), or thymine (T).

nucleotide pair See **base pair**.

nucleus (pl. nuclei) 1. The dense core of an atom, which contains protons and neutrons. 2. The control center of the eukaryotic cell, containing all of the cell's DNA and occupying up to 10 percent of the space inside the cell.

nutrient A chemical element that is required by a living organism.

O

observation A description, measurement, or record of any object or phenomenon.

one-child policy The policy, adopted by China in 1979, that restricts most couples to having only one child and requires expensive fines, up to years' worth of salary, from families that have more than one child.

oogenesis The series of cell divisions that results in a human egg. Compare *spermatogenesis*.

opposable Able to be placed opposite other digits of the hand or foot. For example, opposable thumbs can be placed opposite each of the other four fingers.

oral cavity The mouth.

order The unit of classification in the Linnaean hierarchy above family and below class.

organ A collection of different types of tissues that form a functional unit with a distinctive shape and location in the body.

organ system A network of organs that work in a closely coordinated manner to perform a distinct set of functions in the body.

organelle Any of the membrane-enclosed subcellular compartments found in eukaryotic cells.

organic molecule A molecule that includes at least one carbon-hydrogen bond.

organism An individual living thing composed of interdependent parts.

osmosis A form of simple diffusion in which water moves in and out of cells (and compartments inside cells).

osteocyte A specialized bone cell that surrounds itself with a hard, nonliving mineral matrix composed largely of calcium and phosphate.

out-of-Africa hypothesis The idea that anatomically modern humans first evolved in Africa about 130,000 years ago from a unique population of archaic *Homo sapiens*, and then eventually spread into other continents to live alongside other hominins.

ovary Either of a pair of female reproductive organs that produce eggs and estrogens in vertebrates. Compare *testis*.

oviduct Also called *fallopian tube*. The tube through which an egg travels from the ovary to the uterus.

ovum (pl. ova) See **egg**.

oxidative phosphorylation The third of three stages of cellular respiration. During this process, the chemical energy of NADH and $FADH_2$ is converted into the chemical energy of ATP, while electrons and hydrogen atoms removed from NADH and $FADH_2$ are handed over to molecular O_2, creating water (H_2O). A large amount of ATP is generated. Compare *glycolysis* and *Krebs cycle*.

oxytocin A hormone—secreted by the fetus and, later in the birth process, by the mother's pituitary gland—that stimulates the uterine muscles and causes the placenta to secrete prostaglandins, which reinforce the contractions.

P

P generation The first set of parents in a series of genetic crosses. Compare F_1 generation and F_2 generation.

pacemaker See **sinoatrial node**.

pancreas A gland that produces insulin and secretes fluids that aid in the digestion of food.

parasite An organism that lives in or on another species and harms it by stealing resources. For example, some parasites suck blood or live off the food in our intestines. Compare *host* (definition 1).

partial negative charge A small electrically negative charge on the end of a polar molecule. Compare *partial positive charge*.

partial positive charge A small electrically positive charge on the end of a polar molecule. Compare *partial negative charge*.

passive carrier protein A transport protein that helps specific substances move passively across the plasma membrane. See also **channel protein**. Compare *active carrier protein*.

passive immunity Immunity to a particular pathogen that is conferred by antibodies not made by the body, but received from an outside source. Compare *active immunity*.

passive transport The movement of a substance without the addition of energy. Compare *active transport*.

pathogen An infectious agent. Compare *host* (definition 2).

PCR See **polymerase chain reaction**.

pedigree A chart similar to a family tree that shows genetic relationships among family members over two or more generations of a family's medical history.

peer-reviewed publication The publishing of original research only after it has passed the scrutiny of experts who have no direct involvement in the research under review, or a scientific journal that follows this standard.

penis The male reproductive organ that introduces sperm into a female or hermaphrodite sexual partner. The penis is also involved in urination in mammals.

perennial A plant that lives three or more years. Compare *annual* and *biennial*.

peripheral nervous system (PNS) One of two main pars of the nervous system, consisting of the sensory nerves plus all the nerves that are not part of the *central nervous system*.

PGD See **preimplantation genetic diagnosis**.

pH scale A logarithmic scale that indicates the concentration of hydrogen ions. The scale goes from 0 to 14, with 0 representing an extremely high concentration of free H^+ ions and 14 representing the lowest concentration.

phagocyte A type of white blood cell that functions as part of the innate immune system to mark and destroy foreign invaders by engulfing and digesting them.

phagocytosis Literally, "cellular eating." A large-scale version of endocytosis in which particles considerably larger than macromolecules are ingested. Compare *pinocytosis*.

pharynx Also called *throat*. An area where the back of the mouth and the two nasal cavities join together into a single passageway.

phenotype The physical, biochemical, or behavioral expression of a particular version of a trait. Compare *genotype*.

phloem One of two types of vascular tissue in plants. Phloem transports sugars from the leaves, where they are produced, to living cells in every part of the plant. Compare *xylem*.

photoperiodism The ability of plants to sense the duration of light and dark in a 24-hour cycle.

phospholipid An organic molecule with a hydrophilic head and a hydrophobic tail.

phospholipid bilayer A double layer of phospholipids in which the heads face out and the tails face in. Plasma membranes are phospholipid bilayers.

photosynthesis The process by which organisms capture energy from the sun and use it to create sugars from carbon dioxide and water, thereby transforming light energy into chemical energy stored in the covalent bonds of sugar molecules. Compare *cellular respiration*.

phylum (pl. phyla) The unit of classification in the Linnaean hierarchy above class and below kingdom.

physical trait An anatomical or physiological characteristic of an individual, such as the shape of an animal's head. Compare *behavioral trait* and *biochemical trait*.

physiology The science that focuses on the functions of anatomical structures. Compare *anatomy*.

pinocytosis Literally, "cellular drinking." A large-scale version of endocytosis in which fluids are ingested. Compare *phagocytosis*.

Plantae One of the six kingdoms of life, encompassing all plants.

plasma The fluid portion of the blood.

plasma membrane A barrier consisting of a phospholipid bilayer that separates a cell from its external environment.

platelet A type of sticky cell fragment found in circulating blood that helps form blood clots.

pleiotropy The pattern of inheritance in which a single gene influences a number of different traits. Compare *epistasis*.

PNS See **peripheral nervous system**.

point mutation A mutation in which only a single base is altered.

polar molecule A molecule whose electrical charge is shared unevenly, with some regions being electrically negative and others electrically positive.

pollinator An animal that transports pollen from one plant to another.

polygenic trait A genetic trait that is governed by the action of more than one gene.

polymer A long strand of repeating units of small molecules called *monomers*. For example, proteins are polymers made up of amino acid monomers.

polymerase chain reaction (PCR) A technique for replicating DNA that can produce millions of copies of a DNA sequence in just a few hours from a small initial amount of DNA.

polyploidy The condition in which an individual's somatic cells have more than two sets of chromosomes.

population A group of individuals of the same species living and interacting in a shared environment.

population density The number of individuals per unit area.

population doubling time The time it takes a population to double in size, as a measure of how fast a population is growing.

population ecology The study of the number of organisms in a particular place.

population size The total number of individuals in a population.

positive feedback The steps of a process that increase its output. Compare *negative feedback*. See also **feedback loop**.

postzygotic barrier A barrier that prevents a zygote from developing into a healthy and fertile individual; that is, a reproductive barrier that acts after a zygote exists. Compare *prezygotic barrier*.

predator A consumer that eats either plants or animals. Compare *prey*.

preimplantation genetic diagnosis (PGD) The removal of one or two cells from an embryo developing in a petri dish, usually 3 days after fertilization, followed by testing for genetic disorders. One or more embryos that are free of disorders are then implanted into a woman's uterus.

prey An animal that is eaten by a *predator*.

prezygotic barrier A barrier that prevents a male gamete and a female gamete from fusing to form a zygote; that is, a reproductive barrier that acts before a zygote exists. Compare *postzygotic barrier*.

primary consumer An organism that eats producers. Compare *secondary consumer*, *tertiary consumer*, and *quaternary consumer*.

primary growth In plants, an increase in length by the repeated, modular addition of a bud-stem-leaf unit aboveground or new lateral roots belowground. Compare *secondary growth*.

primary immune response The slow response mounted by the adaptive immune system against an invading pathogen the very first time a person is exposed to that pathogen. Compare *secondary immune response*.

primary literature Scientific literature where research is first published. Compare *secondary literature*.

primary oocyte An immature egg cell.

primary succession Succession that occurs in a newly created habitat, usually from bare rock or sand. Compare *secondary succession*.

primates The order of mammals to which humans belong. All primates have flexible shoulder and elbow joints, five functional fingers and toes, opposable thumbs, flat nails, and brains that are large in relation to the body.

process of science See **scientific method**.

producer An organism at the bottom of a food chain that uses energy from the sun to produce its own food. Compare *consumer*.

product A substance that results from a chemical reaction. Compare *reactant*.

progesterone The primary progestogen. Compare *estradiol* and *testosterone*.

progestogens Any of the hormones produced in the ovaries that have a number of functions in the female body, including thickening the lining of the uterus and increasing the blood supply to it to create a suitable environment for a developing fetus. Compare *estrogens* and *androgens*.

prokaryote Any of one of the two major groups of living organisms. Only bacteria and archaeans are prokaryotes. Compare *eukaryote*.

promoter A segment of DNA near the beginning of a gene that RNA polymerase recognizes and binds to begin transcription. Compare *terminator*.

protein Any of a major class of macromolecules built of amino acids.

Protista One of the six kingdoms of life, a diverse group that includes amoebas and algae.

proton A positively charged particle found in the nucleus of an atom. Compare *electron* and *neutron*.

pseudoscience Scientific-sounding statements, beliefs, or practices that are not actually based on the scientific method.

pulmonary circuit The circuit in the heart, consisting of the two chambers on the right side, that receives blood low in oxygen and pumps it to the lungs for gas exchange. Compare *systemic circuit*.

Punnett square A grid-like diagram showing all possible ways that two alleles can be brought together through fertilization.

Q

quaternary consumer An organism that eats *tertiary consumers*. Compare *primary consumer* and *secondary consumer*.

R

reasorption The removal of valuable solutes such as sodium, chloride, and sugars from the fluid filtered by the kidneys so that they don't exit the body in the urine. Compare *filtration* and *secretion*.

reactant A substance that undergoes change in a chemical reaction. Compare *product*.

receptor-mediated endocytosis A form of specific endocytosis in which receptor proteins embedded in the membrane recognize specific surface characteristics of substances that will be incorporated into the cell.

receptor protein A site where a molecule from another cell can bind.

recessive allele An allele that has no effect on the phenotype when paired with a dominant allele. Compare *dominant allele*.

recessive genetic disorder A genetic disorder that is inherited as a recessive trait on an autosome. Compare *dominant genetic disorder*.

red blood cell A cell in the blood that greatly increases the oxygen-carrying capacity of blood.

relative species abundance How common one species is when compared to another.

reproduction The making of a new individual like oneself.

reproductive barrier A barrier that prevents two species from interbreeding, making them reproductively isolated.

reproductive isolation The condition in which barriers prevent populations from interbreeding. Compare *ecological isolation* and *geographic isolation*.

reproductive system The organ system that generates gametes and may also support fertilization and prenatal development.

respiratory system The organ system that brings in oxygen and expels carbon dioxide to support cellular respiration.

ribosomal RNA (rRNA) A type of RNA that is an important component of ribosomes. Compare *messenger RNA* and *transfer RNA*.

ribosome The site of protein synthesis (translation) in the cytoplasm. Ribosomes are embedded in the rough endoplasmic reticulum.

RNA polymerase An enzyme that recognizes and binds a gene's promoter sequence and then separates the two strands of DNA during transcription.

RNA splicing Processing of mRNA in which the introns are snipped out of a pre-mRNA and the remaining pieces of mRNA—the exons—are joined to generate the mature mRNA.

root system One of two plant organ systems. It anchors the plant, absorbs water and nutrients from the soil, transports food and water, and may store food. Compare *shoot system*.

rough ER A part of the endoplasmic reticulum, having a knobby appearance because of embedded ribosomes, where proteins are assembled. Compare *smooth ER*.

rRNA See **ribosomal RNA**.

rubisco The enzyme that catalyzes the first step in the Calvin cycle, fixing a carbon molecule from CO_2.

S

S phase The "synthesis" phase of the cell cycle, in which preparations for cell division begin. A critical event during this phase is the replication of all the cell's DNA molecules.

S-shaped growth curve The type of graphical curve that represents logistic growth. Compare *J-shaped growth curve*.

SA node See **sinoatrial node**.

saliva A fluid secreted into the oral cavity to aid in the digestion of food.

sarcomere Any of the contractile units of the muscular system that make up each myofibril.

science A body of knowledge about the natural world, and an evidence-based process for acquiring that knowledge.

scientific claim A statement about how the world works that can be proved or disproved using the scientific method.

scientific literacy An understanding of the basics of science and the scientific process.

scientific method Also called *process of science*. The practices that produce scientific knowledge.

scientific name The unique two-word Latin name, consisting of the genus and species names, that is assigned to a species in the Linnaean hierarchy.

secondary consumer An organism that eats *primary consumers*. Compare *tertiary consumer* and *quaternary consumer*.

secondary growth In plants, an increase in thickness in either stems or roots. Compare *primary growth*.

secondary immune response The rapid response mounted by the adaptive immune system against an invading pathogen the second and subsequent times a person is exposed to that pathogen. Compare *primary immune response*.

secondary literature Scientific literature that summarizes and synthesizes an area of research. Compare *primary literature*.

secondary succession Succession that occurs after a disturbance in a community. Compare *primary succession*.

secretion The active transport by the kidneys of excess quantities of substances such as potassium and hydrogen ions, and some toxins, from the blood into the liquid passing through the kidney and out of the body in the urine. Compare *filtration* and *reabsorption*.

selective breeding The process by which humans allow only individuals with certain inherited characteristics to mate.

selective permeability The quality of plasma membranes by which some substances are allowed to cross the membrane at all times, others are excluded at all times, and still others can pass through the membrane when they are aided by transport proteins.

semiconservative replication The mode of replication by which DNA is duplicated, where one "old" strand (the template strand) is retained (conserved) in each new double helix.

sense To perceive the world through a sensory system such as sight, touch, or smell.

sensory input Signals that are received, transmitted, and processed by the central nervous system.

sensory organ An organ of the body, such as the eyes or ears, that receives sensory input.

set point The genetically determined normal state of any physical or chemical characteristic of the body's internal environment.

sex chromosome One of the two chromosomes (X and Y) that determine gender. Compare *autosome*.

sex-linked Found solely on the X or Y chromosome. See also **X-linked** and **Y-linked**.

sexual dimorphism A distinct difference in appearance between the males and females of a species.

sexual reproduction The process by which genetic information from two individuals is combined to produce offspring. It has two steps: cell division through meiosis, followed by fertilization. Compare *asexual reproduction*.

sexual selection Natural selection in which a trait increases an individual's chance of mating even if it decreases the individual's chance of survival.

shared derived trait A unique feature common to all members of a group that originated in the group's most recent common ancestor and then were passed down in the group.

shoot system One of two plant organ systems, consisting of stems and leaves. Stems provide the plant with structural support, transport food and water, and hold leaves up to intercept light so that they can perform photosynthesis. Compare *root system*.

simple diffusion Diffusion in which substances such as the small, uncharged molecules of water, oxygen, or carbon dioxide, slip between the large molecules in the phospholipid bilayer without much hindrance. Compare *facilitated diffusion*.

sinoatrial (SA) node Also called *pacemaker*. The part of the heart that sends a signal telling the atria to contract. Compare *atrioventricular node*.

sister chromatids The two identical DNA molecules produced by the replication of a chromosome.

skeletal muscle The specialized muscle that is associated with the skeleton. It has a banded appearance. Compare *cardiac muscle* and *smooth muscle*.

skeletal system The organ system, consisting of bones, cartilage, and ligaments, that provides an internal framework to support the body of vertebrates.

small intestine The highly coiled tube, specialized for absorption, into which food moves from the stomach during digestion.

smooth ER A part of the endoplasmic reticulum, having a smooth appearance, where lipids and hormones are manufactured. Compare *rough ER*.

smooth muscle The specialized muscle found in the walls of the digestive system and blood vessels. It has no bands, and its contractions are entirely involuntary. Compare *cardiac muscle* and *skeletal muscle*.

soluble Able to mix completely with water.

solute A dissolved substance, such as sugar in water. Compare *solvent*.

solution Any combination of a solute and a solvent.

solvent The fluid, such as water, into which a substance has dissolved. Compare *solute*.

somatic cell A non–sex cell. Compare *gamete*.

speciation The process by which one species splits to form two species or more.

species 1. Members of a group that can mate with one another to produce fertile offspring. 2. The smallest unit of classification in the Linnaean hierarchy.

species interaction Any of four ecological interactions—mutualism, commensalism, exploitation, and competition—that occur between different species.

species richness The total number of different species that live in a community.

specific response A response mounted by the adaptive immune system against a specific strain of pathogen. Compare *nonspecific response*.

sperm The male gamete. Compare *egg*.

spermatogenesis The series of cell divisions that results in human sperm. Compare *oogenesis*.

spinal cord A thick central nerve cord that is continuous with the brain, acting as a filter between the brain and the sensory neurons.

spongy bone One of the two major types of bone tissue. Honeycombed with numerous tiny cavities, it lies inside the *compact bone*. Spongy bone is most abundant at the knobby ends of our long bones.

spore In plants, a haploid cell produced by meiosis that divides mitotically to produce the gametophyte.

sporophyte A diploid, multicellular individual that arises from the zygote in plants. Compare *gametophyte*.

stabilizing selection The pattern of natural selection in which individuals with intermediate values of an inherited phenotypic trait have an advantage over other individuals in the population. Compare *directional selection* and *disruptive selection*.

start codon The codon AUG; the point on an mRNA strand at which the ribosomes begin translation. Compare *stop codon*.

statistics A branch of mathematics that estimates the reliability of data.

stoma (**pl. stomata**) An air pore in the leaf of a plant that controls gas exchange.

stomach An organ of the digestive system, located between the esophagus and intestines, in which most digestion occurs, through mechanical and chemical means.

stop codon The codon UAA, UAG, or UGA; the point on an mRNA strand at which the ribosomes end translation. Compare *start codon*.

substitution A mutation in which one base is substituted for another in the DNA sequence of a gene. Compare *deletion* and *insertion*.

substrate A molecule that will react to form a new product.

succession The process by which the species in a community change over time.

sustainable Able to be continued indefinitely without causing serious damage to the environment.

swallowing reflex The reaction of the digestive system when food comes into contact with the pharynx, in which the epiglottis seals off the entry into the trachea and food is pushed into the esophagus.

sympatric speciation The formation of new species in the absence of geographic isolation. Compare *allopatric speciation*.

synovial sac The cavity inside a joint that is filled with a lubricating fluid that reduces friction between two bony surfaces.

systemic circuit The circuit in the heart, consisting of the two chambers on the left side, that receives oxygenated blood from the lungs and pumps it to the body. Compare *pulmonary circuit*.

T

T cell A type of lymphocyte that matures in the thymus. T cells are involved in cell-mediated immunity. Compare *B cell*.

technology The practical application of scientific techniques and principles.

temperate forest A terrestrial biome characterized by snowy winters and moist, warm summers, and dominated by trees and shrubs adapted to relatively rich soil.

template strand The strand of DNA that is used as a template to make a new strand of DNA.

tendon A specialized, flexible band of tissue, rich in collagen, that joins muscle to bone. Compare *ligament*.

terminator A segment of DNA that, when reached by RNA polymerase, stops transcription. Compare *promoter*.

tertiary consumer An organism that eats *secondary consumers*. Compare *primary consumer* and *quaternary consumer*.

testis (**pl. testes**) Either of a pair of male reproductive organs that produce sperm and androgens in vertebrates. Compare *ovary*.

testosterone The primary androgen. Compare *estradiol* and *progesterone*.

theory A hypothesis, or a group of related hypotheses, that has received substantial confirmation through diverse lines of investigation by independent researchers. Compare *fact*.

throat See **pharynx**.

thymine (**T**) One of the four nucleotides that make up DNA. The other three are adenine (A), guanine (G), and cytosine (C).

tissue A group of cells that function in an integrated manner to perform a unique set of tasks in the body.

trachea Also called *windpipe*. The structure that connects the pharynx to the bronchi.

transcription The synthesis of RNA based on a DNA template. Compare *translation*.

transfer RNA (**tRNA**) A type of RNA that facilitates translation by delivering specific amino acids to the ribosomes as codons are read off of an mRNA. Compare *messenger RNA* and *ribosomal RNA*.

translation The process by which ribosomes convert the information in mRNA into proteins. Compare *transcription*.

transposon Also called *jumping gene*. A DNA sequence that can move from one position on a chromosome to another, or even from one chromosome to another.

transport protein A protein that acts like a gate, channel, or pump that allows molecules to move into and out of a cell.

treatment group Also called *experimental group*. The group of subjects in an experiment that is maintained under the same standard set of conditions as the control group but is subjected to manipulation of the independent variable. Compare *control group*.

trimester Any of the three defined stages of human pregnancy. Each trimester is about 3 months long.

tRNA See **transfer RNA**.

trophic level Each level of the energy pyramid, corresponding to a step in a food chain.

tropical forest A terrestrial biome characterized by warm temperatures, about 12 hours of daylight, and seasonally heavy or year-round rains, and containing a rich diversity of organisms.

tundra A terrestrial biome containing permafrost that is found at the poles and on mountaintops. Tundra is dominated by low-growing flowering plants and a lack of trees.

U

up-regulation The speeding up of gene expression. Compare *down-regulation*.

upper respiratory system The part of the respiratory system that includes airways in the nose, mouth, and throat. Compare *lower respiratory system*.

urinary system The organ system that removes excess fluid from the body, along with waste products, toxins, and other water-soluble substances that are not needed.

urine The waste-carrying watery solution that is expelled from our bodies.

uterus The female reproductive organ in which a fertilized egg implants and develops until birth.

V

vacuole An organelle in plant cells that acts as a garbage or recycling center and that stores water. Compare *lysosome*.

vagina The female reproductive organ that connects the uterus to the external genitalia.

variable A characteristic of any object or individual organism that can change.

variable trait A trait that is different in different individuals of a species. Compare *invariant trait*.

vascular tissue One of three types of plant tissue. Forming the innermost layer of the plant, vascular tissue (consisting of phloem and xylem) contains stacks of long cells forming continuous tubes that run throughout the plant body and transport materials throughout the plant. Compare *dermal tissue* and *ground tissue*.

vein Any of the large vessels that carry blood back to the heart. Compare *artery* and *capillary*.

ventricle Either of the two lower chambers of the heart. Compare *atrium*.

vertebrate An animal with an internal vertebral column (backbone) composed of a series of strong, hollow, cylindrical vertebrae that enclose and protect a major nerve cord.

vesicle A sac, formed by the bulging inward or outward of a section of the plasma membrane, that moves molecules from place to place inside a cell but also may transport substances into and out of the cell.

vestigial trait A trait that is inherited from a common ancestor but no longer used. Vestigial traits may appear as reduced or degenerated parts whose function is hard to discern.

villus (**pl. villi**) Any of the large number of fingerlike projections in the small intestine that are specialized for nutrient absorption.

virus A small, infectious agent that can replicate only inside a living cell.

vitamin Any of various small, organic nutrients needed by the human body, but only in tiny amounts.

voice box See **larynx**.

W

warning coloration Bright coloring of an animal that alerts a potential predator to dangerous chemicals in the animal's tissues. Compare *camouflage* and *mimicry*.

water-soluble vitamin A vitamin that can dissolve in water and therefore tends not to accumulate in body tissues because it can be easily excreted in urine. Compare *fat-soluble vitamin*.

weather Short-term atmospheric conditions, such as today's temperature, precipitation, wind, humidity, and cloud cover. Compare *climate*.

windpipe See **trachea**.

X

X-linked Found solely on the X chromosome. Compare *Y-linked*.

xylem One of two types of vascular tissue in plants. Xylem transports water and minerals, absorbed from the soil, upward from the roots and outward from the central stem to the leaves. Compare *phloem*.

Y

Y-linked Found solely on the Y chromosome. Compare *X-linked*.

Z

zygote The single cell that results from fertilization.

Credits

Sources for Infographics

Chapter 1
p. 15, "Bats in the Barn": Data from Justin G. Boyles, Paul M. Cryan, Gary F. McCracken, and Thomas H. Kunz, "Economic Importance of Bats in Agriculture," *Science*, April 1, 2001, *www.sciencemag.org/content/332/6025/41.*

Chapter 2
p. 29, "What's It All Made Of?": Earth's Atmosphere data from Wikipedia, *en.wikipedia.org/wiki/Abundance_of_the_chemical_elements.* The Universe data from WebElements, *www.webelements.com/periodicity/abundance_universe/.* The Human Body data from Wikipedia, *en.wikipedia.org/wiki/Chemical_makeup_of_the_human_body.* Earth's Crust data from Wikipedia, *en.wikipedia.org/wiki/Abundance_of_elements_in_Earth%27s_crust.*

Chapter 3
p. 51, "Sizing Up Life": Data from Genetic Science Learning Center at the University of Utah, "Cell Size and Scale," learn.*genetics.utah.edu/content/cells/scale/.*

Chapter 4
p. 67, "Growing Demand": Total world energy consumption data from REN21, "Renewables 2012 Global Status Report," *www.map.ren21.net/GSR/GSR2012.pdf.* Biodiesel and Bioethanol data from the Organization for Economic Co-operation and Development (OECD) and the Food and Agriculture Organization (FAO) of the United Nations, "OECD-FAO Agricultural Outlook 2011–2020," *www.oecd.org/site/oecd-faoagriculturaloutlook/48178823.pdf.*

Chapter 5
p. 87, "Plastic by the Numbers": Data from U.S. Environmental Protection Agency, Office of Resource Conservation and Recovery, "Municipal Solid Waste Generation, Recycling, and Disposal in the United States, Tables and Figures for 2012," *www.epa.gov/epawaste/nonhaz/municipal/pubs/2012_msw_dat_tbls.pdf.* Recycling categories and uses from New York State Department of Environmental Conservation, "Recycling Plastics Is as Easy as. . . 1, 2, 3 (4, 5, 6, 7)!" *www.dec.ny.gov/docs/materials_minerals_pdf/plasticpam.pdf,* and from American Chemistry Council, "What Does That Chasing Arrow Symbol on Plastic Products Mean?", *plastics makeitpossible.com/2011/06/what-does-that-chasing-arrow-symbol-on-plastic-products-mean/.*

Chapter 6
p. 103, "Does Bigger Mean Better?": Data from National Center for Biotechnology Information, U.S. National Library of Medicine, "Genome Information by Organism," *www.ncbi.nlm.nih.gov/genome/browse/.*

Chapter 7
p. 117, "Genetic Diseases Affecting Americans": Data from Cold Spring Harbor Laboratory, DNA Learning Center, "Your Genes, Your Health," *www.ygyh.org.*

Chapter 8
p. 138, "Grocery Cart Outbreaks": *Salmonella* data from Centers for Disease Control and Prevention, "Reports of Selected *Salmonella* Outbreak Investigations," *www.cdc.gov/salmonella/outbreaks.html.* E. coli data from U.S. Food and Drug Administration, "FDA Finalizes Report on 2006 Spinach Outbreak," *www.fda.gov/NewsEvents/Newsroom/PressAnnouncements/2007/ucm108873.htm.* *Listeria* data from Centers for Disease Control and Prevention, "*Listeria* Outbreaks," *www.cdc.gov/listeria/outbreaks/index.html.*

Chapter 9
p. 154, "The Deadly Price of a Pandemic": Data from U.S. Department of Health and Human Services, "Pandemic

Flu History," *www.flu.gov/pandemic/history/index.html*, and from World Health Organization, "Global Influenza Programme," *www.who.int/influenza/en/*.

Chapter 10
p. 167, "Watching Evolution Happen": From Oliver Tenaillon, Alejandra Rodríguez-Verdugo, Rebecca L. Gaut, Pamela McDonald, Albert F. Bennett, Anthony D. Long, and Brandon S. Gaut, "The Molecular Diversity of Adaptive Convergence," *Science*, Jan. 27, 2012, *www.ncbi.nlm.nih.gov /pubmed/22282810*.

Chapter 11
p. 193, "Race Against Resistance": Antibacterial drugs data from Infectious Diseases Society of America, "Bad Bugs, No Drugs: As Antibiotic Discovery Stagnates... A Public Health Crisis Brews," *www.idsociety.org/uploadedfiles/idsa /policy_and_advocacy/current_topics_and_issues/antimi crobial_resistance/10x20/images/bad%20bugs%20no%20 drugs.pdf*. Incidence of resistant strains data from Centers for Disease Control and Prevention, "Antibiotic Resistance Threats in the United States, 2013," *www.cdc.gov/drugresis tance/threat-report-2013/pdf/ar-threats-2013-508.pdf*.

Chapter 12
p. 211, "On the Diversity of Species": Data from A.D. Chapman, "Numbers of Living Species in Australia and the World, 2nd Edition, A Report for the Australian Biological Resources Study, September 2009," *www.environment.gov .au/node/13875*.

Chapter 13
p. 233, "The Sixth Extinction": Data from International Union for Conservation of Nature and Natural Resources, *The IUCN Red List of Threatened Species*, Summary Statistics, *www.iucnredlist.org/about/summary-statistics*.

Chapter 14
p. 253, "Hereditary Heirlooms": Data from *National Geographic*, "Genes Are Us. And Them," *ngm.nationalgeo graphic.com/2013/07/125-explore/shared-genes*.

Chapter 15
p. 271, "Forest Devastation": Deforestation and forest degradation data from G. Kissinger, M. Herold, and V. De Sy, "Drivers of Deforestation and Forest Degradation: A Synthesis Report for REDD+ Policymakers," Lexeme Consulting, Vancouver, Canada, August 2012, *www.gov.uk/government /uploads/system/uploads/attachment_data/file/65505 /6316-drivers-deforestation-report.pdf*. Top 5 country data

from Food and Agriculture Organization of the United Nations, "Change in Extent of Forest and Other Wooded Land 1990–2005," *www.fao.org/forestry/32033/en/*.

Chapter 16
p. 289, "The Cost of a Kid": Data from U.S. Department of Agriculture Center for Nutrition Policy and Promotion, "Parents Projected to Spend $245,340 to Raise a Child Born in 2013, According to USDA Report," *www.cnpp.usda.gov /reports-publications*.

Chapter 17
p. 306, "Cause and Effect": Data from William J. Ripple and Robert L. Beschta, "Restoring Yellowstone's Aspen with Wolves," *Biological Conservation* 138 (2007), *www.cof.orst .edu/leopold/papers/Restoring%20Yellowstone%20 aspen%20with%20wolves.pdf*.

Chapter 18
p. 324, "Productive Plants": Data from Oak Ridge National Laboratory Distributed Active Archive Center, "NPP and the Global Carbon Cycle," *daac.ornl.gov/NPP/html_docs/npp _ccycle.html*.

Chapter 19
p. 341, "Preventing Pregnancy": Data from James Trussell, "Contraceptive failure in the United States," *Contraception*, May 2011; 83(5): 397–404, *www.ncbi.nlm.nih.gov/pmc/arti-cles/PMC3638209/*.

Chapter 20
p. 361, "Nutritional Needs": Multivitamin data from One A Day® Men's Pro Edge and Women's Pro Edge vitamins. RDA data from Jennifer J. Otten, Jennifer Pitzi Hellwig, and Linda D. Meyers (editors), *Dietary Reference Intakes: The Essential Guide to Nutrient Requirements*, The National Academies Press, Washington, DC, 2006, *www.nap.edu/catalog/11537.html*; *www.nal.usda.gov/fnic/DRI/Essential_Guide/DRIEssential GuideNutReq.pdf*. Food data from National Institutes of Health, Office of Dietary Supplements, "Dietary Supplements Fact Sheets," *ods.od.nih.gov/factsheets/list-all/*.

Chapter 21
p. 383, "Have a Heart": Waiting list registration data from United Network for Organ Sharing, "Data Slides for Spring Regional Meetings: 2013 Data," *www.unos.org/docs/ DataSlides_Spring_2014.pdf*. Completed transplants data from U.S. Department of Health and Human Services, Organ Procurement and Transplantation Network, *optn.transplant. hrsa.gov/*.

Chapter 22

p. 401, "Driven by Hormones": Melatonin data from B. Selmaoui and Y. Touitou, "Reproducibility of the circadian rhythms of serum cortisol and melatonin in healthy subjects: A study of three different 24-h cycles over six weeks," *Life Sci.*, 73 (2003): 3339–3349. Epinephrine data from University of Mississippi Medical Center, "Physical Exercise - Epinephrine," *www.umc.edu/Education/Schools/Medicine/Basic_Science/ Physiology_and_Biophysics/Core_Facilities(Physiology)/ Physical_Exercise_-_Epinephrine.aspx*. Cortisol data from J.H. Wilmore, D.L. Costill, and W.L. Kenney, *Physiology of Sport and Exercise*, Human Kinetics, Champaign, IL, 2007.

Chapter 23

p. 415, "GM Crops Take Root": Data from Clive James, "Global Status of Commercialized Biotech/GM Crops: 2012," ISAAA Brief No. 44, ISAAA, Ithaca, NY, 2012, *www.isaaa. org/resources/publications/briefs/44/download/isaaa- brief-44-2012.pdf*.

Chapter 24

p. 437, "Safety in Numbers": Data from Caroline L. Trotter and Martin C.J. Maiden, "Meningococcal Vaccines and Herd Immunity: Lessons Learned from Serogroup C Conjugate Vaccination Programs," *Expert Review of Vaccines*, July 2009, *www.ncbi.nlm.nih.gov/pubmed/19538112*.

Text permissions

Figure 15.9: Graph reprinted by permission of Earth Policy Institute from Janet L. Larsen, "2013 Marked the Thirty-seventh Consecutive Year of Above-Average Temperature," Eco-Economy Indicator (Washington, DC: 4 February 2014); **Figure 24.5:** Graph: "The real cause of increasing autism prevalence?" by J. Emory Parker. Reprinted by permission of the author.

Photos

Chapter 1

p.3: U.S. Fish and Game; **p.4:** © Gerrit Vyn; **p.6 (far-left):** Alamy; **p.6 (left):** Scott Camazine/Science Source; **p.6 (center):** Shutterstock; **p.6(right):** Shutterstock; **p.6 (far-right):** Shutterstock; **p.79 (top):** Photo by Dr. Kimberli Miller, courtesy USGS National Wildlife Health Center; **p,7 (bottom):** © Stephen Alvarez/National Geographic Creative; **p.8(top):** Photo used with permission from New York State Department of Environmental Conservation; **p.8 (bottom):** © Dr. David Blehert; **p.8 (right):** Kevin Wenner/Pennsylvania

Game Commission; **p.9 (top):** Photo by Apic/Getty Images; **p.9 (bottom):** photograph by Steve Grodsky; **p.10 (all):** © Amy Smotherman Burgess/Knoxville News Sentinel /ZUMAPRESS.com; **p.13 (top-left):** Photo used with permission from New York State Department of Environmental Conservation; **p.13 (top-right):** Deborah Springer; **p.13 (center-left):** Greg Turner/Pennsylvania Game Commission; **p.13 (center-right):** Alamy; **p.13 (bottom-left):** Photo courtesy Ryan von Linden/New York Department of Environmental Conservation; **p.13 (bottom-right):** Dr. Mary Hausbeck; **p.17:** Lindsey Hefferman, Pennsylvania Game Commission.

Chapter 2

p.18-19: Shutterstock; **p.20:** © DAVID MCNEW/X00184/ Reuters/Corbis; **p.21:** Time & Life Pictures/Getty Image; **p.23 (top):** Courtesy of Henderson James Cleaves III; **p.23 (bottom):** Scripps Institution of Oceanography, UC San Diego; **p.25:** Scripps Institution of Oceanography, UC San Diego; **p.26:** Yosemite/Wikimedia Commons; **p.27:** © imagebroker/Alamy; **p.28 (top):** EUMETSAT/NASA; **p.28 (bottom):** NASA Ames Research Center/SETI/Peter Jenniskens; **p.30 (both):** Alamy.

Chapter 3

p.37: George Diebold/Getty; Shutterstock; **p.38:** Photo courtesy J. Craig Venter Institute; **p.39 (top-left):** Hazel Appleton, Health Protection Agency Centre for Infections/ Science Source; **p.39 (bottom-left):** EMBL/Johanna Höög; **p.39 (top-right):** Biophoto Associates/Science Source, Colorization by: Mary Martin; **p.39 (bottom-right):** CNRI/ Science Source; **p.39 (bottom):** © JESSICA RINALDI/ Reuters/Corbis; **p.40:** Roger Harris/Science Source; **p.41(top):** Photo courtesy J. Craig Venter Institute; **p.41 (bottom):** Courtesy of Neal Devaraj; **p.44:** Photo Researchers, Inc./Science Source; **p.46 (top-left):** Don W. Fawcett/Science Source; **p.46 (bottom-left):** © Dr. Dennis Kunkel/Visuals Unlimited/Corbis; **p.46 (bottom-right):** Dennis Kunkel/ Phototake/Alamy; **p.47 (top):** Hybrid Medical Animation/ Science Source; **p.47 (center):** Biophoto Associates/Science Source; **p.47 (bottom):** Biophoto Associates/Science Source; **p.48 (top):** © Dennis Kunkel Microscopy, Inc./Visuals Unlimited/Corbis; **p.48 (top-left):** © Russell Kightley; **p.48 (center-left):** © Russell Kightley; **p.48 (bottom-left):** University of Edinburgh, Wellcome Images; **p.48 (bottom):** © Russell Kightley; **p.49 (top):** Biophoto Associates/Science Source; **p.49 (top-center):** Dr. Torsten Wittmann/Science Source; **p.49 (bottom-center):** Dr. George Chapman/Visuals Unlimited, Inc.; **p.49 (bottom):** Bill Longcore/Science Source/Photo Researchers.

Index

Page numbers set in *italic* type refer to materials in figures or tables.

amino acids, *continued*
    essential amino acids, 350
    genetic code, *152*
    left- and right-handed forms, 28
    Miller experiments, 21–22, 23–24, 25, 27, 30
    in proteins, 21, 27, 31
    protein synthesis, *145*, 150, *151*, *152*, 153
amniocentesis, 122
Amoeba proteus, 51
amphibians, 222, 230, 242
amygdala, 380
anabolism, 59, *60*, 68
anaerobic organisms, 64, *65*
anaerobic processes, 64, *65*
analogous traits, 188
analytical methods, 8, *10*
anaphase in meiosis, *83*, *84*
anaphase in mitosis, *79*
anatomy, definition, 331
anchorage dependence, 77
androgen insensitivity syndrome (AIS), 115
androgens, 340
angiogenesis, 77
Animalia, 226, *227*, *229*, 230, 241
annuals (plants), 408, 409, 410, 414, 416, 418–19
anthers, *412*, *417*
antibacterial drugs approved by FDA, *193*
antibiotic resistance. *See also* VRSA (vancomycin-resistant *Staphylococcus aureus*)
    Escherichia coli, 137
    incidence of strains of resistant bacteria, *193*
    methicillin-resistant *Staphylococcus aureus* (MRSA), 182, *183*, 185, 186, *193*, 197
    natural selection and, *183*, 185–86, 194–95
    penicillin resistant *Staphylococcus aureus*, 182, 183
    prevention, 197
antibodies, *145*, 399–400
antibody-mediated immunity, 399, *400*
anticodons, *151*, 153
antigens, 399, 400
aorta, 371, *373*
Aphytis chrysomphali wasp, *305*
Aphytis lingnanensis wasp, *305*
appendicular skeleton, 356
applied research, 429
Arabidopsis, 39
Archaea (domain and kingdom), 221–22, 226, *227*, 241
Archaeopteryx, 220, 223, 224, 225–26, 231, 232
Ardipithecus ramidus, 245, *251*
arrector pili muscle, 352
arteries, 371
artificial life. *See* synthetic biology

artificial selection, 163, *164*, 165, *167*
asexual reproduction. *See also* mitosis
    bacteria, 191, 194, 204, *221*
    clone production, 76, 81, *335*
    definition, 76, 335
    eukaryotes, 78, *335*
    plants, 416
aspen trees. *See also* Yellowstone National Park ecology
    consumption by elk, 298, 299, 302–3, *306*
    decline in Yellowstone, 296–97, 298, 300
    ecological community, *297*, 298
    relationship with wolves, 296–97, 298, 299–300, 303, *306*, 307–8
    restoration in Yellowstone, 303, 307–8
    wolf-elk-aspen food chain, 298–99, 302
asteroids, 27–28, 232
atomic mass number, 20, *21*
atomic number, 20, *21*
atomic structure, 20, *21*
atoms, definitions, *14*, 20
ATP (adenosine triphosphate)
    Calvin cycle, 61, *62*
    from cellular respiration, 50, 64, *65*, *66*
    electron transport chain, 61, *62*, *66*
    as energy carrier, 60, *61*
    from glycolysis, 64, *65*, *66*
    high-energy phosphate bonds, 60, *61*
    from Krebs cycle or citric acid cycle, *66*, 68
    from oxidative phosphorylation, *66*, 68
    production in mitochondria, 50, 64
ATP synthase, 61, *62*, *66*
atrioventricular (AV) node, *373*
atrium (plural atria), 372, *373*
Autier, Philippe, 359, 361
autism and autism spectrum disorder
    arguments against claimed link to vaccines, 428, 431–34
    correlation with organic food sales, 431
    correlation with vaccine use, 430–31
    gene in Dobermans, *99*
    incidence of, 430
    original vaccine article in *Lancet*, 429, 433, 434
    scientific claims of vaccines as cause, 426, 427, 429–30
    symptoms, 429
autoimmune diseases, 395–96, 402
autonomic nervous system, 390, 402
autosomal dominant disorders, 120–21
autosomal recessive disorders, *118*, 119–20
autosomes, 113, *114*, *118*, 119
auxins, *413*
axial skeleton, 356
axons, 380, *382*

B

Bacteria (domain and kingdom), 221–22, 226, *227*, 241
bacteriophages ("phages"), 136
Bada, Jeffrey, 20, 23–24, 25, 27, 28, 30–31
Balch, Jennifer, 260, 261, 262, 267–68, 270, 273
barnacles, *302*, *307*
base pair, definition, 129
bases, definition, 28, 30
basic research, 428–29
Batrachochytruim dendrobatidis, *13*
bats. *See also* white-nose syndrome
    bat caves, conditions inside, 6
    big brown bat (*Eptesicus fuscus*), *15*
    Indiana bats, 12
    insect consumption, *15*
    little brown bats (*Myotis lucifugus*), 4, 12, *15*
    researchers in bat caves, 4, 5–6, 7, *10*, 14
B cells, 399, *400*
beak size of birds, natural selection, 164–65, 188, *189*
beavers, 298, 302, 304, 307, 308
Bedau, Mark, 40
behavioral isolation as reproductive barrier, 212, *213*
behavioral traits, 92
Behr, Melissa, 6–7, 8, 9
Belyaev, Dmitry, 102
benign tumors, 77
Beschta, Robert, 296, 299, 300, 303, 307
beta cells (pancreas), 75, 76
bias in science, 12, 428–29, 433
biennial plants, 408–9
big brown bat (*Eptesicus fuscus*), *15*
bile, *353*, 354
binary fission, 76
biochemical traits, 92
bioengineering. *See* tissue engineering
biofuels
    algae, 56–57, 63–64, 68, *69*
    from corn and soybeans, 56, 63
    definition, 56
    from intermediate wheatgrass, 420
    Prius algae biofuel vehicle, *69*
    world consumption, *67*
biogeography, 174–75
biological classification. *See also* species
    amphibians, 242
    Animalia, 226, *227*, *229*, 230, 241
    birds, 242
    bony fish, 242
    Chordata phylum, 242
    classes, 241
    domains, 221–22, 226, *227–28*, 241
    evolutionary trees, 223–26, 231, *232*

    families, 241
    Fungi, 226, *227*, 230, 241
    genus (plural genera), 241
    Homo genus, 242, 245
    jawless fish, 242
    kingdoms, 226, *227–29*, 241
    Linnaean hierarchy, 241
    mammals, 242
    orders, 241
    phylum (plural phyla), 241
    Plantae (plants), 226, *227*, *228*, 241
    Protista (protists), 226, *227*, *228*, 241
    reptiles, 242
    scientific names, 241
    vertebrates, 242
biological communities, *14*. *See also* ecological communities;
      Yellowstone National Park ecology
biological hierarchy, 13, *14*
biology, synthetic. *See* synthetic biology
biomass, 63, 273, 317, 318, 322–23, 325
biomes
    boreal forest biome, *270*, *320*
    chaparral biome, *320*
    definition, *14*, 319
    desert biome, 319, *321*, 322, *324*
    estuaries, *321*, 322
    freshwater biome, *320*, *324*
    global maps, *270*, *320–21*
    grassland biome, *270*, *321*, *324*
    marine biome, *321*
    net primary productivity (NPP), 322, *323*, *324*
    role of sunlight in creating, 268, 319
    temperate deciduous forest biome, *270*, *321*
    tropical forest biome, *270*, 319, *320*, 322
    tundra biome, *270*, 319, *320*, 322, *324*
    types of biomes, *320–21*, *324*
biopharming, 144, 145, *146*, 156
bioreactors, 57, *59*, 60, 63, 68
biosphere, *14*, 260, 261
biotic environmental factors, 261, 272, 314, 316–17, *318*
bipedalism and upright posture, 242–43, *244*, 245, *246*
birds and evolution
    Archaeopteryx, 220, 223, 224, 225–26, 231, 232
    Aurornis xui, *224*, 231–32
    biological classification, 242
    dinosaur-bird evolutionary tree, 223–26, 231, *232*, 242
    embryonic development, 242
    Eosinopteryx brevipenna, *224*, 231, *232*
    evolution of flight, 226
    theropods, 223, 224, 226
    Xiaotingia zhengi, 224–25, 231, *232*
birth control methods, comparison, *339*

human immunodeficiency virus (HIV), 395
hummingbirds, coevolution with flowers, 210
Hunt, Patricia, 74–75, 76, 79–80, 82, 85–86
Huntington disease, *118*, 121, 122, 196, 198
hydrogen, atomic structure, *21*
hydrogen bonds, 24–26
hydrologic cycle (water cycle), 268, *269*, 270, 317
hydrophilic molecules, 25
hydrophobic molecules, 25
hypertonic solutions, 44
hypodermis, 352
hypothalamus, 380, *391*, *393*
hypotheses. *See also* scientific method
   accept, reject, or modify hypothesis, *5*
   definition, 7, *13*
   falsifiable or refutable hypotheses, 8
   formation of, 4, *5*, 7–8
   support *vs.* proof, 8, *9*, 11
   testing, *5*, 8–11, *12*
hypotonic solutions, 44

I

immune memory, 399
immune system
   active immunity, 400
   adaptive immune system, 399–400
   adrenaline effect, 394
   antibodies, *145*, 399–400
   antibody-mediated immunity, 399, *400*
   antigens, 399, 400
   B cells, 399, *400*
   cell-mediated immunity, 399, *400*
   external defenses, 395, *396*, 397, 399
   functions, *333*, 390
   immune memory, 399
   innate immune system, 395–97, *398*, 399
   nonspecific responses, 397, 399
   passive immunity, 400
   primary immune response, 399
   response to endotoxin, 391, 393, 394
   response to vaccines, 144, *145*
   secondary immune response, 399
   specific responses, 399
incomplete dominance, 100
independent variables, 9–10
indeterminate growth, 412, 413
Indiana bats, 12
Indohyus. See also whales
   adaptation to aquatic life, 169–70
   body shape and size, *168*, *169*
   ear bones, 162

fossils, 162, 165, 169–71, 173, 176–77
induced fit, 58, *59*
inflammation, 361, 397, 398, 399
influenza vaccine, 144, 146–48, 149–50, *151*, 155–56
influenza virus, size, *51*
influenza virus H1N1 (swine flu), 146–47, *147*, *154*
ingestion, 352
innate immune system, 395–97, *398*, 399
insulin, 75, 173, *174*, 334–35, 396
integumentary system (skin), *333*, 350, 352, *355*
interference competition, 307
intermediate wheatgrass (*Thinopyrum intermedium*)
   breeding, 412, 413–14, 416, 419
   disease and pest resistance, 412
   domestication, 410, 414, 416, 419–20
   dormancy, 412
   general characteristics, 410
   gene sequencing, 419–20
   Kernza, 416, 418, 420
   leaves, 410
   roots, 410, 412, 414, 416
   seed size and yield, 412, 414, 416, 419
   uses, 410, 420
   water stress, 412
interphase. *See also* cell cycle
   G_0 phase, *75*, 76, 79
   G_1 phase, *75*, 76, 77–78, 79, *81*
   G_2 phase, *75*, 79, *81*
   overview, *75*, 77, *78*, *79*
   S phase, 75–76, 77, *78*, 79, *81*
intrauterine device, *339*
introns, 149, *150*
invariant traits, 92
in vitro fertilization (IVF), 122
ionic bonds, 24
ions, definition, 24
iron, nutritional requirements and sources, *362*
Irschick, Duncan, 202–3, 204
isotonic solutions, 43
isotopes, 20

J

J. Craig Venter Institute (JCVI), 38–39, 40, 50
jawless fish, 242
Jenner, Edward, *426*
Jensen, Tina Kold, 338, 340, 341
joints, 357, *358*
Jones, Kimberly, 56, 57, 59, 63
Jørgensen, Niels, 340, 341
J-shaped growth curve, 281, *282*
Jurassic period, *223*, 226, *234*

metaphase, 77, *78*

mitosis, 76, 77–78, 80–81

overview, 77–78

prophase, 76, *78*

telophase, *79*

mitotic spindle, *78–79, 83*

molecules, definition, *14*, 20

monarch butterfly, 303–4

monocots, 409

monomers, definition, 31

monosaccharides, structure, *33*

monotremes, 242

morphology, definition, 205, 207

motor neurons, *379*

motor output, *379*, 381

mouse oocytes and egg abnormalities, 74–75, 79, 80, 85–86

MRSA (methicillin-resistant *Staphylococcus aureus*), 182, *183*, 185, 186, *193*, 197

muscle fibers, 358, *359, 360*

muscle tissue

function and types, *332*

microscopic structure, 358, *359, 360*

muscle fibers, 358, *359, 360*

myofibrils, 358, *360*

sarcomeres, 358, *359, 360*

vitamin D and muscle health, 358–59

muscular system

cardiac muscle, *332, 359*

overview, *333*, 358–59

skeletal muscle, *332*, 358, *359*, 394

smooth muscle, *332, 352, 359*, 370

voluntary contractions, 358

in voluntary contractions, 358

musk ox, *305*

mutation

definition, 92, 135, 153, 189

deletions, 135, 153

DNA replication errors, 134–36

and evolution, 184, 189, 191, 192, 194, 206

frameshift mutations, 153

horizontal gene transfer, 136, 137, 139, 191–92, 194

insertions, 135, 153

mismatch errors, 135

point mutations, 135–36, *153*

silent mutations, 135–36

substitution point mutations, 135, *136*, 153

mutualism, 209, 300–301, *302*

Mycoplasma capricolum (*M. capricolum*), 40, *41*

Mycoplasma mycoides (*M. mycoides*)

synthetic cell creation, 38, 40–41

synthetic genome construction, 39–40, 50

transformation from *M. capricolum*, 40, *41*

myofibrils, 358, *360*

N

NADH, 60, 64, *65, 66*, 68

NADPH, 60, 61, *62*

natural selection

adaptive traits, 185

antibiotic resistance, *183*, 185–86, 194–95

beak size of birds, 164–65, 188, *189*

directional selection, 186, *187*

disruptive selection, 186, 188, *189*

overview, 163–65, 185

sexual selection, 194

stabilizing selection, 186, *188*

natural transformation, 191

Neanderthals (*Homo neanderthalensis*). *See also* human evolution

biological classification (Linnaean hierarchy), 241, 242

discovery, 240

fossils, 240, 245, 254

genome sequencing, 240, 246–47, *253*

interbreeding with modern humans, 240, 245–46, 254

mitochondrial DNA, 246, 254

Neanderthal DNA in human genomes, 248–49, 250

at Riparo Mezzena rock shelter, 240, 245, 254, *255*

negative feedback, 334–35

Neogene period, *223, 234*

nephrons, 377

nerve cells (neurons), 39, 44, 75, 76

nervous system. *See also* neurons

anatomy, *379*

autonomic nervous system modulation, 390, 402

brain, function and anatomy, 245, 379, 380

central nervous system (CNS), 379, 380–81

functions, *333*, 369

peripheral nervous system (PNS), 379, 380–81

sensory input, *379*, 380, 381

sensory organs, 379–80

sensory receptors, 352, 380, *381*

spinal cord, *332*, 379, *380*, 382

nervous tissue, *332, 352*

net primary productivity (NPP), 322, *323, 324*, 416

neurons. *See also* nervous system

anatomy, *39*, 381, *382*

axons, 380, *382*

dendrites, 380, *382*

function, 75, 381, *382*

gray matter, 380

motor neurons, *379*

nondividing state, 76

cottonwoods, 296, 299, 300, 302, 303, 307
coyote, 298, *300*
ecological community changes since 1920s, 298
elk, *297*, 298–300, 302–3, 304, *306*, 307
grizzly bears, 302, 307
Lamar River, 296, 303
secondary succession, 308, *309*
willows, 296, 298, *300*, 304
Y-linked genes, 115